An Introduction to Basic Fourier Series

Developments in Mathematics

VOLUME 9

Series Editor:

Krishnaswami Alladi, *University of Florida, U.S.A.*

Aims and Scope

Developments in Mathematics is a book series publishing

(i) Proceedings of Conferences dealing with the latest research advances,

(ii) Research Monographs, and

(iii) Contributed Volumes focussing on certain areas of special interest.

Editors of conference proceedings are urged to include a few survey papers for wider appeal. Research monographs which could be used as texts or references for graduate level courses would also be suitable for the series. Contributed volumes are those where various authors either write papers or chapters in an organized volume devoted to a topic of special/current interest or importance. A contributed volume could deal with a classical topic which is once again in the limelight owing to new developments.

An Introduction to Basic Fourier Series

by

Sergei K. Suslov

Department of Mathematics and Statistics,
Arizona State University, Tempe, Arizona, U.S.A.

KLUWER ACADEMIC PUBLISHERS
DORDRECHT / BOSTON / LONDON

A C.I.P. Catalogue record for this book is available from the Library of Congress.

ISBN 1-4020-1221-7

Published by Kluwer Academic Publishers,
P.O. Box 17, 3300 AA Dordrecht, The Netherlands.

Sold and distributed in North, Central and South America
by Kluwer Academic Publishers,
101 Philip Drive, Norwell, MA 02061, U.S.A.

In all other countries, sold and distributed
by Kluwer Academic Publishers,
P.O. Box 322, 3300 AH Dordrecht, The Netherlands.

The author was supported by NSF grant # DMS 9803443

Printed on acid-free paper

Printed in the Netherlands.

Dedicated to Dick Askey on his 70th birthday

Contents

Foreword

It was with the publication of Norbert Wiener's book "The Fourier Integral and Certain of Its Applications"[**165**] in 1933 by Cambridge University Press that the mathematical community came to realize that there is an alternative approach to the study of classical Fourier Analysis, namely, through the theory of classical orthogonal polynomials. Little would he know at that time that this little idea of his would help usher in a new and exiting branch of classical analysis called q-Fourier Analysis. Attempts at finding q-analogs of Fourier and other related transforms were made by other authors, but it took the mathematical insight and instincts of none other then Richard Askey, the grand master of Special Functions and Orthogonal Polynomials, to see the natural connection between orthogonal polynomials and a systematic theory of q-Fourier Analysis. The paper that he wrote in 1993 with N. M. Atakishiyev and S. K Suslov, entitled "An Analog of the Fourier Transform for a q-Harmonic Oscillator"[**13**], was probably the first significant publication in this area. The Poisson kernel for the continuous q-Hermite polynomials plays a role of the q-exponential function for the analog of the Fourier integral under consideration; see also [**14**] for an extension of the q-Fourier transform to the general case of Askey–Wilson polynomials. (Another important ingredient of the q-Fourier Analysis, that deserves thorough investigation, is the theory of q-Fourier series.)

Up until that time one of the stumbling blocks was the lack of a suitable q-exponential (and q-trigonometric) function that had the right analytic properties for use as an appropriate kernel. Fortunately, at about the same time in early 1990's Atakishiyev and Suslov, in their work on the solutions of the difference equation of hypergeometric type [**17**] and, independently, Ismail and Zhang, in their work on an inverse the Askey–Wilson q-difference operator [**76**], came up with just the right function. So all the ingredients were in place for a smooth take-off. There were important pieces of work by several authors, notably Bustoz, Ismail, Stanton and Zhang, but the one person who took up the challenge with a great deal of enthusiasm and passion is the author of this book, S. K. Suslov. He, along with Bustoz, found the orthogonality relations for a suitable pair of q-trigonometric functions and introduced corresponding basic analog of Fourier series [**26**]. He established the addition theorems [**140**], [**142**] and found the right normalization for the above q-exponential functions, enabling him to introduce exact analog of the elementary identity $\sin^2\theta + \cos^2\theta = 1$. In fact, in his most recent work he was able to show (it is included in this book) that these q-trigonometric functions are indeed the sines and cosines of integrals that depend on the angle.

This book has a whole wealth of material, some quite elementary, but others not so. The author has touched on q-analogs of many areas of Fourier analysis as the present state of knowledge allowed him to. He has found reasonable analogs of most of the basic theorems of classical Fourier Analysis. In other words, a pretty complete book for a field that is, and will probably remain, in a state of rapid flux for quite some time.

Some experts may ask if this was the right time to come up with a more-than-300 page book in this subject. But I think compiling all the scattered results in a single volume can only add to the excitement of getting involved in this new area of research. I hope the reader will agree with me that S. K. Suslov has done a marvelous job. I hope this book will help researchers not only to understand the subject but also to come up with new ideas.

Mizan Rahman

Carleton University
Ottawa, Canada

Preface

The subject of Special Functions was an active research area during the last two decades. A few books were published on different aspects of the theory including [**10**], [**35**], [**46**], [**111**] and several conference proceedings. The areas of so-called q-series and their numerous applications were among particularly active research topics. This book summarizes the results obtained during the last several years by a host of authors including Bustoz, Gosper, Ismail, Rahman, Stanton, Suslov, and Zhang in the theory of basic hypergeometric functions and classical analysis — a novel theory of basic Fourier series.

Although the theory of basic hypergeometric functions was originated in the works of Euler, Jacobi, Heine and other classical authors (see Gasper and Rahman's book [**46**] for an excellent account on the theory and a historic review), the basic analogs of Fourier series were introduced only recently during the current "renaissance period" of the Theory of Special Functions. In a way, they represent an interesting missing topic in classical analysis and q-series, which has been discovered only recently.

Dick Askey [**12**] characterized special functions as those functions that are "useful"; he wrote that "one studies special functions not for their own sake, but to be able to use them to solve problems". His lectures show how special functions can be used in various mathematical problems. Constructing q-analogs of the important classical results itself may be considered as one of those interesting problems in analysis. The experience of the last several years has definitely indicated that the new basic exponential and basic trigonometric functions, which have recently appeared in the theory of basic Fourier series, are very useful in discovering new formulas that have been overlooked at previous stages of the investigation. A further study of their properties and applications seems very promising.

Finding a q-analog of a classical expression is usually an "exception" rather then a "rule". Indeed, it is generally much harder to do and this is why analogs of many important classical formulas are not known yet. The theory of Fourier series has many beautiful results and it is surprising that in some cases their q-analogs can be constructed explicitly and investigated in detail. This does not appear to be true at the moment for other q-special functions that are higher then basic exponential functions, say for q-Bessel functions on a q-quadratic grid (the author would be happy to be wrong on that!). That makes the emerging theory of basic Fourier series more precious.

Our main motivation for writing this book was to present the new results on q-Fourier series so that they are readily available to students and researchers — the author made a goal to discuss the main facts of the theory of the basic Fourier series in a unified form and to emphasize, as much as possible, the connections with application, many of those still remain unknown. He hopes that the book can be used as a reference book in a future investigation of the q-Fourier series.

In his book on Special Functions [**126**], Earl Rainville wrote at the beginning of the chapter devoted to the theory of the theta functions "It seems safe to say that no topic in mathematics is more replete with beautiful formulas than that on which we now embark". For the last several years the author has had the same feeling about the theory of basic Fourier series. This research utilizes methods of classical analysis, approximation theory, orthogonal polynomials, entire functions, and numerical analysis to study the branch of q-special functions dealing with the q-analog of Fourier series and its applications. One of the main reasons for writing this book was a strong desire to share my understanding of the beauty of the subject with the reader. We hope that the reader will have the same impression reading this book and that it will stimulate further advances in q-special functions and their applications.

Most of the material is elementary enough so that persons with a good background in analysis should be able to read this book. We assume, nonetheless, that the reader is familiar with the first three chapters of Gasper and Rahman's book [**46**] on basic hypergeometric functions. In our study of q-Fourier series we shall frequently use summation and transformation formulas from this book and, for the benefit of the reader, we have included some of these formulas in Appendix A in order to make our presentation as self-contained as possible. We shall also need to know basic properties of the fundamental Askey–Wilson polynomials [**15**] and some of their special and limiting cases. These are classical topics in the theory of orthogonal polynomials and q-series and nowadays this material can be easily found in several excellent books and papers including [**9**], [**10**], [**46**], [**16**], [**59**], and [**86**].

We use the common numbering system of letting (m,n,p) refer to the p-th numbered display in Section n of Chapter m, letting Ex. m.p refer to the p-th Exercise of Chapter m, and letting (A,n,p) refer to the p-th numbered display in Section n of Appendix A and so on.

Due to time limitations, the author was unable to be as comprehensive in the coverage of basic Fourier series, their extensions and applications as he would have liked. In particular, the reader will not be able to find here a systematic treatment of an analog of the Bernoulli polynomials introduced recently by Ismail and Rahman [**67**] by means of formal q-Fourier series expansions. A systematic study of the q-Bessel functions on a q-quadratic grid [**63**], orthogonal very-well-poised ${}_8\varphi_7$ functions [**141**], [**143**], and the Askey–Wilson function transform [**90**], [**91**], which are natural extensions of the basic trigonometric systems, deserves a separate book in the future when the research in this area will be completed. A list of interesting topics for future investigation is provided at the end of the book.

With a great deal of respect I dedicate this book to Dick Askey on his 70th birthday, who has inspired so many people in the areas of orthogonal polynomials and special functions including the author, and who helped many of us so many times during different stages of our careers. It would be fair to say that this book wouldn't be written without this support.

I would like to thank the publisher for their cooperation and patience during the preparation of this book and Krishnaswami Alladi for the possibility to publish it in the Kluwer series "Developments in Mathematics". The National Science Foundation has provided financial aid for the research in 1998–2002 under the grant # DMS 9803443; the author thanks the Analysis Program Director Carlos Berenstein for this support. The Department of Mathematics and Statistics of Arizona State University has provided encouragement for the project.

Ocean Beach, San Diego, California, where an essential part of this book was written, has given an inspiration and energy to complete the project.

I would like to express my deep gratitude and appreciation to Joaquin Bustoz, a friend and a colleague in this mathematical endeavor for the last seven years, who read the entire manuscript starting from a very preliminary version and made many valuable remarks and suggestions which were taken into account in the final version of the book. I also thank all my friends and colleagues, including Christian Berg, George Gasper, Kevin Kadell, Erik Koelink, John McDonald, Mizan Rahman, Andreas Ruffing, and Dennis Stanton, who read different drafts of the manuscript, for their valuable comments and suggestions — I took them very seriously and appreciate your help and support very much. I wish to thank Bill Gosper for an introduction to the "world of Macsyma" and his help on many occasions — all numerical calculations in this book were performed with the help of his Macsyma program "`namesum`" which allows effective evaluation of infinite sums and infinite products. I am grateful to Christian Krattenthaler for an introduction to his Mathematica packages `HYP` and `HYPQ` [**96**] which helped to check many formulas in this book.

Sergei K. Suslov

CHAPTER 1

Introduction

Our main objective in this book is to summarize the results obtained during the last several years in a novel theory of basic Fourier series. We define basic exponential functions and basic trigonometric functions on a q-quadratic grid, introduce the corresponding analogs of Fourier series, and then study some of their properties and applications.

1.1. Some Basic Exponential Functions

Classical Fourier series are expansions in trigonometric systems. There is a great deal of interest these days in q-analogs of important classical results. A few q-analogs of the exponential and trigonometric functions are known. The F. H. Jackson q-analogs of the exponential function,

$$e_q(x) := \sum_{n=0}^{\infty} \frac{x^n}{(q;q)_n} = \frac{1}{(x;q)_\infty}, \qquad |x| < 1, \tag{1.1.1}$$

$$E_q(x) := \sum_{n=0}^{\infty} \frac{q^{n(n-1)/2}\, x^n}{(q;q)_n} = (-x;q)_\infty, \tag{1.1.2}$$

were introduced at the beginning of the last century (the summation formulas are due to Euler). These basic exponential functions (1.1.1)–(1.1.2) were studied in detail and have been found useful in many applications; see papers by Koornwinder and Swarttouw [**95**], Kalnins and Miller [**82**]–[**83**], Floreanini and Vinet [**40**], Koornwinder [**94**], and references therein.

Jackson's q-exponential functions are very special cases of the *basic hypergeometric series* defined by

$$\begin{aligned} &{}_r\varphi_s\left(\begin{matrix} a_1, a_2, \ldots, a_r \\ b_1, b_2, \ldots, b_s \end{matrix} ; q, t \right) \\ &\quad = \sum_{n=0}^{\infty} \frac{(a_1, a_2, \ldots, a_r; q)_n}{(q, b_1, b_2, \ldots, b_s; q)_n} \left((-1)^n q^{n(n-1)/2} \right)^{1+s-r} t^n, \end{aligned} \tag{1.1.3}$$

where the standard notations for the *q-shifted factorials* are

$$(a;q)_n := \prod_{k=0}^{n-1} (1 - aq^k), \qquad (a_1, a_2, \ldots, a_r; q)_n := \prod_{k=1}^{r} (a_k; q)_n, \tag{1.1.4}$$

$$(a;q)_\infty := \lim_{n\to\infty} (a;q)_n, \qquad (a_1,a_2,\ldots,a_r;q)_\infty := \prod_{k=1}^{r} (a_k;q)_\infty, \tag{1.1.5}$$

$$(a;q)_\alpha := \frac{(a;q)_\infty}{(aq^\alpha;q)_\infty}, \qquad (a_1,a_2,\ldots,a_r;q)_\alpha := \prod_{k=1}^{r} (a_k;q)_\alpha, \tag{1.1.6}$$

provided $|q| < 1$. If $0 < |q| < 1$, the ${}_r\varphi_s$ series converges absolutely for all t if $r \le s$ and for $|t| < 1$ if $r = s+1$. The reader can find an excellent account on the theory of basic hypergeometric series in Gasper and Rahman's book **[46]**. See also Gasper's **[45]** and Ismail's **[59]** lecture notes for an elementary introduction.

Some new q-exponential and q-trigonometric functions have been introduced in **[17]** and **[76]**. They have been under intense investigation during the last few years and appear from different contexts. Atakishiyev and Suslov **[17]**, **[140]** constructed solutions of a q-analog of the equation for harmonic motion. Basic analogs of the expansion formulas of the plane wave were found by Ismail and Zhang **[76]**, Ismail, Rahman and Zhang **[69]**, Floreanini, LeTourneux and Vinet **[43]**–**[44]**, and Ismail, Rahman and Stanton **[68]**.

"Addition" theorems for the basic exponential and trigonometric functions were established by Suslov **[140]**, **[144]** and by Ismail and Stanton **[71]** who also evaluated several important integrals involving these basic exponential functions. Floreanini, LeTourneux and Vinet gave an algebraic proof of Suslov's addition formula **[43]**.

The *basic exponential function* on a q-quadratic grid is given by

$$\begin{aligned} \mathcal{E}_q(x,y;\alpha) &= \frac{(\alpha^2;q^2)_\infty}{(q\alpha^2;q^2)_\infty} \\ &\times \sum_{n=0}^{\infty} \frac{q^{n^2/4}\,\alpha^n}{(q;q)_n} e^{-in\varphi} \left(-q^{(1-n)/2}e^{i\theta+i\varphi}, -q^{(1-n)/2}e^{i\varphi-i\theta};q\right)_n, \end{aligned} \tag{1.1.7}$$

where $x = \cos\theta$, $y = \cos\varphi$ and $|\alpha| < 1$. This function was originally introduced by Ismail and Zhang **[76]** with different notation and normalization; see also **[17]**, **[140]** for the corresponding solutions of a q-analog of the equation for harmonic motion. The above notation is due to Suslov **[140]**. Analytic continuation in a larger domain will be discussed in the next chapters. The function $\mathcal{E}_q(x,y;\alpha)$ is an analog of $\exp\alpha(x+y)$; one can easily show that

$$\lim_{q\to1^-} \mathcal{E}_q(x,y;\alpha(1-q)/2) = \exp\alpha(x+y). \tag{1.1.8}$$

A special case

$$\mathcal{E}_q(x;\alpha) = \mathcal{E}_q(x,0;\alpha) \tag{1.1.9}$$

is an analog of $\exp\alpha x$. The recent study of the basic exponential functions (1.1.7) and (1.1.9) has shown that their theory is very rich; see, for example, our review paper **[144]** and references therein; natural q-extensions of many classical results have already been established and many more extensions of important classical formulas are remained to be found.

1.2. Basic Fourier Series

A periodic function, $f(x) = f(x+2\pi)$, with period 2π can be expanded in the *Fourier series*

$$f(x) = \sum_{n=-\infty}^{\infty} c_n e^{inx} \tag{1.2.1}$$

with

$$c_n = \frac{1}{2\pi} \int_{-\pi}^{\pi} f(x) e^{-inx} \, dx. \tag{1.2.2}$$

A detailed account on the extensive theory of these series can be found in **[3]**, **[20]**, **[156]**, **[164]**, and **[166]**. The expression for the coefficients (1.2.2) follows from the orthogonality relation

$$\int_{-\pi}^{\pi} e^{inx} e^{-imx} \, dx = 0, \qquad m \neq n, \tag{1.2.3}$$

which can be easily verified.

Bustoz and Suslov **[26]** established the following *orthogonality property* of the basic exponential functions (1.1.9) with respect to an absolutely continuous measure

$$\begin{aligned} \int_0^{\pi} \mathcal{E}_q(\cos\theta; i\omega_m)\, \mathcal{E}_q(\cos\theta; -i\omega_n) \left(e^{2i\theta},\, e^{-2i\theta};\, q\right)_{1/2} \, d\theta \\ = 2k(\omega_n)\, \delta_{mn}. \end{aligned} \tag{1.2.4}$$

Here

$$k(\omega) = \pi \frac{\left(q^{1/2}; q\right)_\infty^2}{(q;q)_\infty^2} \frac{\left(-\omega^2; q^2\right)_\infty}{\left(-q\omega^2; q^2\right)_\infty} \sum_{k=0}^{\infty} \frac{q^{k/2}}{1+\omega^2 q^k}; \tag{1.2.5}$$

$\omega_m, \omega_n = 0, \pm\omega_1, \pm\omega_2, \pm\omega_3, \ldots$ and $\omega_0 = 0, \omega_1, \omega_2, \omega_3, \ldots$, are nonnegative solutions of the transcendental equation

$$\left(-i\omega; q^{1/2}\right)_\infty = \left(i\omega; q^{1/2}\right)_\infty \tag{1.2.6}$$

arranged in ascending order of magnitude. The corresponding *basic analog of Fourier series* is given by

$$f(\cos\theta) = \sum_{n=-\infty}^{\infty} c_n\, \mathcal{E}_q(\cos\theta; i\omega_n) \tag{1.2.7}$$

with

$$\begin{aligned} c_n = \frac{1}{2k(\omega_n)} \int_0^{\pi} & f(\cos\theta)\, \mathcal{E}_q(\cos\theta; -i\omega_n) \\ & \times \left(e^{2i\theta},\, e^{-2i\theta};\, q\right)_{1/2} \, d\theta. \end{aligned} \tag{1.2.8}$$

Explicit expansions in basic Fourier series lead to a new class of formulas in q-special functions never investigated before from the analytical or numerical point of view. For example, expansion of $\mathcal{E}_q(x;\alpha)$ in q-Fourier series has the form

$$\mathcal{E}_q(x;\alpha) = \frac{(-\alpha;q^{1/2})_\infty - (\alpha;q^{1/2})_\infty}{(q\alpha^2;q^2)_\infty} \tag{1.2.9}$$
$$\times \left[\frac{1}{2\kappa(0)\,\alpha} + \sum_{n=1}^{\infty} \frac{(-1)^n}{\kappa(\omega_n)(\alpha^2+\omega_n^2)} \right.$$
$$\left. \times \sqrt{\frac{(-q\omega_n^2;q^2)_\infty}{(-\omega_n^2;q^2)_\infty}}\,(\alpha C_q(x;\omega_n) - \omega_n S_q(x;\omega_n)) \right],$$

where

$$\kappa(\omega) = \sum_{k=0}^{\infty} \frac{q^{k/2}}{1+\omega^2 q^k}. \tag{1.2.10}$$

The series (1.2.9) is clearly a q-analog of the classical expansion

$$e^{\alpha x} = \frac{e^{\alpha\pi} - e^{-\alpha\pi}}{\pi} \left[\frac{1}{2\alpha} + \sum_{n=1}^{\infty} \frac{(-1)^n}{\alpha^2+n^2} (\alpha\cos nx - n\sin nx) \right] \tag{1.2.11}$$

and this gives us an opportunity to introduce, in a natural way, extensions of the classical Bernoulli and Euler polynomials and the Riemann zeta function [**146**].

Namely, a q-version of the classical zeta function

$$\zeta(z) = \sum_{n=1}^{\infty} n^{-z}, \qquad \text{Re } z > 1 \tag{1.2.12}$$

can be defined as

$$\zeta_q(z) = \sum_{n=1}^{\infty} \frac{1}{\kappa(\omega_n)\ \omega_n^z}, \qquad \text{Re } z > 1 \tag{1.2.13}$$

and the following summation formulas hold

$$\zeta_q(2) = \frac{q^{1/2}}{2\left(1-q^{3/2}\right)}, \qquad \zeta_q(4) = \frac{q^2}{2\left(1-q^{3/2}\right)^2\left(1-q^{5/2}\right)} \tag{1.2.14}$$

as analogs of Euler's results

$$\zeta(2) = \frac{\pi^2}{6}, \qquad \zeta(4) = \frac{\pi^4}{90}. \tag{1.2.15}$$

A thorough investigation of the basic, or q-Fourier series (1.2.7)–(1.2.8) and some of their applications is our main objective in this book. They have been under intense study during the last several years; different aspects of the theory were developed in [**26**], [**48**], [**56**], [**67**], [**68**], [**69**], [**71**], [**76**], [**140**], [**142**], [**144**], [**145**], [**146**], [**147**], and [**148**].

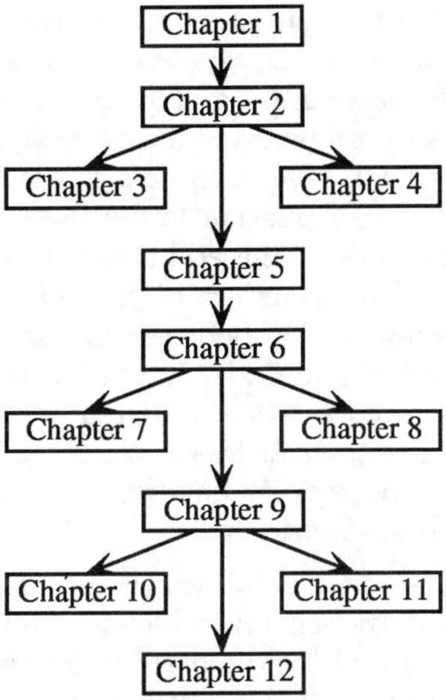

FIGURE 1.1. Main connections among the chapters.

1.3. About this Book

In this book we intend to lay a sound foundation for the study of basic Fourier series. In Chapter 2 we introduce basic exponential functions and basic trigonometric functions on a q-quadratic grid and study some of their elementary properties. Addition theorems are discussed in Chapter 3. Some important expansions and integrals involving basic exponential functions under consideration are derived in Chapter 4. Chapter 5 gives a formal introduction to the theory of basic Fourier series which are the main topic of this book. In Chapter 6 we investigate some elementary properties of these series, including completeness of the basic trigonometric system, uniform convergence and analytic continuation, and consider several examples. More advanced results on the completeness of the q-trigonometric systems are discussed in Chapter 7. Chapter 8 deals with the improved asymptotics of the zeros of basic trigonometric functions which are important for investigation of the convergence of these series. In Chapter 9 we give some examples of expansions of elementary and q-functions in basic Fourier series. Extensions of the Bernoulli and Euler polynomials and numbers, as well as the Riemann zeta function are introduced in Chapter 10. Numerical investigation of the basic Fourier series is discussed in Chapter 11. In conclusion, some suggestions for further work are discussed in Chapter 12.

A more detailed information about the structure of the book can be obtained from the Table of Contents. The diagram on Figure 1.1 presents main connections among the chapters.

When it was possible the author tried to give different proofs of the main results and to emphasize connections between different parts of the theory. See, for example, Sections 4.4 and 5.2 below, where two independent proofs of the orthogonality property of the basic trigonometric system are presented; Chapter 4, where different proofs of the Ismail and Zhang and Ismail, Rahman and Zhang expansion formulas are given; or Chapter 3, where several different proofs of the addition formula are provided. This gives more "insights" on the structure of the theory and emphasizes the relations among various topics. The book contains about a hundred and fifty exercises of different level of complexity located at the end of each chapter. Primarily, these exercises cover results related to the main topics of the corresponding chapter and their further extensions; they may include also some information that is potentially important for the future investigation in the areas related to basic Fourier series.

Summarizing all this together it seems important to say that at the moment we understand only "general contours" of a rather extensive theory of the basic Fourier series and the discovery of an "entire picture" is still a subject for the future investigations. The author hopes that this book will provide some guidelines in this journey.

1.4. Exercises for Chapter 1

(1) Verify the following identities for the q-shifted factorials

$$(a;q)_n = \left(q^{1-n}/a;q\right)_n (-a)^n q^{\binom{n}{2}}, \tag{1.4.1}$$

$$\frac{(a;q)_n}{(b;q)_n} = \frac{\left(q^{1-n}/a;q\right)_n}{\left(q^{1-n}/b;q\right)_n}\left(\frac{a}{b}\right)^n, \tag{1.4.2}$$

$$\left(aq^{-n};q\right)_n = (q/a;q)_n \left(-\frac{a}{q}\right)^n q^{-\binom{n}{2}}, \tag{1.4.3}$$

$$\frac{(aq^{-n};q)_n}{(bq^{-n};q)_n} = \frac{(q/a;q)_n}{(q/b;q)_n}\left(\frac{a}{b}\right)^n, \tag{1.4.4}$$

$$(a;q)_{n-k} = \frac{(a;q)_n}{\left(q^{1-n}/a;q\right)_k}\left(-\frac{q}{a}\right)^k q^{\binom{k}{2}-nk}, \tag{1.4.5}$$

$$\frac{(a;q)_{n-k}}{(b;q)_{n-k}} = \frac{(a;q)_n}{(b;q)_n}\frac{\left(q^{1-n}/b;q\right)_n}{\left(q^{1-n}/a;q\right)_n}\left(\frac{b}{a}\right)^n, \tag{1.4.6}$$

$$\left(q^{-n};q\right)_k = \frac{(q;q)_n}{(q;q)_{n-k}}(-1)^k q^{\binom{k}{2}-nk}, \tag{1.4.7}$$

$$\begin{aligned}\left(aq^{-n};q\right)_k &= \frac{(a;q)_k\,(q/a;q)_n}{\left(q^{1-k}/a;q\right)_n} q^{-nk} \\ &= \frac{(q/a;q)_n}{(q/a;q)_{n-k}}(-a)^k q^{\binom{k}{2}-nk},\end{aligned} \tag{1.4.8}$$

$$\left(aq^{-n};q\right)_{n-k} = \frac{(q/a;q)_n}{(q/a;q)_k}\left(-\frac{a}{q}\right)^{n-k} q^{\binom{k}{2}-\binom{n}{2}}, \tag{1.4.9}$$

$$\frac{(aq^{-n};q)_{n-k}}{(bq^{-n};q)_{n-k}} = \frac{(q/a;q)_n}{(q/b;q)_n}\frac{(q/b;q)_k}{(q/a;q)_k}\left(\frac{a}{b}\right)^{n-k}, \tag{1.4.10}$$

$$(a;q)_{n+k} = (a;q)_n\,(aq^n;q)_k\,, \tag{1.4.11}$$

$$(a;q)_{2n} = \left(a, aq; q^2\right)_n, \tag{1.4.12}$$

$$\left(a^2;q^2\right)_n = (a,-a;q)_n\,, \tag{1.4.13}$$

$$\frac{\left(qa^{1/2},-qa^{1/2};q\right)_n}{\left(a^{1/2},-a^{1/2};q\right)_n} = \frac{\left(aq^2;q^2\right)_n}{\left(a;q^2\right)_n} = \frac{1-aq^{2n}}{1-a}, \tag{1.4.14}$$

$$\left(aq^{-n};q\right)_{2n} = (-a)^n\,q^{-n(n+1)/2}\,(a,q/a;q)_n\,,$$

$$\left(aq^{-n};q\right)_{2n+1} = (1-a)\,(-a)^n\,q^{-n(n+1)/2}\,(aq,q/a;q)_n\,,$$

where $n = 0, 1, 2, \ldots$ and

$$\binom{n}{2} = \frac{n\,(n-1)}{2}.$$

A more complete list of useful identities is given in Appendix I of the book [**46**].

(2) Prove the identity $(-q;q)_\infty\left(q;q^2\right)_\infty = 1$.

(3) Prove that

$$e_q(x) = {}_1\varphi_0\,(0;-;q,z) = \frac{1}{(x;q)_\infty}, \qquad |x|<1,$$

$$E_q(x) = {}_0\varphi_0\,(0;-;q,-z) = (-x;q)_\infty\,.$$

(4) Prove the *q-binomial theorem*

$${}_1\varphi_0\,(a;-;q,z) = \sum_{n=0}^{\infty}\frac{(a;q)_n}{(q;q)_n}\,z^n = \frac{(az;q)_\infty}{(z;q)_\infty},$$

where $|z|<1$ and $|q|<1$, and changing $z \to -z$ show that

$$\sum_{k=0}^{\infty}\frac{(a;q)_{2k}}{(q;q)_{2k}}\,z^{2k} = \frac{1}{2}\left[\frac{(az;q)_\infty}{(z;q)_\infty} + \frac{(-az;q)_\infty}{(-z;q)_\infty}\right],$$

$$\sum_{k=0}^{\infty}\frac{(a;q)_{2k+1}}{(q;q)_{2k+1}}\,z^{2k+1} = \frac{1}{2}\left[\frac{(az;q)_\infty}{(z;q)_\infty} - \frac{(-az;q)_\infty}{(-z;q)_\infty}\right].$$

(5) Show that

$$\lim_{q\to 1^-}\frac{(q^\alpha z;q)_\infty}{(z;q)_\infty} = (1-z)^{-\alpha}, \qquad |z|<1,$$

$$\lim_{q\to 1^-} e_q\left((1-q)\,x\right) = \lim_{q\to 1^-} E_q\left((1-q)\,x\right) = e^x.$$

(6) The *q-gamma function*

$$\Gamma_q(z) = (1-q)^{1-z}\,\frac{(q;q)_\infty}{(q^z;q)_\infty}, \qquad 0<q<1$$

was introduced by Thomae [**153**] and later by Jackson [**79**]. Show that

$$\Gamma_q(z+1) = \frac{1-q^z}{1-q}\,\Gamma_q(z)$$

and

$$\lim_{q\to 1^-} \Gamma_q(z) = \Gamma(z).$$

See [**10**], [**46**] and references therein for a detailed investigation of the properties of the q-gamma function.

(7) Show that

$$\mathcal{E}_q(x,y;\alpha) = \mathcal{E}_q(y,x;\alpha), \qquad \mathcal{E}_q(0,0;\alpha)=1,$$
$$\lim_{q\to 1^-} \mathcal{E}_q\left(x,y;\alpha\left(1-q^{1/2}\right)\right) = \exp\alpha(x+y).$$

(8) Introduce the *basic cosine* $C_q(x;\omega)$ and the *basic sine* $S_q(x;\omega)$ functions by the following analog of Euler's formula

$$\mathcal{E}_q(x;i\omega) = C_q(x;\omega) + iS_q(x;\omega)$$

and show that

$$C_q(x;\omega) = \frac{\left(-\omega^2;q^2\right)_\infty}{\left(-q\omega^2;q^2\right)_\infty}\,{}_2\varphi_1\left(\begin{matrix} -qe^{2i\theta},\ -qe^{-2i\theta} \\ q \end{matrix}\ ;\ q^2,\ -\omega^2\right)$$

and

$$S_q(x;\omega) = \frac{\left(-\omega^2;q^2\right)_\infty}{\left(-q\omega^2;q^2\right)_\infty}\,\frac{2q^{1/4}\omega}{1-q}\,\cos\theta$$
$$\times\ {}_2\varphi_1\left(\begin{matrix} -q^2e^{2i\theta},\ -q^2e^{-2i\theta} \\ q^3 \end{matrix}\ ;\ q^2, -\omega^2\right).$$

(9) Show that the real form of the basic Fourier series (1.2.7)–(1.2.8) is

$$f(\cos\theta) = a_0 + \sum_{n=1}^{\infty}\left(a_n C_q(\cos\theta;\omega_n) + b_n S_q(\cos\theta;\omega_n)\right),$$

where $\omega_0 = 0, \omega_1, \omega_2, \omega_3, \ldots$, are nonnegative zeros of are nonnegative solutions of the transcendental equation

$$\left(-i\omega;q^{1/2}\right)_\infty = \left(i\omega;q^{1/2}\right)_\infty$$

arranged in ascending order of magnitude and the q-Fourier coefficients are given by

$$a_0 = \frac{1}{2k(0)} \int_0^\pi f(\cos\theta) \left(e^{2i\theta},\ e^{-2i\theta};\ q\right)_{1/2} d\theta,$$

$$a_n = \frac{1}{k(\omega_n)} \int_0^\pi f(\cos\theta)\, C_q(\cos\theta;\omega_n) \left(e^{2i\theta}, e^{-2i\theta}; q\right)_{1/2} d\theta,$$

$$b_n = \frac{1}{k(\omega_n)} \int_0^\pi f(\cos\theta)\, S_q(\cos\theta;\omega_n) \left(e^{2i\theta}, e^{-2i\theta}; q\right)_{1/2} d\theta.$$

(10) Jackson [**77**], [**78**], [**79**] introduced the q-*Bessel functions* by

$$J_\nu^{(1)}(x;q) = \frac{\left(q^{\nu+1};q\right)_\infty}{(q;q)_\infty} \left(\frac{x}{2}\right)^\nu {}_2\varphi_1\left(\begin{matrix} 0,\ 0 \\ q^{\nu+1} \end{matrix};\ q,\ -\frac{x^2}{4}\right),$$

$$J_\nu^{(2)}(x;q) = \frac{\left(q^{\nu+1};q\right)_\infty}{(q;q)_\infty} \left(\frac{x}{2}\right)^\nu {}_0\varphi_1\left(\begin{matrix} - \\ q^{\nu+1} \end{matrix};\ q,\ -\frac{x^2}{4}q^{\nu+1}\right),$$

where $0 < q < 1$. The above notations are due to Ismail [**55**], [**56**]. Show that

$$J_\nu^{(2)}(x;q) = \left(-x^2/4;q\right)_\infty J_\nu^{(1)}(x;q), \qquad |x| < 2$$

and

$$\lim_{q\to 1^-} J_\nu^{(k)}((1-q)x;q) = J_\nu(x), \qquad k = 1, 2.$$

See [**46**], [**55**], [**56**] and references therein for a detailed account of the properties of Jackson's q-Bessel functions.

CHAPTER 2

Basic Exponential and Trigonometric Functions

In this chapter we shall introduce the basic exponential and basic trigonometric functions as solutions of a difference analog of the equation for harmonic motion on a q-quadratic grid. Some of their elementary properties will be derived in order to form the basis for developing the theory of basic Fourier series and study some of their applications in the subsequent chapters.

2.1. Differential Equation for Harmonic Motion

The exponential and trigonometric functions can be introduced in elementary calculus in many different ways, in particular, on the basis of the second order differential equation

$$\sigma u'' + \lambda u = 0, \tag{2.1.1}$$

where σ and λ are arbitrary constants. This equation is obviously invariant under the shift of an independent variable $x \to x + y$ and its power series solution is

$$u = \sum_{n=0}^{\infty} \frac{(-\lambda/\sigma)^{n/2}}{n!} (x+y)^n . \tag{2.1.2}$$

If $\lambda/\sigma = -\alpha^2 < 0$, the solutions of (2.1.1) can be found in terms of the exponential functions

$$u_1 = e^{\alpha x} = \sum_{n=0}^{\infty} \frac{(\alpha x)^n}{n!}, \qquad u_2 = e^{-\alpha x}, \tag{2.1.3}$$

$$u_3 = e^{\alpha(x+y)}, \qquad u_4 = e^{-\alpha(x+y)}, \tag{2.1.4}$$

$$u_5 = \cosh \alpha x = \frac{1}{2}\left(e^{\alpha x} + e^{-\alpha x}\right), \tag{2.1.5}$$

$$u_6 = \sinh \alpha x = \frac{1}{2}\left(e^{\alpha x} - e^{-\alpha x}\right), \tag{2.1.6}$$

$$u_7 = \cosh \alpha\,(x+y)\,, \qquad u_8 = \sinh \alpha\,(x+y)\,. \tag{2.1.7}$$

When $\lambda/\sigma = \omega^2 > 0$, Eq. (2.1.1) is called the *differential equation for harmonic motion* and its solutions of are related to the trigonometric functions

$$u_1 = e^{i\omega x} = \cos \omega x + i \sin \omega x, \tag{2.1.8}$$

$$u_2 = e^{-i\omega x} = \cos \omega x - i \sin \omega x, \tag{2.1.9}$$

$$u_3 = e^{i\omega(x+y)}, \qquad u_4 = e^{-i\omega(x+y)}, \tag{2.1.10}$$

$$u_5 = \cos \alpha x, \qquad u_6 = \sin \omega x, \tag{2.1.11}$$

$$u_7 = \cos \omega (x+y), \qquad u_8 = \sin \omega (x+y). \tag{2.1.12}$$

Although from the point of view of real numbers the exponential function $\exp(\alpha x)$ and the trigonometric functions $\cos(\omega x)$ and $\sin(\omega x)$ seem to be derived from completely different sources, they are closely related in the complex plane and in what follows we will not usually distinguish the difference between solutions of the forms (2.1.3)–(2.1.7) and (2.1.8)–(2.1.12).

Many properties of the exponential and trigonometric functions can be derived directly from the differential equation (2.1.1) they satisfy. In particular, a second order differential equation may have only two linearly independent solutions, which leads, in a natural way, to the addition theorems for the exponential and trigonometric functions. For example, from (2.1.3) and (2.1.4)

$$e^{\alpha(x+y)} = Ae^{\alpha x} + Be^{-\alpha x}, \tag{2.1.13}$$

where A and B are some constants independent of x. Differentiating (2.1.13) with respect to x,

$$e^{\alpha(x+y)} = Ae^{\alpha x} - Be^{-\alpha x}, \tag{2.1.14}$$

and therefore

$$e^{\alpha(x+y)} = Ae^{\alpha x}. \tag{2.1.15}$$

Letting $x = 0$, we arrive at the addition theorem for the exponential function

$$e^{\alpha(x+y)} = e^{\alpha x} e^{\alpha y}. \tag{2.1.16}$$

One can prove (2.1.16) by direct series manipulations with the help of the binomial theorem. It is also well-known that the exponential function $e^{\alpha(x+y)}$ is the only measurable solution of the functional equation

$$f(x+y) = f(x)\, f(y). \tag{2.1.17}$$

Similar arguments result in the addition theorem for the trigonometric functions

$$\cos \omega (x+y) = \cos \omega x \cos \omega y - \sin \omega x \sin \omega y, \tag{2.1.18}$$

$$\sin \omega (x+y) = \sin \omega x \cos \omega y + \cos \omega x \sin \omega y, \tag{2.1.19}$$

which are the key tools in trigonometry. The Sturm–Liouville method for the equation of harmonic motion leads, in a natural way, to the orthogonality property of the trigonometric system, which is a starting point of the investigation of the classical Fourier series.

There is a great deal of interest these days in q-analogs of the exponential and trigonometric functions. In this chapter we shall introduce the basic exponential functions and the basic trigonometric functions as solutions of a difference analog of the equation for harmonic motion (2.1.1) in a complete analogy with the classical case. In the next chapter we shall study the addition theorems and related q-trigonometric identities for these functions in order to form a basis for investigation of the basic Fourier series, the main object of this book, in the subsequent chapters.

2.2. Difference Analog of Equation for Harmonic Motion

For twice differentiable functions u and x, defined in a neighborhood of the point z in the complex plane, the Taylor formula implies the following approximations of the first and the second derivatives[1]

$$\frac{u(z+h)-u(z)}{x(z+h)-x(z)}=\frac{du}{dx}+\mathrm{O}(h), \tag{2.2.1}$$

$$\frac{u(z)-u(z-h)}{x(z)-x(z-h)}=\frac{du}{dx}+\mathrm{O}(h), \tag{2.2.2}$$

and

$$\frac{1}{2}\left[\frac{u(z+h)-u(z)}{x(z+h)-x(z)}+\frac{u(z)-u(z-h)}{x(z)-x(z-h)}\right]=\frac{du}{dx}+\mathrm{O}\left(h^2\right), \tag{2.2.3}$$

$$\frac{1}{x(z+h/2)-x(z-h/2)} \times\left[\frac{u(z+h)-u(z)}{x(z+h)-x(z)}-\frac{u(z)-u(z-h)}{x(z)-x(z-h)}\right]=\frac{d^2u}{dx^2}+\mathrm{O}\left(h^2\right) \tag{2.2.4}$$

as $h \to 0$. For $h=1$ let functions $u(z)$ and $x(z)$ be defined at the points z, $z \pm 1/2$, and $z \pm 1$ in the complex z-plane and consider the *forward*, *backward* and *symmetric* difference operators given by

$$\Delta f(z)=f(z+1)-f(z), \tag{2.2.5}$$

$$\nabla f(z)=f(z)-f(z-1), \tag{2.2.6}$$

$$\delta f(z)=f(z+1/2)-f(z-1/2), \tag{2.2.7}$$

respectively; see Ex. 2.2 for elementary properties of these operators. In view of (2.2.1)–(2.2.3), the difference operators

$$\frac{\Delta u}{\Delta x}=\frac{u(z+1)-u(z)}{x(z+1)-x(z)}, \qquad \frac{\nabla u}{\nabla x}=\frac{u(z)-u(z-1)}{x(z)-x(z-1)} \tag{2.2.8}$$

and

$$\frac{1}{2}\left(\frac{\Delta u}{\Delta x}+\frac{\nabla u}{\nabla x}\right)=\frac{1}{2}\left[\frac{u(z+1)-u(z)}{x(z+1)-x(z)}+\frac{u(z)-u(z-1)}{x(z)-x(z-1)}\right] \tag{2.2.9}$$

can be viewed as difference analogs of the first derivative du/dx on a nonuniform lattice $x(z)$. In a similar fashion, due to (2.2.4) the difference operator

$$\frac{\Delta}{\nabla x_1}\left(\frac{\nabla u}{\nabla x}\right)=\frac{1}{\nabla x_1}\Delta\left(\frac{\nabla u}{\nabla x}\right)=\frac{1}{\delta x}\delta\left(\frac{\delta u}{\delta x}\right) \tag{2.2.10}$$
$$=\frac{1}{x(z+1/2)-x(z-1/2)}\left[\frac{u(z+1)-u(z)}{x(z+1)-x(z)}-\frac{u(z)-u(z-1)}{x(z)-x(z-1)}\right],$$

where $x_1(z)=x(z+1/2)$, is an analog of the second derivative d^2u/dx^2 on a nonuniform lattice $x(z)$.

[1] We use the symbol O in a usual sense; see Appendix B.4 for the definition.

Replacing the second derivative u'' by the difference operator (2.2.10), one can write the difference analog of Eq. (2.1.1) as

$$\sigma \frac{\Delta}{\nabla x_1(z)} \left(\frac{\nabla u(z)}{\nabla x(z)} \right) + \lambda u(z) = 0 \tag{2.2.11}$$

with the forward and backward difference operators defined by (2.2.5)–(2.2.6). Eq. (2.2.11) is the simplest case of a general *difference equation of hypergeometric type* on nonuniform lattices; see [**113**], [**111**], [**139**] and Ex. 2.20 for more details.

We shall consider the following *lattices*

$$x(z) = \begin{cases} C_1 q^{-z} + C_2 q^z + C_3, & \text{if } q \neq 1; \\ C_1 z^2 + C_2 z + C_3, & \text{if } q = 1. \end{cases} \tag{2.2.12}$$

where σ, λ, C_1, C_2, C_3, and q are constants. One can assume, without any lost of generality, that $C_3 = 0$. When $q = 1$ and $C_1 \neq 0$, the lattice in (2.2.12) is called *quadratic*; in the case $q = 1$ and $C_1 = 0$ this lattice is *linear*. In a similar fashion, when $q \neq 1$ and $C_1, C_2 \neq 0$ the lattice in (2.2.12) is called *q-quadratic*; and in the case $q \neq 1$ and C_1 (or C_2 but not both) $= 0$, this lattice is *q-linear*. Some properties of these *classical lattices* are discussed in Exs. 2.3–2.7.

The power series solutions of the differential equation (2.1.1) were constructed above in the form

$$u = \sum_{n=0}^{\infty} \frac{(-\lambda/\sigma)^{n/2}}{n!} (x+y)^n . \tag{2.2.13}$$

Solutions of the difference equation of hypergeometric type can be found by an extended power series method [**17**]. The most general formal "power series" solution of Eq. (2.2.11) has a somewhat similar form

$$u(z,s) = \sum_{n=0}^{\infty} \frac{(-\lambda/\sigma)^{n/2}}{\gamma(n)!} \left[x(z) - x\left(s + \frac{n-1}{2} \right) \right]^{(n)}, \tag{2.2.14}$$

where s is an additional "free" parameter, representing an analog of the shift in the classical case; an analog of the factorial is defined by

$$\gamma(n)! := \gamma(1)\,\gamma(2)\cdots\gamma(n), \qquad \gamma(k) := \frac{q^{k/2} - q^{-k/2}}{q^{1/2} - q^{-1/2}}, \tag{2.2.15}$$

and a "generalized power" is given by

$$[x(z) - x(v)]^{(n)} := \prod_{k=0}^{n-1} [x(z) - x(v-k)] . \tag{2.2.16}$$

A more formal definition and further properties of the generalized power functions are discussed in Exs. 2.9–2.11. The difference-differentiation formula

$$\frac{\delta}{\delta x(z)}\left[x(z)-x(v)\right]^{(n)}=\gamma(n)\ \left[x(z)-x(v-1/2)\right]^{(n-1)}, \tag{2.2.17}$$

where the symmetric difference operator is defined by (2.2.7), can be established by a straightforward calculation on a general q-quadratic grid (2.2.12); see **[139]** and Ex. 2.9. Using the formula (2.2.17) one can easily show that the series (2.2.14) satisfies the following equation

$$\frac{\delta u(z,s)}{\delta x(z)}=\left(-\frac{\lambda}{\sigma}\right)^{1/2}u(z,s). \tag{2.2.18}$$

Applying the divided difference operator $\delta/\delta x(z)$ once again, we arrive at Eq. (2.2.11) in a symmetric form,

$$\frac{\delta}{\delta x(z)}\left(\frac{\delta u(z,s)}{\delta x(z)}\right)+\frac{\lambda}{\sigma}\,u(z,s)=0, \tag{2.2.19}$$

provided that the series (2.2.14) converges at the points z, $z\pm 1/2$, and $z\pm 1$ in the complex z-plane.

The "even" and "odd" parts of (2.2.14), namely,

$$u_+(z,s)=\sum_{k=0}^{\infty}\frac{(-\lambda/\sigma)^k}{\gamma(2k)!}\left[x(z)-x(s+k-1/2)\right]^{(2k)}, \tag{2.2.20}$$

$$u_-(z,s)=\sum_{k=0}^{\infty}\frac{(-\lambda/\sigma)^{k+1/2}}{\gamma(2k+1)!}\left[x(z)-x(s+k)\right]^{(2k+1)}, \tag{2.2.21}$$

in turn, both satisfy Eq. (2.2.11). Also,

$$\frac{\delta u_+(z,s)}{\delta x(z)}=\left(-\frac{\lambda}{\sigma}\right)^{1/2}u_-(z,s), \tag{2.2.22}$$

$$\frac{\delta u_-(z,s)}{\delta x(z)}=\left(-\frac{\lambda}{\sigma}\right)^{1/2}u_+(z,s) \tag{2.2.23}$$

in view of (2.2.17).

Solutions (2.2.14) and (2.2.22)–(2.2.23) can be represented in terms of the basic hypergeometric series (1.1.3). Indeed, a simple calculation shows that

$$\left[x(z)-x\left(s+\frac{n-1}{2}\right)\right]^{(n)} \tag{2.2.24}$$

$$=(-C_1)^n\,q^{-ns}\left(q^{s-z-(n-1)/2},\mu q^{s+z-(n-1)/2};q\right)_n,$$

or

$$\left[x(z)-x(s+k-1/2)\right]^{(2k)}=(C_1C_2)^k\,q^{-k^2} \tag{2.2.25}$$

$$\times\left(q^{1/2+z-s}, q^{1/2-z+s}, \mu q^{1/2+z+s}, q^{1/2-z-s}/\mu; q\right)_k,$$

and

$$[x(z)-x(s+k)]^{(2k+1)} = [x(z)-x(s)] \tag{2.2.26}$$
$$\times (C_1C_2)^k q^{-k(k+1)} \left(q^{1+z-s}, q^{1-z+s}, \mu q^{1+z+s}, q^{1-z-s}/\mu; q\right)_k,$$

where $\mu = C_2/C_1$. Therefore solution (2.2.14) can be written as the sum of two ${}_4\varphi_3$'s,

$$\begin{aligned} &u(z,s) \qquad\qquad (2.2.27)\\ &= {}_4\varphi_3\left(\begin{matrix} q^{1/2+z-s}, q^{1/2-z+s}, \mu q^{1/2+z+s}, q^{1/2-z-s}/\mu \\ -q,\quad q^{1/2},\quad -q^{1/2}\end{matrix}; q, -\frac{\lambda C_1C_2(1-q)^2}{\sigma q^{1/2}}\right)\\ &+\left(-\frac{\lambda}{\sigma}\right)^{1/2}[x(z)-x(s)]\\ &\times {}_4\varphi_3\left(\begin{matrix} q^{1+z-s}, q^{1-z+s}, \mu q^{1+z+s}, q^{1-z-s}/\mu \\ -q,\quad q^{3/2},\quad -q^{3/2}\end{matrix}; q, -\frac{\lambda C_1C_2(1-q)^2}{\sigma q^{1/2}}\right).\end{aligned}$$

Both the ${}_4\varphi_3$'s converge absolutely when $\left|C_1C_2(1-q)^2\lambda/\sigma\right| < 1$. Analytic continuation in a larger domain will be discussed later.

2.3. Basic Exponential Functions

It is natural to consider the function (2.2.27) as a general q-analog of the exponential function on q-linear and q-quadratic grids (2.2.12). Let us specify the parameters in a convenient way, change the normalization, and introduce the *basic exponential function* on a q-quadratic grid as follows

$$\begin{aligned} \mathcal{E}_q(x,y;\alpha) &= \frac{(\alpha^2;q^2)_\infty}{(q\alpha^2;q^2)_\infty} \qquad\qquad (2.3.1)\\ &\times \sum_{n=0}^{\infty}\frac{q^{n^2/4}\,\alpha^n}{(q;q)_n} e^{-in\varphi}\left(-q^{(1-n)/2}e^{i\theta+i\varphi}, -q^{(1-n)/2}e^{i\varphi-i\theta}; q\right)_n\\ &= \frac{(\alpha^2;q^2)_\infty}{(q\alpha^2;q^2)_\infty}\\ &\times\left[{}_4\varphi_3\left(\begin{matrix} -q^{1/2}e^{i\theta+i\varphi}, -q^{1/2}e^{i\theta-i\varphi}, -q^{1/2}e^{i\varphi-i\theta}, -q^{1/2}e^{-i\theta-i\varphi} \\ -q,\quad q^{1/2},\quad -q^{1/2}\end{matrix}; q, \alpha^2\right)\right.\\ &+\frac{2q^{1/4}}{1-q}\alpha(\cos\theta+\cos\varphi)\\ &\times\left.{}_4\varphi_3\left(\begin{matrix} -qe^{i\theta+i\varphi}, -qe^{i\theta-i\varphi}, -qe^{i\varphi-i\theta}, -qe^{-i\theta-i\varphi} \\ -q,\quad q^{3/2},\quad -q^{3/2}\end{matrix}; q, \alpha^2\right)\right],\end{aligned}$$

where $x = \cos\theta$, $y = \cos\varphi$ and $|\alpha| < 1$. Analytic continuation in a larger domain will be discussed later. The $\mathcal{E}_q(x, y; \alpha)$, which is a function of three variables x, y, α and the base q, is an analog of $\exp\alpha(x + y)$,

$$\lim_{q \to 1^-} \mathcal{E}_q(x, y; \alpha(1-q)/2) = \exp\alpha(x+y). \tag{2.3.2}$$

The basic exponential functions on a q-quadratic grid were originally introduced by Ismail and Zhang **[76]** with a different notation and normalization; see also **[17]**, **[140]** for the corresponding solutions of a q-analog of the equation for harmonic motion. The above notation (2.3.1) is due to Suslov **[140]**.

We shall also consider the function

$$\begin{aligned} \mathcal{E}_q(x; \alpha) &= \mathcal{E}_q(x, 0; \alpha) \qquad (2.3.3)\\ &= \frac{(\alpha^2; q^2)_\infty}{(q\alpha^2; q^2)_\infty} \sum_{n=0}^\infty \frac{q^{n^2/4}\,\alpha^n}{(q;q)_n} (-i)^n \left(-iq^{(1-n)/2}e^{i\theta}, -iq^{(1-n)/2}e^{-i\theta}; q\right)_n \\ &= \frac{(\alpha^2; q^2)_\infty}{(q\alpha^2; q^2)_\infty} \left[{}_2\varphi_1\left(\begin{matrix} -qe^{2i\theta}, -qe^{-2i\theta} \\ q \end{matrix} ; q^2, \alpha^2 \right) \right. \\ &\quad \left. + \frac{2q^{1/4}}{1-q} \alpha\cos\theta \; {}_2\varphi_1\left(\begin{matrix} -q^2e^{2i\theta}, -q^2e^{-2i\theta} \\ q^3 \end{matrix} ; q^2, \alpha^2 \right) \right] \end{aligned}$$

as a q-analog of $\exp(\alpha x)$.

The following simple properties hold

$$\mathcal{E}_q(0, 0; \alpha) = \mathcal{E}_q(0; \alpha) = 1, \tag{2.3.4}$$

$$\mathcal{E}_q(x, y; \alpha) = \mathcal{E}_q(y, x; \alpha). \tag{2.3.5}$$

Indeed,

$$\begin{aligned} \mathcal{E}_q(0, 0; \alpha) = \mathcal{E}_q(0; \alpha) &= \frac{(\alpha^2; q^2)_\infty}{(q\alpha^2; q^2)_\infty} \; {}_2\varphi_1\left(\begin{matrix} q, \; q \\ q \end{matrix} ; \; q^2, \; \alpha^2 \right) \\ &= \frac{(\alpha^2; q^2)_\infty}{(q\alpha^2; q^2)_\infty} \; {}_1\varphi_0\left(\begin{matrix} q \\ - \end{matrix} ; \; q^2, \; \alpha^2 \right) = 1 \end{aligned} \tag{2.3.6}$$

by the q-binomial theorem. The symmetry relation (2.3.5) is due to the following representations for the q-shifted factorials

$$\begin{aligned} & e^{-2ik\varphi} \left(-q^{(1-2k)/2}e^{i\theta+i\varphi}, -q^{(1-2k)/2}e^{i\varphi-i\theta}; q\right)_{2k} \qquad (2.3.7)\\ &= e^{-2ik\varphi} \left(-q^{(1-2k)/2}e^{i\theta+i\varphi}, -q^{(1-2k)/2}e^{i\varphi-i\theta}; q\right)_k \\ &\times \left(-q^{1/2}e^{i\theta+i\varphi}, -q^{1/2}e^{i\varphi-i\theta}; q\right)_k \\ &= q^{-k^2} \left(-q^{1/2}e^{i\theta+i\varphi}, -q^{1/2}e^{i\varphi-i\theta}, -q^{1/2}e^{i\theta-i\varphi}, -q^{1/2}e^{-i\theta-i\varphi}; q\right)_k \end{aligned}$$

and

$$e^{-i(2k+1)\varphi} \left(-q^{-k}e^{i\theta+i\varphi}, -q^{-k}e^{i\varphi-i\theta}; q\right)_{2k+1} \tag{2.3.8}$$

$$= e^{-i(2k+1)\varphi}\left(1+e^{i\theta+i\varphi}\right)\left(1+e^{i\varphi-i\theta}\right)$$
$$\times\left(-q^{-k}e^{i\theta+i\varphi},-q^{-k}e^{i\varphi-i\theta},-qe^{i\theta+i\varphi},-qe^{i\theta-i\varphi};q\right)_k$$
$$= 4q^{-k(k+1)}\cos\left(\left(\theta+\varphi\right)/2\right)\cos\left(\left(\theta-\varphi\right)/2\right)$$
$$\times\left(-qe^{i\theta+i\varphi},-qe^{i\varphi-i\theta},-qe^{i\theta-i\varphi},-qe^{-i\theta-i\varphi};q\right)_k,$$

which are obviously symmetric in θ and φ. Another symmetry property

$$\mathcal{E}_q\left(x;\alpha\right)=\mathcal{E}_q\left(-x;-\alpha\right) \tag{2.3.9}$$

follows directly from (2.3.3) if one replaces $x\to -x$, or $\theta\to\pi-\theta$, and $\alpha\to-\alpha$.

The following special values of the q-exponential function are important in the theory of basic Fourier series:

$$\mathcal{E}_q\left(\eta;\alpha\right)=\mathcal{E}_q\left(-\eta;-\alpha\right)=\frac{\left(-\alpha;q\right)_\infty}{\left(q^{1/2}\alpha;q\right)_\infty}=\frac{\left(-\alpha;q^{1/2}\right)_\infty}{\left(q\alpha^2;q^2\right)_\infty} \tag{2.3.10}$$

with $\eta=\left(q^{1/4}+q^{-1/4}\right)/2$. They can be derived from (2.3.3). Indeed, when $x=\eta$, or $e^{i\theta}=q^{1/4}$, one gets

$$\frac{\left(q\alpha^2;q^2\right)_\infty}{\left(\alpha^2;q^2\right)_\infty}\,\mathcal{E}_q\left(\eta;\alpha\right)={}_2\varphi_1\left(\begin{matrix}-q^{1/2},\ -q^{3/2}\\ q\end{matrix};\ q^2,\ \alpha^2\right) \tag{2.3.11}$$
$$+\alpha\frac{1+q^{1/2}}{1-q}\,{}_2\varphi_1\left(\begin{matrix}-q^{3/2},\ -q^{5/2}\\ q^3\end{matrix};\ q^2,\ \alpha^2\right)$$
$$=\sum_{k=0}^{\infty}\frac{\left(-q^{1/2},-q^{3/2};q^2\right)_k}{\left(q,\ q^2;q^2\right)_k}\,\alpha^{2k}+\alpha\frac{1+q^{1/2}}{1-q}\sum_{k=0}^{\infty}\frac{\left(-q^{3/2},-q^{5/2};q^2\right)_k}{\left(q^2,\ q^3;q^2\right)_k}\,\alpha^{2k}$$
$$=\sum_{k=0}^{\infty}\frac{\left(-q^{1/2};q\right)_{2k}}{\left(q;q\right)_{2k}}\,\alpha^{2k}+\sum_{k=0}^{\infty}\frac{\left(-q^{1/2};q\right)_{2k+1}}{\left(q;q\right)_{2k+1}}\,\alpha^{2k+1}$$
$$=\sum_{n=0}^{\infty}\frac{\left(-q^{1/2};q\right)_n}{\left(q;q\right)_n}\,\alpha^n=\frac{\left(-q^{1/2}\alpha;q\right)_\infty}{\left(\alpha;q\right)_\infty}$$

by (1.4.12) and the q-binomial theorem, which implies (2.3.10) and the following relation with Jackson's q-exponential function:

$$E_{q^{1/2}}\left(\alpha\right)=\left(q\alpha^2;q^2\right)_\infty\mathcal{E}_q\left(\eta;\alpha\right). \tag{2.3.12}$$

The basic exponential functions $\mathcal{E}_q\left(x,y;\alpha\right)$ and $\mathcal{E}_q\left(x;\alpha\right)$ satisfy the difference equation

$$\frac{\delta u}{\delta x}=\frac{2q^{1/4}\alpha}{1-q}\,u, \tag{2.3.13}$$

which is a q-version of

$$\frac{d}{dx} e^{\alpha(x+y)} = \alpha\, e^{\alpha(x+y)} \tag{2.3.14}$$

on a q-quadratic grid; see (2.2.18). The operator $\delta/\delta x$ is the standard Askey–Wilson divided difference operator

$$\frac{\delta u(z)}{\delta x(z)} = \frac{u(z+1/2) - u(z-1/2)}{x(z+1/2) - x(z-1/2)} \tag{2.3.15}$$

with $x(z) = (q^z + q^{-z})/2 = \cos\theta$, $q^z = e^{i\theta}$; see Ex. 2.11. Applying this operator to (2.3.13) once again, we obtain the second order difference equation

$$\frac{\delta^2 u}{\delta x^2} = \left(\frac{2q^{1/4}}{1-q}\alpha\right)^2 u, \tag{2.3.16}$$

which is equivalent to (2.2.11).

A double series extension of the q-exponential function (2.3.1) was recently introduced in **[142]** and **[144]** as

$$\begin{aligned} &\mathcal{E}_q(x, y; \alpha, \beta) \qquad (2.3.17)\\ &= \frac{(\beta^2; q^2)_\infty}{(q\alpha^2; q^2)_\infty} \sum_{n=0}^{\infty} \frac{q^{n^2/4}\, \beta^n}{(q;q)_n} \\ &\times e^{-in\varphi} \left(-q^{(1-n)/2} e^{i\theta+i\varphi}\alpha/\beta,\ -q^{(1-n)/2} e^{i\varphi-i\theta}\alpha/\beta; q\right)_n \\ &\times {}_2\varphi_2\left(\begin{matrix} q^{-n}, & \alpha^2/\beta^2 \\ -q^{(1-n)/2}e^{i\theta+i\varphi}\alpha/\beta, & -q^{(1-n)/2}e^{i\varphi-i\theta}\alpha/\beta \end{matrix}; q, qe^{2i\varphi}\right). \end{aligned}$$

This function, which depends on four variables x, y, α, β and the base q, naturally appears in an "addition" theorem for the q-exponential functions discussed in the next chapter; see Section 3.5, Eq. (3.5.1). If $\beta = \alpha$, the second series terminates and we obtain the basic exponential function (2.3.1). Changing the order of summation gives an alternate form

$$\begin{aligned} &\mathcal{E}_q(x, y; \alpha, \beta) \qquad (2.3.18)\\ &= \frac{(\beta^2; q^2)_\infty}{(q\alpha^2; q^2)_\infty} \sum_{k=0}^{\infty} \frac{(\alpha^2/\beta^2; q)_k}{(q;q)_k} q^{k^2/4} \left(\beta e^{i\varphi}\right)^k \\ &\times \sum_{n=0}^{\infty} \frac{q^{n(n-2k)/4}}{(q;q)_n} \beta^n e^{-in\varphi} \\ &\times \left(-q^{(1-n+k)/2} e^{i\theta+i\varphi}\alpha/\beta, -q^{(1-n+k)/2} e^{i\varphi-i\theta}\alpha/\beta; q\right)_n. \end{aligned}$$

When $\beta = \alpha$ the first sum terminates and we obtain (2.3.1) once again. The second sum can be reduced to the sum of two ${}_4\varphi_3$ series similar to those in

Eq. (2.3.1):

$$\mathcal{E}_q(x,y;\alpha,\beta) = \frac{(\beta^2;q^2)_\infty}{(q\alpha^2;q^2)_\infty}\sum_{k=0}^{\infty}\frac{(\alpha^2/\beta^2;q)_k}{(q;q)_k}q^{k^2/4}\left(\beta e^{i\varphi}\right)^k \tag{2.3.19}$$

$$\times\left[{}_4\varphi_3\left(\begin{matrix}-q^{(1+k)/2}e^{i\theta+i\varphi}\alpha/\beta, -q^{(1-k)/2}e^{-i\theta-i\varphi}\beta/\alpha, \\ -q, \\ -q^{(1+k)/2}e^{i\varphi-i\theta}\alpha/\beta, -q^{(1-k)/2}e^{i\theta-i\varphi}\beta/\alpha \\ q^{1/2}, \quad -q^{1/2}\end{matrix}\;;q,\,\alpha^2\right)\right.$$

$$+\frac{q^{1/4}\alpha}{1-q}\left(q^{k/2}e^{i\varphi}\alpha/\beta + q^{-k/2}e^{-i\varphi}\beta/\alpha + 2\cos\theta\right)$$

$$\left.\times{}_4\varphi_3\left(\begin{matrix}-q^{1+k/2}e^{i\theta+i\varphi}\alpha/\beta, -q^{1-k/2}e^{-i\theta-i\varphi}\beta/\alpha, \\ -q, \\ -q^{1+k/2}e^{i\varphi-i\theta}\alpha/\beta, -q^{1-k/2}e^{i\theta-i\varphi}\beta/\alpha \\ q^{3/2}, \quad -q^{3/2}\end{matrix}\;;q,\,\alpha^2\right)\right].$$

The q-exponential function $\mathcal{E}_q(x,y;\alpha,\beta)$ given by (2.3.17) is an analog of $\exp(\alpha x+\beta y)$. Indeed, in the limit $q\to 1^-$ one gets

$$\lim_{q\to 1^-}\mathcal{E}_q(x,y;(1-q)\alpha/2,(1-q)\beta/2) \tag{2.3.20}$$

$$=\sum_{n=0}^{\infty}\frac{(\beta/2)^n}{n!}\,e^{-in\varphi}\left(1+e^{i\theta+i\varphi}\alpha/\beta\right)^n\left(1+e^{i\varphi-i\theta}\alpha/\beta\right)^n$$

$$\times\sum_{k=0}^{n}\frac{(-n)_k}{k!}\left(-\frac{\left(1-\alpha^2/\beta^2\right)e^{2i\varphi}}{\left(1+e^{i\theta+i\varphi}\alpha/\beta\right)\left(1+e^{i\varphi-i\theta}\alpha/\beta\right)}\right)^k$$

$$=\sum_{n=0}^{\infty}\frac{(\beta/2)^n}{n!}\,e^{-in\varphi}\left(1+e^{i\theta+i\varphi}\alpha/\beta\right)^n\left(1+e^{i\varphi-i\theta}\alpha/\beta\right)^n$$

$$\times\left(1+\frac{\left(1-\alpha^2/\beta^2\right)e^{2i\varphi}}{\left(1+e^{i\theta+i\varphi}\alpha/\beta\right)\left(1+e^{i\varphi-i\theta}\alpha/\beta\right)}\right)^n$$

$$=\sum_{n=0}^{\infty}\frac{(\alpha x+\beta y)^n}{n!} = \exp(\alpha x+\beta y)$$

by the binomial theorem. We shall also see in the next chapter that the function $u = \mathcal{E}_q(x,y;\alpha,\beta)$ in (2.3.17)–(2.3.18) is a double series solution of the difference equations (2.3.13) and (2.3.16). Methods of solving of the difference equations of hypergeometric type are discussed in [**17**], [**139**] and [**140**]; see also Exs. 2.20–2.26. They do not involve this double series solution.

The function $\mathcal{E}_q(x,y;\alpha,\beta)$ satisfies the following simple properties

$$\mathcal{E}_q(x,y;\alpha,\alpha) = \mathcal{E}_q(x,y;\alpha), \quad \mathcal{E}_q(x,y;\alpha,\beta) = \mathcal{E}_q(y,x;\beta,\alpha), \tag{2.3.21}$$

$$\mathcal{E}_q(x,0;\alpha,\beta)=\mathcal{E}_q(x;\alpha), \qquad \mathcal{E}_q(0,y;\alpha,\beta)=\mathcal{E}_q(y;\beta). \tag{2.3.22}$$

These properties are closely related to a q-addition theorem discussed in the next chapter; see Section 3.5, Eq. (3.5.1).

Near the end of this section we shall derive convenient representations for the $\mathcal{E}_q(x;\alpha)$. Using the transformation formula (III.3) of [**46**] or (A.3.3), one can rewrite (2.3.3) as

$$\begin{aligned}\mathcal{E}_q(x;\alpha) &= {}_2\varphi_1\left(\begin{matrix}-e^{2i\theta},-e^{-2i\theta}\\ q\end{matrix};q^2,q\alpha^2\right)\\ &\quad+\frac{2q^{1/4}}{1-q}\alpha\cos\theta\ {}_2\varphi_1\left(\begin{matrix}-qe^{2i\theta},-qe^{-2i\theta}\\ q^3\end{matrix};q^2,q\alpha^2\right)\\ &=\sum_{n=0}^{\infty}\frac{q^{n/2}\alpha^n}{(q;q)_n}\ \phi_n(x;q),\end{aligned} \tag{2.3.23}$$

where by the definition

$$\phi_{2k}(x;q)=\prod_{p=0}^{k-1}\left(4x^2q^{2p}+\left(1-q^{2p}\right)^2\right), \tag{2.3.24}$$

$$\phi_{2k+1}(x;q)=2q^{-1/4}\ x\prod_{p=0}^{k-1}\left(4x^2q^{2p+1}+\left(1-q^{2p+1}\right)^2\right). \tag{2.3.25}$$

(We assume that the empty products when $k=0$ here are equal to 1.) The right side in (2.3.23) provides Taylor's expansion of the $\mathcal{E}_q(x;\alpha)$ with respect to α in the larger domain $q\,|\alpha|^2<1$; see also Ex. 3.4.

In the recent papers [**72**]–[**73**] Ismail and Stanton pointed out the following representations for the q-exponential function

$$\begin{aligned}\mathcal{E}_q(x;\alpha) &= \frac{\left(-\alpha;q^{1/2}\right)_\infty}{\left(q\alpha^2;q^2\right)_\infty}\ {}_2\varphi_1\left(\begin{matrix}q^{1/4}e^{i\theta},\ q^{1/4}e^{-i\theta}\\ -q^{1/2}\end{matrix};q^{1/2},-\alpha\right)\\ &=\frac{\left(\alpha;q^{1/2}\right)_\infty}{\left(q\alpha^2;q^2\right)_\infty}\ {}_2\varphi_1\left(\begin{matrix}-q^{1/4}e^{i\theta},\ -q^{1/4}e^{-i\theta}\\ -q^{1/2}\end{matrix};q^{1/2},\alpha\right),\end{aligned} \tag{2.3.26}$$

or

$$\begin{aligned}\mathcal{E}_q(x;\alpha) &= \frac{\left(-\alpha;q^{1/2}\right)_\infty}{\left(q\alpha^2;q^2\right)_\infty}\sum_{n=0}^{\infty}\frac{(-\alpha)^n}{(q;q)_n}\varphi_n(x;q)\\ &=\frac{\left(\alpha;q^{1/2}\right)_\infty}{\left(q\alpha^2;q^2\right)_\infty}\sum_{n=0}^{\infty}\frac{\alpha^n}{(q;q)_n}\varphi_n(-x;q),\end{aligned} \tag{2.3.27}$$

where $|\alpha|<1$ and by the definition

$$\varphi_n(\cos\theta;q)=\left(q^{1/4}e^{i\theta},q^{1/4}e^{-i\theta};q^{1/2}\right)_n. \tag{2.3.28}$$

One can see that these representations follow directly from (2.3.1) and (2.3.3) when $e^{i\varphi} = q^{1/4}$, with the help of the simple identity

$$q^{n(n-1)/4} \left(q^{(1-n)/2+1/4}e^{i\theta}, q^{(1-n)/2+1/4}e^{-i\theta}; q\right)_n = \left(q^{1/4}e^{i\theta}, q^{1/4}e^{-i\theta}; q^{1/2}\right)_n, \tag{2.3.29}$$

the addition theorem (3.1.3) discussed in the next chapter, and the special values of the basic exponential function (2.3.10); see Ex. 2.28. Another independent proof of the representations (2.3.26)–(2.3.27) will be given on the basis of the difference equation (2.2.11) in Section 3.4 of the next chapter.

Eqs. (2.3.26) imply the symmetry relation (2.3.9) for the basic exponential function in the form

$$\mathcal{E}_q(x; -\alpha) = \mathcal{E}_q(-x; \alpha), \tag{2.3.30}$$

which follows also from the generating relation (3.4.15) derived in the next chapter, and the special values

$$\mathcal{E}_q(-x; 1) = \mathcal{E}_q(x; -1) = \frac{\left(q^{1/2}, q^{1/4}e^{i\theta}, q^{1/4}e^{-i\theta}; q^{1/2}\right)_\infty}{(q;q)_\infty (q;q^2)_\infty} \tag{2.3.31}$$

due to (III.3) and (III.1) of [**46**] or (A.3.3) and (A.3.1), respectively.

Eqs. (2.3.26) imply also the following representation

$$\mathcal{E}_q(x; \alpha) = \frac{\left(-\alpha q^{1/4}e^{i\theta}, -\alpha q^{1/4}e^{-i\theta}; q^{1/2}\right)_\infty}{(q\alpha^2;q^2)_\infty \left(-q^{1/2}; q^{1/2}\right)_\infty} \times {}_2\varphi_2\left(\begin{matrix} \alpha, \ -\alpha \\ -\alpha q^{1/4}e^{i\theta}, \ -\alpha q^{1/4}e^{-i\theta} \end{matrix}; q^{1/2}, \ -q^{1/2}\right) \tag{2.3.32}$$

in view of the transformation formula (A.3.5) from Appendix A. Eq. (2.3.32) shows that the function $(q\alpha^2;q^2)_\infty \mathcal{E}_q(x;\alpha)$ is an entire function in α and z when $e^{i\theta} = q^z$. On the other hand, from (2.3.26),

$$\begin{aligned} \mathcal{E}_q(x; \alpha) = &\frac{\left(q^{1/4}e^{-i\theta}, -q^{1/4}e^{-i\theta}, -\alpha q^{1/4}e^{i\theta}, -q^{1/4}e^{-i\theta}/\alpha; q^{1/2}\right)_\infty}{(q\alpha^2;q^2)_\infty \left(-q^{1/2}, e^{-2i\theta}, -q^{1/2}/\alpha; q^{1/2}\right)_\infty} \\ &\times {}_2\varphi_1\left(\begin{matrix} q^{1/4}e^{i\theta}, -q^{1/4}e^{i\theta} \\ q^{1/2}e^{2i\theta} \end{matrix}; q^{1/2}, \frac{q^{1/2}}{\alpha}\right) \\ &+ \frac{\left(q^{1/4}e^{i\theta}, -q^{1/4}e^{i\theta}, -\alpha q^{1/4}e^{-i\theta}, -q^{1/4}e^{i\theta}/\alpha; q^{1/2}\right)_\infty}{(q\alpha^2;q^2)_\infty \left(-q^{1/2}, e^{2i\theta}, -q^{1/2}/\alpha; q^{1/2}\right)_\infty} \\ &\times {}_2\varphi_1\left(\begin{matrix} q^{1/4}e^{-i\theta}, -q^{1/4}e^{-i\theta} \\ q^{1/2}e^{-2i\theta} \end{matrix}; q^{1/2}, \frac{q^{1/2}}{\alpha}\right), \end{aligned} \tag{2.3.33}$$

when $|\alpha| > q^{1/2}$ by (III.32) of [**46**] or (A.3.8).

2.4. Basic Trigonometric Functions

The *basic cosine* $C_q(x,y;\omega)$ and *basic sine* $S_q(x,y;\omega)$ functions can be introduced by the following analog of the Euler formula

$$\mathcal{E}_q(x,y;i\omega)=C_q(x,y;\omega)+iS_q(x,y;\omega). \tag{2.4.1}$$

Hence,

$$\begin{aligned} C_q(x,y;\omega) &= \frac{(-\omega^2;q^2)_\infty}{(-q\omega^2;q^2)_\infty} \\ &\times {}_4\varphi_3\left(\begin{matrix} -q^{1/2}e^{i\theta+i\varphi}, -q^{1/2}e^{i\theta-i\varphi}, -q^{1/2}e^{i\varphi-i\theta}, -q^{1/2}e^{-i\theta-i\varphi} \\ -q, \quad q^{1/2}, \quad -q^{1/2} \end{matrix} ;q,-\omega^2\right) \end{aligned} \tag{2.4.2}$$

and

$$\begin{aligned} S_q(x,y;\omega) &= \frac{(-\omega^2;q^2)_\infty}{(-q\omega^2;q^2)_\infty}\frac{2q^{1/4}}{1-q}\,\omega(\cos\theta+\cos\varphi) \\ &\times {}_4\varphi_3\left(\begin{matrix} -qe^{i\theta+i\varphi}, -qe^{i\theta-i\varphi}, -qe^{i\varphi-i\theta}, -qe^{-i\theta-i\varphi} \\ -q, \quad q^{3/2}, \quad -q^{3/2} \end{matrix} ;q,-\omega^2\right). \end{aligned} \tag{2.4.3}$$

They are q-analogs of $\cos\omega(x+y)$ and $\sin\omega(x+y)$, respectively. Indeed, by (2.3.2) and (2.4.1),

$$\lim_{q\to1^-} C_q(x,y;\omega(1-q)/2)=\cos\omega(x+y), \tag{2.4.4}$$

$$\lim_{q\to1^-} S_q(x,y;\omega(1-q)/2)=\sin\omega(x+y). \tag{2.4.5}$$

The special cases $y=0$ $(\varphi=\pi/2)$ of (2.4.2)–(2.4.3) are

$$C_q(x;\omega)=\frac{(-\omega^2;q^2)_\infty}{(-q\omega^2;q^2)_\infty}\,{}_2\varphi_1\left(\begin{matrix} -qe^{2i\theta}, \ -qe^{-2i\theta} \\ q \end{matrix} ;\ q^2,\ -\omega^2\right) \tag{2.4.6}$$

and

$$\begin{aligned} S_q(x;\omega) &= \frac{(-\omega^2;q^2)_\infty}{(-q\omega^2;q^2)_\infty}\frac{2q^{1/4}\omega}{1-q}\cos\theta \\ &\times {}_2\varphi_1\left(\begin{matrix} -q^2e^{2i\theta}, \ -q^2e^{-2i\theta} \\ q^3 \end{matrix} ;\ q^2,-\omega^2\right). \end{aligned} \tag{2.4.7}$$

The basic cosine and sine functions (2.4.6)–(2.4.7) are defined here for $|\omega|<1$ only. An analytic continuation of these functions in a larger domain was discussed in **[26]**, **[63]**, **[76]**, and **[140]**. For example,

$$\begin{aligned} C_q(x;\omega) &= \frac{(q\omega^2e^{2i\theta},\ q\omega^2e^{-2i\theta};\ q^2)_\infty}{(q,\ -q\omega^2;\ q^2)_\infty} \\ &\times {}_2\varphi_2\left(\begin{matrix} -\omega^2, & -q\omega^2 \\ q\omega^2e^{2i\theta}, & q\omega^2e^{-2i\theta} \end{matrix} ;\ q^2,\ q\right) \end{aligned} \tag{2.4.8}$$

and

$$S_q(x;\omega) = \frac{\left(q^2\omega^2 e^{2i\theta},\ q^2\omega^2 e^{-2i\theta};\ q^2\right)_\infty}{\left(q^3,\ -q\omega^2;\ q^2\right)_\infty} \frac{2q^{1/4}\omega}{1-q} \cos\theta \quad (2.4.9)$$
$$\times\ {}_2\varphi_2\left(\begin{matrix} -\omega^2, & -q\omega^2 \\ q^2\omega^2 e^{2i\theta}, & q^2\omega^2 e^{-2i\theta} \end{matrix}\ ;\ q^2,\ q^3\right)$$

by (A.3.5). One can see from (2.4.8) and (2.4.9) that the basic trigonometric functions (2.4.6) and (2.4.7) are entire functions in z when $e^{i\theta} = q^z$. On the other hand, from (2.4.6) and (2.4.7) we get

$$C_q(\cos\theta;\omega) = {}_2\varphi_1\left(\begin{matrix} -e^{2i\theta}, -e^{-2i\theta} \\ q \end{matrix}\ ; q^2, -q\omega^2\right) \quad (2.4.10)$$
$$= \frac{\left(q\omega^2 e^{2i\theta},\ e^{-2i\theta}/\omega^2; q^2\right)_\infty}{\left(e^{-2i\theta}; q\right)_\infty \left(q, -q\omega^2, -q/\omega^2; q^2\right)_\infty}\ {}_2\varphi_1\left(\begin{matrix} -e^{2i\theta}, -qe^{2i\theta} \\ q^2 e^{4i\theta} \end{matrix}\ ; q^2, -\frac{q^2}{\omega^2}\right)$$
$$+ \frac{\left(q\omega^2 e^{-2i\theta},\ e^{2i\theta}/\omega^2; q^2\right)_\infty}{\left(e^{2i\theta}; q\right)_\infty \left(q, -q\omega^2, -q/\omega^2; q^2\right)_\infty}\ {}_2\varphi_1\left(\begin{matrix} -e^{-2i\theta}, -qe^{-2i\theta} \\ q^2 e^{-4i\theta} \end{matrix}\ ; q^2, -\frac{q^2}{\omega^2}\right)$$

and

$$S_q(\cos\theta;\omega) \quad (2.4.11)$$
$$= \frac{2q^{1/4}\omega}{1-q} \cos\theta\ {}_2\varphi_1\left(\begin{matrix} -qe^{2i\theta}, -qe^{-2i\theta} \\ q^3 \end{matrix}\ ; q^2, -q\omega^2\right)$$
$$= \frac{q^{1/4}\omega e^{i\theta} \left(q^2\omega^2 e^{2i\theta}, e^{-2i\theta}/\omega^2; q^2\right)_\infty}{\left(e^{-2i\theta}; q\right)_\infty \left(q, -q\omega^2, -q/\omega^2; q^2\right)_\infty}\ {}_2\varphi_1\left(\begin{matrix} -e^{2i\theta}, -qe^{2i\theta} \\ q^2 e^{4i\theta} \end{matrix}\ ; q^2, -\frac{q^2}{\omega^2}\right)$$
$$+ \frac{q^{1/4}\omega e^{-i\theta} \left(q^2\omega^2 e^{-2i\theta}, e^{2i\theta}/\omega^2; q^2\right)_\infty}{\left(e^{2i\theta}; q\right)_\infty \left(q, -q\omega^2, -q/\omega^2; q^2\right)_\infty}\ {}_2\varphi_1\left(\begin{matrix} -e^{-2i\theta}, -qe^{-2i\theta} \\ q^2 e^{-4i\theta} \end{matrix}\ ; q^2, -\frac{q^2}{\omega^2}\right)$$

by (III.3) and (III.32) of [**46**] or (A.3.3) and (A.3.8). Eqs. (2.4.10) and (2.4.11) determine the large ω-asymptotics of the basic cosine $C_q(x;\omega)$ and basic sine $S_q(x;\omega)$ functions, respectively.

The basic trigonometric functions can also be written as

$$C_q(\cos\theta;\omega) \quad (2.4.12)$$
$$= \frac{\left(q\omega^2 e^{2i\theta}; q^2\right)_\infty}{\left(e^{-2i\theta}; q\right)_\infty \left(q, -q\omega^2; q^2\right)_\infty} \sum_{n=0}^{\infty} \frac{q^{2n}\left(-e^{2i\theta}; q\right)_{2n}}{\left(q^2, q^2 e^{4i\theta}; q^2\right)_n} \left(q\omega^2 e^{2i\theta}; q^2\right)_n^{-1}$$
$$+ \frac{\left(q\omega^2 e^{-2i\theta}; q^2\right)_\infty}{\left(e^{2i\theta}; q\right)_\infty \left(q, -q\omega^2; q^2\right)_\infty} \sum_{n=0}^{\infty} \frac{q^{2n}\left(-e^{-2i\theta}; q\right)_{2n}}{\left(q^2, q^2 e^{-4i\theta}; q^2\right)_n} \left(q\omega^2 e^{-2i\theta}; q^2\right)_n^{-1}$$

and

$$S_q(\cos\theta;\omega) \quad (2.4.13)$$
$$= \frac{e^{i\theta}\omega \left(q^2\omega^2 e^{2i\theta}; q^2\right)_\infty}{\left(e^{-2i\theta}; q\right)_\infty \left(q, -q\omega^2; q^2\right)_\infty} \sum_{n=0}^{\infty} \frac{q^{2n+1/4}\left(-qe^{2i\theta}; q\right)_{2n}}{\left(q^2, q^2 e^{4i\theta}; q^2\right)_n} \left(q^2\omega^2 e^{2i\theta}; q^2\right)_n^{-1}$$

$$+\frac{e^{-i\theta}\omega\left(q^2\omega^2e^{-2i\theta};q^2\right)_\infty}{\left(e^{2i\theta};q\right)_\infty\left(q,-q\omega^2;q^2\right)_\infty}\sum_{n=0}^{\infty}\frac{q^{2n+1/4}\left(-qe^{-2i\theta};q\right)_{2n}}{\left(q^2,q^2e^{-4i\theta};q^2\right)_n}\left(q^2\omega^2e^{-2i\theta};q^2\right)_n^{-1}$$

due to (A.3.6). It is worth noting that the expansions (2.4.12)–(2.4.13) are not in terms of the usual inverse power sequence $\left\{(x\omega)^{-n}\right\}_{n=0}^{\infty}$, but are sums of two expansions in terms of the "inverse generalized powers" $\left(q^2\omega^2e^{\pm 2i\theta};q^2\right)_n^{-1}$; cf. [**54**].

Formulas (2.4.8)–(2.4.13) give analytic continuation of the basic trigonometric functions (2.4.6)–(2.4.7) in the entire complex ω-plane. The analytic continuation of the q-trigonometric functions (2.4.1)–(2.4.2) can be obtained on the basis of the following addition theorems

$$C_q(x,y;\omega)=C_q(x;\omega)\,C_q(y;\omega)-S_q(x;\omega)\,S_q(y;\omega),$$
$$S_q(x,y;\omega)=S_q(x;\omega)\,C_q(y;\omega)+C_q(x;\omega)\,S_q(y;\omega),$$

discussed in the next chapter; see Eqs. (3.2.7)–(3.2.8). On the other hand,

$$\begin{aligned}
&C_q(x,y;\omega) \qquad\qquad (2.4.14)\\
&=\frac{\left(-q^{1/2}e^{i\theta-i\varphi},-q^{1/2}e^{i\varphi-i\theta},q^{1/2}\omega^2e^{i\theta+i\varphi},q^{1/2}e^{-i\theta-i\varphi}/\omega^2;q\right)_\infty}{\left(e^{-2i\theta},e^{-2i\varphi},-q/\omega^2;q\right)_\infty\left(-q\omega^2;q^2\right)_\infty^2}\\
&\times {}_4\varphi_3\left(\begin{matrix}q^{1/2}e^{i\theta+i\varphi},-q^{1/2}e^{i\theta+i\varphi},qe^{i\theta+i\varphi},-qe^{i\theta+i\varphi}\\ qe^{2i\theta},\ qe^{2i\varphi},\ qe^{2i\theta+2i\varphi}\end{matrix};q,-\frac{q}{\omega^2}\right)\\
&+\frac{\left(-q^{1/2}e^{i\theta+i\varphi},-q^{1/2}e^{-i\theta-i\varphi},q^{1/2}\omega^2e^{i\theta-i\varphi},q^{1/2}e^{i\varphi-i\theta}/\omega^2;q\right)_\infty}{\left(e^{-2i\theta},e^{2i\varphi},-q/\omega^2;q\right)_\infty\left(-q\omega^2;q^2\right)_\infty^2}\\
&\times {}_4\varphi_3\left(\begin{matrix}q^{1/2}e^{i\theta-i\varphi},-q^{1/2}e^{i\theta-i\varphi},qe^{i\theta-i\varphi},-qe^{i\theta-i\varphi}\\ qe^{2i\theta},\ qe^{-2i\varphi},\ qe^{2i\theta-2i\varphi}\end{matrix};q,-\frac{q}{\omega^2}\right)\\
&+\frac{\left(-q^{1/2}e^{i\theta+i\varphi},-q^{1/2}e^{-i\theta-i\varphi}q^{1/2}\omega^2e^{i\varphi-i\theta},q^{1/2}e^{i\theta-i\varphi}/\omega^2;q\right)_\infty}{\left(e^{2i\theta},e^{-2i\varphi},-q/\omega^2;q\right)_\infty\left(-q\omega^2;q^2\right)_\infty^2}\\
&\times {}_4\varphi_3\left(\begin{matrix}q^{1/2}e^{i\varphi-i\theta},-q^{1/2}e^{i\varphi-i\theta},qe^{i\varphi-i\theta},-qe^{i\varphi-i\theta}\\ qe^{-2i\theta},\ qe^{2i\varphi},\ qe^{2i\varphi-2i\theta}\end{matrix};q,-\frac{q}{\omega^2}\right)\\
&+\frac{\left(-q^{1/2}e^{i\theta-i\varphi},-q^{1/2}e^{i\varphi-i\theta},q^{1/2}\omega^2e^{-i\theta-i\varphi},q^{1/2}e^{i\theta+i\varphi}/\omega^2;q\right)_\infty}{\left(e^{2i\theta},e^{2i\varphi},-q/\omega^2;q\right)_\infty\left(-q\omega^2;q^2\right)_\infty^2}\\
&\times {}_4\varphi_3\left(\begin{matrix}q^{1/2}e^{-i\theta-i\varphi},-q^{1/2}e^{-i\theta-i\varphi},qe^{-i\theta-i\varphi},-qe^{-i\theta-i\varphi}\\ qe^{-2i\theta},\ qe^{-2i\varphi},\ qe^{-2i\theta-2i\varphi}\end{matrix};q,-\frac{q}{\omega^2}\right)
\end{aligned}$$

and

$$\begin{aligned}
&S_q(x,y;\omega)=2q^{1/4}\omega\left(\cos\theta+\cos\varphi\right) \qquad\qquad (2.4.15)\\
&\times\left[\frac{\left(-qe^{i\theta-i\varphi},-qe^{i\varphi-i\theta},q\omega^2e^{i\theta+i\varphi},e^{-i\theta-i\varphi}/\omega^2;q\right)_\infty}{\left(1+e^{-i\theta-i\varphi}\right)\left(e^{-2i\theta},e^{-2i\varphi},-q/\omega^2;q\right)_\infty\left(-q\omega^2;q^2\right)_\infty^2}\right.
\end{aligned}$$

$$\times\ {}_4\varphi_3\left(\begin{matrix} q^{1/2}e^{i\theta+i\varphi}, -q^{1/2}e^{i\theta+i\varphi}, qe^{i\theta+i\varphi}, -qe^{i\theta+i\varphi} \\ qe^{2i\theta},\quad qe^{2i\varphi},\quad qe^{2i\theta+2i\varphi} \end{matrix} ; q, -\frac{q}{\omega^2}\right)$$

$$+\frac{\left(-qe^{i\theta+i\varphi}, -qe^{-i\theta-i\varphi}, q\omega^2 e^{i\theta-i\varphi}, e^{i\varphi-i\theta}/\omega^2; q\right)_\infty}{\left(1+e^{i\varphi-i\theta}\right)\left(e^{-2i\theta}, e^{2i\varphi}, -q/\omega^2; q\right)_\infty \left(-q\omega^2; q^2\right)_\infty^2}$$

$$\times\ {}_4\varphi_3\left(\begin{matrix} q^{1/2}e^{i\theta-i\varphi}, -q^{1/2}e^{i\theta-i\varphi}, qe^{i\theta-i\varphi}, -qe^{i\theta-i\varphi} \\ qe^{2i\theta},\quad qe^{-2i\varphi},\quad qe^{2i\theta-2i\varphi} \end{matrix} ; q, -\frac{q}{\omega^2}\right)$$

$$+\frac{\left(-qe^{i\theta+i\varphi}, -qe^{-i\theta-i\varphi} q\omega^2 e^{i\varphi-i\theta}, e^{i\theta-i\varphi}/\omega^2; q\right)_\infty}{\left(1+e^{i\theta-i\varphi}\right)\left(e^{2i\theta}, e^{-2i\varphi}, -q/\omega^2; q\right)_\infty \left(-q\omega^2; q^2\right)_\infty^2}$$

$$\times\ {}_4\varphi_3\left(\begin{matrix} q^{1/2}e^{i\varphi-i\theta}, -q^{1/2}e^{i\varphi-i\theta}, qe^{i\varphi-i\theta}, -qe^{i\varphi-i\theta} \\ qe^{-2i\theta},\quad qe^{2i\varphi},\quad qe^{2i\varphi-2i\theta} \end{matrix} ; q, -\frac{q}{\omega^2}\right)$$

$$+\frac{\left(-qe^{i\theta-i\varphi}, -qe^{i\varphi-i\theta}, q\omega^2 e^{-i\theta-i\varphi}, e^{i\theta+i\varphi}/\omega^2; q\right)_\infty}{\left(1+e^{i\theta+i\varphi}\right)\left(e^{2i\theta}, e^{2i\varphi}, -q/\omega^2; q\right)_\infty \left(-q\omega^2; q^2\right)_\infty^2}$$

$$\times\ {}_4\varphi_3\left(\begin{matrix} q^{1/2}e^{-i\theta-i\varphi}, -q^{1/2}e^{-i\theta-i\varphi}, qe^{-i\theta-i\varphi}, -qe^{-i\theta-i\varphi} \\ qe^{-2i\theta},\quad qe^{-2i\varphi},\quad qe^{-2i\theta-2i\varphi} \end{matrix} ; q, -\frac{q}{\omega^2}\right)\Bigg]$$

by the special case $r = 3$ of the transformation formula (4.5.2) of [**46**] or (A.3.18).

The basic trigonometric functions (2.4.1)–(2.4.2) and (2.4.6)–(2.4.7) are solutions of difference equation (2.2.11) on a q-quadratic lattice, where $x(z) = \left(q^z + q^{-z}\right)/2$, $q^z = e^{i\theta}$, $x_1(z) = x(z+1/2)$, $\lambda/\sigma = 4q^{1/2}\omega^2/\left(1-q\right)^2$. See [**17**], [**76**], [**111**], [**139**], [**140**], and Exs. 2.20–2.26 for more details on the theory of q-difference equations. Eq. (2.2.11) can be rewritten in a more symmetric form

$$\frac{\delta}{\delta x(z)}\left(\frac{\delta u(z)}{\delta x(z)}\right) + \frac{4q^{1/2}}{(1-q)^2}\,\omega^2\,u(z) = 0, \tag{2.4.16}$$

where $\delta f(z) = f(z+1/2) - f(z-1/2)$.

The q-trigonometric functions satisfy the difference-differentiation formulas [**76**], [**140**]:

$$\frac{\delta}{\delta x}\,C_q(x,y;\omega) = -\frac{2q^{1/4}}{1-q}\,\omega\,S_q(x,y;\omega) \tag{2.4.17}$$

and

$$\frac{\delta}{\delta x}\,S_q(x,y;\omega) = \frac{2q^{1/4}}{1-q}\,\omega\,C_q(x,y;\omega) \tag{2.4.18}$$

in view of (2.2.22) and (2.2.23), respectively. Applying the operator $\delta/\delta x$ to the both sides of (2.4.17) or (2.4.18) we obtain equation (2.4.16) again.

Near the end of this section we shall discuss special cases of the basic trigonometric functions which are very important in the theory of q-Fourier series. Ismail and Stanton [26] have found that both ${}_2\varphi_1$ series in (2.4.6)–(2.4.7) can be summed by a consequence of the q-binomial theorem when $x = \eta = \left(q^{1/4}+q^{-1/4}\right)/2$. Their observation gives the simple "infinite product" representations for $S_q(\eta;\omega)$ and $C_q(\eta;\omega)$:

$$S_q(\eta;\omega) = \frac{(-i\omega;q^{1/2})_\infty - (i\omega;q^{1/2})_\infty}{2i(-q\omega^2;q^2)_\infty} \tag{2.4.19}$$

and

$$C_q(\eta;\omega) = \frac{(-i\omega;q^{1/2})_\infty + (i\omega;q^{1/2})_\infty}{2(-q\omega^2;q^2)_\infty}. \tag{2.4.20}$$

These representations are consequences of the q-Euler formula (2.4.1) due to

$$S_q(\eta;\omega) = \frac{1}{2i}\left(\mathcal{E}_q(\eta;i\omega) - \mathcal{E}_q(\eta;-i\omega)\right), \tag{2.4.21}$$

$$C_q(\eta;\omega) = \frac{1}{2}\left(\mathcal{E}_q(\eta;i\omega) + \mathcal{E}_q(\eta;-i\omega)\right), \tag{2.4.22}$$

and the special values of the basic exponential function (2.3.10). Using (II.2) of [46] or (A.2.2), we obtain from (2.4.19)–(2.4.20) the following series expansions

$$S_q(\eta;\omega) = \frac{1}{(-q\omega^2;q^2)_\infty}\sum_{n=0}^{\infty}(-1)^n\frac{q^{n(n+1/2)}}{\left(q^{1/2};q^{1/2}\right)_{2n+1}}\,\omega^{2n+1} \tag{2.4.23}$$

and

$$C_q(\eta;\omega) = \frac{1}{(-q\omega^2;q^2)_\infty}\sum_{n=0}^{\infty}(-1)^n\frac{q^{n(n-1/2)}}{\left(q^{1/2};q^{1/2}\right)_{2n}}\,\omega^{2n}, \tag{2.4.24}$$

which are very convenient for numerical investigation of these functions.

There are also important relations with Jackson's basic Bessel function $J_\nu^{(2)}(r;q)$ considered in Ex. 1.10, namely, the basic sine $S_q(\eta;\omega)$ and basic cosine $C_q(\eta;\omega)$ functions at $\eta = \left(q^{1/4}+q^{-1/4}\right)/2$ are just multiples of $J_{\pm 1/2}^{(2)}(2\omega;q)$:

$$S_q(\eta;\omega) = \frac{(q;q)_\infty}{\left(q^{1/2};q\right)_\infty}\frac{\omega^{1/2}}{\left(-q\omega^2;q^2\right)_\infty}J_{1/2}^{(2)}(2\omega;q), \tag{2.4.25}$$

$$C_q(\eta;\omega) = \frac{(q;q)_\infty}{\left(q^{1/2};q\right)_\infty}\frac{\omega^{1/2}}{\left(-q\omega^2;q^2\right)_\infty}J_{-1/2}^{(2)}(2\omega;q). \tag{2.4.26}$$

This is an analog of the well-known fact that the classical Bessel functions of the form $J_{m+1/2}(z)$ can be reduced to the elementary functions [163]. Relations (2.4.25)–(2.4.26) follow from the following quadratic transformation formula [68]

$${}_2\varphi_1\left(\begin{matrix}-q^{\nu+1}, & -q^{\nu+2} \\ & q^{2\nu+2}\end{matrix}\;;\;q^2,\;-r^2/4\right) \tag{2.4.27}$$

$$= \frac{(q;q)_\infty}{(q^{\nu+1};q)_\infty} \frac{(2/r)^\nu}{(-r^2/4;q^2)_\infty} J_\nu^{(2)}(r;q),$$

where $|r| < 2$, relating the ${}_2\varphi_1$ of a given structure with Jackson's basic Bessel functions $J_\nu^{(2)}(r;q)$ given by

$$J_\nu^{(2)}(x;q) = \frac{(q^{\nu+1};q)_\infty}{(q;q)_\infty} \sum_{n=0}^\infty q^{(\nu+n)n} \frac{(-1)^n (x/2)^{\nu+2n}}{(q;q)_n (q^{\nu+1};q)_n}. \tag{2.4.28}$$

A similar to (2.4.27) relation was earlier found by Rahman [**116**].

It is worth noting also the relations with the q-trigonometric functions $\mathrm{Sin}_q(z)$ and $\mathrm{Cos}_q(z)$, defined in Ex. 1.14 of [**46**]. Namely,

$$S_q(\eta;\omega) = E_{q^2}\left(q\omega^2\right) \ \mathrm{Sin}_{q^{1/2}}(\omega), \tag{2.4.29}$$

$$C_q(\eta;\omega) = E_{q^2}\left(q\omega^2\right) \ \mathrm{Cos}_{q^{1/2}}(\omega), \tag{2.4.30}$$

where $E_q(z) = e_q^{-1}(-z)$ are the q-exponential functions defined by (1.3.15)–(1.3.17) of [**46**] or (A.2.1)–(A.2.2).

2.5. q-Linear and Linear Grids

In this section we introduce analogs of exponential and trigonometric functions on q-linear and linear grids. Taking the limit $C_2 \to 0$ with $C_1 = 1$ and $\alpha^2 = q^{-1/2}(1-q)^2 \lambda/\sigma$ in (2.2.27), one gets

$$\begin{aligned} \varepsilon_q(x,y;\alpha) &= {}_3\varphi_3\left(\begin{matrix} -q^{1/2}x/y,\ -q^{1/2}y/x,\ 0 \\ -q,\ q^{1/2},\ -q^{1/2} \end{matrix} ;\ q,\ -\alpha^2 xy \right) \\ &\quad + \frac{\alpha}{1-q}(x+y)\ {}_3\varphi_3\left(\begin{matrix} -qx/y,\ -qy/x,\ 0 \\ -q,\ q^{3/2},\ -q^{3/2} \end{matrix} ;\ q,\ -\alpha^2 q^{1/2} xy \right) \\ &= \sum_{n=0}^\infty \frac{q^{n(n-1)/4}}{(q;q)_n} \left(-q^{(1-n)/2} y/x; q\right)_n (\alpha x)^n \end{aligned} \tag{2.5.1}$$

as a q-version of the $\exp \alpha (x+y)$ on a q-linear grid. This function has appeared in [**118**]. Let us introduce also

$$\begin{aligned} \varepsilon_q(\alpha x) = \varepsilon_q(x,0;\alpha) &= {}_1\varphi_1\left(\begin{matrix} 0 \\ q \end{matrix} ;\ q^2,\ -q^{1/2}\alpha^2 x^2 \right) \\ &\quad + \frac{\alpha}{1-q} x\ {}_1\varphi_1\left(\begin{matrix} 0 \\ q^3 \end{matrix} ;\ q^2,\ -q^{3/2}\alpha^2 x^2 \right) \\ &= \sum_{n=0}^\infty \frac{q^{n(n-1)/4}}{(q;q)_n} (\alpha x)^n \end{aligned} \tag{2.5.2}$$

as an analog of the $\exp(\alpha x)$; see [**17**], [**38**], [**118**].

The basic trigonometric functions on a q-linear grid can be introduced by letting $\alpha = i\omega$ in (2.5.1)–(2.5.2) and using an analog of Euler's formula,

$$\varepsilon_q(x,y;i\omega) = c_q(x,y;\omega) + i s_q(x,y;\omega), \tag{2.5.3}$$

where

$$c_q(x,y;\omega) = {}_3\varphi_3\left(\begin{matrix} -q^{1/2}x/y,\ -q^{1/2}y/x,\ 0 \\ -q,\ q^{1/2},\ -q^{1/2} \end{matrix} ;\ q,\ \omega^2 xy \right) \tag{2.5.4}$$

and

$$s_q(x,y;\omega) = \frac{\omega}{1-q}(x+y) \tag{2.5.5}$$
$$\times {}_3\varphi_3\left(\begin{matrix} -q^{1/2}x/y,\ -q^{1/2}y/x,\ 0 \\ -q,\ q^{1/2},\ -q^{1/2} \end{matrix} ;\ q,\ q^{1/2}\omega^2 xy \right)$$

are q-analogs of the $\cos\omega(x+y)$ and $\sin\omega(x+y)$ on a q-linear grid, respectively.

In a similar fashion,

$$\varepsilon_q(i\omega x) = c_q(\omega x) + i s_q(\omega x), \tag{2.5.6}$$

where

$$c_q(\omega x) = c_q(x,0;\omega) = {}_1\varphi_1\left(\begin{matrix} 0 \\ q \end{matrix} ;\ q^2,\ q^{1/2}\omega^2 x^2 \right) \tag{2.5.7}$$

and

$$s_q(\omega x) = s_q(x,0;\omega) \tag{2.5.8}$$
$$= \frac{\omega}{1-q} x\ {}_1\varphi_1\left(\begin{matrix} 0 \\ q^3 \end{matrix} ;\ q^2,\ q^{3/2}\omega^2 x^2 \right)$$

are q-analogs of the $\cos\omega x$ and $\sin\omega x$ on a q-linear grid, respectively.

In the case of the linear lattice, $x(z) = z$, the function (2.2.27) takes a simple form

$$u(z,s) = u(z-s) \tag{2.5.9}$$

with

$$u(z) = {}_2F_1\left(\begin{matrix} 1/2+z,\ \ 1/2-z \\ 1/2 \end{matrix} ;\ -\frac{\alpha^2}{4} \right) \tag{2.5.10}$$
$$+\alpha z\ {}_2F_1\left(\begin{matrix} 1+z,\ \ 1-z \\ 3/2 \end{matrix} ;\ -\frac{\alpha^2}{4} \right)$$
$$= \sum_{n=0}^{\infty} \frac{\alpha^n}{n!} \left(z - \frac{n-1}{2} \right)_n .$$

Here $(a)_n = a(a+1)(a+2)\cdots(a+n-1) = \Gamma(a+n)/\Gamma(a)$ and ${}_2F_1$ is the hypergeometric function. Also,

$$u(0) = \sum_{n=0}^{\infty} \frac{(1/2)_n}{n!} \left(-\frac{\alpha^2}{4} \right)^n = \left(1 + \frac{\alpha^2}{4} \right)^{-1/2}, \tag{2.5.11}$$

and one can introduce the function

$$e(z) = u(z)/u(0) \tag{2.5.12}$$

$$= \left[{}_2F_1 \left(\begin{matrix} 1/2+z, \quad 1/2-z \\ 1/2 \end{matrix} ; -\frac{\alpha^2}{4} \right) \right.$$
$$\left. + \alpha z \; {}_2F_1 \left(\begin{matrix} 1+z, \quad 1-z \\ 3/2 \end{matrix} ; -\frac{\alpha^2}{4} \right) \right]$$

as a difference analog of the exponential function. It will be shown in the next chapter that this function satisfies the functional equation

$$e(z+s) = e(z)\; e(s), \tag{2.5.13}$$

whose only differentiable solution is

$$e(z) = \exp(\beta z). \tag{2.5.14}$$

To find the value of the constant β, one can choose $z = 1/2$, which gives

$$\beta = 2\log\left(\left(1+\frac{\alpha^2}{4}\right)^{1/2} + \frac{\alpha}{2}\right). \tag{2.5.15}$$

Thus

$$e(z) := e(z;\alpha) = \left(\left(1+\frac{\alpha^2}{4}\right)^{1/2} + \frac{\alpha}{2}\right)^{2z}. \tag{2.5.16}$$

The direct transformation from (2.5.12) to (2.5.16) can be done with the help of relations (15.1.16) and (15.1.18) of [1].

Changing $\alpha = i\omega$ in (2.5.12) and (2.5.16), we obtain

$$\cos\left(2z\,\arctan\left(\frac{\omega/2}{(1-\omega^2/4)^{1/2}}\right)\right) \tag{2.5.17}$$
$$= \left(1-\frac{\omega^2}{4}\right)^{1/2} {}_2F_1 \left(\begin{matrix} 1/2+z, \quad 1/2-z \\ 1/2 \end{matrix} ; \frac{\omega^2}{4} \right)$$

and

$$\sin\left(2z\,\arctan\left(\frac{\omega/2}{(1-\omega^2/4)^{1/2}}\right)\right) \tag{2.5.18}$$
$$= \left(1-\frac{\omega^2}{4}\right)^{1/2} \omega z \; {}_2F_1 \left(\begin{matrix} 1+z, \quad 1-z \\ 3/2 \end{matrix} ; \frac{\omega^2}{4} \right)$$

as difference analogs of $\cos\omega z$ and $\sin\omega z$.

2.6. Exercises for Chapter 2

(1) *Divided difference operators and approximations.* Using Taylor's expansions show that for the twice differentiable functions $u = u(z)$ and $x = x(z)$, defined in a neighborhood of the point z in the complex plane, the following approximation holds

$$\frac{u(z+h)-u(z)}{x(z+h)-x(z)}$$

$$= \frac{u'(z)}{x'(z)} + \frac{h}{2}\left(\frac{u''(z)}{x'(z)} - \frac{u'(z)\, x''(z)}{(x'(z))^2}\right) + \mathrm{O}\left(h^2\right)$$

as $h \to 0$, where

$$\frac{df}{dx} = \frac{u'(z)}{x'(z)}, \qquad \frac{d^2 f}{dx^2} = \frac{1}{(x'(z))^2}\left(u''(z) - u'(z)\,\frac{x''(z)}{x'(z)}\right)$$

for $u = f(x(z))$ when f is twice differentiable in x, and then verify (2.2.1)–(2.2.4).

(2) *Properties of difference operators.* Verify that

$$\begin{aligned}
&\nabla f(z) = \Delta f(z-1) = \delta f(z-1/2), \\
&\Delta\nabla f(z) = \nabla\Delta f(z) = \delta^2 f(z) = f(z+1) - 2f(z) + f(z-1), \\
&\Delta\left[f(z)\ g(z)\right] = f(z)\ \Delta g(z) + g(z+1)\ \Delta f(z) \\
&= \frac{f(z+1)+f(z)}{2}\,\Delta g(z) + \frac{g(z+1)+g(z)}{2}\,\Delta f(z), \\
&\Delta\left[f(z-1)\ g(z)\right] = f(z)\ \Delta g(z) + g(z)\ \nabla f(z), \\
&\Delta\left(\frac{1}{f(z)}\right) = -\frac{\Delta f(z)}{f(z+1)\, f(z)}, \quad \nabla\left(\frac{1}{f(z)}\right) = -\frac{\nabla f(z)}{f(z)\, f(z-1)},
\end{aligned}$$

where the forward Δ, backward ∇ and symmetric δ difference operators are defined by (2.2.5), (2.2.6) and (2.2.7), respectively.

(3) *Classical lattices and their properties.* Show that the difference equation

$$\frac{1}{2}\left[x(z+1) + x(z)\right] = \alpha x\left(z + \frac{1}{2}\right) + \beta,$$

where α and β are constants, has the following solutions

$$x(z) = \begin{cases} C_1 q^{-z} + C_2 q^{z} + \beta/(1-\alpha), & \text{if } \alpha \neq \pm 1; \\ 4\beta z^2 + C_1 z + C_2, & \text{if } \alpha = 1; \end{cases}$$

where $q = q_1^{1/2} = q_1^{-1/2}$, q_1 and q_2 are the roots of the equation $q_i^2 - 2\alpha q_i + 1 = 0$; cf. (2.2.12). We shall call these solutions *nonuniform lattices of the classical type,* or simply, *classical lattices* **[113]**, **[111]**. Verify that for these lattices the following properties hold

$$\begin{aligned}
&\frac{1}{2}\left[x(z+\mu) + x(z)\right] = \alpha(\mu)\, x\left(z + \frac{\mu}{2}\right) + \beta(\mu), \\
&x(z+\mu) - x(z) = \gamma(\mu)\, \nabla x\left(z + \frac{\mu+1}{2}\right), \\
&\frac{1}{4}\left[x(z+\mu) - x(z)\right]^2 = \gamma^2(\mu)\left[\left(\alpha^2 - 1\right) x^2\left(z + \frac{\mu}{2}\right)\right. \\
&\qquad\qquad \left. + 2(\alpha+1)\,\beta x\left(z + \frac{\mu}{2}\right) + C_3\right],
\end{aligned}$$

where

$$\alpha(\mu) = \begin{cases} \dfrac{q^{\mu/2} + q^{-\mu/2}}{2}; \\ 1; \end{cases} \qquad \beta(\mu) = \begin{cases} \beta \dfrac{1-\alpha(\mu)}{1-\alpha}; \\ \beta\mu^2; \end{cases}$$

$$\gamma(\mu) = \begin{cases} \dfrac{q^{\mu/2} - q^{-\mu/2}}{q^{1/2} - q^{-1/2}}; \\ \mu; \end{cases}$$

and

$$C_3 = \begin{cases} \dfrac{\alpha+1}{\alpha-1}\beta^2 - 4\left(\alpha^2-1\right)C_1C_2; \\ \dfrac{1}{4}C_1^2 - 4\beta C_2; \end{cases}$$

for $\alpha \neq \pm 1$ and $\alpha = 1$, respectively.

(4) Verify the following identities

$$\alpha(\mu)\gamma(\mu) = \frac{1}{2}\gamma(2\mu),$$
$$\alpha^2(\mu) + \left(\alpha^2-1\right)\gamma^2(\mu) = \alpha(2\mu),$$
$$\alpha(\mu+\nu)\alpha(2\mu) - \left(\alpha^2-1\right)\gamma(\mu+\nu)\gamma(2\mu) = \alpha(\mu-\nu),$$
$$\alpha(\mu+\nu)\gamma(2\mu) - \gamma(\mu+\nu)\alpha(2\mu) = \gamma(\mu-\nu),$$
$$\alpha(-\mu) = \alpha(\mu), \qquad \gamma(-\mu) = -\gamma(\mu),$$
$$\alpha(\mu-1)\gamma(\mu) - \alpha(\nu-1)\gamma(\nu) = \alpha(\mu+\nu-1)\gamma(\mu-\nu),$$
$$\gamma(\mu-1)\gamma(\mu) - \gamma(\nu-1)\gamma(\nu) = \gamma(\mu+\nu-1)\gamma(\mu-\nu),$$

where the functions $\alpha(\mu)$ and $\gamma(\mu)$ are defined in the previous exercise.

(5) Verify that the one-parameter subgroup property,

$$A(\mu+\nu) = A(\mu)A(\nu),$$

holds for the following matrix

$$A(\mu) = \begin{pmatrix} \alpha(\mu) & \gamma(\mu) \\ \left(\alpha^2-1\right)\gamma(\mu) & \alpha(\mu) \end{pmatrix}, \qquad \det A(\mu) = 1.$$

(6) Show that

$$\sum_{k=0}^{n} \alpha(2k) = \alpha(n-1)\gamma(n), \quad \sum_{k=0}^{n} \gamma(2k) = \gamma(n-1)\gamma(n).$$

(7) *Characterization of classical lattices.* Prove that the only solutions of the functional equation

$$x(u) - x(v) = \gamma(u-v)\,\delta(u+v),$$

where the functions x and γ are differentiable, and δ is continuous are

$$x(z) = \begin{cases} Ae^{2\sqrt{\gamma'''(0)}z} + Be^{-2\sqrt{\gamma'''(0)}z} + C, & \text{if } \gamma'''(0) \neq 0; \\ Az^2 + Bz + C, & \text{if } \gamma'''(0) = 0 \end{cases}$$

[*Hint*: Derive the following third order differential equation $x'''(z) - 4\gamma'''(0)\,x'(z) = 0$ for these solutions [**33**].]

(8) Show that the divided difference operator of the form

$$\frac{\delta f(x(z))}{\delta x(z)} = \frac{f(x(z+1/2)) - f(x(z-1/2))}{x(z+1/2) - x(z-1/2)}$$

on the classical lattices (2.2.12) takes polynomials of degree n in $x(z)$ to polynomials of degree $n-1$ in $x(z)$ [**111**], [**105**]. [*Hint*: Consider the cases $n = 2$ and $n = 3$ first and then apply the mathematical induction.]

(9) *Generalized powers and their properties.* Introduce the generalized powers $[x_\nu(s) - x_\nu(z)]^{(\mu)}$ on the nonuniform lattices (2.2.12) as functions that satisfy the following "*product rules*" [**139**]

$$\begin{aligned} &[x_\nu(s) - x_\nu(z)]\,[x_\nu(s) - x_\nu(z-1)]^{(\mu)} \\ &= [x_\nu(s) - x_\nu(z)]^{(\mu)}\,[x_{\nu-1}(s) - x_\nu(z-\mu)] \\ &= [x_\nu(s) - x_\nu(z)]^{(\mu+1)}, \end{aligned}$$

$$\begin{aligned} &[x_{\nu-1}(s+1) - x_{\nu-1}(z)]^{(\mu)}\,[x_{\nu-\mu}(s) - x_{\nu-\mu}(z)] \\ &= [x_{\nu-\mu}(s+\mu) - x_{\nu-\mu}(z)]\,[x_{\nu-1}(s) - x_\nu(z)]^{(\mu)} \\ &= [x_\nu(s) - x_\nu(z)]^{(\mu+1)}, \end{aligned}$$

where $x_\mu(z) = x(z+\mu/2)$.
[*Note*: These relations are analogs of the trivial identities

$$(s-z)(s-z)^\mu = (s-z)^\mu (s-z) = (s-z)^{\mu+1}$$

for the power function.]
Show that the following *difference-differentiation formulas* hold [**139**]

$$\begin{aligned} \frac{\Delta_z [x_\nu(s) - x_\nu(z)]^{(\mu)}}{\Delta x_{\nu-\mu+1}(z)} &= -\frac{\Delta_s [x_{\nu+1}(s) - x_{\nu+1}(z)]^{(\mu)}}{\Delta x_{\nu-\mu+1}(z)} \\ &= -\gamma(\mu)\,[x_\nu(s) - x_\nu(z)]^{(\mu-1)}, \end{aligned}$$

$$\begin{aligned} &\frac{\nabla_z}{\nabla x_{\nu-\mu+1}(z)}\left(\frac{1}{[x_\nu(s) - x_\nu(z)]^{(\mu)}}\right) \\ &= -\frac{\Delta_z}{\Delta x_{\nu-1}(z)}\left(\frac{1}{[x_{\nu-1}(s) - x_{\nu-1}(z)]^{(\mu)}}\right) \end{aligned}$$

$$= \frac{\gamma(\mu)}{[x_\nu(s) - x_\nu(z)]^{(\mu+1)}}.$$

[*Note*: These relations can be viewed as analogs of the familiar differentiation formulas

$$\frac{d}{dz}(s-z)^\mu = -\frac{d}{ds}(s-z)^\mu = -\mu(s-z)^{\mu-1},$$
$$\frac{d}{dz}\left(\frac{1}{(s-z)^\mu}\right) = -\frac{d}{ds}\left(\frac{1}{(s-z)^\mu}\right) = \frac{\mu}{(s-z)^{\mu+1}}$$

for the power function.]

(10) Show that the generalized power functions can be defined as [**139**]

$$[x_\nu(s) - x_\nu(z)]^{(\alpha)}$$
$$= C_1^\alpha \frac{\Gamma(s-z+\alpha)\,\Gamma(s+z+\nu+\mu+1)}{\Gamma(s-z)\,\Gamma(s+z+\nu-\alpha+\mu+1)}, \qquad \mu = \frac{C_2}{C_1},$$

where $\Gamma(z)$ is the gamma function, for the lattices $x(z) = C_1 z^2 + C_2 z + C_3$ and as

$$[x_\nu(s) - x_\nu(z)]^{(\alpha)}$$
$$= C_1^\alpha q^{-\alpha(s+\nu/2)} \frac{\left(q^{s-z}, \mu q^{s+z+\nu-\alpha+1}; q\right)_\infty}{\left(q^{s-z+\alpha}, \mu q^{s+z+\nu+1}; q\right)_\infty}, \qquad \mu = \frac{C_2}{C_1}$$

for the lattices $x(z) = C_1 q^{-z} + C_2 q^z + C_3$ when $|q| < 1$.

(11) The *Askey–Wilson divided difference operator* of the first order [**15**] is given by

$$\mathcal{D}_q f = \frac{\delta u(z)}{\delta x(z)} = \frac{u(z+1/2) - u(z-1/2)}{x(z+1/2) - x(z-1/2)}$$

where $u(z) = f(q^z)$, $x(z) = (q^z + q^{-z})/2 = \cos\theta$, and $q^z = e^{i\theta}$. Show that

$$\mathcal{D}_q \left(ae^{i\theta}, ae^{-i\theta}; q\right)_\alpha = -2a\frac{1-q^\alpha}{1-q}\left(aq^{1/2}e^{i\theta}, aq^{1/2}e^{-i\theta}; q\right)_{\alpha-1},$$

$$\mathcal{D}_q^n \left(ae^{i\theta}, ae^{-i\theta}; q\right)_\alpha = (-2a)^n \frac{\left(q^{\alpha-n+1}; q\right)_n}{(1-q)^n} q^{n(n-1)/4}$$
$$\times \left(aq^{n/2}e^{i\theta}, aq^{n/2}e^{-i\theta}; q\right)_{\alpha-n}.$$

Here $|q| < 1$, α is a complex number,

$$(a;q)_\alpha = \frac{(a;q)_\infty}{(aq^\alpha;q)_\infty},$$

and the principal value of q^α is taken [**46**].

(12) The *differential equation of hypergeometric type* has the form [**113**]

$$\sigma(z)\, y'' + \tau(z)\, y' + \lambda y = 0,$$

where $\sigma(z)$, $\tau(z)$ are polynomials of at most second and first degree, respectively, and λ is a constant. Verify that the function $v_1 = y'(z)$ satisfies the differential equation of the same type

$$\sigma(z)\, v_1'' + \tau_1(z)\, v_1' + \mu_1 v_1 = 0,$$

where

$$\tau_1(z) = \tau(z) + \sigma'(z), \qquad \mu_1 = \lambda + \tau'.$$

Show by the induction that the functions $v_n = y^{(n)}(z)$ are solutions of the differential equations of hypergeometric type

$$\sigma(z)\, v_n'' + \tau_n(z)\, v_n' + \mu_n v_n = 0,$$

where

$$\tau_n(z) = \tau(z) + n\sigma'(z), \qquad \mu_n = \lambda + n\tau' + \frac{1}{2}n(n-1)\,\sigma''.$$

(13) Rewrite the differential equation of hypergeometric type in *self-adjoint form*

$$(\sigma\rho y')' + \lambda\rho y = 0, \qquad (\sigma\rho)' = \tau\rho$$

and show that

$$\frac{d}{dz}\left(\sigma(z)\,\rho(z)\, W(y_1,\, y_2)\right) = (\lambda_1 - \lambda_2)\, y_1(z)\, y_2(z)\, \rho(z),$$

where $y_1 = y_1(z)$ and $y_2 = y_2(z)$ are solutions corresponding to the eigenvalues λ_1 and λ_2, respectively, and

$$W(y_1,\, y_2) = \begin{vmatrix} y_1(z) & y_2(z) \\ y_1'(z) & y_2'(z) \end{vmatrix}$$

is the *Wronskian.*

(14) *Classical orthogonal polynomials.* Prove that the differential equation of hypergeometric type has a particular solution of the form $y(z) = y_n(z)$, where $y_n(z)$ is a polynomial of degree n, if

$$\lambda_n = -n\tau' - n(n-1)\,\sigma''/2, \qquad n = 0, 1, 2, \ldots .$$

Find these polynomial solutions in the form of the *Rodrigues formula* [**1**], [**37**], [**111**], [**113**]:

$$y_n(x) = \frac{B_n}{\rho(z)} \left[\sigma^n(z)\,\rho(z)\right]^{(n)},$$

where the B_n are constants, and $\rho(z)$ is a solution of the *Pearson equation*

$$\left[\sigma(z)\,\rho(z)\right]' = \tau(z)\,\rho(z).$$

Show that under the conditions $\sigma(x)\rho(x)x^k\big|_{s=a,b} = 0$ with $k = 0, 1, 2, \dots$, the polynomials $y_n(x)$ and $y_m(x)$ are *orthogonal*:

$$\int_a^b y_n(x)\, y_m(x)\, \rho(x)\, dx = 0, \qquad m \neq n.$$

[*Note*: Among polynomial solutions of the differential equation of hypergeometric type are the *Jacobi*, *Laguerre* and *Hermite* polynomials [**1**], [**10**], [**37**], [**113**], [**151**]. They are often called (*very*) *classical orthogonal polynomials* due to Koornwinder.]

(15) *Integral representations.* Let $\rho(z)$ satisfy the Pearson equation

$$[\sigma(z)\rho(z)]' = \tau(z)\rho(z)$$

and let ν be a root of the equation

$$\lambda + \nu\tau' + \frac{1}{2}\nu(\nu - 1)\sigma'' = 0.$$

Prove that the differential equation of hypergeometric type has a particular solution of the form

$$y = y_\nu(z) = \frac{C_\nu}{\rho(z)} \int_C \frac{\sigma^\nu(s)\rho(s)}{(s - z)^{\nu+1}}\, ds,$$

where the C_ν is a constant and C is a contour in the complex z-plane, if the following conditions are satisfied:

(a) the derivative of the integral

$$\varphi_{\nu\mu}(z) = \int_C \frac{\rho_\nu(s)}{(s - z)^{\mu+1}}\, ds$$

with $\rho_\nu(s) = \sigma^\nu(s)\rho(s)$ can be evaluated by the formula

$$\varphi'_{\nu\mu}(z) = (\mu + 1)\,\varphi_{\nu,\ \mu+1}(z)$$

when $\mu = \nu - 1$ and $\mu = \nu$;

(b) the contour C is chosen so that the equality

$$\left.\frac{\sigma^\nu(s)\rho(s)}{(s - z)^{\nu+1}}\right|_{s_2}^{s_1} = 0$$

holds, where s_1 and s_2 are the end points of the contour C.

[*Note*: This integral representation contains the well-known integral representations for the *hypergeometric* and *confluent hypergeometric functions*, *Hermite function*, and *Poisson's integral representations* for the *Bessel function* of the first kind and for *Hankel functions* among others [**113**].]

(16) *Power series method.* Let a be a root of the equation $\sigma(z) = 0$. Show that the differential equation of hypergeometric type has a power series solution of the form

$$y(z) = \sum_{n=0}^{\infty} c_n (z-a)^n,$$

where

$$\frac{c_{n+1}}{c_n} = -\frac{\lambda + n(\tau' + (n-1)\sigma''/2)}{(n+1)(\tau(a) + n\sigma'(a))},$$

if:

(a) $\lim_{m\to\infty} \frac{d^k}{dx^k} y_m(z) = \frac{d^k}{dx^k} y(z)$ for $k = 0, 1, 2$;

(b) $\lim_{m\to\infty} (\lambda - \lambda_m) c_m (z-a)^m = 0$;

with $y_m(z) = \sum_{n=0}^{m} c_n (z-a)^n$, $\lambda_m = -m\tau' - m(m-1)\sigma''/2$. In the case $\sigma(z) =$ constant $\neq 0$ show that the equation of hypergeometric type has a power series solution of the same form when a is a root of the equation $\tau(z) = 0$,

$$\frac{c_{n+2}}{c_n} = -\frac{\lambda + n\tau'}{(n+1)(n+2)\sigma},$$

and the convergence conditions (a)–(b) are satisfied **[17]**, **[149]**. [*Hint*: Use the identity

$$\rho^{-1}\frac{d}{dz}\left[\sigma\rho\frac{d}{dz}(z-\xi)^n\right] = n(n-1)\sigma(\xi)(z-\xi)^{n-2}$$
$$+ n\tau_{n-1}(\xi)(z-\xi)^{n-1} - \lambda_n (z-\xi)^n,$$

where $\tau_m(\xi) = \tau_m(\xi) + m\sigma'(\xi)$, which can be easily verified.]

(17) *Generalized power series method.* Using the expansion

$$y(z) = \sum_{n=0}^{\infty} c_n (z-\xi)^{\alpha+n}, \quad \frac{c_{n+1}}{c_n} = \frac{\lambda - \lambda_{\alpha+n}}{(\alpha+n+1)\tau_{\alpha+n}(a)},$$

find formal generalized power series solutions of the differential equation of hypergeometric type in the form

$$y(z) = c_0 (z-a)^{\alpha} \sum_{n=0}^{\infty} \prod_{k=0}^{n-1} \frac{(\lambda - \lambda_{\alpha+k})(a-z)}{\tau_{\alpha+k}(a)(\alpha+k+1)}$$

provided $\sigma(a) = 0$ and $\alpha\tau_{\alpha-1}(a) = 0$.
In a similar fashion, using the expansion

$$y(z) = \sum_{n=0}^{\infty} \frac{c_n}{(z-\xi)^{\alpha+n}}, \quad \frac{c_{n+1}}{c_n} = \frac{(\alpha+n)\tau_{-\alpha-n-1}(a)}{\lambda - \lambda_{-\alpha-n-1}},$$

find formal solutions of the form

$$y(z) = \frac{c_0}{(z-a)^\alpha} \sum_{n=0}^{\infty} \prod_{k=0}^{n-1} \frac{(\alpha+k)\,\tau_{-\alpha-k-1}(a)}{(\lambda-\lambda_{-\alpha-k-1})(z-a)}$$

when $\sigma(a) = 0$ and $\lambda = \lambda_{-\alpha}$.
In the case $\sigma(z) = \text{constant} \neq 0$ find the following formal solutions

$$y(z) = \sum_n \frac{c_n}{(z-a)^{\alpha+n}}, \qquad \frac{c_{n+2}}{c_n} = \frac{(\alpha+n)(\alpha+n+1)\,\sigma}{\lambda-\lambda_{-\alpha-n-2}}$$

when $\tau(a) = 0$ and $\lambda = \lambda_{-\alpha}$ for even nonnegative integer n or $\lambda = \lambda_{-\alpha-1}$ for odd positive integer n. Establish when these formal solutions satisfy the equation of hypergeometric type [**149**].

(18) Use the power series method to show that the elementary functions

$$u_0 = e^z = \sum_{n=0}^{\infty} \frac{z^n}{n!},$$

$$u_+ = \cos z = \sum_{n=0}^{\infty} (-1)^n \frac{z^{2n}}{(2n)!},$$

$$u_- = \sin z = \sum_{n=0}^{\infty} (-1)^n \frac{z^{2n+1}}{(2n+1)!}$$

satisfy the differential equations $u'' = u$ and $u'' = -u$, respectively.

(19) Use the power series method to classify all solutions of the differential equation of hypergeometric type depending on degrees of polynomials $\sigma(z)$ and $\tau(z)$. [*Note*: In addition to the *Jacobi*, *Laguerre* and *Hermite* polynomials, these solutions include the *hypergeometric function* of Euler and Gauss, the *confluent hypergeometric function*, the *Hermite function*, and the *Bessel* type functions, which are often called *special functions of mathematical physics* [**1**], [**10**], [**113**], [**149**], [**152**], [**163**], [**164**].]

(20) We replace the differential equation of hypergeometric type by a difference equation on a lattice $x = x(z)$ with variable step $\Delta x(z) = x(z+h) - x(z)$:

$$\frac{\widetilde{\sigma}(x(z))}{x(z+h/2) - x(z-h/2)} \left[\frac{y(z+h)-y(z)}{x(z+h)-x(z)} - \frac{y(z)-y(z-h)}{x(z)-x(z-h)}\right]$$
$$+\frac{\widetilde{\tau}(x(z))}{2} \left[\frac{y(z+h)-y(z)}{x(z+h)-x(z)} + \frac{y(z)-y(z-h)}{x(z)-x(z-h)}\right] + \lambda u(z) = 0.$$

After the linear change of variable $z \to hz$, this *difference equation of hypergeometric type* on a nonuniform lattice takes the form [**113**], [**111**], [**112**]:

$$\widetilde{\sigma}(x(z)) \frac{\Delta}{\nabla x_1(z)} \left(\frac{\nabla y(z)}{\nabla x(z)}\right)$$

$$+\frac{1}{2}\widetilde{\tau}\left(x\left(z\right)\right)\left(\frac{\Delta y\left(z\right)}{\Delta x\left(z\right)}+\frac{\nabla y\left(z\right)}{\nabla x\left(z\right)}\right)+\lambda y\left(z\right)=0,$$

where $x_1(z) = x(z+1/2)$, $\widetilde{\sigma}(x)$ and $\widetilde{\tau}(x)$ are polynomials of at most second and first degree in x, respectively, and λ is a constant. Prove that if the lattice $x(z)$ satisfies the difference equation

$$\frac{1}{2}\left[x\left(z+1\right)+x\left(z\right)\right]=\alpha x\left(z+\frac{1}{2}\right)+\beta,$$

where α and β are constants, and $x^2(z+1)+x^2(z)$ is a polynomials of at most second degree in $x_1(z) = x(z+1/2)$, then the functions $v_k(z) = \Delta v_{k-1}(z)/\Delta x_{k-1}(z)$, $k = 1, 2, \ldots$ with $v_0(z) = y(z)$ and $x_k(z) = x(z+k/2)$ satisfy an equation of the hypergeometric form,

$$\widetilde{\sigma}_k\left(x_k\left(z\right)\right)\frac{\Delta}{\nabla x_{k+1}\left(z\right)}\left(\frac{\nabla v_k\left(z\right)}{\nabla x_k\left(x\right)}\right)$$

$$+\frac{1}{2}\widetilde{\tau}_k\left(x_k\left(z\right)\right)\left(\frac{\Delta v_k\left(z\right)}{\Delta x_k\left(z\right)}+\frac{\nabla v_k\left(z\right)}{\nabla x_k\left(z\right)}\right)+\mu_k v_k\left(z\right)=0,$$

where $\widetilde{\sigma}_k(x_k)$ and $\widetilde{\tau}_k(x_k)$ are polynomials of at most second and first degree in x_k, respectively, and μ_k are constants. Moreover,

$$\widetilde{\sigma}_k\left(x_k\left(z\right)\right)-\frac{1}{2}\widetilde{\tau}_k\left(x_k\left(z\right)\right)\nabla x_{k+1}\left(z\right)$$

$$=\widetilde{\sigma}\left(x\left(z\right)\right)-\frac{1}{2}\widetilde{\tau}\left(x\left(z\right)\right)\nabla x_1\left(z\right):=\sigma\left(z\right),$$

$$\widetilde{\tau}_k\left(x_k\left(z\right)\right)\nabla x_{k+1}\left(z\right)$$

$$=\sigma\left(z+k\right)-\sigma\left(z\right)+\widetilde{\tau}\left(x\left(z+k\right)\right)\nabla x_1\left(z+k\right),$$

and

$$\mu_k=\lambda+\sum_{m=0}^{k-1}\widetilde{\tau}'_m.$$

(21) Rewrite the difference equation of hypergeometric type in *self-adjoint form* [**111**], [**113**]:

$$\frac{\Delta}{\nabla x_1\left(z\right)}\left(\sigma\left(z\right)\rho\left(z\right)\frac{\nabla y\left(z\right)}{\nabla x\left(z\right)}\right)+\lambda\rho\left(z\right)y\left(z\right)=0,$$

$$\frac{\Delta}{\nabla x_1\left(z\right)}\left(\sigma\left(z\right)\rho\left(z\right)\right)=\tau\left(z\right)\rho\left(z\right),$$

where

$$\sigma\left(z\right)=\widetilde{\sigma}\left(x\left(z\right)\right)-\frac{1}{2}\widetilde{\tau}\left(x\left(z\right)\right)\nabla x_1\left(z\right),\quad \tau\left(z\right)=\widetilde{\tau}\left(x\left(z\right)\right).$$

Then show that

$$\Delta\left(\sigma\left(z\right)\rho\left(z\right)W\left(y_1,\ y_2\right)\right)$$

$$=\left(\lambda_1-\lambda_2\right)y_1\left(z\right)y_2\left(z\right)\ \rho\left(z\right)\nabla x_1\left(z\right),$$

where $y_1 = y_1(z)$ and $y_2 = y_2(z)$ are solutions corresponding to the eigenvalues λ_1 and λ_2, respectively, and

$$W(y_1,\ y_2) = \begin{vmatrix} y_1(z) & y_2(z) \\ \dfrac{\nabla y_1(z)}{\nabla x(z)} & \dfrac{\nabla y_2(z)}{\nabla x(z)} \end{vmatrix} = \frac{y_2(z)\, y_1(z-1) - y_1(z)\, y_2(z-1)}{x(z) - x(z-1)}$$

is a *difference analog of the Wronskian.*

(22) *Classical orthogonal polynomials in a discrete variable.* Let $x(z)$ be a nonuniform lattice of the classical type considered in Ex. 2.3. Prove that the difference equation of hypergeometric type has a particular solution of the form $y(z) = y_n(x(z))$, where $y_n(x)$ is a polynomial of degree n in x, if

$$\lambda_n = -\alpha(n)\,\gamma(n)\,\widetilde{\tau}' - \gamma(n)\,\gamma(n-1)\,\widetilde{\sigma}''/2, \quad n = 0, 1, 2, \ldots$$

and functions $\alpha(\mu)$ and $\gamma(\mu)$ are defined in Ex. 2.3.
Find these polynomial solutions in the form of the difference analog of the *Rodrigues type formula* [**113**], [**111**]:

$$y_n(x) = \frac{B_n}{\rho(z)} \nabla^{(n)} \left[\rho_n(z)\right],$$

where the B_n are constants, and $\rho(z)$ is a solution of the difference analog of the *Pearson equation*

$$\Delta\left[\sigma(z)\,\rho(z)\right] = \tau(z)\,\rho(z)\,\nabla x_1(z)$$

with $\sigma(z) = \widetilde{\sigma}(x(z)) - \frac{1}{2}\widetilde{\tau}(x(z))\,\nabla x_1(z)$ and $\tau(z) = \widetilde{\tau}(x(z))$,

$$\rho_n(z) = \rho(z+n) \prod_{k=1}^{n-1} \sigma(z+k)$$

and the n-th order divided difference operator $\nabla^{(n)}$ is given by

$$\nabla^{(n)} = \left(\frac{\nabla}{\nabla x_1}\right)\left(\frac{\nabla}{\nabla x_2}\right)\cdots\left(\frac{\nabla}{\nabla x_n}\right), \quad x_k = x(z + k/2).$$

Show that under the conditions $\sigma(s)\,\rho(s)\,x^k(s - 1/2)\big|_{s=a,b} = 0$ with $k = 0, 1, 2, \ldots$, the polynomial solutions $y_n(x)$ and $y_m(x)$ are *orthogonal* in the following sense:

$$\sum_{s_i=a}^{b-1} y_n(x_i)\ y_m(x_i)\ \rho_i \Delta x_{i-1/2} = 0, \qquad m \neq n,$$

where $x_i = x(s_i)$, $\rho_i = \rho(s_i)$, and $\Delta x_{i-1/2} = \Delta x(s_i - 1/2)$. [*Note*: These polynomial solutions of the difference equation of hypergeometric type corresponding to the positive weight function ρ_i are called *classical orthogonal polynomials in a discrete variable* [**37**], [**113**], [**111**].]

Show that under the conditions

$$\int_C \nabla \left[\rho_1(z)\, x_1^k(z)\right] dz = 0, \qquad k = 0, 1, 2, \ldots$$

the polynomial solutions $y_n(x)$ and $y_m(x)$ are *orthogonal* on a contour C in the complex z-plane:

$$\int_C y_n(x(z))\; y_m(x(z))\; \rho(z)\,\nabla x_1(z)\; dz = 0, \qquad m \neq n.$$

[*Note*: In many cases a special choice of the contour C in this relation gives rise to the continuous orthogonality property for a real-valued system of polynomials. The most general case corresponds to the Askey–Wilson polynomials [**15**], [**18**]; see other examples in [**9**], [**10**], [**16**], [**46**], [**86**], [**111**], [**113**]. The special and/or limiting cases of the Askey–Wilson polynomials are often called *classical orthogonal polynomials* due to Andrews and Askey [**9**].]

(23) Show that for the nonuniform lattices of the classical type

$$\nabla^{(n)} f(z) = \sum_{k=0}^{n} (-1)^{n-k} \frac{\gamma(n)!}{\gamma(k)!\gamma(n-k)!} \times \frac{\nabla x_1(z - n/2 + k)}{\prod_{m=0}^{n} \nabla x_1(z - (m-k)/2)}\, f(z - n + k),$$

where $\gamma(n)! = \gamma(1)\,\gamma(2)\cdots\gamma(n)$ and $\gamma(\mu)$ is defined in Ex. 2.3 [**139**]. [*Hint*: Use the mathematical induction or Cauchy's integral formula and the identity

$$\nabla_z^{(n)} \left(\frac{1}{x_n(s) - x_n(z)} \right) = \frac{\gamma(n)!}{\left[x_n(s) - x_n(z)\right]^{(n+1)}},$$

where $\left[x_n(s) - x_n(z)\right]^{(m)} = \prod_{k=0}^{m-1} \left[x_n(s) - x_n(z-k)\right]$.]

(24) *Analogs of integral representations.* Let $x(z)$ be a nonuniform lattice of the classical type considered in Ex. 2.3. Prove that the difference equation of hypergeometric type has particular solutions of the form [**139**]

$$y = y_\nu(z) = \frac{C_\nu}{\rho(z)} \varphi_{\nu\nu}(z),$$

where the C_ν are constants, $\varphi_{\nu\nu}(z) = \varphi_{\nu\mu}(z)\big|_{\mu=\nu}$ with

$$\varphi_{\nu\mu}(z) = \begin{cases} \displaystyle\sum_{s=a}^{b-1} \frac{\rho(s)\,\nabla x_{\nu+1}(s)}{\left[x_\nu(s) - x_\nu(z)\right]^{(\mu+1)}}, \\[2ex] \displaystyle\int_C \frac{\rho(s)\,\nabla x_{\nu+1}(s)\; ds}{\left[x_\nu(s) - x_\nu(z)\right]^{(\mu+1)}}, \end{cases}$$

and C is a contour in the complex z-plane, if:

(a) $\rho(z)$ and $\rho_\nu(z)$ satisfy the Pearson type equations

$$\Delta\left[\sigma(z)\,\rho(z)\right] = \tau(z)\,\rho(z)\,\nabla x_1(z),$$
$$\Delta\left[\sigma(z)\,\rho_\nu(z)\right] = \tau_\nu(z)\,\rho_\nu(z)\,\nabla x_{\nu+1}(z),$$

and ν is a root of the equation

$$\lambda_\nu + \alpha(\nu)\,\gamma(\nu)\,\widetilde{\tau}' - \gamma(\nu-1)\,\gamma(\nu)\,\widetilde{\sigma}''/2 = 0;$$

(b) the generalized powers $[x(s) - x(z)]^{(\mu)}$ have the properties

$$\begin{aligned}
&[x_\nu(s) - x_\nu(z)]\,[x_\nu(s) - x_\nu(z-1)]^{(\mu)} \\
&= [x_\nu(s) - x_\nu(z)]^{(\mu)}\,[x_{\nu-1}(s) - x_\nu(z-\mu)] \\
&= [x_\nu(s) - x_\nu(z)]^{(\mu+1)},
\end{aligned}$$

$$\begin{aligned}
&[x_{\nu-1}(s+1) - x_{\nu-1}(z)]^{(\mu)}\,[x_{\nu-\mu}(s) - x_{\nu-\mu}(z)] \\
&= [x_{\nu-\mu}(s+\mu) - x_{\nu-\mu}(z)]\,[x_{\nu-1}(s) - x_\nu(z)]^{(\mu)} \\
&= [x_\nu(s) - x_\nu(z)]^{(\mu+1)},
\end{aligned}$$

(c) the difference-differentiation formula

$$\frac{\nabla\varphi_{\nu\mu}(z)}{\nabla x_{\nu-\mu}(z)} = \gamma(\mu+1)\ \varphi_{\nu,\ \mu+1}(z)$$

holds for $\mu = \nu$ and $\mu = \nu - 1$;

(d) the equations

$$\left.\frac{\sigma(s)\,\rho_\nu(s)}{[x_{\nu-1}(s) - x_{\nu-1}(z+1)]^{(\nu+1)}}\right|_a^b = 0$$

and

$$\int_C \Delta\left(\frac{\sigma(s)\,\rho_\nu(s)}{[x_{\nu-1}(s) - x_{\nu-1}(z+1)]^{(\nu+1)}}\right) ds = 0$$

are satisfied in the cases of the sum and the integral, respectively.

[*Hint*: Use the identity

$$\begin{aligned}
&\Delta_s\left(\frac{\sigma(s)\,\rho_\nu(s)}{[x_{\nu-1}(s) - x_{\nu-1}(z+1)]^{(\mu+1)}}\right) \\
&+\frac{\gamma(\mu+1)\,\sigma(s)\,\rho_\nu(s)\,\nabla x_{\nu+1}(s)}{[x_\nu(s) - x_\nu(z+1)]^{(\mu+2)}} \\
&= \frac{\tau_\nu(s)\,\rho_\nu(s)\,\nabla x_{\nu+1}(s)}{[x_{\nu-1}(s+1) - x_{\nu-1}(z+1)]^{(\mu+1)}}
\end{aligned}$$

and properties of the generalized powers.]

(25) *Analog of the power series method.* Under the same conditions as in the previous exercise show that the function

$$y(z) = c_0 [x(z) - x(a+\alpha-1)]^{(\alpha)} \times \sum_{n=0}^{\infty} \prod_{k=0}^{n-1} \frac{(\lambda - \lambda_{\alpha+k}) [x(a+\alpha+k) - x(z)]}{\tau_{\alpha+k}(a)\, \gamma(\alpha+k+1)},$$

where c_0 is a constant, satisfy a nonhomogeneous difference equation of hypergeometric type

$$\widetilde{\sigma}(x(z)) \frac{\Delta}{\nabla x_1(z)} \left(\frac{\nabla y(z)}{\nabla x(z)} \right) + \frac{1}{2} \widetilde{\tau}(x(z)) \left(\frac{\Delta y(z)}{\Delta x(z)} + \frac{\nabla y(z)}{\nabla x(z)} \right) + \lambda y(z) = G(z)$$

with the right-hand side of the form

$$G(z) = c_0 [x(z) - x(a+\alpha-1)]^{(\alpha)} \times \lim_{m\to\infty} (\lambda - \lambda_{\alpha+m}) \prod_{k=0}^{m-1} \frac{(\lambda - \lambda_{\alpha+k}) [x(a+\alpha+k) - x(z)]}{\tau_{\alpha+k}(a)\, \gamma(\alpha+k+1)},$$

if $\sigma(a) = 0$, $\gamma(\alpha)\tau_{\alpha-1}(a) = 0$, and the series converges at the points z, $z \pm 1$. In a similar fashion, show that the series

$$y(z) = c_0[x(z) - x(b)]^{(\alpha)} \times \sum_{n=0}^{\infty} \prod_{k=0}^{n-1} \frac{(\lambda - \lambda_{a+k})[x(b-\alpha-k) - x(z)]}{\tau_{\alpha+k}(b-\alpha-k)\gamma(\alpha+k+1)}$$

is a solution of the difference equation above with the right-hand side

$$G(z) = c_0[x(z) - x(b)]^{(\alpha)} \times \lim_{m\to\infty} (\lambda - \lambda_{\alpha+m}) \prod_{k=0}^{m-1} \frac{(\lambda - \lambda_{a+k})[x(b-\alpha-k) - x(z)]}{\tau_{\alpha+k}(b-\alpha-k)\gamma(\alpha+k+1)},$$

if $\sigma(b) + \tau(b)\nabla x_1(b) = 0$, $\gamma(\alpha)\tau_{\alpha-1}(b-\alpha+1) = 0$, and the series converges at z, $z \pm 1$. [*Hint:* Use the identity

$$\begin{aligned} \rho^{-1}(z) \frac{\Delta}{\nabla x_1(z)} \left(\sigma(z)\rho(z) \frac{\nabla}{\nabla x(z)} [x(z) - x(\xi)]^{(\mu)} \right) &= \gamma(\mu)\gamma(\mu-1)\sigma(\xi-\mu+1)[x(z) - x(\xi-1)]^{(\mu-2)} \\ &+ \gamma(\mu)\tau_{\mu-1}(\xi-\mu+1)[x(z) - x(\xi-1)]^{(\mu-1)} \\ &- \lambda_\mu [x(z) - x(\xi)]^{(\mu)}, \end{aligned}$$

which can be established with the help of the properties of the generalized powers on classical grids [**17**].]

Show that the series

$$y(z) = \frac{c_0}{[x(z) - x(a-1)]^{(\alpha)}} \times \sum_{n=0}^{\infty} \prod_{k=0}^{n-1} \frac{\gamma(\alpha+k)\tau_{-\alpha-k-1}(a)}{(\lambda - \lambda_{-\alpha-k-1})[x(z) - x(a-\alpha-k-1)]}$$

is a solution of the difference equation above with the right-hand side

$$G(z) = -\frac{c_0}{[x(z) - x(a-1)]^{(\alpha)}} \lim_{m\to\infty} (\lambda - \lambda_{-\alpha-m-1}) \times \prod_{k=0}^{m} \frac{\gamma(\alpha+k)\tau_{-\alpha-k-1}(a)}{(\lambda - \lambda_{-\alpha-k-1})[x(z) - x(a-\alpha-k-1)]},$$

if $\sigma(a) = 0$, $\lambda = \lambda_{-\alpha}$, and the series converges at z, $z \pm 1$. In a similar manner, show that the series

$$y(z) = \frac{c_0}{[x(z) - x(b+\alpha)]^{(\alpha)}} \times \sum_{n=0}^{\infty} \prod_{k=0}^{n-1} \frac{\gamma(\alpha+k)\tau_{-\alpha-k-1}(b+\alpha+k+1)}{(\lambda - \lambda_{-\alpha-k-1})[x(z) - x(b+\alpha+k+1)]}$$

is a solution of the difference equation above with

$$G(z) = -\frac{c_0}{[x(z) - x(b+\alpha)]^{(\alpha)}} \lim_{m\to\infty} (\lambda - \lambda_{-\alpha-m-1}) \times \prod_{k=0}^{m} \frac{\gamma(\alpha+k)\tau_{-\alpha-k-1}(b+\alpha+k+1)}{(\lambda - \lambda_{-\alpha-k-1})[x(z) - x(b+\alpha+k+1)]},$$

if $\sigma(b) + \tau(b)\nabla x_1(b) = 0$, $\lambda = \lambda_{-\alpha}$, and the series converges at z, $z \pm 1$. [*Hint*: Use the identity

$$\begin{aligned}\rho^{-1}(z)\frac{\Delta}{\nabla x_1(z)}\left(\sigma(z)\rho(z)\frac{\nabla}{\nabla x(z)}\left(\frac{1}{[x(z)-x(\xi)]^{(\mu)}}\right)\right) &= \frac{\gamma(\mu)\gamma(\mu+1)\sigma(\xi+1)}{[x(z)-x(\xi+1)]^{(\mu+2)}} - \frac{\gamma(\mu)\tau_{-\mu-1}(\xi+1)}{[x(z)-x(\xi)]^{(\mu+1)}} \\ &\quad - \frac{\lambda_{-\mu}}{[x(z)-x(\xi)]^{(\mu)}},\end{aligned}$$

which can be verified with the help of the properties of the generalized powers [**17**].]

[*Note*: In many cases the above extension of the generalized power series method gives us an opportunity to construct more then one solution of the nonhomogeneous difference equation of hypergeometric type corresponding to the same right-hand side [**17**] and, therefore, to construct a solution of the homogeneous equation as a linear combination. This consideration explains the structure of many transformation formulas for basic hypergeometric series up to the transformations of very-well-poised ${}_8\varphi_7$ functions [**46**].]

(26) Classify all solutions of the difference equation of hypergeometric type on the classical nonuniform lattices [**17**].

(27) Verify (2.2.24)–(2.2.27).

(28) Verify (2.3.29) and prove (2.3.26) with the help of the addition theorem (3.1.3) from the next chapter. [*Hint*: Use the addition formula

$$\mathcal{E}_q(x, y; \alpha) = \mathcal{E}_q(x; \alpha)\ \mathcal{E}_q(y; \alpha)$$

with $y = \eta$, or $e^{i\varphi} = q^{1/4}$, and (2.3.29) with $x \to -x$ in order to obtain

$$\mathcal{E}_q(x; \alpha)\ \mathcal{E}_q(\eta; \alpha) = \frac{(\alpha^2; q^2)_\infty}{(q\alpha^2; q^2)_\infty}\ {}_2\varphi_1\left(\begin{matrix} -q^{1/4}e^{i\theta},\ -q^{1/4}e^{-i\theta} \\ -q^{1/2} \end{matrix}; q^{1/2}, \alpha\right).$$

To complete the proof find the value of $\mathcal{E}_q(\eta; \alpha)$ by setting $x = 0$.]

(29) *Positivity property.* Prove that if either $0 \le \alpha < 1$, $x \ge -\eta$ or $-1 < \alpha \le 0$, $x \le \eta = \left(q^{1/4} + q^{-1/4}\right)/2$, then $\mathcal{E}_q(x; \alpha) > 0$ [**72**].

(30) Show that

$$\mathcal{E}_q\left(i\frac{1-q}{2q^{1/2}}; \alpha\right) = \frac{(\alpha^2; q^2)_\infty}{(q\alpha^2; q^2)_\infty} + i\alpha q^{-1/4}.$$

(31) Derive (2.4.10)–(2.4.11) as special cases of (2.4.14)–(2.4.15).

(32) Derive the quadratic transformation formula (2.4.27).

(33) Show that the formal solution of the difference equation (2.2.11) on a quadratic lattice $x(z) = C_1z^2 + C_2z + C_3$ has the form

$$u(z, s) = \sum_{n=0}^{\infty} \frac{(-\lambda/\sigma)^{n/2}}{n!}(-C_1)^n \times \left(s - z - \frac{n-1}{2}\right)_n \left(s + z + \mu - \frac{n-1}{2}\right)_n,$$

where s is a shift parameter and $\mu = C_2/C_1$. Express this formal difference analog of the exponential on a quadratic grid as a sum of two divergent ${}_4F_1$'s. Obtain the difference analog of the exponential function on a linear grid (2.5.9) as a formal limit $C_1 \to 0$.

(34) Show that

$$\frac{\delta}{\delta x}\ \mathcal{E}_q(x, y; \alpha, \beta) = \frac{2q^{1/4}\alpha}{1-q}\ \mathcal{E}_q(x, y; \alpha, \beta),$$

$$\frac{\delta}{\delta y}\ \mathcal{E}_q(x, y; \alpha, \beta) = \frac{2q^{1/4}\beta}{1-q}\ \mathcal{E}_q(x, y; \alpha, \beta).$$

Introduce the following analogs of the cosine and sine functions

$$C_q(x,y;\omega,\varkappa) = \frac{1}{2}\left(\mathcal{E}_q(x,y;i\omega,i\varkappa) + \mathcal{E}_q(x,y;-i\omega,-i\varkappa)\right),$$

$$S_q(x,y;\omega,\varkappa) = \frac{1}{2i}\left(\mathcal{E}_q(x,y;i\omega,i\varkappa) - \mathcal{E}_q(x,y;-i\omega,-i\varkappa)\right)$$

and derive the difference-differentiation formulas

$$\frac{\delta}{\delta x}\, C_q(x,y;\omega,\varkappa) = -\frac{2q^{1/4}\omega}{1-q}\, S_q(x,y;\omega,\varkappa),$$

$$\frac{\delta}{\delta x}\, S_q(x,y;\omega,\varkappa) = \frac{2q^{1/4}\omega}{1-q}\, C_q(x,y;\omega,\varkappa),$$

$$\frac{\delta}{\delta y}\, C_q(x,y;\omega,\varkappa) = -\frac{2q^{1/4}\varkappa}{1-q}\, S_q(x,y;\omega,\varkappa),$$

$$\frac{\delta}{\delta y}\, S_q(x,y;\omega,\varkappa) = \frac{2q^{1/4}\varkappa}{1-q}\, C_q(x,y;\omega,\varkappa)$$

as q-analogs of

$$\frac{d}{dx}\cos(\omega x + \varkappa y) = -\omega \sin(\omega x + \varkappa y),$$

$$\frac{d}{dx}\sin(\omega x + \varkappa y) = \omega \cos(\omega x + \varkappa y),$$

$$\frac{d}{dy}\cos(\omega x + \varkappa y) = -\varkappa \sin(\omega x + \varkappa y),$$

$$\frac{d}{dy}\sin(\omega x + \varkappa y) = \varkappa \cos(\omega x + \varkappa y),$$

respectively.
Show that

$$C_q(x;\omega)\, C_q(y;\varkappa) = \frac{1}{2}\left(C_q(x,y;\omega,\varkappa) + C_q(x,-y;\omega,\varkappa)\right),$$

$$S_q(x;\omega)\, S_q(y;\varkappa) = \frac{1}{2}\left(C_q(x,-y;\omega,\varkappa) - C_q(x,y;\omega,\varkappa)\right),$$

$$S_q(x;\omega)\, C_q(y;\varkappa) = \frac{1}{2}\left(S_q(x,y;\omega,\varkappa) + S_q(x,-y;\omega,\varkappa)\right),$$

$$C_q(x;\omega)\, S_q(y;\varkappa) = \frac{1}{2}\left(S_q(x,y;\omega,\varkappa) - S_q(x,-y;\omega,\varkappa)\right).$$

[*Hint*: Use the addition formulas (3.5.16)–(3.5.17).]

CHAPTER 3

Addition Theorems

In this chapter we shall discuss q-analogs of the addition formula for the exponential function and the corresponding results on q-trigonometry.

3.1. Introduction

The addition theorem for the exponential function can be written as

$$\exp \alpha \left(x+y\right) = \exp \alpha x \ \exp \alpha y. \tag{3.1.1}$$

It is also well-known that $\exp\left(\alpha x\right)$ is the only measurable solution of the functional equation $f\left(x+y\right) = f\left(x\right) f\left(y\right)$. Jackson's q-exponential functions (1.1.1)–(1.1.2) satisfy the following q-addition formulas

$$e_q\left(x+y\right) = e_q\left(y\right)\ e_q\left(x\right), \qquad E_q\left(x+y\right) = E_q\left(x\right)\ E_q\left(y\right), \tag{3.1.2}$$

provided that $xy = qyx$. See [**94**] for a nice review on special functions of q-commuting variables and the bibliography. A commutative q-analog of the addition theorem (3.1.1) involving the q-exponential functions (2.3.1) and (2.3.3) has been established by Suslov [**140**].

THEOREM 3.1.

$$\mathcal{E}_q\left(x,y;\alpha\right) = \mathcal{E}_q\left(x;\alpha\right)\ \mathcal{E}_q\left(y;\alpha\right). \tag{3.1.3}$$

Formula (3.1.3) has attracted some attention and different proofs of this relation were given in [**43**], [**71**], and [**140**]. Floreanini, LeTourneux, and Vinet [**43**] have used group theoretical methods. Ismail and Stanton gave two different proofs on the basis of the connection coefficients for the q-Hermite polynomials [**71**] and yet another proof in [**73**]. See also [**140**] for the original proof by direct series manipulations and [**72**]–[**73**]. In this chapter we shall discuss several different proofs of this result and its extension.

3.2. First Proof of Addition Theorem: Analytic Functions

We shall follow here the idea of one of the original proofs [**140**] using arguments of the theory of analytic functions. The functions $\mathcal{E}_q\left(x,y;\alpha\right)$, $\mathcal{E}_q\left(x;\alpha\right)$ and $\mathcal{E}_q\left(x;-\alpha\right)$ satisfy the second order difference equation (2.3.16). Also, the q-exponential functions $\mathcal{E}_q\left(x;\pm\alpha\right)$ are linearly independent when $\alpha \neq 0$. Therefore, see for example Ruijsenaar's paper [**132**], Section 2.2, or Section 3.4 of this chapter,

$$\mathcal{E}_q\left(x,y;\alpha\right) = A\ \mathcal{E}_q\left(x;\alpha\right) + B\ \mathcal{E}_q\left(x;-\alpha\right), \tag{3.2.1}$$

where $A = A(z,s)$ and $B = B(z,s)$ are, generally speaking, some functions of period 1 in z. Here $x = \cos\theta$, $e^{i\theta} = q^z$ and $y = \cos\varphi$, $e^{i\varphi} = q^s$. Applying $\delta/\delta x$ to both sides of (3.2.1), with the help of (2.3.13)

$$\mathcal{E}_q(x,y;\alpha) = A\ \mathcal{E}_q(x;\alpha) - B\ \mathcal{E}_q(x;-\alpha), \tag{3.2.2}$$

or

$$\mathcal{E}_q(x,y;\alpha) = A\ \mathcal{E}_q(x;\alpha). \tag{3.2.3}$$

The functions $\mathcal{E}_q(x,y;\alpha)$ and $\mathcal{E}_q(x;\alpha)$, which are analytic when $|\alpha| < 1$ by their definitions (2.3.1) and (2.3.3), transparently have a natural purely imaginary period $T = 2\pi i/\log q^{-1}$ in z. Hence, the coefficient

$$A(z,s) = \mathcal{E}_q(x,y;\alpha)/\mathcal{E}_q(x;\alpha) \tag{3.2.4}$$

is a doubly periodic analytic function in z. In view of (2.3.23) the following inequality holds

$$\begin{aligned}|\mathcal{E}_q(x;\alpha) - 1| &\le \sum_{n=1}^{\infty} \frac{\left|\left(-q^{2z}, -q^{-2z}; q^2\right)_n\right|}{\left(q, q^2; q^2\right)_n} \left(q|\alpha|^2\right)^n \\ &+ \frac{2q^{1/4}}{1-q}|\alpha|\,|x| \sum_{n=0}^{\infty} \frac{\left|\left(-q^{1+2z}, -q^{1-2z}; q^2\right)_n\right|}{\left(q^2, q^3; q^2\right)_n} \left(q|\alpha|^2\right)^n.\end{aligned}$$

Moreover,

$$\begin{aligned}\left|\left(-q^{2z}, -q^{-2z}; q^2\right)_n\right| &\le \left(-q^{2\operatorname{Re} z}, -q^{-2\operatorname{Re} z}; q^2\right)_n \le \left(-q, -q^{-1}; q^2\right)_n, \\ \left|\left(-q^{1+2z}, -q^{1-2z}; q^2\right)_n\right| &\le \left(-q^{1+2\operatorname{Re} z}, -q^{1-2\operatorname{Re} z}; q^2\right)_n \le \left(-q^2, -1; q^2\right)_n,\end{aligned}$$

when $-1/2 \le \operatorname{Re} z \le 1/2$ and, therefore,

$$\begin{aligned}|\mathcal{E}_q(x;\alpha) - 1| &\le \sum_{n=1}^{\infty} \frac{\left(-q, -q^{-1}; q^2\right)_n}{\left(q, q^2; q^2\right)_n} \left(q|\alpha|^2\right)^n \\ &+ q^{-1/4}\frac{1+q}{1-q}|\alpha| \sum_{n=0}^{\infty} \frac{\left(-q^2, -1; q^2\right)_n}{\left(q^2, q^3; q^2\right)_n} \left(q|\alpha|^2\right)^n \\ &\le |\alpha| \left(\mathcal{E}_q\left(\left(q^{1/2} + q^{-1/2}\right)/2; 1\right) - 1\right)\end{aligned}$$

by (2.3.23) when $|\alpha| < 1$. As a result,

$$|\mathcal{E}_q(x;\alpha) - 1| \le C\,|\alpha| \tag{3.2.5}$$

for all $x = \left(q^z + q^{-z}\right)/2$, $-1/2 \le \operatorname{Re} z \le 1/2$ and $|\alpha| < 1$. Here, the positive constant $C = \mathcal{E}_q\left(\left(q^{1/2} + q^{-1/2}\right)/2; 1\right) - 1$ depends on q only. Hence,

$$|\mathcal{E}_q(x;\alpha)| \ge 1 - C\,|\alpha| > 0 \tag{3.2.6}$$

for all $|\alpha| < \delta = \min\left(C^{-1}, 1\right)$ and for all z in the parallelogram with the vertices at $-1/2 + i\pi/\log q$, $1/2 + i\pi/\log q$, $1/2 + i\pi/\log q^{-1}$ and $-1/2 + i\pi/\log q^{-1}$ in Figure 3.1. The last inequality shows that for sufficiently small $|\alpha|$ the $\mathcal{E}_q(x;\alpha)$ does not have zeros in z in the parallelogram.

Thus, $A(z,s)$ given by (3.2.4) is an entire doubly periodic function in z when $|\alpha| < \delta$ that is a constant by Liouville's theorem. One can find this

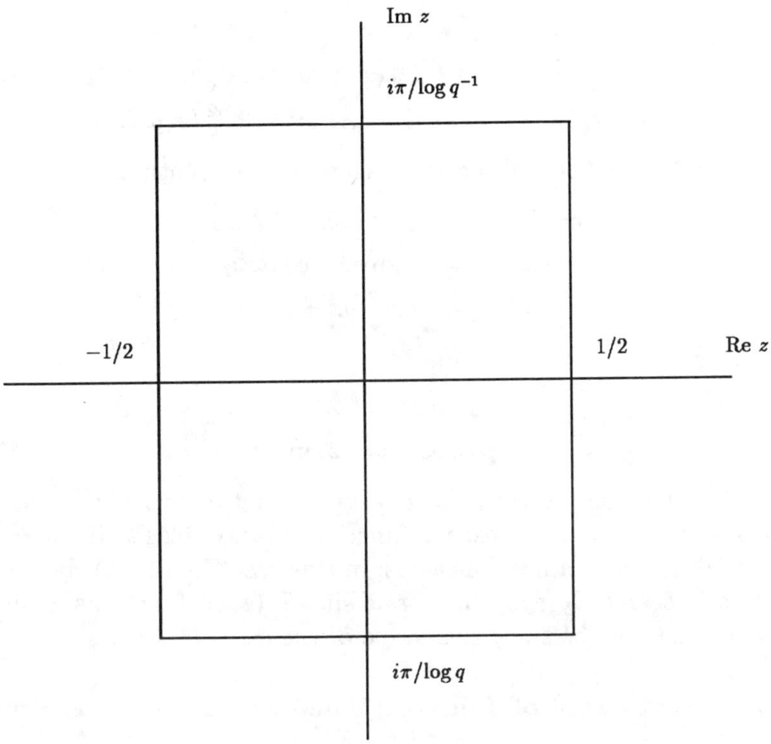

FIGURE 3.1. The parallelogram of the double periodicity in the complex z-plane.

constant choosing $x = 0$ (or $\theta = \pi/2$) in (3.2.3) which results in the addition formula (3.1.3) in the open disk $|\alpha| < \delta$. Analytic continuation of (3.1.3) in the parameter α to the entire complex α-plane completes the proof.

The following q-addition formulas for the basic trigonometric functions have been established in [**140**].

THEOREM 3.2.

$$C_q\,(x, y; \omega) = C_q\,(x; \omega)\, C_q\,(y; \omega) - S_q\,(x; \omega)\, S_q\,(y; \omega)\,, \tag{3.2.7}$$
$$S_q\,(x, y; \omega) = S_q\,(x; \omega)\, C_q\,(y; \omega) + C_q\,(x; \omega)\, S_q\,(y; \omega)\,. \tag{3.2.8}$$

PROOF. Use

$$\mathcal{E}_q\,(x, y; i\omega) = \mathcal{E}_q\,(x; i\omega)\;\; \mathcal{E}_q\,(y; i\omega) \tag{3.2.9}$$

and the analog of Euler formula (2.4.1). □

The special cases of (3.2.7)–(3.2.9) are

$$C_q\,(x, x; \omega) = C_q^2\,(x; \omega) - S_q^2\,(x; \omega)\,, \tag{3.2.10}$$
$$S_q\,(x, x; \omega) = 2S_q\,(x; \omega)\;\; C_q\,(x; \omega)\,, \tag{3.2.11}$$
$$C_q\,(x, -x; \omega) = C_q^2\,(x; \omega) + S_q^2\,(x; \omega)\,, \tag{3.2.12}$$

and

$$C_q(x,-x;\omega)+C_q(x,x;\omega)=2C_q^2(x;\omega), \tag{3.2.13}$$

$$C_q(x,-x;\omega)-C_q(x,x;\omega)=2S_q^2(x;\omega), \tag{3.2.14}$$

which are analogs of the well-known trigonometric identities

$$\cos 2\omega x = \cos^2\omega x-\sin^2\omega x, \tag{3.2.15}$$

$$\sin 2\omega x = 2\sin\omega x\ \cos\omega x, \tag{3.2.16}$$

$$1 = \cos^2\omega x+\sin^2\omega x, \tag{3.2.17}$$

and

$$1+\cos 2\omega x = 2\cos^2\omega x, \tag{3.2.18}$$

$$1-\cos 2\omega x = 2\sin^2\omega x, \tag{3.2.19}$$

respectively. One can see that $C_q(x,-x;\omega)$ is not a constant, what we get in the classical case, but a positive function due to (3.2.12) if the $C_q(x;\omega)$ and $S_q(x;\omega)$ are real valued functions; in this case Eq. (5.8.6) below shows that the basic cosine $C_q(x;\omega)$ and basic sine $S_q(x;\omega)$ functions do not have common x-zeros for the same real value of the ω.

3.3. Second Proof of Addition Theorem: Product Formula

In this section we shall give an alternative proof of the addition theorem (3.1.3) on the basis of a more general product formula for functions (2.2.14).

LEMMA 3.1. *The following symmetry relation holds*

$$\sum_{k=0}^{n}\frac{\left[x(z)-x\left(s+\frac{k-1}{2}\right)\right]^{(k)}}{\gamma(k)!}\frac{\left[x(r)-x\left(t+\frac{n-k-1}{2}\right)\right]^{(n-k)}}{\gamma(n-k)!} \tag{3.3.1}$$

$$=\sum_{k=0}^{n}\frac{\left[x(z)-x\left(t+\frac{k-1}{2}\right)\right]^{(k)}}{\gamma(k)!}\frac{\left[x(r)-x\left(s+\frac{n-k-1}{2}\right)\right]^{(n-k)}}{\gamma(n-k)!}$$

on the general q-quadratic lattice (2.2.12).

PROOF. Relation (3.3.1) is obviously true in two symmetric cases when $t=s$ and $z=r$. If $n=1$ one gets

$$[x(r)-x(t)]+[x(z)-x(s)]=[x(r)-x(s)]+[x(z)-x(t)],$$

which is also correct.

Now we can prove this lemma by mathematical induction. Denote the left side of (3.3.1) as

$$L_n(z):=L_n(z,r,s,t) \tag{3.3.2}$$

$$=\sum_{k=0}^{n}\frac{\left[x(z)-x\left(s+\frac{k-1}{2}\right)\right]^{(k)}}{\gamma(k)!}\frac{\left[x(r)-x\left(t+\frac{n-k-1}{2}\right)\right]^{(n-k)}}{\gamma(n-k)!}.$$

In view of (2.2.17)

$$\frac{\delta L_n(z)}{\delta x(z)} = L_{n-1}(z). \tag{3.3.3}$$

Denote also the difference between the left and right sides of (3.3.1) as

$$D_n(z) := L_n(z, r, s, t) - L_n(z, r, t, s) \tag{3.3.4}$$

and suppose that $D_n(z) = 0$, which means that relation (3.3.1) holds for some positive integer n by the assumption. Then due to (3.3.3)

$$\frac{\delta D_{n+1}(z)}{\delta x(z)} = D_n(z) = 0, \tag{3.3.5}$$

or

$$D_{n+1}(z) = C = \text{constant}, \tag{3.3.6}$$

because $D_{n+1}(z)$ is a polynomial in x by the definition. In order to find this constant we choose $z = r$ when $C = D_{n+1}(r) = 0$. Thus $D_{n+1} = 0$, which proves our lemma by induction. □

We can derive now the following product formula.

THEOREM 3.3. *The functions $u(z,s)$ given by (2.2.14) satisfy*

$$u(z,s)\ u(r,t) = u(z,t)\ u(r,s) \tag{3.3.7}$$

provided that the series converge.

PROOF. In view of (2.2.14) and (3.3.1)

$$\begin{aligned} &u(z,s)\ u(r,t) \\ &= \sum_{k=0}^{\infty} \xi^k \frac{\left[x(z) - x\left(s + \frac{k-1}{2}\right)\right]^{(k)}}{\gamma(k)!} \sum_{k=0}^{\infty} \xi^m \frac{\left[x(r) - x\left(t + \frac{m-1}{2}\right)\right]^{(m)}}{\gamma(m)!} \\ &= \sum_{n=0}^{\infty} \xi^n \sum_{k=0}^{n} \frac{\left[x(z) - x\left(s + \frac{k-1}{2}\right)\right]^{(k)}}{\gamma(k)!} \frac{\left[x(r) - x\left(t + \frac{n-k-1}{2}\right)\right]^{(n-k)}}{\gamma(n-k)!} \\ &= \sum_{n=0}^{\infty} \xi^n \sum_{k=0}^{n} \frac{\left[x(z) - x\left(t + \frac{k-1}{2}\right)\right]^{(k)}}{\gamma(k)!} \frac{\left[x(r) - x\left(s + \frac{n-k-1}{2}\right)\right]^{(n-k)}}{\gamma(n-k)!} \\ &= u(z,t)\ u(r,s), \end{aligned}$$

where $\xi = (-\lambda/\sigma)^{1/2}$. □

COROLLARY 3.1. *The addition theorem* (3.1.3) *is a special case of the product formula* (3.3.7).

PROOF. In order to obtain (3.1.3) from (3.3.7) choose $r = t = z_0$, where $q^{z_0} = i\sqrt{\mu}$ and $\mu = C_2/C_1$:

$$u(z,s)\ u(z_0,z_0) = u(z,z_0)\ u(z_0,s). \tag{3.3.8}$$

But from (2.2.27),

$$u(z,z_0) \tag{3.3.9}$$

$$= {}_2\varphi_1\left(\begin{matrix} -\mu q^{1+2z}, \ -q^{1-2z}/\mu \\ q \end{matrix} ; q^2, -\frac{\lambda C_1 C_2 (1-q)^2}{\sigma q^{1/2}}\right)$$

$$+ \left(-\frac{\lambda}{\sigma}\right)^{1/2} x(z)$$

$$\times \ {}_2\varphi_1\left(\begin{matrix} -\mu q^{2+2z}, \ -q^{2-2z}/\mu \\ q^3 \end{matrix} ; q^2, -\frac{\lambda C_1 C_2 (1-q)^2}{\sigma q^{1/2}}\right),$$

$$u(z_0, s) \tag{3.3.10}$$

$$= {}_2\varphi_1\left(\begin{matrix} -\mu q^{1+2s}, \ -q^{1-2s}/\mu \\ q \end{matrix} ; q^2, -\frac{\lambda C_1 C_2 (1-q)^2}{\sigma q^{1/2}}\right)$$

$$- \left(-\frac{\lambda}{\sigma}\right)^{1/2} x(s)$$

$$\times \ {}_2\varphi_1\left(\begin{matrix} -\mu q^{2+2s}, \ -q^{2-2s}/\mu \\ q^3 \end{matrix} ; q^2, -\frac{\lambda C_1 C_2 (1-q)^2}{\sigma q^{1/2}}\right),$$

and

$$u(z_0, s) = {}_1\varphi_0\left(\begin{matrix} q \\ - \end{matrix} ; q^2, -\frac{\lambda C_1 C_2 (1-q)^2}{\sigma q^{1/2}}\right) \tag{3.3.11}$$

$$= \frac{\left(-C_1 C_2 (1-q)^2 q^{1/2} \lambda/\sigma; q^2\right)_\infty}{\left(-C_1 C_2 (1-q)^2 q^{-1/2} \lambda/\sigma; q^2\right)_\infty}.$$

We have used the q-binomial theorem when $\left|C_1 C_2 (1-q)^2 q^{-1/2} \lambda/\sigma\right| < 1$. Comparing (2.2.27), (2.3.1), (2.3.3), and (3.3.9)–(3.3.11), one gets (3.1.3) from (3.3.7). □

On the other hand, it is easy to verify that (3.3.7) follows from (3.3.8). Thus, the product formula (3.3.7) and the addition theorem (3.3.8) are, in fact, equivalent.

COROLLARY 3.2. *Relations* (3.3.8)–(3.3.11) *provide the analytic continuation of the function* (2.2.27).

PROOF. Indeed, both sides of the equation (3.3.8) are well-defined when $\left|C_1 C_2 (1-q)^2 q^{-1/2} \lambda/\sigma\right| < 1$. But the analytic continuation of the ${}_2\varphi_1$s on the right side in to the larger domain can be easily given [**46**]. For example, with the help of (A.3.3), (A.3.5), and (A.3.8) one gets

$${}_2\varphi_1\left(\begin{matrix} -\mu q^{1+2z}, \ -q^{1-2z}/\mu \\ q \end{matrix} ; q^2, \zeta\right)$$

$$= \frac{\left(-\mu q^{1+2z}\zeta, -q^{1-2z}\zeta/\mu; q^2\right)_\infty}{\left(q, \ \zeta; q^2\right)_\infty}$$

$$\times\ {}_2\varphi_2\left(\begin{matrix}\zeta, & q\zeta \\ -\mu q^{1+2z}\zeta, & -q^{1-2z}\zeta/\mu\end{matrix}\ ;\ q^2,\ q\right) \tag{3.3.12}$$

$$= \frac{\left(-q^{-2z}/\mu, -q^{1-2z}/\mu, -\mu q^{1+2z}\zeta, -q^{1-2z}/\mu\zeta; q^2\right)_\infty}{\left(q,\ q^{-4z}/\mu^2, \zeta, q/\zeta; q^2\right)_\infty}$$

$$\times\ {}_2\varphi_1\left(\begin{matrix}-\mu q^{2z},\ -\mu q^{1+2z} \\ \mu^2 q^{2+4z}\end{matrix}\ ;\ q^2, \frac{q^2}{\zeta}\right)$$

$$+ \frac{\left(-\mu q^{2z}, -\mu q^{1+2z}, -q^{1-2z}\zeta/\mu, -\mu q^{1+2z}/\zeta; q^2\right)_\infty}{\left(q,\ q^{4z}/\mu^2, \zeta, q/\zeta; q^2\right)_\infty}$$

$$\times\ {}_2\varphi_1\left(\begin{matrix}-q^{-2z}/\mu,\ -q^{1-2z}/\mu \\ q^{2-4z}/\mu^2\end{matrix}\ ;\ q^2, \frac{q^2}{\zeta}\right) \tag{3.3.13}$$

and

$${}_2\varphi_1\left(\begin{matrix}-\mu q^{2+2z},\ -q^{2-2z}/\mu \\ q^3\end{matrix}\ ;\ q^2, \zeta\right)$$

$$= \frac{\left(-\mu q^{2+2z}\zeta, -q^{2-2z}\zeta/\mu; q^2\right)_\infty}{\left(q^3,\ \zeta; q^2\right)_\infty}$$

$$\times\ {}_2\varphi_2\left(\begin{matrix}\zeta, & q\zeta \\ -\mu q^{2+2z}\zeta, & -q^{2-2z}\zeta/\mu\end{matrix}\ ;\ q^2,\ q^3\right) \tag{3.3.14}$$

$$= \frac{\left(-q^{1-2z}/\mu, -q^{2-2z}/\mu, -\mu q^{2+2z}\zeta, -q^{-2z}/\mu\zeta; q^2\right)_\infty}{\left(q^3,\ q^{-4z}/\mu^2, \zeta, q/\zeta; q^2\right)_\infty}$$

$$\times\ {}_2\varphi_1\left(\begin{matrix}-\mu q^{1+2z},\ -\mu q^{2z} \\ \mu^2 q^{2+4z}\end{matrix}\ ;\ q^2, \frac{q^2}{\zeta}\right)$$

$$+ \frac{\left(-\mu q^{1+2z}, -\mu q^{2+2z}, -q^{2-2z}\zeta/\mu, -\mu q^{2z}/\zeta; q^2\right)_\infty}{\left(q^3,\ \mu^2 q^{4z}, \zeta, q/\zeta; q^2\right)_\infty}$$

$$\times\ {}_2\varphi_1\left(\begin{matrix}-q^{1-2z}/\mu,\ -q^{-2z}/\mu \\ q^{2-4z}/\mu^2\end{matrix}\ ;\ q^2, \frac{q^2}{\zeta}\right), \tag{3.3.15}$$

where $\zeta = -C_1C_2\,(1-q)^2\,q^{-1/2}\lambda/\sigma$. □

Equations (3.3.8), (3.3.13), and (3.3.15) determine also the asymptotic behavior of the function (2.2.27) for the large values of its argument.

3.4. Third Proof of Addition Theorem: Difference Equation

In this section shall give an independent proof of the addition formula (3.1.3) and will derive several related results on the basis of a general approach using the theory of analytic difference equations discussed by Ruijsenaars [**132**].

3.4.1. Proof of Addition Theorem. Consider the difference equation of hypergeometric type in self-adjoint form

$$\frac{\Delta}{\nabla x_1(z)}\left(\sigma(z)\,\rho(z)\,\frac{\nabla y(z)}{\nabla x(z)}\right)+\lambda\rho(z)\,y(z)=0, \tag{3.4.1}$$

where $\Delta f(z) = \nabla f(z+1) = f(z+1) - f(z)$, $x_1(z) = x(z+1/2)$ and $x(z) = (q^z + q^{-z})/2$. Suppose that y_1 and y_2 are two solutions of (3.4.1) corresponding to the same eigenvalue λ. Then

$$\Delta\left[\sigma(z)\,\rho(z)\,W(y_1, y_2)\right] = 0, \tag{3.4.2}$$

where

$$\begin{aligned} W(y_1,y_2) &= \begin{vmatrix} y_1(z) & y_2(z) \\ \dfrac{\nabla y_1(z)}{\nabla x(z)} & \dfrac{\nabla y_2(z)}{\nabla x(z)} \end{vmatrix} \\ &= \frac{y_2(z)\,y_1(z-1) - y_1(z)\,y_2(z-1)}{x(z)-x(z-1)} \end{aligned} \tag{3.4.3}$$

is the difference analog of the Wronskian. See, for example, [**111**], [**132**], [**139**], [**141**]–[**142**], and Ex. 2.21. If y_3 is another solution of (3.4.1) corresponding to the same eigenvalue, then a straightforward calculation shows

$$y_3(z) = -\frac{W(y_2,y_3)}{W(y_1,y_2)}\,y_1(z) + \frac{W(y_1,y_3)}{W(y_1,y_2)}\,y_2(z). \tag{3.4.4}$$

It follows also from (3.4.2) that the ratios $W(y_k, y_3)/W(y_1, y_2)$, $k = 1, 2$ are periodic functions in z of period 1. The last equation gives us the possibility to find relations between different solutions of (3.4.1). We apply this method here only at the level of the q-exponential functions in order to derive the addition formula (3.1.3).

For the q-exponential functions equation (3.4.1) reduces to (2.2.11) and we can choose $\sigma = \rho = 1$. Consider three different solutions discussed in Section 3.2,

$$y_1(z) = \mathcal{E}_q(x;\alpha), \quad y_2(z) = \mathcal{E}_q(x;-\alpha), \quad y_3(z) = \mathcal{E}_q(x,y;\alpha). \tag{3.4.5}$$

Then, by (3.4.2), $\Delta\left[W(y_k, y_l)\right] = 0$ and for $|\alpha| < 1$ all the q-Wronskians are constants as doubly periodic entire functions. One can find these constants using special values of x. Equations (2.3.13) and (3.4.3) for $W(y_1, y_3)$ give

$$\mathcal{E}_q(x;\alpha)\ \mathcal{E}_q(x(z-1/2),y;\alpha) - \mathcal{E}_q(x,y;\alpha)\ \mathcal{E}_q(x(z-1/2);\alpha) = A. \tag{3.4.6}$$

Letting $x = \eta = x(\pm 1/4) = \left(q^{1/4} + q^{-1/4}\right)/2$,

$$A = \mathcal{E}_q(\eta;\alpha)\ \mathcal{E}_q(\eta,y;\alpha) - \mathcal{E}_q(\eta,y;\alpha)\ \mathcal{E}_q(\eta;\alpha) = 0 \tag{3.4.7}$$

and

$$W(y_1, y_3) = 0. \tag{3.4.8}$$

In the case of $W(y_1, y_2)$, in the same fashion,

$$\mathcal{E}_q(x;\alpha)\ \mathcal{E}_q(x(z-1/2);-\alpha) + \mathcal{E}_q(x;-\alpha)\ \mathcal{E}_q(x(z-1/2);\alpha) = B, \tag{3.4.9}$$

and for $x = \eta$,

$$B = 2\,\mathcal{E}_q(\eta;\alpha)\,\mathcal{E}_q(\eta;-\alpha) = 2\frac{(\alpha^2;q^2)_\infty}{(q\alpha^2;q^2)_\infty}, \tag{3.4.10}$$

in view of

$$\mathcal{E}_q(\eta;\alpha) = \frac{(-\alpha;q^{1/2})_\infty}{(q\alpha^2;q^2)_\infty}, \tag{3.4.11}$$

see (2.3.10). As a result,

$$W(y_1,y_2) = -\frac{4q^{1/4}\alpha}{1-q}\,\frac{(\alpha^2;q^2)_\infty}{(q\alpha^2;q^2)_\infty}. \tag{3.4.12}$$

Finally, substituting (3.4.8) in (3.4.4) and letting $x = 0$ we get

$$\frac{W(y_2,y_3)}{W(y_1,y_2)} = -\mathcal{E}_q(y;\alpha) \tag{3.4.13}$$

and our proof of the addition formula (3.1.3) is complete. We have also evaluated the q-Wronskian

$$W(y_2,y_3) = \frac{4q^{1/4}\alpha}{1-q}\,\frac{(\alpha^2;q^2)_\infty}{(q\alpha^2;q^2)_\infty}\,\mathcal{E}_q(y;\alpha) \tag{3.4.14}$$

which is of an independent interest.

3.4.2. Generating Function. Our next goal is to derive the following generating relation

$$(q\alpha^2;q^2)_\infty\,\mathcal{E}_q(x;\alpha) = \sum_{n=0}^{\infty}\frac{q^{n^2/4}}{(q;q)_n}\,\alpha^n\,H_n(x|q) \tag{3.4.15}$$

for the continuous q-Hermite polynomials of Rogers given by

$$H_n(\cos\theta|q) = \sum_{k=0}^{n}\frac{(q;q)_n}{(q;q)_k\,(q;q)_{n-k}}\,e^{i(n-2k)\theta}, \tag{3.4.16}$$

see, for example, [**46**]. Different proofs of (3.4.15) are given in [**76**], [**41**], [**43**], [**68**], [**71**], and [**140**]. It is worth noting that the series in (3.4.15) analytically continues the left side to an entire function in x and α.

Consider the series

$$F(x,\alpha) := \sum_{n=0}^{\infty}\frac{q^{n^2/4}}{(q;q)_n}\,\alpha^n\,H_n(x|q)\,. \tag{3.4.17}$$

Using the difference-differentiation formula

$$\frac{\delta}{\delta x}\,H_n(x|q) = 2q^{(1-n)/2}\,\frac{1-q^n}{1-q}\,H_{n-1}(x|q) \tag{3.4.18}$$

for the continuous q-Hermite polynomials; see, for example, [**140**]; one can show that the function

$$y_4(z) = F(x,\alpha) \tag{3.4.19}$$

is a solution of (2.3.13) and (3.3.9). Thus,

$$y_4(z) = -\frac{W(y_2,y_4)}{W(y_1,y_2)}\, y_1(z) + \frac{W(y_1,y_4)}{W(y_1,y_2)}\, y_2(z). \tag{3.4.20}$$

Evaluation of $W(y_1,y_4)$ gives

$$\mathcal{E}_q(x;\alpha)\;F(x(z-1/2),\alpha) - F(x,\alpha)\;\mathcal{E}_q(x(z-1/2);\alpha) = C \tag{3.4.21}$$

and letting $x=\eta$ we obtain $C=0$, or

$$W(y_1,y_4) = 0. \tag{3.4.22}$$

Substituting this in (3.4.20) and choosing $x=0$ one gets

$$F(0,\alpha) = -\frac{W(y_2,y_4)}{W(y_1,y_2)}. \tag{3.4.23}$$

So, we need to evaluate

$$F(0,\alpha) = \sum_{n=0}^{\infty} \frac{q^{n^2/4}}{(q;q)_n}\, \alpha^n\, H_n(0|q), \tag{3.4.24}$$

where

$$H_{2k+1}(0|q) = 0, \qquad H_{2k}(0|q) = (-1)^k \left(q;q^2\right)_k. \tag{3.4.25}$$

These special values can be found from (3.4.16); see also Ex. 7.17 of [**46**] and Ex. 3.10 here; or from the Rogers generating function given, for example, in Ex. 1.28 of the same book; see also (4.7.3) here. Therefore

$$F(0,\alpha) = \sum_{n=0}^{\infty} (-1)^k \frac{q^{k^2}}{(q^2;q^2)_k}\, \alpha^{2k} = \left(q\alpha^2;q^2\right)_\infty \tag{3.4.26}$$

by (1.1.2) and

$$F(x,\alpha) = \left(q\alpha^2;q^2\right)_\infty\, \mathcal{E}_q(x;\alpha), \tag{3.4.27}$$

which is the generating relation (3.4.15).

In a similar fashion one can derive the generating formula found in [**71**]

$$\mathcal{E}_q(x;\beta) = \left(\beta^2;q^2\right)_\infty \sum_{n=0}^{\infty} \frac{q^{n^2/4}}{(q;q)_n}\, \beta^n\, H_n\left(x|q^{-1}\right), \tag{3.4.28}$$

where

$$H_n\left(\cos\theta|q^{-1}\right) = \sum_{k=0}^{n} q^{k^2-kn} \frac{(q;q)_n}{(q;q)_k\,(q;q)_{n-k}}\, e^{i(n-2k)\theta}. \tag{3.4.29}$$

The above generating relations (3.4.15) and (3.4.28) are q-analogs of the generating function

$$\sum_{n=0}^{\infty} \frac{\alpha^n}{n!}\, H_n(x) = e^{2\alpha x-\alpha^2} \tag{3.4.30}$$

for the Hermite polynomials. Indeed, in the case of (3.4.15) rewrite the left side as $E_{q^2}\left(-q\alpha^2\right)\mathcal{E}_q\left(x;\alpha\right)$, replace $\alpha\to\sqrt{2\left(1-q\right)}\ \alpha$, $x\to\sqrt{\left(1-q\right)/2}\ x$, and use the following limiting relations

$$\lim_{q\to1^-}\left(\frac{2}{1-q}\right)^{n/2}H_n\left(\sqrt{\frac{1-q}{2}}\ x\ \middle|\ q\right)=H_n\left(x\right), \tag{3.4.31}$$

$$\lim_{q\to1^-}\mathcal{E}_q\left(\sqrt{\frac{1-q}{2}}\ x,\sqrt{\frac{1-q}{2}}\ y;\sqrt{\frac{1-q}{2}}\ \alpha\right)=e^{\alpha(x+y)}, \tag{3.4.32}$$

$$\lim_{q\to1^-}E_{q^2}\left(-2q\left(1-q\right)\alpha^2\right)=e^{-\alpha^2}. \tag{3.4.33}$$

Eq. (3.4.31) can be proven with the help of the three term recurrence relation for the continuous q-Hermite polynomials, see [**75**] and Ex. 3.2; Eq. (3.4.32) follows from the definition (2.3.1), see Ex. 3.3; and (3.4.33) is a consequence of the limiting relation for Jackson's q-exponential function, see Ex. 1.5. The case of (3.4.28) is similar.

3.4.3. Representation for Basic Exponential Function. We shall derive the representation (2.3.27) for the basic exponential function as another application of the method under consideration. The generalized power function (2.3.28) has the following difference-differentiation formula

$$\frac{\delta}{\delta x}\ \varphi_n\left(x;q\right)=-2q^{1/4}\ \frac{1-q^n}{1-q}\ \varphi_{n-1}\left(x;q\right). \tag{3.4.34}$$

This follows from (2.3.29) or can be verified by a direct calculation. Consider the series

$$G\left(x,\alpha\right):=\sum_{n=0}^{\infty}\frac{\left(-\alpha\right)^n}{\left(q;q\right)_n}\ \varphi_n\left(x;q\right). \tag{3.4.35}$$

Using (3.4.34) one can easily show that the function

$$y_5\left(z\right)=G\left(x,\alpha\right) \tag{3.4.36}$$

is a solution of Eqs. (2.3.13) and (2.3.16). Therefore

$$y_5\left(z\right)=-\frac{W\left(y_2,y_5\right)}{W\left(y_1,y_2\right)}\ y_1\left(z\right)+\frac{W\left(y_1,y_5\right)}{W\left(y_1,y_2\right)}\ y_2\left(z\right). \tag{3.4.37}$$

Evaluation of $W\left(y_2,y_5\right)$ gives

$$\mathcal{E}_q\left(x;-\alpha\right)\ G\left(x\left(z-1/2\right),\alpha\right)-G\left(x,\alpha\right)\ \mathcal{E}_q\left(x\left(z-1/2\right);-\alpha\right)=D \tag{3.4.38}$$

and letting $x=\eta$ we obtain

$$W\left(y_2,y_5\right)=\frac{4q^{1/4}\alpha}{1-q}\ \frac{\left(\alpha;q^{1/2}\right)_\infty}{\left(q\alpha^2;q^2\right)_\infty}. \tag{3.4.39}$$

In a similar fashion,

$$W\left(y_1,y_5\right)=0 \tag{3.4.40}$$

and

$$G(x,\alpha) = \frac{\left(\alpha; q^{1/2}\right)_\infty}{\left(\alpha^2; q^2\right)_\infty} \, \mathcal{E}_q(x;\alpha), \tag{3.4.41}$$

which implies (2.3.27).

This method can be used to derive functional relations between different solutions of the difference equation of the hypergeometric type (3.4.1) up to the level of very-well-poised ${}_8\varphi_7$ functions.

3.5. Another Addition Theorem

Although $\mathcal{E}_q(x;\alpha)$ is an analog of $\exp \alpha x$, the function $\mathcal{E}_q(x;\alpha)$ is not symmetric in x and α, so one would expect $\mathcal{E}_q(x;\alpha)$ to have two different addition theorems. Eq. (3.1.3) gives the addition theorem in the variable x. Recently Ismail and Stanton [**71**] have found an important expansion formula (4.1.1) and called it the addition theorem in the variable α because it becomes $\exp \alpha x \ \exp \beta x = \exp(\alpha+\beta)\,x$ when $q \to 1^-$. In [**142**] the author was able to find another version of the "addition" formula with respect to both variables x and α which extends Theorem 3.1.

THEOREM 3.4.

$$\mathcal{E}_q(x,y;\alpha,\beta) = \mathcal{E}_q(x;\alpha) \ \mathcal{E}_q(y;\beta) \tag{3.5.1}$$

This formula can be thought of as a general analog of the relation

$$\exp(\alpha x + \beta y) = \exp \alpha x \ \exp \beta y. \tag{3.5.2}$$

Clearly, Theorem 3.4 gives the addition formula (3.1.3) when $\beta = \alpha$ in view of (2.3.17). The case of the addition theorem in the variable α, raised by Ismail and Stanton, arises when $y = x$. Theorem 3.1 simplifies the product of two single series to a similar single series, while Theorem 3.4 allows to factor the double series into the product of two single series. It is worth mentioning also that the "addition" formulas (3.1.3) and (3.5.1) give an analytic continuation of the q-exponential functions $\mathcal{E}_q(x,y;\alpha)$ and $\mathcal{E}_q(x,y;\alpha,\beta)$ into the entire complex α and β-planes because the analytic continuation of the ${}_2\varphi_1$ functions in (2.3.3) is well-known; see, for example, [**46**], [**26**], and [**140**], or (3.3.12)–(3.3.15).

Let us establish first an auxiliary result [**26**].

LEMMA 3.2. *The following connection relation holds*

$$\begin{aligned} &\frac{\left(q\alpha^2 r^2; q^2\right)_\infty}{\left(q\alpha^2; q^2\right)_\infty} \, \mathcal{E}_q(\cos\theta; \alpha r) \\ &\quad = \frac{1}{2\pi}\int_0^\pi \frac{\left(q, r^2, e^{2i\varphi}, \ e^{-2i\varphi};\ q\right)_\infty}{\left(re^{i\theta+i\varphi}, re^{i\theta-i\varphi}, re^{i\varphi-i\theta}, re^{-i\theta-i\varphi}; q\right)_\infty} \\ &\qquad \times \mathcal{E}_q(\cos\varphi;\alpha) \ d\varphi \end{aligned} \tag{3.5.3}$$

where $|r| < 1$.

PROOF. The continuous q-Hermite polynomials (3.4.16) have the following bilinear generating function, the Poisson kernel,

$$\sum_{n=0}^{\infty} \frac{r^n}{(q;q)_n} H_n(\cos\theta|q)\, H_n(\cos\varphi|q) \tag{3.5.4}$$
$$= \frac{(r^2;\, q)_\infty}{(re^{i\theta+i\varphi}, re^{i\theta-i\varphi}, re^{i\varphi-i\theta}, re^{-i\theta-i\varphi}; q)_\infty}$$

where $|r| < 1$. The orthogonality relation for these polynomials is

$$\int_0^\pi H_m(\cos\theta|q)\, H_n(\cos\theta|q) \left(e^{2i\theta},\, e^{-2i\theta};\, q\right)_\infty d\theta \tag{3.5.5}$$
$$= 2\pi \frac{(q;q)_n}{(q;q)_\infty}\, \delta_{mn},$$

see, for example, [46]. Expanding $\mathcal{E}_q(\cos\varphi;\alpha)$ in the right side of (3.5.3) in the uniformly convergent series of the continuous q-Hermite polynomials with the aid of the generating relation (3.4.15) one gets

$$\frac{1}{2\pi}\int_0^\pi \frac{(q, r^2, e^{2i\varphi},\, e^{-2i\varphi};\, q)_\infty}{(re^{i\theta+i\varphi}, re^{i\theta-i\varphi}, re^{i\varphi-i\theta}, re^{-i\theta-i\varphi}; q)_\infty} \tag{3.5.6}$$
$$\times \left(q\alpha^2; q^2\right)_\infty \mathcal{E}_q(\cos\varphi;\alpha)\; d\varphi$$
$$= \sum_{n=0}^{\infty} \frac{q^{n^2/4}}{(q;q)_n}\, \alpha^n$$
$$\times \frac{1}{2\pi}\int_0^\pi \frac{(q, r^2, e^{2i\varphi},\, e^{-2i\varphi};\, q)_\infty}{(re^{i\theta+i\varphi}, re^{i\theta-i\varphi}, re^{i\varphi-i\theta}, re^{-i\theta-i\varphi}; q)_\infty}$$
$$\times H_n(\cos\varphi|q)\; d\varphi.$$

The series in (3.5.4) converges uniformly when $-1 \le x, y \le 1$ and $|r| < 1$. Therefore, using the orthogonality relation (3.5.5),

$$\frac{1}{2\pi}\int_0^\pi \frac{(q, r^2, e^{2i\varphi},\, e^{-2i\varphi};\, q)_\infty\; H_n(\cos\varphi|q)}{(re^{i\theta+i\varphi}, re^{i\theta-i\varphi}, re^{i\varphi-i\theta}, re^{-i\theta-i\varphi}; q)_\infty}\, d\varphi \tag{3.5.7}$$
$$= r^n\, H_n(\cos\theta|q).$$

From (3.5.6), (3.5.7), and (3.4.15) we finally arrive at the connection relation (3.5.3). □

Now we can prove the addition formula (3.5.1).

PROOF. The proof of (3.5.1) uses the connection relation (3.5.3) which we rewrite as

$$\frac{(q\alpha^2;q^2)_\infty}{(q\beta^2;q^2)_\infty}\, \mathcal{E}_q(\cos\theta;\alpha) \tag{3.5.8}$$
$$= \frac{1}{2\pi}\int_0^\pi \frac{(q, \alpha^2/\beta^2, e^{2i\psi},\, e^{-2i\psi};\, q)_\infty\; \mathcal{E}_q(\cos\psi;\beta)\; d\psi}{(e^{i\theta+i\psi}\alpha/\beta, e^{i\theta-i\psi}\alpha/\beta, e^{-i\theta+i\psi}\alpha/\beta, e^{-i\theta-i\psi}\alpha/\beta; q)_\infty}$$

provided that $\alpha < \beta$. Multiplying both sides of (3.5.8) by $\mathcal{E}_q(\cos\varphi;\beta)$ and then using the addition formula (3.1.3), the symmetry relation (2.3.5), and the definition (2.3.1) one gets

$$\frac{(q\alpha^2;q^2)_\infty}{(q\beta^2;q^2)_\infty}\,\mathcal{E}_q(\cos\theta;\alpha)\ \mathcal{E}_q(\cos\varphi;\beta) \tag{3.5.9}$$
$$=\frac{1}{2\pi}\int_0^\pi \frac{(q,\alpha^2/\beta^2,e^{2i\psi},\ e^{-2i\psi};\ q)_\infty\ \mathcal{E}_q(\cos\varphi,\cos\psi;\beta)}{(e^{i\theta+i\psi}\alpha/\beta,e^{i\theta-i\psi}\alpha/\beta,e^{-i\theta+i\psi}\alpha/\beta,e^{-i\theta-i\psi}\alpha/\beta;q)_\infty}\,d\psi$$
$$=\frac{(\beta^2;q^2)_\infty}{(q\beta^2;q^2)_\infty}\,(q,\alpha^2/\beta^2;\ q)_\infty\sum_{n=0}^\infty\frac{q^{n^2/4}}{(q;q)_n}\left(\beta e^{-i\varphi}\right)^n$$
$$\times\frac{1}{2\pi}\int_0^\pi\frac{(e^{2i\psi},\ e^{-2i\psi},-q^{(1-n)/2}e^{i\varphi+i\psi},-q^{(1-n)/2}e^{i\varphi-i\psi};\ q)_\infty}{\left(-q^{(1+n)/2}e^{i\varphi+i\psi},-q^{(1+n)/2}e^{i\varphi-i\psi};q\right)_\infty}$$
$$\times\frac{d\psi}{(e^{i\theta+i\psi}\alpha/\beta,e^{i\theta-i\psi}\alpha/\beta,e^{-i\theta+i\psi}\alpha/\beta,e^{-i\theta-i\psi}\alpha/\beta;q)_\infty}.$$

The last integral can be evaluated as the special case $a=b=0$ of the Nassrallah–Rahman integral,

$$\frac{1}{2\pi}\int_0^\pi\frac{(e^{2i\psi},\ e^{-2i\psi},ge^{i\psi},ge^{-i\psi};\ q)_\infty\ d\psi}{(ce^{i\psi},ce^{-i\psi},de^{i\psi},de^{-i\psi},fe^{i\psi},fe^{-i\psi};q)_\infty} \tag{3.5.10}$$
$$=\frac{(cg;q)_\infty}{(q,cd,cf;q)_\infty}\,{}_2\varphi_1\left(\begin{matrix} g/d,\ g/f\\ cg\end{matrix};q,\ df\right),$$

see (6.3.2) and (6.3.8) of [**46**] or (A.4.3). Therefore

$$\mathcal{E}_q(\cos\theta;\alpha)\ \mathcal{E}_q(\cos\varphi;\beta) \tag{3.5.11}$$
$$=\frac{(\beta^2;q^2)_\infty}{(q\alpha^2;q^2)_\infty}\sum_{n=0}^\infty\frac{q^{n^2/4}}{(q;q)_n}\,\beta^n e^{-in\varphi}\left(-q^{(1-n)/2}e^{i\theta+i\varphi}\alpha/\beta;q\right)_n$$
$$\times\,{}_2\varphi_1\left(\begin{matrix}-q^{(1-n)/2}e^{i\theta+i\varphi}\beta/\alpha,\ q^{-n}\\ -q^{(1-n)/2}e^{i\theta+i\varphi}\alpha/\beta\end{matrix};q,\ -q^{(1+n)/2}e^{i\varphi-i\theta}\alpha/\beta\right).$$

Use of the transformation (III.3) of [**46**] or (A.3.3) completes the proof. □

The function $\mathcal{E}_q(x,y;\alpha,\beta)$ on the left side of (3.5.1) is an analog of the exponential function $\exp(\alpha x+\beta y)$; see (2.3.20). Due to the addition theorem (3.5.1) this function $u=\mathcal{E}_q(x,y;\alpha,\beta)$ in (2.3.17)–(2.3.18) is a double series solution of the difference equations (2.3.13) and (2.3.16). Equation (3.5.1) leads also to the product formula

$$\mathcal{E}_q(x,y;\alpha,\beta)\ \mathcal{E}_q(z,w;\gamma,\delta)=\mathcal{E}_q(x,z;\alpha,\gamma)\ \mathcal{E}_q(y,w;\beta,\delta), \tag{3.5.12}$$

which is obviously a q-analog of

$$\exp(\alpha x+\beta y)\exp(\gamma z+\delta w)=\exp(\alpha x+\gamma z)\exp(\beta y+\delta w). \tag{3.5.13}$$

Equation (3.5.12) is an extension of the product formula (3.3.7) in the case of the q-quadratic lattice under consideration.

Two limiting cases of (3.5.1) are of interest. When $\beta \to 0$, we obtain the generating relation (3.4.15) for the continuous q-Hermite polynomials. Another limiting case, $\alpha \to 0$, results in the generating relation (3.4.28).

Further extension of Theorem 3.2 can be given. Let us introduce the basic trigonometric functions

$$C_q(x, y; \omega, \varkappa) = \frac{1}{2}\left(\mathcal{E}_q(x, y; i\omega, i\varkappa) + \mathcal{E}_q(x, y; -i\omega, -i\varkappa)\right), \tag{3.5.14}$$

$$S_q(x, y; \omega, \varkappa) = \frac{1}{2i}\left(\mathcal{E}_q(x, y; i\omega, i\varkappa) - \mathcal{E}_q(x, y; -i\omega, -i\varkappa)\right) \tag{3.5.15}$$

as analogs of $\cos(\omega x + \varkappa y)$ and $\sin(\omega x + \varkappa y)$, respectively. The following addition formulas hold [**142**].

THEOREM 3.5.

$$C_q(x, y; \omega, \varkappa) = C_q(x; \omega)\, C_q(y; \varkappa) - S_q(x; \omega)\, S_q(y; \varkappa), \tag{3.5.16}$$

$$S_q(x, y; \omega, \varkappa) = S_q(x; \omega)\, C_q(y; \varkappa) + C_q(x; \omega)\, S_q(y; \varkappa). \tag{3.5.17}$$

These formulas are obviously q-analogs of

$$\cos(\omega x + \varkappa y) = \cos\omega x \cos\varkappa y - \sin\omega x \sin\varkappa y, \tag{3.5.18}$$

$$\sin(\omega x + \varkappa y) = \sin\omega x \cos\varkappa y + \cos\omega x \sin\varkappa y. \tag{3.5.19}$$

Clearly, our Theorem 3.2 is the special case $\varkappa = \omega$ of Theorem 3.5.

3.6. Addition Theorems on q-Linear and Linear Grids

In the case of a q-linear grid the addition theorem has the form.

THEOREM 3.6. *The following addition formula holds*

$$\varepsilon_q(x, y; \alpha) = \varepsilon_q(\alpha x)\ \varepsilon_q(\alpha y) \tag{3.6.1}$$

for the basic exponential functions given by (2.5.1)–(2.5.2).

This relation was found by Rahman [**118**]. The proof can be given, for example, on the basis of the difference equation as in Section 3.2. We shall give here a direct proof by using the following version of the q-binomial theorem.

LEMMA 3.3. *If $x = x(z) = C_1 q^{-z}$ and $y = x(s) = C_1 q^{-s}$, then*

$$\left[x(z) - x\left(s + \frac{n-1}{2}\right)\right]^{(n)} = \sum_{k=0}^{n} \frac{\gamma(n)!}{\gamma(k)!\,\gamma(n-k)!}\, x^k (-y)^{n-k}. \tag{3.6.2}$$

PROOF. In the case of the q-linear grid $x = C_1 q^{-z}$ from the definition of the generalized power (2.2.16) one gets

$$\left[x(z) - x\left(s + \frac{n-1}{2}\right)\right]^{(n)} = \sum_{k=0}^{n} c_k\, x^k, \tag{3.6.3}$$

where c_k are some constants independent of x. Let us apply the divided difference operator $\delta/\delta x$ to the both sides of (3.6.4) m times. With the help of the difference-differentiation formula (2.2.17),

$$\frac{\gamma(n)!}{\gamma(n-m)!}\left[x(z)-x\left(s+\frac{n-m-1}{2}\right)\right]^{(n-m)} \tag{3.6.4}$$

$$=\sum_{k=m}^{n} c_k\,\gamma(k)\gamma(k-1)\cdots\gamma(k-m+1)\;x^{k-m}(z).$$

Substituting $x(z)=0$ (or letting $z\to-\infty$ when $|q|<1$),

$$\frac{\gamma(n)!}{\gamma(n-m)!}\left[0-x\left(s+\frac{n-m-1}{2}\right)\right]^{(n-m)}=\gamma(m)!\;c_m. \tag{3.6.5}$$

But from (2.2.16),

$$\left[0-x\left(s+\frac{n-m-1}{2}\right)\right]^{(n-m)} \tag{3.6.6}$$

$$=(-1)^{n-m}\prod_{k=0}^{n-m-1} x\left(s+\frac{n-m-1}{2}-k\right)$$

$$=(-1)^{n-m}x\left(s+\frac{n-m-1}{2}\right)\cdots x\left(s-\frac{n-m-1}{2}\right)$$

$$=(-1)^{n-m}y^{n-m},$$

which gives

$$c_m=\frac{\gamma(n)!}{\gamma(m)!\gamma(n-m)!}\;(-y)^{n-m} \tag{3.6.7}$$

and our proof is complete. □

We gave the proof of (3.6.2) in terms of generalized powers. Using (2.2.15) and (2.2.24) one can reduce (3.6.2) to a terminating form of the q-binomial theorem.

THEOREM 3.7. *Let*

$$u(x,y)=\sum_{n=0}^{\infty}\frac{\xi^n}{\gamma(n)!}\left[x(z)-x\left(s+\frac{n-1}{2}\right)\right]^{(n)} \tag{3.6.8}$$

and

$$u(x)=u(x,0)=\sum_{n=0}^{\infty}\frac{\xi^n}{\gamma(n)!}x^n \tag{3.6.9}$$

with $x=C_1q^{-z}$ *and* $y=C_1q^{-s}$. *Then*

$$u(x,y)=u(x)\;u(-y). \tag{3.6.10}$$

PROOF. By (3.6.2)

$$\begin{aligned} u(x,y) &= \sum_{n=0}^{\infty} \frac{\xi^n}{\gamma(n)!} \left[x(z) - x\left(s + \frac{n-1}{2}\right)\right]^{(n)} \\ &= \sum_{n=0}^{\infty} \xi^n \sum_{k=0}^{n} \frac{x^k (-y)^{n-k}}{\gamma(n)!\,\gamma(n-k)!} \\ &= \sum_{k=0}^{\infty} \xi^k \frac{x^n}{\gamma(k)!} \sum_{m=0}^{\infty} \xi^m \frac{(-y)^m}{\gamma(m)!}, \end{aligned}$$

provided that all series converge. □

The addition formula (3.6.1) is equivalent to the relation (3.6.10) when $y \to -y$.

As a consequence of (3.6.1), we obtain the addition formulas for the q-trigonometric functions (2.5.4)–(2.5.5) and (2.5.7)–(2.5.8).

THEOREM 3.8.

$$c_q(x,y;\omega) = c_q(\omega x)\, c_q(\omega y) - s_q(\omega x)\, s_q(\omega y), \tag{3.6.11}$$

$$s_q(x,y;\omega) = s_q(\omega x)\, c_q(\omega y) + c_q(\omega x)\, s_q(\omega y). \tag{3.6.12}$$

Special cases of these relations similar to (3.2.10)–(3.2.14) are valid.

In the case of the linear lattice, $x(z) = z$, the general product formula (3.3.7) takes the form

$$u(z-s)\, u(r-t) = u(z-t)\, u(r-s) \tag{3.6.13}$$

for the function (2.5.10), which implies the usual addition formula

$$e(z+s) = e(z)\ e(s) \tag{3.6.14}$$

for the exponential function (2.5.12).

3.7. Application: Continuous q-Hermite Polynomials

Let us introduce the following polynomials $H_n(x,y|q)$ in two variables,

$$H_{2n}(x,y|q) = (-1)^n \frac{(q;q^2)_n}{q^n} \tag{3.7.1}$$

$$\times {}_6\varphi_5 \left(\begin{matrix} q^{-n}, -q^{-n}, -q^{1/2}e^{i\theta+i\varphi}, -q^{1/2}e^{i\theta-i\varphi}, -q^{1/2}e^{i\varphi-i\theta}, -q^{1/2}e^{-i\theta-i\varphi} \\ -q,\ \ q^{1/2},\ \ -q^{1/2},\ \ 0,\ \ 0 \end{matrix} ; q, q^2 \right)$$

and

$$H_{2n+1}(x,y|q) = (-1)^n\, 2 \frac{(q^3;q^2)_n}{q^{2n}} (\cos\theta + \cos\varphi) \tag{3.7.2}$$

$$\times {}_6\varphi_5 \left(\begin{matrix} q^{-n}, -q^{-n}, -qe^{i\theta+i\varphi}, -qe^{i\theta-i\varphi}, -qe^{i\varphi-i\theta}, -qe^{-i\theta-i\varphi} \\ -q,\ \ q^{3/2},\ \ -q^{3/2},\ \ 0,\ \ 0 \end{matrix} ; q, q^2 \right)$$

with $x=\cos\theta$ and $y=\cos\varphi$, as an extension of the continuous q-Hermite polynomials (3.4.16). One can see that

$$H_n(x,0|q)=H_n(x|q) \tag{3.7.3}$$

and we can interpret the $H_n(x,y|q)$ as a q-version of the Hermite polynomials $H_n(x+y)$ with the shifted argument [**140**].

It is easy to see that the q-exponential function (2.3.1) is the generating function for the polynomials $H_n(x,y|q)$.

THEOREM 3.9. *We have*

$$\sum_{n=0}^{\infty}\frac{q^{n^2/4}}{(q;q)_n}\,\alpha^n\,H_n(x,y|q)=\left(q\alpha^2;q^2\right)_\infty\mathcal{E}_q(x,y;\alpha) \tag{3.7.4}$$
$$=\left(\alpha^2;q^2\right)_\infty\sum_{n=0}^{\infty}\frac{q^{n^2/4}\,\alpha^n}{(q;q)_n}e^{-in\varphi}\left(-q^{(1-n)/2}e^{i\theta+i\varphi},-q^{(1-n)/2}e^{i\varphi-i\theta};q\right)_n,$$

when $q|\alpha|^2<1$.

PROOF. Substitute (3.7.1) and (3.7.2) into the left side of (3.7.4) and interchange the order of summation with the help of (I.12) and (II.2) of [**46**] or (1.4.7) and (A.2.2). As a result one gets the right side of (3.7.4) due to (2.3.1). The series in the left side of (3.7.4) converges absolutely and uniformly when $-1\le x,y\le 1$ and $\left|q\alpha^2\right|<1$. This can be shown with the help of the uniform bound for the polynomials $H_n(x,y|q)$ found in Ex. 6.2. We leave the details to the reader. □

Several properties of polynomials $H_n(x,y|q)$ follow from the generating relation (3.7.4). Eq. (3.4.15) is clearly its special case when $y=0$, which implies (3.7.3).

COROLLARY 3.3. *The continuous q-Hermite polynomials* $H_n(x,y|q)$ *are given by*

$$H_n(x,y|q)=\sum_{k=0}^{[n/2]}(-1)^k\frac{(q;q)_n}{\left(q^2;q^2\right)_k(q;q)_{n-2k}}q^{k(2k-n-1)} \tag{3.7.5}$$
$$\times e^{-i(n-2k)\varphi}\left(-q^{(1-n+2k)/2}e^{i\theta+i\varphi},-q^{(1-n+2k)/2}e^{i\varphi-i\theta};q\right)_{n-2k}$$

and an "inverse" of this expansion has the form

$$e^{-in\varphi}\left(-q^{(1-n)/2}e^{i\theta+i\varphi},-q^{(1-n)/2}e^{i\varphi-i\theta};q\right)_n \tag{3.7.6}$$
$$=\sum_{k=0}^{[n/2]}q^{k(k-n)}\frac{(q;q)_n}{\left(q^2;q^2\right)_k(q;q)_{n-2k}}H_{n-2k}(x,y|q).$$

Here $x=\cos\theta$ *and* $y=\cos\varphi$.

PROOF. By (1.1.2)

$$\left(\alpha^2;q^2\right)_\infty = \sum_{k=0}^{\infty} \frac{q^{k(k-1)}}{\left(q^2;q^2\right)_k} \left(-\alpha^2\right)^k . \tag{3.7.7}$$

Substituting this expansion in the right side of (3.7.4) and equating coefficients of α^n on both sides, we get (3.7.5). To derive (3.7.6), multiply both sides of (3.7.4) by

$$\sum_{k=0}^{\infty} \frac{\alpha^{2k}}{\left(q^2;q^2\right)_k} = \frac{1}{\left(\alpha^2;q^2\right)_\infty} \tag{3.7.8}$$

due to (1.1.1) and equate coefficients of α^n. □

Relations (3.7.5) and (3.7.6) are clearly q-extensions of the classical formula **[126]**, **[151]**

$$H_n(x+y) = \sum_{k=0}^{[n/2]} (-1)^k \frac{n!}{k!\,(n-k)!} (2x+2y)^{n-2k} , \tag{3.7.9}$$

and its "inverse",

$$(2x+2y)^n = \sum_{k=0}^{[n/2]} \frac{n!}{k!\,(n-k)!} H_{n-2k}(x+y) , \tag{3.7.10}$$

respectively.

COROLLARY 3.4. *The continuous q-Hermite polynomials* $H_n(x,y|q)$ *have the following difference-differentiation formulas*

$$\frac{\delta H_n(x,y|q)}{\delta x} = \frac{\delta H_n(x,y|q)}{\delta y} = 2q^{(1-n)/2} \frac{1-q^n}{1-q} H_{n-1}(x,y|q) . \tag{3.7.11}$$

PROOF. Apply the divided difference operators $\delta/\delta x$ and $\delta/\delta y$ to both sides of (3.7.4) with the help of (2.3.13), and compare coefficients of α^n. □

The addition theorem (3.1.3) gives us the possibility to express polynomials $H_n(x,y|q)$ in terms of the continuous q-Hermite polynomials $H_n(x|q)$.

THEOREM 3.10. *The following expansion in the continuous q-Hermite polynomials holds*

$$H_n(x,y|q) \tag{3.7.12}$$

$$= \sum_{k=0}^{n} q^{(k-n)(k+n-2)/4} \frac{(q;q)_n}{(q;q)_k\,(q;q)_{n-k}} H_k(x|q)\ \phi_{n-k}(y;q) ,$$

where $\phi_m(y;q)$ *are given by* (2.3.24)–(2.3.25).

PROOF. Due to (3.7.4), (3.1.3) and (2.3.23):

$$\sum_{n=0}^{\infty} \frac{q^{n^2/4}}{(q;q)_n} \alpha^n H_n(x,y|q) = \left(q\alpha^2;q^2\right)_\infty \mathcal{E}_q(x;\alpha)\,\mathcal{E}_q(y;\alpha)$$

$$= \sum_{k=0}^{\infty} \frac{q^{k^2/4}\alpha^k}{(q;q)_k} H_k(x|q) \sum_{m=0}^{\infty} \frac{q^{m/2}\alpha^m}{(q;q)_m} \phi_m(y;q)$$
$$= \sum_{n=0}^{\infty} \alpha^n \sum_{k=0}^{n} \frac{q^{k^2/4+(n-k)/2}}{(q;q)_k (q;q)_{n-k}} H_k(x|q)\ \phi_{n-k}(y;q).$$

Equating coefficients of α^n on both sides, we obtain (3.7.12). □

In a similar fashion, expansions (3.4.15), (2.3.23) and (2.3.27) give rise to the following summation formulas

$$\phi_n(x;q) = q^{n(n-2)/4} \sum_{k=0}^{[n/2]} \frac{q^{k(k-n+1)} (q;q)_n}{(q^2;q^2)_k (q;q)_{n-2k}} H_{n-2k}(x|q), \tag{3.7.13}$$

$$H_n(x|q) = q^{-n(n-2)/4} \sum_{k=0}^{[n/2]} \frac{(-1)^k q^{k(k-1)} (q;q)_n}{(q^2;q^2)_k (q;q)_{n-2k}} \phi_{n-2k}(x;q), \tag{3.7.14}$$

and

$$\varphi_n(-x;q) = \sum_{k=0}^{n} \frac{q^{k^2/4} (q;q)_n}{(q;q)_k \left(q^{1/2};q^{1/2}\right)_{n-k}} H_k(x|q), \tag{3.7.15}$$

$$H_n(x|q) = \sum_{k=0}^{n} \frac{(-1)^{n-k} q^{(n-k)(n-k-1)/4} (q;q)_n}{(q;q)_k \left(q^{1/2};q^{1/2}\right)_{n-k}} \varphi_k(-x;q), \tag{3.7.16}$$

respectively.

3.8. Exercises for Chapter 3

(1) Verify that

$$\gamma(n)! = q^{-n(n-1)/4} \frac{(q;q)_n}{(1-q)^n},$$
$$\frac{\gamma(n)!}{\gamma(k)!\gamma(n-k)!} = q^{-(n-k)k/2} \frac{(q;q)_n}{(q;q)_k (q;q)_{n-k}}$$
$$= (-1)^k q^{(n+1)k/2} \frac{(q^{-n};q)_k}{(q;q)_k}$$

and show that for the general q-quadratic lattice (2.2.12) relation (3.3.1) can be rewritten as the following sum of q-factorials

$$q^{-nt} \sum_{k=0}^{n} (-1)^k q^{(t-s+(n+1)/2)k} \frac{(q^{-n};q)_k}{(q;q)_k}$$
$$\times \left(q^{s-z-(k-1)/2},\ \mu q^{s+z-(k-1)/2};q\right)_k$$
$$\times \left(q^{t-r-(n-k-1)/2},\ \mu q^{t+r-(n-k-1)/2};q\right)_{n-k}$$
$$= q^{-ns} \sum_{k=0}^{n} (-1)^k q^{(s-t+(n+1)/2)k} \frac{(q^{-n};q)_k}{(q;q)_k}$$

$$\times \left(q^{t-z-(k-1)/2},\ \mu q^{t+z-(k-1)/2}; q\right)_k$$
$$\times \left(q^{s-r-(n-k-1)/2},\ \mu q^{s+r-(n-k-1)/2}; q\right)_{n-k},$$

where $\mu = C_2/C_1$.

(2) Prove (3.4.31). [*Hint:* Use the three-term recurrence relation

$$2xH_n(x|q) = H_{n+1}(x|q) + (1-q^n) H_{n-1}(x|q)$$

for the continuous q-Hermite polynomials.]

(3) Prove (3.4.32).

(4) Consider Taylor's expansion (2.3.23) of the $\mathcal{E}_q(x;\alpha)$ with respect to α:

$$\mathcal{E}_q(x;\alpha) = \sum_{n=0}^{\infty} \frac{q^{n/2}\alpha^n}{(q;q)_n}\ \phi_n(x;q), \quad q|\alpha|^2 < 1, \tag{3.8.1}$$

where

$$\phi_{2k}(x;q) = 4x^2 \prod_{p=1}^{k-1} \left(4x^2q^{2p} + \left(1-q^{2p}\right)^2\right),$$

$$\phi_{2k+1}(x;q) = 2q^{-1/4}\ x \prod_{p=0}^{k-1} \left(4x^2q^{2p+1} + \left(1-q^{2p+1}\right)^2\right).$$

Verify that

$$\phi_0(x;q) = 1, \qquad \phi_1(x;q) = 2q^{-1/4}x,$$
$$\phi_2(x;q) = 4x^2, \quad \phi_3(x;q) = 2q^{-1/4}x\left(4x^2q + (1-q)^2\right),$$
$$\phi_4(x;q) = 4x^2\left(4x^2q^2 + \left(1-q^2\right)^2\right),$$
$$\phi_5(x;q) = 2q^{-1/4}x\left(4x^2q + (1-q)^2\right)\left(4x^2q^3 + \left(1-q^3\right)^2\right).$$

Show that

$$|\phi_{2k}(x;q)| \le \left(-1;q^2\right)_k^2, \qquad |\phi_{2k+1}(x;q)| \le 2q^{-1/4}\left(-q;q^2\right)_k^2$$

and

$$|\phi_n(x;q)| < C = \max(C_1, C_2),$$

where $-1 \le x \le 1$, $0 < q < 1$, and $C_1 = 2q^{-1/4}\left(-q;q^2\right)_\infty^2$, $C_2 = \left(-1;q^2\right)_\infty^2$.

Show that

$$\phi_n(x;q) = q^{n(n/2-1)/2}\left(1+e^{2i\theta}\right)\left(-q^{2-n}e^{2i\theta};q^2\right)_{n-1} e^{-in\theta},$$

$$\frac{\delta}{\delta x}\ \phi_n(x;q) = 2q^{-1/4}\frac{1-q^n}{1-q}\ \phi_{n-1}(x;q),$$

$$\left(\frac{\delta}{\delta x}\right)^m \phi_n(x;q) = 2^m q^{-m/4} \frac{\left(q^{n-m+1};q\right)_m}{(1-q)^m} \phi_{n-m}(x;q),$$

$$\lim_{q\to 1} \phi_n(x;q) = (2x)^n$$

and verify that [**73**]

$$\mathcal{E}_q(x;\alpha) = \sum_{n=0}^{\infty} \frac{\left(1+e^{2i\theta}\right)\left(-q^{-n}e^{2i\theta};q^2\right)_n e^{-in\theta}}{(q;q)_n \left(1+q^{-n}e^{2i\theta}\right)} q^{n^2/4} \alpha^n.$$

(5) Consider expansions (2.3.27):

$$\mathcal{E}_q(x;\alpha) = \frac{\left(-\alpha;q^{1/2}\right)_\infty}{\left(q\alpha^2;q^2\right)_\infty} \sum_{n=0}^{\infty} \frac{(-\alpha)^n}{(q;q)_n} \varphi_n(x;q)$$

$$= \frac{\left(\alpha;q^{1/2}\right)_\infty}{\left(q\alpha^2;q^2\right)_\infty} \sum_{n=0}^{\infty} \frac{\alpha^n}{(q;q)_n} \varphi_n(-x;q), \quad |\alpha|<1,$$

where

$$\varphi_n(x;q) = \left(q^{1/4}e^{i\theta}, q^{1/4}e^{-i\theta}; q^{1/2}\right)_n, \quad x = \cos\theta.$$

Verify that

$$\varphi_0(x;q) = 1, \qquad \varphi_1(x;q) = 1+q^{1/2}-2q^{1/4}x,$$

$$\varphi_2(x;q) = \left(1+q^{1/2}-2q^{1/4}x\right)\left(1+q^{3/2}-2q^{3/4}x\right),$$

$$\varphi_3(x;q) = \left(1+q^{1/2}-2q^{1/4}x\right)\left(1+q^{3/2}-2q^{3/4}x\right) \times \left(1+q^{5/2}-2q^{5/4}x\right).$$

Show that

$$\varphi_n(x;q) = \prod_{k=0}^{n-1}\left(1+q^{k+1/2}-2q^{k/2+1/4}x\right),$$

$$\left(q^{1/4};q^{1/2}\right)_n^2 \le |\varphi_n(x;q)| \le \left(-q^{1/4};q^{1/2}\right)_n^2$$

and

$$|\varphi_n(x;q)| < \left(-q^{1/4};q^{1/2}\right)_\infty^2,$$

when $-1 \le x \le 1$ and $0<q<1$.
Show that

$$\frac{\delta}{\delta x}\varphi_n(x;q) = -2q^{1/4}\frac{1-q^n}{1-q}\varphi_{n-1}(x;q),$$

$$\left(\frac{\delta}{\delta x}\right)^m \varphi_n(x;q) = (-2)^m q^{m/4} \frac{\left(q^{n-m+1};q\right)_m}{(1-q)^m} \varphi_{n-1}(x;q),$$

$$\lim_{q\to 1}\varphi_n(x;q) = 2^n(1-x)^n.$$

(6) Consider Taylor's expansion of the basic exponential $\mathcal{E}_q(x, y; \alpha)$ in α:

$$\mathcal{E}_q(x, y; \alpha) = \sum_{n=0}^{\infty} \frac{q^{n/2}\alpha^n}{(q;q)_n} \phi_n(x, y; q)$$
$$= \frac{(\alpha^2; q^2)_\infty}{(q\alpha^2; q^2)_\infty} \sum_{n=0}^{\infty} \frac{q^{n^2/4}\,\alpha^n}{(q;q)_n} \varphi_n(x, y; q),$$

where $q|\alpha|^2 < 1$ and by the definition

$$\varphi_n(x, y; q) = e^{-in\varphi}\left(-q^{(1-n)/2}e^{i\theta+i\varphi}, -q^{(1-n)/2}e^{i\varphi-i\theta}; q\right)_n$$

with $x = \cos\theta$ and $y = \cos\varphi$. Verify that

$$\phi_n(x, y; q) = q^{n(n-2)/4}(q;q)_n$$
$$\times \sum_{k=0}^{[n/2]} \frac{q^{k(k-n+1)}(q^{-1}; q^2)_k}{(q^2;q^2)_k (q;q)_{n-2k}} \varphi_{n-2k}(x, y; q)$$

and show that

$$\frac{\delta}{\delta x}\phi_n(x, y; q) = \frac{\delta}{\delta y}\phi_n(x, y; q) = 2q^{-1/4}\frac{1-q^n}{1-q}\phi_{n-1}(x, y; q),$$
$$\phi_n(x, y; q) = \phi_n(y, x; q), \qquad \phi_n(x, 0; q) = \phi_n(x; q).$$

(7) *Variants of the q-binomial formula.* The addition formula $e^{\alpha(x+y)} = e^{\alpha x}e^{\alpha y}$ implies

$$(x+y)^n = \sum_{k=0}^{n} \frac{n!}{k!\,(n-k)!} x^k y^{n-k}.$$

With the same notations as in the previous exercises, use the addition theorem $\mathcal{E}_q(x, y; \alpha) = \mathcal{E}_q(x; \alpha)\,\mathcal{E}_q(y; \alpha)$ in order to derive the following q-analogs of this binomial formula:

$$\varphi_n(x, y; q) = \sum_{k=0}^{n} (-1)^k q^{-n^2/4} \frac{(q;q)_n}{(q;q)_k (q;q)_{n-k}}$$
$$\times \varphi_k(x; q)\,\varphi_{n-k}(-y; q),$$

$$\varphi_n(x, y; q) = \sum_{k=0}^{n} q^{(k-n)(k+n-2)//4} \frac{(q;q)_n}{(q;q)_k (q;q)_{n-k}}$$
$$\times \varphi_k(x, 0; q)\,\phi_{n-k}(y; q),$$

$$\phi_n(x, y; q) = \sum_{k=0}^{n} \frac{(q;q)_n}{(q;q)_k (q;q)_{n-k}} \phi_k(x; q)\,\phi_{n-k}(y; q).$$

[*Remark*: In our notations $q^{n^2/4}\varphi_n(x, -\eta; q) = (-1)^n\ \varphi_n(x; q)$, where $\eta = \left(q^{1/4} + q^{-1/4}\right)/2$.]

(8) The *Taylor formula for polynomials* $f(x)$ evaluates the coefficients f_k in the expansion

$$f(x)=\sum_{k=0}^{\infty} f_k\ (x-c)^k, \qquad f_k=\frac{f^{(k)}(c)}{k!}.$$

This formula can be extended by considering other polynomial bases and suitable operators [**58**], [**73**]. Derive the following generalizations using the Askey–Wilson first order divided difference operator $\mathcal{D}_q=\delta/\delta x$ defined in Ex. 2.11 and three q-polynomial bases:

(a)

$$f(x)=\sum_{k=0}^{n} f_k\ \psi_k(x,a;q),$$

where

$$f_k=\frac{(q-1)^k}{(2a)^k(q;q)_k}q^{-k(k-1)/4}\left(\mathcal{D}_q^k f(x)\right)\Big|_{x=x_k},$$

$$\psi_k(x,a;q)=\left(ae^{i\theta},ae^{-i\theta};q\right)_k, \qquad x=\cos\theta,$$

and x_k are given by

$$x_k=\frac{1}{2}\left(aq^{k/2}+q^{-k/2}/a\right), \quad k=0,1,2,\dots,n;$$

(b)

$$f(x)=\sum_{k=0}^{n} f_k\ \phi_k(x;q),$$

where

$$f_k=\frac{(1-q)^k}{2^k(q;q)_k}q^{k/2}\left(\mathcal{D}_q^k f(x)\right)\Big|_{x=0},$$

$$\phi_n(x;q)=q^{n(n/2-1)/2}\left(1+e^{2i\theta}\right)\left(-q^{2-n}e^{2i\theta};q^2\right)_{n-1}e^{-in\theta}$$

with $x=\cos\theta$; see also Ex. 3.4;

(c)

$$f(x)=\sum_{k=0}^{n} f_k\ \varphi_k(x;q),$$

where

$$f_k=\frac{(q-1)^k}{2^k(q;q)_k}q^{-k/2}\left(\mathcal{D}_q^k f(x)\right)\Big|_{x=\eta},$$

$$\varphi_n(x;q)=\left(q^{1/4}e^{i\theta},q^{1/4}e^{-i\theta};q^{1/2}\right)_n$$

with $x=\cos\theta$; see also Ex. 3.5.

(9) Show that the continuous q-Hermite polynomials given by (3.4.16) can be represented in terms of the basic hypergeometric series as

$$H_n(x|q) = q^{-n/4}\left(-q^{1/2};q^{1/2}\right)_n \times {}_3\varphi_2\left(\begin{matrix} q^{-n/2},\ q^{1/4}e^{i\theta},\ q^{1/4}e^{-i\theta} \\ -q^{1/2},\ 0 \end{matrix};q^{1/2},\ q^{1/2}\right);$$

$$H_{2n}(x|q) = (-1)^n \frac{\left(q;q^2\right)_n}{q^n} \times {}_3\varphi_2\left(\begin{matrix} q^{-2n},\ -qe^{2i\theta},\ -qe^{-2i\theta} \\ q,\ 0 \end{matrix};q^2,\ q^2\right),$$

$$H_{2n+1}(x|q) = (-1)^n 2\frac{\left(q^3;q^2\right)_n}{q^{2n}}\cos\theta \times {}_3\varphi_2\left(\begin{matrix} q^{-2n},\ -q^2e^{2i\theta},\ -q^2e^{-2i\theta} \\ q^3,\ 0 \end{matrix};q^2,\ q^2\right);$$

$$H_{2n}(x|q) = q^{n^2}\left(-q^{1-2n}e^{2i\theta},\ -q^{1-2n}e^{-2i\theta};q^2\right)_n \times {}_2\varphi_2\left(\begin{matrix} q^{-2n},\ q^{1-2n} \\ -q^{1-2n}e^{2i\theta},\ -q^{1-2n}e^{-2i\theta} \end{matrix};q^2,\ q\right),$$

$$H_{2n+1}(x|q) = 2q^{n^2+n}\cos\theta\left(-q^{-2n}e^{2i\theta},\ -q^{-2n}e^{-2i\theta};q^2\right)_n \times {}_2\varphi_2\left(\begin{matrix} q^{-2n},\ q^{-1-2n} \\ -q^{-2n}e^{2i\theta},\ -q^{-2n}e^{-2i\theta} \end{matrix};q^2,\ q\right);$$

$$H_n(x|q) = \frac{e^{i\theta}}{\left(e^{-2i\theta};q\right)_\infty}\,{}_2\varphi_1\left(\begin{matrix} 0,\ 0 \\ qe^{2i\theta} \end{matrix};q,\ q^{n+1}\right) + \frac{e^{-i\theta}}{\left(e^{2i\theta};q\right)_\infty}\,{}_2\varphi_1\left(\begin{matrix} 0,\ 0 \\ qe^{-2i\theta} \end{matrix};q,\ q^{n+1}\right),$$

where $x = \cos\theta$.

(10) Show that

$$H_n(-x|q) = (-1)^n H_n(x|q),$$

$$H_{2n+1}(0|q) = 0, \qquad H_{2n}(0|q) = (-1)^n\left(q;q^2\right)_n,$$

$$H_n\left(\frac{1}{2}\left(q^{1/4}+q^{-1/4}\right)\middle|\, q\right) = q^{-n/4}\frac{(q;q)_n}{\left(q^{1/2};q^{1/2}\right)_n}.$$

(11) Verify the following "inversion with respect to the base" relations for the q-shifted factorials

$$\begin{aligned}(a;q)_n &= \left(a^{-1};p\right)_n (-a)^n p^{-n(n-1)/2},\\ (p;p)_n &= (-1)^n q^{-n(n+1)/2} (q;q)_n, \quad p=q^{-1}\end{aligned}$$

and show that

$$\begin{aligned}H_n\left(x|q^{-1}\right) &= \sum_{k=0}^{n} \frac{(p;p)_n}{(p;p)_k (p;p)_{n-k}} e^{i(n-2k)\theta}\\ &= \sum_{k=0}^{n} q^{k^2-kn} \frac{(q;q)_n}{(q;q)_k (q;q)_{n-k}} e^{i(n-2k)\theta}\\ &= (-1)^n q^{-n(n-1)/2} (q;q)_n \lim_{\beta\to\infty} \beta^{-n} C_n(x,\beta|q),\end{aligned}$$

where $x=\cos\theta$ and $C_n(x,\beta|q)$ are the continuous q-ultraspherical polynomials given by

$$C_n(\cos\theta;\beta|q) = \sum_{k=0}^{n} \frac{(\beta;q)_k (\beta;q)_{n-k}}{(q;q)_k (q;q)_{n-k}} e^{i(n-2k)\theta},$$

see, for example, [**46**].
Show that

$$e_q(x) = E_{q^{-1}}\left(-xq^{-1}\right), \qquad E_q(x) = e_{q^{-1}}\left(-xq^{-1}\right)$$

and

$$\mathcal{E}_q(x;\alpha) = \mathcal{E}_{q^{-1}}\left(x;-\alpha q^{-1/2}\right) = \mathcal{E}_{q^{-1}}\left(-x;\alpha q^{-1/2}\right).$$

(12) Show that the continuous q-ultraspherical polynomials $C_n(x,\beta|q)$ have the following representations in terms of the basic hypergeometric series

$$\begin{aligned}C_n(x,\beta|q) &= \beta^{-n/2}\frac{\left(\beta^2;q\right)_n}{(q;q)_n}\\ &\quad\times {}_4\varphi_3\left(\begin{matrix} q^{-n},\ \beta^2q^n,\ \beta^{1/2}e^{i\theta},\ \beta^{1/2}e^{-i\theta} \\ \beta q^{1/2},\ -\beta,\ -\beta q^{1/2}\end{matrix};q,\ q\right)\\ &= \frac{(\beta;q)_n}{(q;q)_n} e^{in\theta}\, {}_2\varphi_1\left(\begin{matrix} q^{-n},\ \beta \\ \beta^{-1}q^{1-n}\end{matrix};q,\ q\beta^{-1}e^{-2i\theta}\right)\\ &= \frac{\left(\beta^2;q\right)_n}{(q;q)_n}\beta^{-n}e^{-in\theta}\, {}_3\varphi_2\left(\begin{matrix} q^{-n},\ \beta,\ \beta e^{2i\theta} \\ \beta^2,\ 0\end{matrix};q,\ q\right)\\ &= q^{-n/4}\frac{\left(\beta;q^{1/2}\right)_n}{\left(q^{1/2};q^{1/2}\right)_n}\end{aligned}$$

$$\times {}_4\varphi_3 \left(\begin{matrix} q^{-n/2},\ \beta q^{n/2},\ q^{1/4}e^{i\theta},\ q^{1/4}e^{-i\theta} \\ -q^{1/2},\ \beta^{1/2}q^{1/4},\ -\beta^{1/2}q^{1/4} \end{matrix} ; q^{1/2}, q^{1/2} \right)$$

$$= \frac{(\beta;q)_n}{(q;q)_n} \left(-e^{2i\theta}q^{2-n};q\right)_{n-1} \left(1+e^{2i\theta}\right) e^{-in\theta}$$

$$\times {}_4\varphi_3 \left(\begin{matrix} q^{-n},\ q^{1-n},\ -q^{1-n}/\beta,\ -q^{2-n}/\beta \\ q^{2-2n}/\beta^2,\ -q^{2-n}e^{2i\theta},\ -q^{2-n}e^{-2i\theta} \end{matrix} ; q^2, q^2 \right),$$

where $x = \cos\theta$.

(13) Show that

$$C_n(-x;\beta|q) = (-1)^n C_n(x;\beta|q),$$

$$C_{2n+1}(0;\beta|q) = 0, \qquad C_{2n}(0;\beta|q) = (-1)^n \frac{(\beta^2;q^2)_n}{(q^2;q^2)_n},$$

$$C_n\left(\frac{1}{2}\left(\beta^{1/2}+\beta^{-1/2}\right);\beta\middle| q\right) = \beta^{-n/2}\frac{(\beta^2;q)_n}{(q;q)_n}.$$

$$C_n\left(\frac{1}{2}\left(q^{1/4}+q^{-1/4}\right);\beta\middle| q\right) = q^{-n/4}\frac{\left(\beta;q^{1/2}\right)_n}{\left(q^{1/2};q^{1/2}\right)_n}.$$

(14) *A q-shift operator.* Ismail introduced the action of the q-translation operator $T_q(y)$ on the continuous q-Hermite polynomials $H_n(x|q)$ as

$$T_q(y)\, H_n(x|q) = H_n(x \dotplus y|q)$$
$$= \sum_{k=0}^{n} q^{(k-n)(k+n-2)/4} \frac{(q;q)_n}{(q;q)_k\,(q;q)_{n-k}}\, H_k(x|q)\ \phi_{n-k}(y;q),$$

where polynomials $\phi_m(y;q)$ are given by (2.3.24)–(2.3.25), see also Ex. 3.4; and then extended the action of $T_q(y)$ to the space of all polynomials by linearity [**61**]:

$$P_n(x) = \sum_{k=0}^{n} c_k\ H_k(x|q),$$

$$P_n(x \dotplus y) = T_q(y)\, P_n(x) = \sum_{k=0}^{n} c_k\ T_q(y)\, H_k(x|q)$$
$$= \sum_{k=0}^{n} c_k\ H_k(x \dotplus y|q).$$

Use (3.7.12) to show that

$$T_q(y)\, H_n(x|q) = H_n(x \dotplus y|q) = H_n(x,y|q),$$

where $H_n(x,y|q)$ are the continuous q-Hermite polynomials in two variables introduced in [**140**] and given by (3.7.1)–(3.7.2) or (3.7.5).

Use (3.7.6) to show that

$$T_q(y)\,\varphi_n(x,0;q) = \varphi_n(x \dotplus y,0;q) = \varphi_n(x,y;q),$$

where the "generalized powers" $\varphi_n(x,y;q)$ are defined in Ex. 3.6. On the other hand, introduce the action of a q-translation operator $\mathcal{T}_q(y)$ on the "powers" $\varphi_n(x,0;q)$ as

$$\mathcal{T}_q(y)\,\varphi_n(x,0;q) = \varphi_n(x,y;q),$$

and then extend this action of $\mathcal{T}_q(y)$ to the space of all polynomials by linearity; see also Ex. 3.7. Show that $\mathcal{T}_q(y) = T_q(y)$ on all polynomials.

Use (3.7.4) to show that

$$T_q(y)\,\mathcal{E}_q(x;\alpha) = \mathcal{E}_q(x \dotplus y;\alpha) = \mathcal{E}_q(x,y;\alpha),$$

when $q\,|\alpha|^2 < 1$.

(15) *A characterization property of the basic exponential function.* Prove that the function $u(x) = \mathcal{E}_q(x,y;\alpha)$ is the only analytic in x solution of the difference equation (2.3.13),

$$\frac{\delta u}{\delta x} = \frac{2q^{1/4}}{1-q}\alpha\; u,$$

on the lattice $x = (q^z + q^{-z})/2$ that satisfies the initial condition

$$u(0) = \mathcal{E}_q(y;\alpha)$$

and does not have poles in the parallelogram on Figure 3.1.
[*Hint*: Follow the proof of the addition theorem in Section 3.1 in order to obtain

$$u(x) = A\mathcal{E}_q(x;\alpha),$$

where A is a constant independent of x. Then let $x = 0$ and use the addition theorem.]

(16) *Separation of variables.* Prove that the function $F(x,y) = \mathcal{E}_q(x,y;\alpha)$ is the only analytic in x and y particular solution of the partial difference equation

$$\frac{\delta F(x,y)}{\delta x} = \frac{\delta F(x,y)}{\delta y}$$

on the q-quadratic lattices $x = (q^z + q^{-z})/2$ and $y = (q^s + q^{-s})/2$ of the form

$$F(x,y) = f(x)\;\; g(y)$$

provided that functions $f(x)$ and $g(y)$ do not have poles in the parallelogram on Figure 3.1.
[*Hint*: Derive $\dfrac{\delta f}{\delta x}/f = \dfrac{\delta g}{\delta y}/g =$ constant and use a special case of the previous exercise.]

CHAPTER 4

Some Expansions and Integrals

Our main objective in this chapter is to derive several important integrals and expansion formulas involving basic trigonometric functions on a q-quadratic grid. These results were established by Ismail, Rahman, Stanton, and Zhang in **[68]**, **[69]**, **[71]**, and **[76]**.

4.1. Main Results

Ismail and Stanton **[71]** have found the following expansion formula

$$\mathcal{E}_q(x;\alpha)\,\mathcal{E}_q(x;\beta) \tag{4.1.1}$$
$$= \sum_{m=0}^{\infty} q^{m^2/4}\alpha^m\, H_m(x|q)\, \frac{\left(-\alpha\beta q^{(m+1)/2};q\right)_\infty}{\left(q\alpha^2, q\beta^2;q^2\right)_\infty}\, \frac{\left(-q^{(1-m)/2}\beta/\alpha;q\right)_m}{(q;q)_m},$$

where $H_m(x|q)$ are the continuous q-Hermite polynomials given by (3.4.16). They call expansion (4.1.1) the "addition" theorem with respect to the parameter α because it becomes $\exp\alpha x\,\exp\beta x = \exp(\alpha+\beta)x$ in the limit $q \to 1^-$.

The expansion (4.1.1) leads to the following integral evaluation

$$\int_0^\pi \mathcal{E}_q(\cos\theta;\alpha)\,\mathcal{E}_q(\cos\theta;\beta)\; H_m(\cos\theta|q)\,\left(e^{2i\theta}, e^{-2i\theta};q\right)_\infty\, d\theta \tag{4.1.2}$$
$$= 2\pi q^{m^2/4}\alpha^m \frac{\left(-\alpha\beta q^{(m+1)/2};q\right)_\infty \left(-q^{(1-m)/2}\beta/\alpha;q\right)_m}{(q;q)_\infty\, \left(q\alpha^2, q\beta^2;q^2\right)_\infty}$$

due to the orthogonality relation for the continuous q-Hermite polynomials (3.5.5). An important extension of the integral (4.1.2), namely,

$$\int_0^\pi \mathcal{E}_q(\cos\theta;\alpha)\,\mathcal{E}_q(\cos\theta;\beta)\; C_m(\cos\theta;\gamma|q)\, \frac{\left(e^{2i\theta},\, e^{-2i\theta};\, q\right)_\infty}{\left(\gamma e^{2i\theta},\, \gamma e^{-2i\theta};\, q\right)_\infty}\, d\theta$$
$$= 2\pi\, \frac{\left(\gamma,\, \gamma q^{m+1}, -\alpha\beta q^{(m+1)/2};q\right)_\infty}{\left(q,\, \gamma^2 q^m;q\right)_\infty \left(q\alpha^2,\, q\beta^2;q^2\right)_\infty}\, \alpha^m q^{m^2/4} \frac{\left(-q^{(1-m)/2}\beta/\alpha;q\right)_m}{(q;q)_m}$$
$$\times\, {}_2\varphi_2\left(\begin{matrix} -q^{(m+1)/2}\alpha/\beta, -q^{(m+1)/2}\beta/\alpha \\ \gamma q^{m+1},\ -q^{(m+1)/2}\alpha\beta \end{matrix};q, -q^{(m+1)/2}\alpha\beta\gamma\right), \tag{4.1.3}$$

had also been evaluated in [**71**]. Here $C_m(x;\gamma|q)$ are the continuous q-ultraspherical polynomials given by

$$C_m(\cos\theta;\beta|q)=\sum_{k=0}^{m}\frac{(\beta;q)_k\,(\beta;q)_{m-k}}{(q;q)_k\,(q;q)_{m-k}}\,e^{i(m-2k)\theta}, \tag{4.1.4}$$

see, for example, [**46**]; the continuous q-Hermite polynomials are their special case

$$H_m(x|q)=(q;q)_m\,C_m(x;0|q) \tag{4.1.5}$$

by (3.4.16) and (4.1.5). The special case $\beta=0$ of (4.1.3) takes the form

$$\int_0^{\pi}\mathcal{E}_q(\cos\theta;\alpha)\,\frac{\left(e^{2i\theta},\ e^{-2i\theta};\ q\right)_\infty}{\left(\gamma e^{2i\theta},\ \gamma e^{-2i\theta};\ q\right)_\infty}\,C_m(\cos\theta;\gamma|q)\,d\theta \tag{4.1.6}$$
$$=\frac{2\pi\left(\gamma,\gamma q^{m+1};q\right)_\infty\,\alpha^m q^{m^2/4}}{(q;q)_m\,(q,\ \gamma^2 q^m;q)_\infty\,(q\alpha^2;q^2)_\infty}\ {}_0\varphi_1\left(\begin{matrix}-\\ \gamma q^{m+1}\end{matrix};\ q,\ \alpha^2\gamma q^{m+1}\right).$$

The expansion formula associated with (4.1.3) is

$$\mathcal{E}_q(\cos\theta;\alpha)\,\mathcal{E}_q(\cos\theta;\beta) \tag{4.1.7}$$
$$=\sum_{m=0}^{\infty}q^{m^2/4}\alpha^m\frac{\left(-\alpha\beta q^{(m+1)/2};q\right)_\infty\left(-q^{(1-m)/2}\beta/\alpha;q\right)_m}{(\gamma;q)_m\left(q\alpha^2,\ q\beta^2;q^2\right)_\infty}$$
$$\times\, C_m(\cos\theta;\gamma|q)$$
$$\times\,{}_2\varphi_2\left(\begin{matrix}-q^{(m+1)/2}\alpha/\beta,-q^{(m+1)/2}\beta/\alpha\\ \gamma q^{m+1},\ -q^{(m+1)/2}\alpha\beta\end{matrix};q,-q^{(m+1)/2}\alpha\beta\gamma\right).$$

The expansions formulas and integrals found by Ismail and Stanton in [**71**] are important contributions in the growing area of q-series. For example, the special case $m=0$ and $\gamma=q^{1/2}$ of the integral (4.1.3) leads to the orthogonality property of the basic trigonometric system originally found in [**26**] by a different method. We shall give this elementary proof of the orthogonality relation in Section 4.4. See also [**146**] and Chapters 9 to 10 for further applications of the Ismail and Stanton expansions and integrals in the theory of basic Fourier series.

In Section 4.5 we shall derive the Ismail and Zhang formula [**76**] as a q-analog of the expansion of the plane wave in the ultraspherical polynomials; this formula can be viewed as the special case $\beta=0$ of (4.1.7). In Section 4.6 we follow [**68**] in order to establish several more expansions of the basic exponential function in classical q-orthogonal polynomials including Ismail, Rahman and Zhang analog of the expansion formula of the plane wave in the Jacobi polynomials [**69**]. More expansions are discussed in Section 4.7 and a simple proof of the Ismail, Rahman and Zhang formula is presented in Section 4.8. A few more integrals are evaluated in Section 4.9.

4.2. Proofs of (4.1.1)

In this section we intend to give two proofs of the expansion (4.1.1). Our first proof is, essentially, the original proof of this result due to Ismail and Stanton [**71**] on the basis of the generating function (3.4.15) and the linearization formula of Rogers [**46**],

$$\frac{H_m(x|q)\, H_n(x|q)}{(q;q)_m\,(q;q)_n} = \sum_{k=0}^{\min(m,n)} \frac{H_{m+n-2k}(x|q)}{(q;q)_k\,(q;q)_{m-k}\,(q;q)_{n-k}}, \tag{4.2.1}$$

for the continuous q-Hermite polynomials.

Use of (3.4.15) and (4.2.1) gives

$$\begin{aligned}
&\left(q\alpha^2, q\beta^2; q^2\right)_\infty \mathcal{E}_q(x;\alpha)\,\mathcal{E}_q(x;\beta) \qquad (4.2.2)\\
&= \sum_{m,n=0}^{\infty} \frac{q^{(m^2+n^2)/4}\alpha^m\beta^n}{(q;q)_m\,(q;q)_n} H_m(x|q)\, H_n(x|q)\\
&= \sum_{m,n,k} q^{(m^2+n^2)/4}\alpha^m\beta^n \frac{H_{m+n-2k}(x|q)}{(q;q)_k\,(q;q)_{m-k}\,(q;q)_{n-k}}\\
&= \sum_{M,N} q^{(M^2+N^2)/4}\alpha^M\beta^N \frac{H_{M+N}(x|q)}{(q;q)_M\,(q;q)_N}\\
&\quad\times \sum_{k=0}^{\infty} \frac{q^{k(k-1)/2}}{(q;q)_k}\left(\alpha\beta q^{(M+N+1)/2}\right)^k\\
&= \sum_{M,N} q^{(M^2+N^2)/4}\alpha^M\beta^N \frac{H_{M+N}(x|q)}{(q;q)_M\,(q;q)_N}\left(-\alpha\beta q^{(M+N+1)/2};q\right)_\infty .
\end{aligned}$$

We substitute $m = M + k$ and $n = N + k$ in the third line here and use the Euler summation formula (1.1.2) in the last but one line. Letting $M+N = m$ in the last expression and using (I.12) of [**46**] or (1.4.7) one can write

$$\begin{aligned}
&\sum_{m=0}^{\infty} q^{m^2/4}\alpha^m\, H_m(x|q)\left(-\alpha\beta q^{(m+1)/2};q\right)_\infty\\
&\quad\times \sum_{N=0}^{\infty} \frac{(q^{-m};q)_N}{(q;q)_N}\left(-\frac{\beta}{\alpha} q^{(m+1)/2}\right)^N\\
&= \sum_{m=0}^{\infty} q^{m^2/4}\alpha^m\, H_m(x|q)\left(-\alpha\beta q^{(m+1)/2};q\right)_\infty \left(-\frac{\beta}{\alpha} q^{(1-m)/2};q\right)_m
\end{aligned}$$

by the q-binomial theorem and the first proof is complete.

Our second proof of (4.1.1) uses also (3.4.15) and (4.2.1), but we evaluate the integral (4.1.2). Obviously, this integral evaluation is equivalent to (4.1.1). Let us start from the evaluation of the following "matrix elements",

$$\int_0^\pi H_m(\cos\theta|q)\,\mathcal{E}_q(\cos\theta;\alpha)\,H_n(\cos\theta|q)\left(e^{2i\theta}, e^{-2i\theta};q\right)_\infty d\theta \tag{4.2.3}$$

$$= 2\pi \frac{\alpha^{m+n} q^{(m+n)^2/4}}{(q\alpha^2; q^2)_\infty (q; q)_\infty} \, {}_2\varphi_1 \left(\begin{matrix} q^{-m}, \, q^{-n} \\ 0 \end{matrix} ; q, \, \frac{q}{\alpha^2} \right),$$

which is of independent interest. Using the linearization formula (4.2.1) in the integrand, one gets

$$\sum_k \frac{(q;q)_m (q;q)_n}{(q;q)_k (q;q)_{m-k} (q;q)_{n-k}}$$
$$\times \int_0^\pi \mathcal{E}_q (\cos\theta; \alpha) H_{m+n-2k} (\cos\theta | q) \left(e^{2i\theta}, e^{-2i\theta}; q \right)_\infty \, d\theta$$
$$= 2\pi \frac{(q;q)_m (q;q)_n}{(q\alpha^2; q^2)_\infty (q;q)_\infty} \sum_k \frac{q^{(m+n-2k)^2/4} \alpha^{m+n-2k}}{(q;q)_k (q;q)_{m-k} (q;q)_{n-k}}$$

with the help of the generating function (3.4.15) and the orthogonality relation (3.5.5). Transformation of the sum of the q-shifted factorials in the last line by (I.12) of [**46**] or (1.4.7) results in (4.2.3).

In order to complete the proof, let us denote the integral in the left side of (4.1.2) as I. Applying (3.4.15) and (4.2.3), one gets

$$I = \frac{1}{(q\alpha^2; q^2)_\infty} \sum_{n=0}^\infty \frac{q^{n^2/4} \, \beta^n}{(q;q)_n} \tag{4.2.4}$$
$$\times \int_0^\pi H_m (\cos\theta|q) \, \mathcal{E}_q (\cos\theta; \alpha) H_n (\cos\theta | q) \left(e^{2i\theta}, e^{-2i\theta}; q \right)_\infty \, d\theta$$
$$= 2\pi \frac{q^{m^2/4} \alpha^m}{\left(q\alpha^2, q\beta^2; q^2\right)_\infty (q;q)_\infty}$$
$$\times \sum_{n=0}^\infty q^{n(n+m)/2} \frac{(\alpha\beta)^n}{(q;q)_n} \sum_k \frac{(q^{-m}, \, q^{-n}; q)_k}{(q;q)_k} \left(\frac{q}{\alpha^2} \right)^k .$$

The change the order of summation and the substitution $n = k + l$ with the help of (I.12) of [**46**] or (1.4.7) give

$$I = 2\pi \frac{q^{m^2/4} \alpha^m}{\left(q\alpha^2, q\beta^2; q^2\right)_\infty (q;q)_\infty}$$
$$\times \sum_k \frac{(q^{-m}; q)_k}{(q;q)_k} \left(-\frac{\alpha}{\beta} q^{(m+1)/2} \right)^k \sum_l \frac{q^{l(l-1)/2}}{(q;q)_l} \left(\alpha\beta q^{(m+1)/2} \right)^l .$$

These series can be summed by the q-binomial theorem and by the Euler summation formula (1.1.2), respectively. This results in (4.1.2) and our second proof is complete.

4.3. Proof of (4.1.3)

This proof is due to Ismail and Stanton [**71**], with somewhat different details. Let us denote the integral in the left side of (4.1.3) as I_m. Using expansion (4.1.1),

$$I_m = \sum_{n=0}^{\infty} q^{n^2/4} \alpha^n \frac{\left(-\alpha\beta q^{(n+1)/2}; q\right)_\infty \left(-q^{(1-n)/2}\beta/\alpha; q\right)_n}{\left(q\alpha^2, q\beta^2; q^2\right)_\infty} \tag{4.3.1}$$
$$\times \int_0^\pi \frac{H_n(\cos\theta|q)\, C_m(\cos\theta;\gamma|q)}{(q;q)_n} \frac{\left(e^{2i\theta},\ e^{-2i\theta};\ q\right)_\infty}{\left(\gamma e^{2i\theta},\ \gamma e^{-2i\theta};\ q\right)_\infty}\, d\theta.$$

In order to evaluate this integral, Ismail and Stanton have used the special case $\gamma = 0$ of the Rogers connection coefficient formula [**46**],

$$C_m(x;\gamma|q) \tag{4.3.2}$$
$$= \sum_{k=0}^{[n/2]} \frac{\beta^k (\gamma/\beta;q)_k (\gamma;q)_{n-k} \left(1-\beta q^{n-2k}\right)}{(q;q)_k (q\beta;q)_{n-k} (1-\beta)} C_{n-2k}(x;\beta|q),$$

namely,

$$\frac{H_n(\cos\theta|q)}{(q;q)_n} = \sum_{k=0}^{[n/2]} \frac{\gamma^k \left(1-\gamma q^{n-2k}\right)}{(q;q)_k (q\gamma;q)_{n-k} (1-\gamma)} C_{n-2k}(\cos\theta;\gamma|q). \tag{4.3.3}$$

This gives

$$\int_0^\pi \frac{H_n(\cos\theta|q)\, C_m(\cos\theta;\gamma|q)}{(q;q)_n} \frac{\left(e^{2i\theta},\ e^{-2i\theta};\ q\right)_\infty}{\left(\gamma e^{2i\theta},\ \gamma e^{-2i\theta};\ q\right)_\infty}\, d\theta \tag{4.3.4}$$
$$= \sum_{k=0}^{[n/2]} \frac{\gamma^k \left(1-\gamma q^{n-2k}\right)}{(q;q)_k (q\gamma;q)_{n-k} (1-\gamma)}$$
$$\times \int_0^\pi C_m(\cos\theta;\gamma|q)\, C_{n-2k}(\cos\theta;\gamma|q)\, \frac{\left(e^{2i\theta},\ e^{-2i\theta};\ q\right)_\infty}{\left(\gamma e^{2i\theta},\ \gamma e^{-2i\theta};\ q\right)_\infty}\, d\theta$$
$$= 2\pi \frac{(\gamma, q\gamma; q)_\infty}{(q,\gamma^2;q)_\infty} \sum_{k=0}^{[n/2]} \frac{\gamma^k \left(\gamma^2;q\right)_m}{(q;q)_k (q\gamma;q)_{n-k} (q;q)_m}\, \delta_{m,n-2k}$$

in view of the orthogonality relation for the continuous q-ultraspherical polynomials [**15**], [**46**]

$$\int_0^\pi C_m(\cos\theta;\beta|q)\, C_n(\cos\theta;\beta|q)\, \frac{\left(e^{2i\theta},\ e^{-2i\theta};\ q\right)_\infty}{\left(\beta e^{2i\theta},\ \beta e^{-2i\theta};\ q\right)_\infty}\, d\theta \tag{4.3.5}$$
$$= 2\pi \frac{(\beta, q\beta; q)_\infty \left(\beta^2;q\right)_n (1-\beta)}{\left(q,\beta^2;q\right)_\infty (q;q)_n (1-\beta q^n)}\, \delta_{mn}.$$

Therefore

$$I_m = 2\pi \frac{(\gamma, q\gamma; q)_\infty}{(q,\gamma^2;q)_\infty \left(q\alpha^2, q\beta^2; q^2\right)_\infty} \frac{\left(\gamma^2;q\right)_m}{(q;q)_m} \tag{4.3.6}$$
$$\times \sum_{k=0}^{\infty} \frac{q^{(m+2k)^2/4} \alpha^{m+2k} \gamma^k}{(q;q)_k (q\gamma;q)_{m+k}}$$

$$\times \left(-\alpha\beta q^{(1+m)/2+k}; q\right)_{\infty} \left(-q^{(1-m)2-k}\beta/\alpha; q\right)_{m+2k}.$$

This series can be transformed to the ${}_2\varphi_2$ in the right side of (4.1.3) with the help of the following relations

$$(q\gamma;q)_{m+k} = (q\gamma;q)_m \left(\gamma q^{m+1};q\right)_k,$$

$$\left(-\alpha\beta q^{(1+m)/2+k};q\right)_{\infty} = \frac{\left(-\alpha\beta q^{(1+m)/2};q\right)_{\infty}}{\left(-\alpha\beta q^{(1+m)/2};q\right)_k},$$

$$\left(-q^{(1-m)/2-k}\beta/\alpha;q\right)_{m+2k}$$
$$= \left(-q^{(1-m)/2-k}\beta/\alpha;q\right)_{m+k} \left(-q^{(1+m)/2}\beta/\alpha;q\right)_k,$$

$$\left(-q^{(1-m)/2-k}\beta/\alpha;q\right)_{m+k}$$
$$= \left(-q^{(1-m)/2-k}\beta/\alpha;q\right)_k \left(-q^{(1-m)/2}\beta/\alpha;q\right)_m,$$

$$\left(-q^{(1-m)/2-k}\beta/\alpha;q\right)_k$$
$$= \left(-q^{(1+m)/2}\alpha/\beta;q\right)_k \left(q^{-(m+1)/2}\beta/\alpha\right)^k q^{-k(k-1)/2},$$

see Appendix I of [**46**] or Ex. 1.1 for the convenient formulas of transformations for the q-shifted factorials. This completes the proof.

4.4. Orthogonality Property

The orthogonality relation for the trigonometric functions on the interval $(-l,l)$ is

$$\frac{1}{2l}\int_{-l}^{l} \exp\left(i\frac{\pi m}{l}x\right) \exp\left(-i\frac{\pi n}{l}x\right)\, dx = \delta_{mn}, \tag{4.4.1}$$

where

$$\delta_{mn} = \begin{cases} 1 & \text{if} \quad m=n, \\ 0 & \text{if} \quad m\neq n. \end{cases} \tag{4.4.2}$$

Bustoz and Suslov [**26**] have established an analog of this important property for the q-exponential function $\mathcal{E}_q(x;i\omega)$ using arguments of the Sturm–Liouville theory for the second order divided difference Askey–Wilson operator; see, for example, [**111**] and [**139**]. A special case of the integral (4.1.3) gives rise to an elementary proof of this property which plays a key role in the theory of basic Fourier series.

THEOREM 4.1. *The following orthogonality property holds*

$$\int_0^{\pi} \mathcal{E}_q(\cos\theta;i\omega_m)\,\mathcal{E}_q(\cos\theta;-i\omega_n)\,\left(e^{2i\theta},\ e^{-2i\theta};\ q\right)_{1/2}\, d\theta \tag{4.4.3}$$
$$= 2k(\omega_n)\ \delta_{mn}.$$

Here

$$k(\omega) = \pi \frac{\left(q^{1/2};q\right)_\infty^2}{(q;q)_\infty^2} \frac{\left(-\omega^2;q^2\right)_\infty}{\left(-q\omega^2;q^2\right)_\infty} \sum_{k=0}^{\infty} \frac{q^{k/2}}{1+\omega^2 q^k}; \tag{4.4.4}$$

$\omega_m, \omega_n = 0, \pm\omega_1, \pm\omega_2, \pm\omega_3, \ldots$ *and* $\omega_0 = 0, \omega_1, \omega_2, \omega_3, \ldots$ *, are nonnegative solutions of the transcendental equation*

$$\left(-i\omega;q^{1/2}\right)_\infty = \left(i\omega;q^{1/2}\right)_\infty \tag{4.4.5}$$

arranged in ascending order of magnitude.

PROOF. Using (III.4) of [**46**] or (A.3.4), let us transform the ${}_2\varphi_2$ function in the right side of (4.1.3) with $m=0$ and $\gamma = q^{1/2}$ to a ${}_2\varphi_1$ which can be summed by a consequence of the q-binomial theorem, see Ex. 1.4,

$$\begin{aligned}
&\frac{\left(\alpha^2;q\right)_\infty}{\left(-q^{1/2}\alpha\beta;q\right)_\infty} {}_2\varphi_1\left(\begin{matrix} -q^{1/2}\beta/\alpha, -q\beta/\alpha \\ q^{3/2}\end{matrix} ; q, \alpha^2\right) \\
&= \frac{\left(\alpha^2;q\right)_\infty \left(1-q^{1/2}\right)}{\left(-q^{1/2}\alpha\beta;q\right)_\infty (\alpha+\beta)} \sum_{k=0}^{\infty} \frac{\left(-\beta/\alpha;q^{1/2}\right)_{2k+1}}{\left(q^{1/2};q^{1/2}\right)_{2k+1}} \alpha^{2k+1} \\
&= \frac{\left(\alpha^2;q\right)_\infty \left(1-q^{1/2}\right)}{2\left(-q^{1/2}\alpha\beta;q\right)_\infty (\alpha+\beta)} \left[\frac{\left(-\beta;q^{1/2}\right)_\infty}{\left(\alpha;q^{1/2}\right)_\infty} - \frac{\left(\beta;q^{1/2}\right)_\infty}{\left(-\alpha;q^{1/2}\right)_\infty}\right].
\end{aligned}$$

Therefore

$$\begin{aligned}
&\int_0^\pi \mathcal{E}_q(\cos\theta;\alpha)\,\mathcal{E}_q(\cos\theta;\beta)\left(e^{2i\theta},\ e^{-2i\theta};\ q\right)_{1/2} d\theta \\
&\qquad = \pi \frac{\left(q^{1/2};q\right)_\infty^2}{(q;q)_\infty^2} \frac{\left(-\alpha,\ -\beta;q^{1/2}\right)_\infty - \left(\alpha,\ \beta;q^{1/2}\right)_\infty}{(\alpha+\beta)\left(q\alpha^2,\ q\beta^2;q^2\right)_\infty}.
\end{aligned} \tag{4.4.6}$$

This basic integral is obviously a q-extension of the elementary integral

$$\int_{-1}^{1} e^{\alpha x}\, e^{\beta x}\, dx = \frac{e^{\alpha+\beta} - e^{-\alpha-\beta}}{\alpha+\beta}. \tag{4.4.7}$$

Since,

$$\begin{aligned}
&\left(-\alpha,\ -\beta;q^{1/2}\right)_\infty - \left(\alpha,\ \beta;q^{1/2}\right)_\infty \\
&\quad = \left[\left(-\alpha;q^{1/2}\right)_\infty - \left(\alpha;q^{1/2}\right)_\infty\right]\left(-\beta;q^{1/2}\right)_\infty \\
&\quad + \left[\left(-\beta;q^{1/2}\right)_\infty - \left(\beta;q^{1/2}\right)_\infty\right]\left(\alpha;q^{1/2}\right)_\infty,
\end{aligned} \tag{4.4.8}$$

the orthogonality relation (4.4.3) follows immediately from (4.4.6) if $\alpha = i\omega_m$, $\beta = -i\omega_n$ and $\omega_n \neq \omega_m$ are the roots of the transcendental equation (4.4.5).

In order to obtain the expression (4.4.4) for the normalization constant $k(\omega)$ one can take the limit $\beta \to -\alpha$ in the right side of (4.4.6),

$$\int_0^{\pi} \mathcal{E}_q(\cos\theta;\alpha)\ \mathcal{E}_q(\cos\theta;-\alpha)\ \left(e^{2i\theta},\ e^{-2i\theta};\ q\right)_{1/2}\ d\theta \tag{4.4.9}$$
$$= \pi \frac{\left(q^{1/2};q\right)_\infty^2}{(q;q)_\infty^2}\ \frac{\left(\alpha^2;q^2\right)_\infty}{\left(q\alpha^2;q^2\right)_\infty}\ \frac{d}{d\alpha} \log \frac{\left(-\alpha;q^{1/2}\right)_\infty}{\left(\alpha;q^{1/2}\right)_\infty},$$

and then substitute $\alpha = i\omega$. This results in (4.4.4) after evaluation of the derivative and our proof is complete. □

Equation (4.4.5) can be rewritten in terms of the basic sine function $S_q(\eta;\omega)$ as

$$S_q\left(\frac{1}{2}\left(q^{1/4}+q^{-1/4}\right);\omega\right) = \frac{(-i\omega;q^{1/2})_\infty - (i\omega;q^{1/2})_\infty}{2i(-q\omega^2;q^2)_\infty} = 0 \tag{4.4.10}$$

due to (2.4.19); we use the notation $\eta = \left(q^{1/4}+q^{-1/4}\right)/2$ throughout the book. One can also see that the orthogonality relation (4.4.3) holds at the zeros of the basic cosine function

$$C_q\left(\frac{1}{2}\left(q^{1/4}+q^{-1/4}\right);\omega\right) = \frac{(-i\omega;q^{1/2})_\infty + (i\omega;q^{1/2})_\infty}{2(-q\omega^2;q^2)_\infty} = 0, \tag{4.4.11}$$

see (2.4.20). We leave the details to the reader.

4.5. Ismail and Zhang Formula

The classical expansion of the plane wave in the ultraspherical polynomials is

$$e^{irx} = \left(\frac{2}{r}\right)^{\nu} \Gamma(\nu) \sum_{m=0}^{\infty} i^m (\nu+m)\, J_{\nu+m}(r)\, C_m^{\nu}(x). \tag{4.5.1}$$

Ismail and Zhang [**76**] had found the following q-analog of this formula

$$\mathcal{E}_q(x;i\omega) = \frac{(q;q)_\infty\,\omega^{-\nu}}{(q^\nu;q)_\infty\,(-q\omega^2;q^2)_\infty} \tag{4.5.2}$$
$$\times \sum_{m=0}^{\infty} i^m \left(1-q^{\nu+m}\right)\ q^{m^2/4}\ J_{\nu+m}^{(2)}(2\omega;q)\ C_m\left(x;q^\nu\,|\,q\right),$$

where $J_{\nu+m}^{(2)}(2\omega;q)$ is Jackson's q-Bessel function defined by

$$J_\nu^{(2)}(x;q) = \frac{\left(q^{\nu+1};q\right)_\infty}{(q;q)_\infty} \sum_{n=0}^{\infty} q^{(\nu+n)n} \frac{(-1)^n\,(x/2)^{\nu+2n}}{(q;q)_n\,(q^{\nu+1};q)_n}. \tag{4.5.3}$$

Different proofs of this important formula were given in [**41**], [**69**], [**68**], and [**71**]. One can obtain (4.5.2) as the limiting case $\beta \to 0$ of the expansion (4.1.7). Ismail and Stanton [**71**] have found the following simple proof of

(4.5.2) using the generation relation (3.4.14) and the connection coefficient formula (4.3.3):

$$\begin{aligned}
&\left(q\alpha^2;q^2\right)_\infty \mathcal{E}_q(x;\alpha) \qquad (4.5.4)\\
&=\sum_{n=0}^{\infty}\frac{q^{n^2/4}}{(q;q)_n}\,\alpha^n\,H_n(x|q)\\
&=\sum_{n,k} q^{n^2/4}\alpha^n\frac{\beta^k\left(1-\beta q^{n-2k}\right)}{(q;q)_k\,(q\beta;q)_{n-k}\,(1-\beta)}\,C_{n-2k}(x;\beta|q)\\
&=\sum_m \frac{1-\beta q^m}{1-\beta}q^{m^2/4}\alpha^m\,C_m(x;\beta|q)\\
&\quad\times\sum_k\frac{\alpha^{2k}\beta^k}{(q;q)_k\,(q\beta;q)_{m+k}}q^{k(m+k)}.
\end{aligned}$$

We have substituted here $n=m+2k$ in the second line. The sum over k gives the q-Bessel function and the infinite products in (4.5.2) when $\alpha=i\omega$ and $\beta=q^\nu$.

The integral evaluation corresponding to (4.5.2) has the form

$$\int_0^\pi \mathcal{E}_q(\cos\theta;i\omega)\;C_m(\cos\theta;q^\nu|q)\;\frac{\left(e^{2i\theta},\,e^{-2i\theta};\,q\right)_\infty}{\left(q^\nu e^{2i\theta},\,q^\nu e^{-2i\theta};\,q\right)_\infty}\,d\theta \qquad (4.5.5)$$

$$=2\pi\;i^m q^{m^2/4}\frac{\omega^{-\nu}}{\left(-q\omega^2;q^2\right)_\infty}\;\frac{\left(q^\nu;q\right)_\infty\left(q^{2\nu};q\right)_m}{\left(q^{2\nu};q\right)_\infty(q;q)_m}\;J^{(2)}_{\nu+m}(2\omega;q).$$

This gives a q-extension of *Gegenbauer's integral representation* for Bessel functions:

$$\int_0^\pi e^{ir\cos\theta}C_m^\nu(\cos\theta)\,(\sin\theta)^{2\nu}\;d\theta=2\pi i^m\frac{\Gamma(2\nu+m)}{(2r)^\nu\,\Gamma(\nu)\,m!}\;J_{\nu+m}(r)\,, \qquad (4.5.6)$$

see [**37**]; Vol. 2, Eq. (7.8.11). The simplest cases $m=0$ correspond to the *Poisson integral representation* and its basic extension.

4.6. q-Exponentials and Connection Coefficient Problems

In this section we derive several more expansions of the basic exponential function in the classical q-orthogonal polynomials following a general approach suggested by Ismail, Rahman and Stanton [**68**].

4.6.1. Expansion Formula. The Askey–Wilson polynomials [**15**] are

$$\begin{aligned}
p_n(x)&=p_n(x;a,b,c,d)=p_n(x;a,b,c,d|q) \qquad (4.6.1)\\
&=a^{-n}\;(ab,\,ac,\,ad;q)_n\;{}_4\varphi_3\left(\begin{matrix}q^{-n},abcdq^{n-1},ae^{i\theta},ae^{-i\theta}\\ ab,ac,ad\end{matrix}\;;\;q,\,q\right),
\end{aligned}$$

where $x=\cos\theta$. These polynomials are the most general known classical orthogonal polynomials; see, for example, [**9**], [**15**], [**16**], [**18**], [**46**], [**86**], [**111**], and Ex. 2.22. In their fundamental memoir [**15**], Askey and Wilson studied in detail many properties of these polynomials, which are symmetric in all four parameters a, b, c, d. The orthogonality relation is

$$\int_0^\pi \frac{p_n(\cos\theta; a,b,c,d)\, p_m(\cos\theta; a,b,c,d)\, \left(e^{2i\theta}, e^{-2i\theta}; q\right)_\infty}{\left(ae^{i\theta}, ae^{-i\theta}, be^{i\theta}, be^{-i\theta}, ce^{i\theta}, ce^{-i\theta}, de^{i\theta}, de^{-i\theta}; q\right)_\infty}\, d\theta$$
$$= \delta_{nm} \frac{2\pi\,(abcd;q)_\infty}{(q, ab, ac, ad, bc, bd, cd; q)_\infty} \tag{4.6.2}$$
$$\times \frac{\left(1 - abcdq^{-1}\right)\, (q, ab, ac, ad, bc, bd, cd; q)_n}{\left(1 - abcdq^{2n-1}\right)\, (abcdq^{-1}; q)_n}$$

when $\max(|a|, |b|, |c|, |d|) < 1$.

The following lemma establishes the expansion of the generalized powers in the Askey–Wilson polynomials [**68**].

LEMMA 4.1. *The coefficients in the expansion*

$$\left(\gamma e^{i\theta}, \gamma e^{-i\theta}; q\right)_n = \sum_{k=0}^{n} a_{n,k}\, p_k(\cos\theta; a,b,c,d) \tag{4.6.3}$$

are given by

$$a_{n,k} = \frac{(q, \gamma d; q)_n (\gamma/d; q)_{n-k} (abcdq^{-1}; q)_k}{(q, \gamma d; q)_k (q;q)_{n-k} (abcdq^{-1}; q)_{2k}} \tag{4.6.4}$$
$$\times (-\gamma)^k\, q^{k(k-1)/2}\, {}_4\varphi_3\left(\begin{matrix} q^{k-n}, adq^k, bdq^k, cdq^k \\ d\gamma q^k, q^{1-n+k} d/\gamma, abcdq^{2k} \end{matrix}; q, q\right).$$

Alternate forms of the coefficients are

$$a_{n,k} = \frac{(q, c\gamma, d\gamma, ab; q)_n (abcdq^{-1}; q)_k \left(1 - abcdq^{2k-1}\right)}{(q, c\gamma, d\gamma, ab; q)_k (q;q)_{n-k} (abcd; q)_{n+k} \left(1 - abcdq^{-1}\right)}$$
$$\times (-\gamma)^k\, q^{k(k-1)/2}\, {}_4\varphi_3\left(\begin{matrix} q^{k-n}, cdq^k, \gamma/a, \gamma/b \\ c\gamma q^k, d\gamma q^k, q^{1-n}/ab \end{matrix}; q, q\right) \tag{4.6.5}$$

and

$$a_{n,k} = \frac{(q, a\gamma, b\gamma, c\gamma, d\gamma; q)_n (\gamma^2, abcdq^{-1}; q)_k \left(1 - abcdq^{2k-1}\right)}{(q, a\gamma, b\gamma, c\gamma, d\gamma; q)_k (\gamma^2; q)_n (q;q)_{n-k} (abcd; q)_{n+k} \left(1 - abcdq^{-1}\right)}$$
$$\times (-\gamma)^k\, q^{k(k-1)/2}\, {}_8W_7\left(\gamma^2 q^{k-1}; q^{k-n}, \gamma/a, \gamma/b, \gamma/c, \gamma/d;\ q,\ abcdq^{n+k}\right). \tag{4.6.6}$$

This is valid if $\max(|a|, |b|, |c|, |d|) < 1$.

PROOF. Multiplying (4.6.3) by $p_m(x; a,b,c,d)\,\rho(x)$, where the weight function $\rho(x)$ is given explicitly by (4.6.2), and integrating over x on $(-1, 1)$, we find that

$$a_{n,m} \frac{2\pi\,(abcd;q)_\infty}{(q, ab, ac, ad, bc, bd, cd; q)_\infty} \tag{4.6.7}$$

$$\times a^m \frac{\left(1-abcdq^{-1}\right)(q,bc,bd,cd;q)_m}{\left(1-abcdq^{2m-1}\right)\left(abcdq^{-1};q\right)_m}$$

$$= \sum_{k=0}^{m} q^k \frac{\left(q^{-m}, abcdq^{m-1};q\right)_k}{(q,ab,ac,ad;q)_k}$$

$$\times \int_0^{\pi} \frac{\left(e^{2i\theta}, e^{-2i\theta};q\right)_\infty}{\left(aq^k e^{i\theta}, aq^k e^{-i\theta}, be^{i\theta}, be^{-i\theta}, ce^{i\theta}, ce^{-i\theta}, de^{i\theta}, de^{-i\theta};q\right)_\infty}$$

$$\times \frac{\left(\gamma e^{i\theta}, \gamma e^{-i\theta};q\right)_\infty}{\left(\gamma q^n e^{i\theta}, \gamma q^n e^{-i\theta};q\right)_\infty} \, d\theta.$$

The last integral can be evaluated by the Nassrallah–Rahman integral (6.3.8) of [**46**] or (A.4.3), and is equal to

$$\frac{2\pi\left(b\gamma, c\gamma, d\gamma, abcdq^k, bcd\gamma q^n;q\right)_\infty}{\left(q,bc,bd,cd,bcd\gamma, abq^k, acq^k, adq^k, b\gamma q^n, c\gamma q^n, d\gamma q^n;q\right)_\infty}$$

$$\times {}_8W_7\left(bcd\gamma q^{-1}; q^{-n}, bc, bd, cd, q^{-k}\gamma/a;\ q,\ a\gamma q^{k+n}\right)$$

$$= \frac{2\pi\,(abcd;q)_\infty}{(q,ab,ac,ad,bc,bd,cd;q)_\infty}\,(ab,ac;q)_k\,(\gamma d,\gamma/d;q)_n$$

$$\times \sum_{l=0}^{n} q^l \frac{\left(q^{-n},bd,cd;q\right)_l (ad;q)_{k+l}}{\left(q,d\gamma,q^{1-n}d/\gamma;q\right)_l (abcd;q)_{k+l}}, \tag{4.6.8}$$

where (III.17) of [**46**] or (A.3.14) was used in the last step.

From (4.6.7)–(4.6.8) we have

$$a^m \frac{\left(1-abcdq^{-1}\right)(q,bc,bd,cd;q)_m}{\left(1-abcdq^{2m-1}\right)\left(abcdq^{-1};q\right)_m} a_{n,m} \tag{4.6.9}$$

$$= (\gamma d,\gamma/d;q)_n \sum_{l=0}^{n} q^l \frac{\left(q^{-n},ad,bd,cd;q\right)_l}{\left(q,abcd,d\gamma,q^{1-n}d/\gamma;q\right)_l}$$

$$\times {}_3\varphi_2\left(\begin{matrix} q^{-m}, abcdq^{m-1}, adq^l \\ abcdq^l, ad \end{matrix};\ q,\ q\right).$$

The ${}_3\varphi_2$ is now summed by (II.12) of [**46**] or (A.2.8),

$${}_3\varphi_2\left(\begin{matrix} q^{-m}, abcdq^{m-1}, adq^l \\ abcdq^l, ad \end{matrix};\ q,\ q\right) = \frac{\left(q^{l-m+1},bc;q\right)_m}{\left(abcdq^l, q^{1-m}/ad;q\right)_m},$$

and after some simplification we obtain (4.6.4). Application of the Sears transformation (III.15) of [**46**] or (A.3.12), gives the alternate representation (4.6.5).

It is clear from (4.6.3) that the coefficients $a_{n,k}$ must be symmetric in a, b, c, d but it is not obvious in either of two forms (4.6.4) or (4.6.5). The symmetry becomes explicit when we apply the Watson transformation (III.17) of [**46**] or (A.3.14) to obtain (4.6.6). □

We shall use Lemma 4.1 in the following way. Let us multiply (4.6.3) by c_n, where $\{c_n\}$ is an arbitrary sequence, and sum over n,

$$\sum_{n=0}^{\infty} c_n \left(\gamma e^{i\theta}, \gamma e^{-i\theta}; q\right)_n = \sum_{k=0}^{n} p_k(\cos\theta; a, b, c, d) \sum_{n=0}^{\infty} a_{n+k,k}\, c_{n+k}, \tag{4.6.10}$$

provided that the left side of (4.6.10) converges and interchanging the sums is justifiable. If one chooses $\gamma = \varepsilon q^{(1-n)/2}$ for some constant ε and c_n in accordance with the definition (2.3.1), then Eq. (4.6.10) is an expansion for the q-exponential function $\mathcal{E}_q(x, y; \alpha)$. We shall specialize the parameters a, b, c, d in such a way that the ${}_4\varphi_3$ series in (4.6.4) and (4.6.5), or the ${}_8\varphi_7$ series in (4.6.6), can be summed. This gives an opportunity to derive several interesting expansion formulas [**68**].

4.6.2. Expansion in Continuous q-Ultraspherical Polynomials. Here we present an independent proof of the Ismail and Zhang formula (4.5.2) on the basis of Lemma 4.1 [**68**]. The continuous q-ultraspherical polynomials $C_n(x; a^2|q)$ are obtained from the Askey–Wilson polynomials by

$$p_n\left(x; a, aq^{1/2}, -a, -aq^{1/2}|q\right) = \frac{(q;q)_n\,(a^4;q)_{2n}}{(a^2, a^4; q)_{2n}}\, C_n\left(x; a^2|q\right). \tag{4.6.11}$$

We consider (4.6.6) for the continuous q-ultraspherical polynomials when $b = aq^{1/2}$, $c = -a$, $d = -aq^{1/2}$ and α is arbitrary. Using (3.4.7) of [**46**] or (A.3.11), we have

$$\begin{aligned}
&{}_8W_7\left(\gamma^2 q^{k-1}; q^{-1/2}\gamma/a, -q^{-1/2}\gamma/a, \gamma/a, -\gamma/a, q^{k-n}; a^4 q^{n+k+1}\right)\\
&= \frac{(\gamma^2 q^k, a^4 q^{2k+1}; q)_\infty}{(a^2 q^{k+1}, \gamma^2 a^2 q^{2k}; q)_\infty}\, {}_2\varphi_1\left(\begin{matrix} q^{-1}\gamma^2/a^2,\ q^{-n}/a^2 \\ \gamma^2 q^n \end{matrix}\ ;\ q,\ a^4 q^{n+k+1}\right)\\
&= \frac{(\gamma^2 q^k, \gamma^2 a^2 q^{2n}, a^4 q^{2k+1}; q)_\infty}{(\gamma^2 q^n, \gamma^2 a^2 q^{2k}, a^4 q^{n+k+1}; q)_\infty}\, {}_2\varphi_1\left(\begin{matrix} q^{k-n},\ q^{-n}/a^2 \\ a^2 q^{k+1} \end{matrix}\ ;\ q,\ \gamma^2 a^2 q^{2n}\right),
\end{aligned}$$

where Eq. (III.2) of [**46**] or (A.3.2) was used in the last step. Thus, in this case Lemma 4.1 reduces to

$$\begin{aligned}
\left(\gamma e^{i\theta}, \gamma e^{-i\theta}; q\right)_n &= \sum_{k=0}^{n} \frac{(q;q)_n\,(-\gamma)^k\, q^{k(k-1)/2}}{(q;q)_{n-k}\,(a^2;q)_k}\, C_k(\cos\theta; a^2|q)\\
&\quad\times {}_2\varphi_1\left(\begin{matrix} q^{k-n},\ q^{-n}/a^2 \\ a^2 q^{k+1} \end{matrix}\ ;\ q,\ \gamma^2 a^2 q^{2n}\right).
\end{aligned} \tag{4.6.12}$$

There are several choices for parameter γ when the ${}_2\varphi_1$ series is summable. In order to derive the Ismail and Zhang formula (4.5.2), we choose $\gamma = -iq^{(1-n)/2}$ and apply the Bailey–Daum formula (II.9) of [**46**] or (A.2.7) to get

$$
{}_2\varphi_1\left(\begin{matrix} q^{k-n},\ q^{-n}/a^2 \\ a^2q^{k+1} \end{matrix}\ ;\ q,\ \gamma^2a^2q^{2n}\right) \tag{4.6.13}
$$

$$
= \begin{cases} 0 & \text{if } n-k \text{ is odd,} \\ (-1)^m\, q^{-m^2}\, \dfrac{(q, -a^2q^{k+1}, -a^2q^{k+2};\ q^2)_m}{(a^4q^{2k+2};\ q^2)_m} & \text{if } n-k=2m, \end{cases}
$$

where the identity $(-q;q)_\infty\,(q;q^2)_\infty = 1$ was used. The choice of c_n corresponding to (2.3.3) leads to

$$
\mathcal{E}_q(x;\alpha) = \frac{(\alpha^2;q^2)_\infty}{(q\alpha^2;q^2)_\infty} \sum_{k=0}^{\infty} \frac{q^{k^2/4}\alpha^k}{(a^2;q)_k}\, C_k(x;a^2|q) \tag{4.6.14}
$$

$$
\times\ {}_2\varphi_1\left(\begin{matrix} -a^2q^{k+1},\ -a^2q^{k+2} \\ a^4q^{2k+2} \end{matrix}\ ;\ q,\ \alpha^2\right).
$$

Setting $\alpha = i\omega$ and $a = q^{\nu/2}$ we obtain (4.5.2) with the help of the quadratic transformation formula (2.4.27). On the other hand, this gives a proof of this quadratic transformation because an independent proof of the Ismail and Zhang formula was given in Section 4.5.

4.6.3. Expansion in Continuous q-Jacobi Polynomials. We evaluate the connection coefficients $a_{n,k}$ using special cases of balanced ${}_4\varphi_3$'s. This leads to expansions of the q-exponential function in special families of Askey–Wilson polynomials including the Ismail, Rahman and Zhang expansion formula in continuous q-Jacobi polynomials [**69**]. The corresponding balanced ${}_4\varphi_3$ evaluation is Andrews' formula (4.3) of [**8**]; see also (A.2.10). So, (4.6.4) is summable if $b = aq^{1/2}$, $c \to -c$, $d = -cq^{1/2}$ and $\gamma = -q^{3/4-n/2}$:

$$
a_{n,k} = q^{-((n-k)(n+k-1)-k)/4}\, \frac{(q;q)_n\,(a^2c^2;q)_k}{(q;q)_k\,(a^2c^2;q)_{2k}} \tag{4.6.15}
$$

$$
\times \frac{\left(-aq^{1/4+k/2},\ cq^{1/4+k/2};q^{1/2}\right)_{n-k}}{\left(q^{1/2},\ acq^{k+1/2};q^{1/2}\right)_{n-k}},
$$

where (I.7) of [**46**] or (1.4.1) was used. From (4.6.10) with

$$
c_n = \frac{q^{n(n-1)/4}}{(q;q)_n}\,\alpha^n \tag{4.6.16}
$$

we obtain that

$$
\frac{(q\alpha^2;q^2)_\infty}{(\alpha^2;q^2)_\infty}\,\mathcal{E}_q(x,\eta;\alpha) \tag{4.6.17}
$$

$$
= \sum_{k=0}^{\infty} \frac{q^{k^2/4}\alpha^k\,(a^2c^2;q)_k}{(q;q)_k\,(a^2c^2;q)_{2k}}\, p_k(x;a,aq^{1/2},-c,-cq^{1/2})
$$

$$\times\ {}_2\varphi_1\left(\begin{matrix} cq^{1/4+k/2},\ -aq^{1/4+k/2} \\ acq^{k+1/2} \end{matrix};\ q^{1/2},\ \alpha\right),$$

where $\eta=\left(q^{1/4}+q^{-1/4}\right)/2$. As a result with the help of the addition theorem

$$\mathcal{E}_q\left(x,\eta;\alpha\right)=\mathcal{E}_q\left(x;\alpha\right)\ \mathcal{E}_q\left(\eta;\alpha\right)=\mathcal{E}_q\left(x;\alpha\right)\ \frac{\left(-\alpha;q^{1/2}\right)_\infty}{\left(q\alpha^2;q^2\right)_\infty} \tag{4.6.18}$$

one gets the Ismail, Rahman and Zhang formula **[69]**,

$$\begin{aligned}\mathcal{E}_q\left(x;\alpha\right) &= \frac{\left(\alpha^2;q^2\right)_\infty}{\left(-\alpha;q^{1/2}\right)_\infty}\sum_{k=0}^{\infty}\frac{q^{k^2/4}\alpha^k\left(a^2c^2;q\right)_k}{(q;q)_k\left(a^2c^2;q\right)_{2k}} \\ &\quad\times p_k(x;a,aq^{1/2},-c,-cq^{1/2}) \\ &\quad\times\ {}_2\varphi_1\left(\begin{matrix} cq^{1/4+k/2},\ -aq^{1/4+k/2} \\ acq^{k+1/2} \end{matrix};\ q^{1/2},\ \alpha\right),\end{aligned} \tag{4.6.19}$$

as an analog of the classical expansion of the plane wave in Jacobi polynomials,

$$\begin{aligned}e^{i\omega x} &= e^{-i\omega}\sum_{n=0}^{\infty}\frac{(\alpha+\beta+1)_n}{(\alpha+\beta+1)_{2n}}\ (2i\omega)^n\ P_n^{(\alpha,\beta)}(x) \\ &\quad\times\ {}_1F_1\left(\begin{matrix}\beta+n+1\\ \alpha+\beta+2n+2\end{matrix};\ 2i\omega\right),\end{aligned} \tag{4.6.20}$$

see **[37]**; Vol. 2, Eq. (10.20.4); independent proofs of (4.6.19) and (4.6.20) will be given in Section 4.8.

Askey and Wilson **[15]** define the continuous q-Jacobi polynomials by

$$\begin{aligned}P_n^{(\alpha,\beta)}(x|q) &= \frac{\left(q^{\alpha+1};q\right)_n}{(q;q)_n} \\ &\quad\times\ {}_4\varphi_3\left(\begin{matrix} q^{-n},\ q^{\alpha+\beta+n+1},\ q^{(2\alpha+1)/4}e^{i\theta},\ q^{(2\alpha+1)/4}e^{-i\theta} \\ q^{\alpha+1},\ -q^{(\alpha+\beta+n+1)/2},\ -q^{(\alpha+\beta+n+2)/2}\end{matrix};\ q,\ q\right)\end{aligned} \tag{4.6.21}$$

(another definition is due to Rahman **[114]**, **[46]**), such that

$$\lim_{q\to 1^-} P_n^{(\alpha,\beta)}(x|q)=P_n^{(\alpha,\beta)}(x). \tag{4.6.22}$$

In this notation, if $\alpha=i\omega$, the expansion formula (4.6.19) has the form

$$\begin{aligned}\mathcal{E}_q\left(x;i\omega\right) &= \frac{\left(-\omega^2;q^2\right)_\infty}{\left(-i\omega;q^{1/2}\right)_\infty}\sum_{n=0}^{\infty}q^{n(n-2\alpha-1)/4}\ (i\omega)^n\ \frac{\left(q^{\alpha+\beta+1};q\right)_n}{\left(q^{\alpha+\beta+1};q\right)_{2n}} \\ &\quad\times\left(-q^{(\alpha+\beta+1)/2};q^{1/2}\right)_n\ P_n^{(\alpha,\beta)}(x|q) \qquad (4.6.23) \\ &\quad\times\ {}_2\varphi_1\left(\begin{matrix} q^{(\beta+n+1)/2},\ -q^{(\alpha+n+1)/2} \\ q^{(\alpha+\beta+2n+2)/2}\end{matrix};\ q^{1/2},\ i\omega\right),\end{aligned}$$

which is evidently a q-analog of (4.6.20).

4.6.4. Expansions in Special Askey–Wilson polynomials. Let us evaluate the connection coefficients $a_{n,k}$ in a few more special cases of the Askey–Wilson polynomials and consider corresponding expansion formulas. Setting $\gamma = -iq^{(1-n)/2}$, $b=-a$, $d=-c$ in (4.6.5), we can use Andrews' terminating q-analog of Watson's ${}_3F_2$ sum, see (II.17) of [**46**] or (A.2.9). As a result

$$
{}_4\varphi_3\left(\begin{matrix} q^{k-n},\ -c^2q^k,\ iq^{(1-n)/2}/a,\ -iq^{(1-n)/2}/a \\ icq^{(1-n)/2+k},\ -icq^{(1-n)/2+k},\ -q^{1-n}/a^2 \end{matrix};\ q,\ q\right) \tag{4.6.24}
$$
$$
= (-1)^m q^{-m^2} \frac{\left(q,\ a^2c^2q^{2k};q^2\right)_m}{\left(-c^2q^{1+k-2m},\ -a^2q^k;q^2\right)_m}\,\delta_{n,k+2m}
$$

and the corresponding connection coefficients are

$$
a_{n,k} = (-1)^m i^k q^{-m^2-mk} \frac{\left(a^2c^2q^{-1};q\right)_k (q;q)_{k+2m}}{(q;q)_k \left(a^2c^2q^{-1};q\right)_{2k}} \tag{4.6.25}
$$
$$
\times \frac{\left(-a^2q^{k+1},\ -c^2q^{k+1};q^2\right)_m}{\left(q^2,\ a^2c^2q^{2k+1};q^2\right)_m}.
$$

Using (4.6.10) with

$$
c_n = \frac{q^{n^2/4}}{(q;q)_n}(-i\alpha)^n \tag{4.6.26}
$$

we arrive at the following expansion formula [**68**]:

$$
\mathcal{E}_q(x;\alpha) = \frac{\left(\alpha^2;q^2\right)_\infty}{\left(q\alpha^2;q^2\right)_\infty} \sum_{k=0}^{\infty} \frac{q^{k^2/4}\alpha^k \left(a^2c^2q^{-1};q\right)_k}{(q;q)_k \left(a^2c^2q^{-1};q\right)_{2k}} \tag{4.6.27}
$$
$$
\times\, p_k(x;a,-a,c,-c)\ {}_2\varphi_1\left(\begin{matrix} -a^2q^{k+1},\ -c^2q^{k+1} \\ a^2c^2q^{2k+1} \end{matrix};\ q^2,\ \alpha^2\right).
$$

In [**63**] it was pointed out that

$$
\left(\frac{r}{2}\right)^{\nu} \frac{\left(q^{\nu+1}, -r^2/4;q\right)_\infty}{(q;q)_\infty}\ {}_2\varphi_1\left(\begin{matrix} q^{(\nu+1)/2}e^{i\theta},\ q^{(\nu+1)/2}e^{-i\theta} \\ q^{\nu+1} \end{matrix};\ q,\ -\frac{r^2}{4}\right)
$$

is a q-analog of the Bessel function $J_\nu(xr)$ when $x=\cos\theta$; see Ex. 5.6. The ${}_2\varphi_1$ in (4.6.27) can be expressed in terms of the above q-Bessel function.

If $c=0$, then the ${}_2\varphi_1$ in (4.6.27) becomes a ${}_1\varphi_0$, which can be summed by the q-binomial theorem

$$
{}_1\varphi_0\left(\begin{matrix} -a^2q^{k+1} \\ - \end{matrix};\ q^2,\ \alpha^2\right) = \frac{\left(-\alpha^2a^2q^{k+1};q^2\right)_\infty}{\left(\alpha^2;q^2\right)_\infty}, \tag{4.6.28}
$$

and the corresponding expansion found in [**68**] is

$$
\left(q\alpha^2;q^2\right)_\infty \mathcal{E}_q(x;\alpha) = \sum_{k=0}^{\infty} \frac{q^{k^2/4}\alpha^k}{(q;q)_k} \left(-\alpha^2a^2q^{k+1};q^2\right)_\infty\ p_k(x;a,-a), \tag{4.6.29}
$$

where $p_k(x;a,-a) = p_k(x;a,-a|q) = p_k(x;a,-a,0,0|q)$ are special Al-Salam and Chihara polynomials. Since (7.5.25) and (7.5.34) of [46] imply

$$p_n\left(x;q^{1/2},-q^{1/2},0,0|q\right) = H_n\left(x|q^2\right), \tag{4.6.30}$$

we arrive also at the following expansion formula [68]

$$\left(q\alpha^2;q^2\right)_\infty \mathcal{E}_q(x;\alpha) = \sum_{k=0}^{\infty} \frac{q^{k^2/4}\alpha^k}{(q;q)_k} \left(-\alpha^2 q^{k+2};q^2\right)_\infty H_k\left(x|q^2\right), \tag{4.6.31}$$

which may be compared with (3.4.15).

4.7. More Expansions and Integrals

The Al-Salam and Chihara polynomials are defined by

$$p_n(x;a,b) = p_n(x;a,b|q) = p_n(x;a,b,0,0|q) \tag{4.7.1}$$

$$= a^{-n}\,(ab;q)_n\ {}_3\varphi_2\left(\begin{matrix} q^{-n}, ae^{i\theta}, ae^{-i\theta} \\ ab, 0 \end{matrix}\ ;\ q,\ q\right), \quad x = \cos\theta,$$

see [7], [15], and [86]. They have the following generating function

$$\sum_{n=0}^{\infty} \frac{t^n}{(q;q)_n}\, p_n(x;a,b) = \frac{(at,bt;q)_\infty}{\left(te^{i\theta},te^{-i\theta};q\right)_\infty}, \tag{4.7.2}$$

which is a two parameter extension of the Rogers generation relation for the continuous q-Hermite polynomials

$$\sum_{n=0}^{\infty} \frac{t^n}{(q;q)_n}\, H_n(x|q) = \frac{1}{\left(te^{i\theta},te^{-i\theta};q\right)_\infty}. \tag{4.7.3}$$

An expansion of the $\mathcal{E}_q(x;\alpha)$ in special Al-Salam and Chihara polynomials $p_n(x;a,-a)$ is given by Eq. (4.6.29). Here we would like to note that this formula can also be derived on the basis of the connection coefficient relation

$$H_n(x|q) = \sum_{k=0}^{[n/2]} \frac{(q;q)_n\ a^{2k}}{\left(q^2;q^2\right)_k (q;q)_{n-2k}}\, p_{n-2k}(x;a,-a|q), \tag{4.7.4}$$

which follows from the generating functions (4.7.2) and (4.7.3); see Ex. 4.5 below. In a complete analogy with (4.5.4),

$$\left(q\alpha^2;q^2\right)_\infty \mathcal{E}_q(x;\alpha) = \sum_{n=0}^{\infty} \frac{q^{n^2/4}}{(q;q)_n}\,\alpha^n\, H_n(x|q) \tag{4.7.5}$$

$$= \sum_{n,k} q^{n^2/4}\alpha^n\, \frac{a^{2k}}{\left(q^2;q^2\right)_k (q;q)_{n-k}}\, p_{n-2k}(x;a,-a|q)$$

$$= \sum_m \frac{q^{m^2/4}\alpha^m}{(q;q)_m}\, p_m(x;a,-a|q) \sum_k q^{k(k-1)} \frac{\left(\alpha^2 a^2 q^{m+1}\right)^k}{(q;q)_k}$$

$$= \sum_{m=0}^{\infty} \frac{q^{m^2/4}\alpha^m}{(q;q)_m} \left(-\alpha^2 a^2 q^{m+1}; q^2\right)_\infty \; p_m(x; a, -a|q),$$

where we have used the Euler summation formula (A.2.2) in the last step.

Ismail and Stanton evaluated the following integral [**71**]

$$\int_0^\pi \mathcal{E}_q(\cos\theta;\alpha)\,\mathcal{E}_q(\cos\theta;\beta)\; p_m(\cos\theta; a, -a|q) \tag{4.7.6}$$

$$\times \frac{\left(e^{2i\theta},\, e^{-2i\theta};\, q\right)_\infty}{\left(a^2e^{2i\theta},\, a^2e^{-2i\theta};\, q^2\right)_\infty}\, d\theta$$

$$= 2\pi q^{m^2/4}\alpha^m \frac{\left(-\alpha\beta q^{(m+1)/2}; q\right)_\infty \left(-q^{(1-m)/2}\beta/\alpha; q\right)_m}{\left(q,\, -a^2q^m; q\right)_\infty \left(q\alpha^2,\, q\beta^2; q^2\right)_\infty}$$

$$\times {}_2\varphi_2\left(\begin{matrix} -q^{(m+1)/2}\alpha/\beta, -q^{(m+1)/2}\beta/\alpha \\ -q,\; -q^{(m+1)/2}\alpha\beta \end{matrix}; q, -q^{(m+1)/2}\alpha\beta a^2\right),$$

which is equivalent to the orthogonal expansion

$$\mathcal{E}_q(\cos\theta;\alpha)\,\mathcal{E}_q(\cos\theta;\beta) \tag{4.7.7}$$

$$= \sum_{m=0}^{\infty} q^{m^2/4}\alpha^m \frac{\left(-\alpha\beta q^{(m+1)/2}; q\right)_\infty \left(-q^{(1-m)/2}\beta/\alpha; q\right)_m}{\left(q\alpha^2,\, q\beta^2; q^2\right)_\infty\; (q;q)_m}$$

$$\times\, p_m(\cos\theta; a, -a|q)$$

$$\times {}_2\varphi_2\left(\begin{matrix} -q^{(m+1)/2}\alpha/\beta, -q^{(m+1)/2}\beta/\alpha \\ -q,\; -q^{(m+1)/2}\alpha\beta \end{matrix}; q, -q^{(m+1)/2}\alpha\beta a^2\right)$$

in view of the orthogonality relation [**15**], [**46**]

$$\int_0^\pi \frac{p_m(\cos\theta; a, b)\; p_n(\cos\theta; a, b)\left(e^{2i\theta},\, e^{-2i\theta};\, q\right)_\infty}{\left(ae^{i\theta}, ae^{-i\theta}, be^{i\theta}, be^{-i\theta}; q\right)_\infty}\, d\theta \tag{4.7.8}$$

$$= 2\pi \frac{(q, ab; q)_n}{(q, ab; q)_\infty}\, \delta_{mn}.$$

The proof of (4.7.6) is similar to the proof of (4.1.3). Denoting the integral in the left side of (4.7.6) as I_m and using expansion (4.1.1), one finds

$$I_m = \sum_{n=0}^{\infty} q^{n^2/4}\alpha^n \frac{\left(-\alpha\beta q^{(n+1)/2}; q\right)_\infty \left(-q^{(1-n)/2}\beta/\alpha; q\right)_n}{\left(q\alpha^2, q\beta^2; q^2\right)_\infty} \tag{4.7.9}$$

$$\times \int_0^\pi \frac{H_n(\cos\theta|q)\, p_m(\cos\theta; a, -a|q)}{(q;q)_n} \frac{\left(e^{2i\theta},\, e^{-2i\theta};\, q\right)_\infty}{\left(a^2e^{2i\theta},\, a^2e^{-2i\theta};\, q^2\right)_\infty}\, d\theta.$$

The last integral can be evaluated with the help of the connection coefficient formula (4.7.4),

$$\int_0^\pi \frac{H_n(\cos\theta|q)\, p_m(\cos\theta; a, -a|q)}{(q;q)_n} \frac{\left(e^{2i\theta},\, e^{-2i\theta};\, q\right)_\infty}{\left(a^2e^{2i\theta},\, a^2e^{-2i\theta};\, q^2\right)_\infty}\, d\theta \tag{4.7.10}$$

$$= \sum_{k=0}^{[n/2]} \frac{a^{2k}}{(q^2;q^2)_k\,(q;q)_{n-2k}}$$

$$\times \int_0^\pi p_m(\cos\theta;a,-a|q)\,p_{n-2k}(\cos\theta;a,-a|q)\,\frac{\left(e^{2i\theta},\ e^{-2i\theta};\ q\right)_\infty}{\left(a^2e^{2i\theta},\ a^2e^{-2i\theta};\ q^2\right)_\infty}\,d\theta$$

$$= 2\pi\frac{\left(q,-a^2;q\right)_m}{\left(q,-a^2;q\right)_\infty}\sum_{k=0}^{[n/2]}\frac{a^{2k}}{(q^2;q^2)_k\,(q;q)_{n-2k}}\,\delta_{m,n-2k}$$

in view of the orthogonality relation (4.7.8). Therefore

$$I_m = \frac{2\pi q^{m^2/4}\alpha^m}{\left(q,-a^2q^m;q\right)_\infty\left(q\alpha^2,q\beta^2;q^2\right)_\infty} \times\sum_{k=0}^{\infty}\frac{q^{(m+2k)^2/4}\alpha^{m+2k}a^{2k}}{(q,-q;q)_k} \times\left(-\alpha\beta q^{(1+m)/2+k};q\right)_\infty\left(-q^{(1-m)2-k}\beta/\alpha;q\right)_{m+2k}. \tag{4.7.11}$$

This series can be transformed to the ${}_2\varphi_2$ in the right side of (4.7.6) with the help of the q-shifted factorial relations at the end of Section 4.3. We leave the details to the reader.

4.8. Second Proof of Ismail, Rahman and Zhang Formula

In this section we shall present an independent proof of the expansion formulas (4.6.19) and (4.6.23) in the continuous q-Jacobi polynomials. Let us start from the classical case $q = 1$.

4.8.1. Expansion in Jacobi Polynomials. A simple proof of the expansion of the plane wave in Jacobi polynomials (4.6.20) can be given in the following manner. Consider

$$e^{ir(x+y)} = \sum_{n=0}^{\infty} c_n\ P_n^{(\alpha,\beta)}(x), \tag{4.8.1}$$

where

$$d_n^2\ c_n = \int_{-1}^{1} e^{ir(x+y)}P_n^{(\alpha,\beta)}(x)\,(1-x)^\alpha\,(1+x)^\beta\ dx \tag{4.8.2}$$

and

$$d_n^2 = \int_{-1}^{1}\left(P_n^{(\alpha,\beta)}(x)\right)^2(1-x)^\alpha\,(1+x)^\beta\ dx = 2^{\alpha+\beta+1}\frac{\Gamma(\alpha+n+1)\,\Gamma(\beta+n+1)}{(\alpha+\beta+2n+1)\,n!\,\Gamma(\alpha+\beta+n+1)} \tag{4.8.3}$$

is the $\mathcal{L}^2$-norm of the Jacobi polynomials $P_n^{(\alpha,\beta)}(x)$ when $\alpha > -1$ and $\beta > -1$. By the Rodrigues formula

$$P_n^{(\alpha,\beta)}(x) = \frac{(-1)^n}{2^n n!}(1-x)^{-\alpha}(1+x)^{-\beta} \times \frac{d^n}{dx^n}\left[(1-x)^{\alpha+n}(1+x)^{\beta+n}\right], \tag{4.8.4}$$

and integrating by parts

$$\begin{aligned} d_n^2\, c_n &= \frac{(-1)^n}{2^n n!}\int_{-1}^{1} e^{ir(x+y)}\left((1-x)^{\alpha+n}(1+x)^{\beta+n}\right)^{(n)} dx \\ &= \frac{(-1)^n}{2^n n!}\left[e^{ir(x+y)}\left((1-x)^{\alpha+n}(1+x)^{\beta+n}\right)^{(n)}\Big|_{-1}^{1} \right. \\ &\left. - \int_{-1}^{1}\left(e^{ir(x+y)}\right)'\left((1-x)^{\alpha+n}(1+x)^{\beta+n}\right)^{(n-1)} dx\right] \\ &= \frac{(-1)^{n-1}}{2^n n!}(ir)\int_{-1}^{1} e^{ir(x+y)}\left((1-x)^{\alpha+n}(1+x)^{\beta+n}\right)^{(n-1)} dx \\ &= \ldots \\ &= \frac{(ir)^n}{2^n n!}\int_{-1}^{1} e^{ir(x+y)}(1-x)^{\alpha+n}(1+x)^{\beta+n}\, dx. \end{aligned} \tag{4.8.5}$$

In order to evaluate the last integral, one can choose $y = 1$ and expand $e^{ir(x+1)}$ in Taylor's series:

$$\begin{aligned} d_n^2\, c_n &= \frac{(ir)^n}{2^n n!}\sum_{k=0}^{\infty}\frac{(ir)^k}{k!}\int_{-1}^{1}(1-x)^{\alpha+n}(1+x)^{\beta+n+k}\, dx \\ &= \frac{(ir)^n}{2^n n!}\frac{\Gamma(\beta+n+1)}{\Gamma(\alpha+\beta+2n+2)}\sum_{k=0}^{\infty}\frac{(\beta+n+1)_k}{(\alpha+\beta+2n+2)_k\, k!}(ir)^k \end{aligned} \tag{4.8.6}$$

with $(a)_k = a(a+1)\cdot \ldots \cdot (a+n-1) = \Gamma(a+n)/\Gamma(a)$, in view of (4.8.3). As a result we arrive at

$$e^{ir(x+1)} = \sum_{n=0}^{\infty}\frac{(\alpha+\beta+1)_n}{(\alpha+\beta+1)_{2n}}(2ir)^n\, P_n^{(\alpha,\beta)}(x) \times {}_1F_1\left(\begin{matrix}\beta+n+1\\ \alpha+\beta+2n+2\end{matrix};\, 2ir\right) \tag{4.8.7}$$

and the proof is complete.

4.8.2. Expansion in Continuous q-Jacobi Polynomials. We can repeat practically the same consideration in the q-case. The corresponding expansion in the Askey–Wilson polynomials has the form

$$\mathcal{E}_q(x,y;\alpha) = \sum_{n=0}^{\infty} c_n\, p_n(x;a,b,c,d) \tag{4.8.8}$$

with

$$d_n^2 c_n = \int_{-1}^{1} \mathcal{E}_q(x, y; \alpha)\ p_n(x; a, b, c, d)\ \rho(x)\ dx, \tag{4.8.9}$$

where $d_n^2 = d_n^2(a, b, c, d)$ and $\rho(x) = \rho(x; a, b, c, d)$ are the $\mathcal{L}^2$-norm and the weight function in the orthogonality relation (4.6.2), respectively. The Rodrigues-type formula for the Askey–Wilson polynomials [**15**], [**18**], [**111**] is

$$\begin{aligned} p_n(x; a, b, c, d) &= \frac{(-1)^n (1-q)^n q^{n(n-1)/4}}{2^n \rho(z; a, b, c, d)} \\ &\times \left(\frac{\delta}{\delta x(z)}\right)^n \rho\left(z; aq^{n/2}, bq^{n/2}, cq^{n/2}, dq^{n/2}\right), \end{aligned} \tag{4.8.10}$$

where

$$\rho(z; a, b, c, d) = \frac{\left(q^{2z}, q^{1-2z}; q\right)_\infty q^{-z}}{\left(aq^z, aq^{-z}, bq^z, bq^{-z}, cq^z, cq^{-z}, dq^z, dq^{-z}; q\right)_\infty} \tag{4.8.11}$$

and $x = (q^z + q^{-z})/2$, $q^z = e^{i\theta}$. Hence, (4.8.9) can be rewritten as

$$\begin{aligned} d_n^2 c_n &= (-1)^n 2^{-n} (1-q)^n q^{n(n-1)/4} \frac{q^{1/2} \log q}{2(1-q)} \\ &\times \int_C \mathcal{E}_q(x, y; \alpha) \left(\frac{\delta}{\delta x(z)}\right)^n \rho\left(z; aq^{n/2}, bq^{n/2}, cq^{n/2}, dq^{n/2}\right)\ \delta x(z)\ dz, \end{aligned} \tag{4.8.12}$$

where the contour C is given in Figure 3.1 such that $z = i\theta / \log q^{-1}$; $-\pi \leq \theta \leq \pi$ and

$$dx = \frac{q^{1/2} \log q}{1-q}\ \delta x(z)\ dz. \tag{4.8.13}$$

Consider

$$\delta\left[f\left(z + \frac{1}{2}\right) g(z)\right] = f(z)\, \delta g(z) + g\left(z + \frac{1}{2}\right) \delta f\left(z + \frac{1}{2}\right) \tag{4.8.14}$$

with

$$f(z) = \mathcal{E}_q(x(z), y; \alpha) \tag{4.8.15}$$

and

$$\begin{aligned} g(z) &= \left(\frac{\delta}{\delta x(z)}\right)^{n-1} \rho\left(z; aq^{n/2}, bq^{n/2}, cq^{n/2}, dq^{n/2}\right) \\ &= \left(\frac{-2}{1-q}\right)^{n-1} q^{-(n-1)(n-2)/4} p_{n-1}\left(x(z); aq^{1/2}, bq^{1/2}, cq^{1/2}, dq^{1/2}\right) \\ &\times \rho\left(z; aq^{1/2}, bq^{1/2}, cq^{1/2}, dq^{1/2}\right) \end{aligned} \tag{4.8.16}$$

in view of (4.8.10). Then

$$\int_C f(z)\, \delta g(z)\ dz = \int_C \delta\left[f\left(z + \frac{1}{2}\right) g(z)\right]\ dz \tag{4.8.17}$$

$$-\int_C g\left(z+\frac{1}{2}\right)\delta f\left(z+\frac{1}{2}\right)\,dz$$

and with the help of the Cauchy theorem and the periodicity property $q^z = q^{z+T}$, $T = 2\pi i/\log q^{-1}$ one gets

$$\int_C \delta\left[f\left(z+\frac{1}{2}\right)g(z)\right]\,dz = \int_D \left[f\left(z+\frac{1}{2}\right)g(z)\right]\,dz = 0, \qquad (4.8.18)$$

$$\int_C g\left(z+\frac{1}{2}\right)\delta f\left(z+\frac{1}{2}\right)\,dz = \int_C g(z)\,\delta f(z)\,dz, \qquad (4.8.19)$$

where D is the boundary of the rectangle on Figure 3.1 oriented counter-clockwise, because the integrands do not have singularities inside this rectangle when $\max(|a|,|b|,|c|,|d|) < 1$. Applying this "integration by parts" formula n times,

$$\begin{aligned}
&\int_C \mathcal{E}_q(x,y;\alpha)\left(\frac{\delta}{\delta x(z)}\right)^n \rho\left(z;aq^{n/2},bq^{n/2},cq^{n/2},dq^{n/2}\right)\,\delta x(z)\,dz \\
&= -\int_C \frac{\delta}{\delta x(z)}\mathcal{E}_q(x,y;\alpha) \\
&\times \left(\frac{\delta}{\delta x(z)}\right)^{n-1} \rho\left(z;aq^{n/2},bq^{n/2},cq^{n/2},dq^{n/2}\right)\,\delta x(z)\,dz \\
&= \dots \\
&= (-1)^n \int_C \left(\frac{\delta}{\delta x(z)}\right)^n \mathcal{E}_q(x,y;\alpha) \qquad (4.8.20)\\
&\times\rho\left(z;aq^{n/2},bq^{n/2},cq^{n/2},dq^{n/2}\right)\,\delta x(z)\,dz.
\end{aligned}$$

By (2.3.13)

$$\left(\frac{\delta}{\delta x}\right)^n \mathcal{E}_q(x,y;\alpha) = \left(\frac{2q^{1/4}\alpha}{1-q}\right)^n \mathcal{E}_q(x,y;\alpha) \qquad (4.8.21)$$

and as a result,

$$d_n^2 c_n = q^{n^2/4}\alpha^n \int_{-1}^{1} \mathcal{E}_q(x,y;\alpha)\,\rho\left(x;aq^{n/2},bq^{n/2},cq^{n/2},dq^{n/2}\right)\,dx. \qquad (4.8.22)$$

In a complete analogy with the classical case, we can evaluate this integral when $y = \pm\eta$ and the choice of the parameters $b = aq^{1/2}$, $c \to -c$, $d = -cq^{1/2}$ corresponds to the case of the continuous q-Jacobi polynomials. On the other hand, in view of (2.3.29), we can choose $y = 0$ and use (2.3.27) instead, which gives

$$\begin{aligned}
d_n^2 c_n\Big|_{y=0} &= q^{n^2/4}\alpha^n \frac{\left(\alpha;q^{1/2}\right)_\infty}{\left(q\alpha^2;q^2\right)_\infty}\sum_{k=0}^{\infty}\frac{\alpha^k}{(q;q)_k} \qquad (4.8.23)\\
&\times \int_0^\pi \frac{\left(-q^{1/4}e^{i\theta},-q^{1/4}e^{-i\theta};q^{1/2}\right)_\infty}{\left(-q^{1/4+k/2}e^{i\theta},-q^{1/4+k/2}e^{-i\theta};q^{1/2}\right)_\infty}
\end{aligned}$$

$$\times \frac{\left(e^{2i\theta}, e^{-2i\theta}; q\right)_\infty}{\left(aq^{n/2}e^{i\theta}, aq^{n/2}e^{-i\theta}, aq^{n/2+1/2}e^{i\theta}, aq^{n/2+1/2}e^{-i\theta}; q\right)_\infty}$$
$$\times \frac{d\theta}{\left(-cq^{n/2}e^{i\theta}, -cq^{n/2}e^{-i\theta}, -cq^{n/2+1/2}e^{i\theta}, -cq^{n/2+1/2}e^{-i\theta}; q\right)_\infty}.$$

After transformation of the infinite products in the integrand to the base $q^{1/2}$ with the help of (1.4.12) and (1.4.13), this integral can be evaluated as a special case of the Askey–Wilson integral (A.4.1):

$$\int_0^\pi \frac{\left(e^{2i\theta}, e^{-2i\theta}; q^{1/2}\right)_\infty}{\left(q^{1/4}e^{i\theta}, q^{1/4}e^{-i\theta}, -q^{1/4+k/2}e^{i\theta}, -q^{1/4+k/2}e^{-i\theta}; q^{1/2}\right)_\infty} \tag{4.8.24}$$
$$\times \frac{d\theta}{\left(aq^{n/2}e^{i\theta}, aq^{n/2}e^{-i\theta}, -cq^{n/2}e^{i\theta}, -cq^{n/2}e^{-i\theta}; q^{1/2}\right)_\infty}$$
$$= \frac{2\pi \left(acq^{n+1/2}; q^{1/2}\right)_\infty \left(-q^{1/2}, cq^{n/2+1/4}, -aq^{n/2+1/4}; q^{1/2}\right)_k}{\left(q, a^2q^{n+1/2}, c^2q^{n+1/2}; q\right)_\infty \left(-acq^n; q^{1/2}\right) \left(acq^{n+1/2}; q\right)_k}.$$

Taking into account that

$$d_n^2 = d_n^2\left(a, aq^{1/2}, -c, -cq^{1/2}\right) \tag{4.8.25}$$
$$= \frac{2\pi \left(a^2c^2q^n; q\right)_\infty}{\left(q^{n+1}, a^2q^{n+1/2}, c^2q^{n+1/2}; q\right)_\infty \left(1 - acq^n\right) \left(-acq^n; q^{1/2}\right)_\infty^2},$$

from (4.8.23) and (4.8.24) we finally obtain

$$c_n\big|_{y=0} = q^{n^2/4}\alpha^n \frac{\left(\alpha^2; q^2\right)_\infty}{\left(-\alpha; q^{1/2}\right)_\infty} \frac{\left(a^2c^2; q\right)_n}{\left(a^2c^2; q\right)_{2n} (q; q)_n} \tag{4.8.26}$$
$$\times {}_2\varphi_1\left(\begin{matrix} cq^{1/4+k/2}, \ -aq^{1/4+k/2} \\ acq^{k+1/2} \end{matrix} ; \ q^{1/2}, \ \alpha \right),$$

which results in the Ismail, Rahman and Zhang expansion formula (4.6.19). It is worth noting that in the current proof of this formula we did not use a single summation or transformation formula for the basic hypergeometric series.

4.9. Miscellaneous Results

In this section we derive a few more integrals involving the basic exponential functions on a q-quadratic grid. Our first integral,

$$\int_0^\pi \mathcal{E}_q(\cos\theta; \alpha)\, \mathcal{E}_q(\cos\theta; \beta)\, \mathcal{E}_q(\cos\theta; \gamma) \left(e^{2i\theta}, e^{-2i\theta}; q\right)_\infty d\theta \tag{4.9.1}$$
$$= 2\pi \sum_{m=0}^\infty q^{m^2/2} (\alpha\gamma)^m \frac{\left(-\alpha\beta q^{(m+1)/2}; q\right)_\infty \left(-q^{(1-m)/2}\beta/\alpha; q\right)_m}{(q; q)_\infty \left(q\alpha^2, q\beta^2, q\gamma^2; q\right)_\infty (q; q)_m},$$

follows directly from (4.1.2) and (3.4.15). The second integral,

$$\int_0^\pi \mathcal{E}_q(\cos\theta;\alpha)\,\mathcal{E}_q(\cos\theta;\beta)\,\frac{\left(e^{2i\theta},e^{-2i\theta};q\right)_\infty}{\left(ae^{i\theta},ae^{-i\theta};q\right)_\infty}\,d\theta \tag{4.9.2}$$

$$= 2\pi\sum_{m=0}^{\infty} q^{m^2/4}(\alpha a)^m\,\frac{\left(-\alpha\beta q^{(m+1)/2};q\right)_\infty\left(-q^{(1-m)/2}\beta/\alpha;q\right)_m}{(q;q)_\infty\left(q\alpha^2,q\beta^2;q\right)_\infty(q;q)_m},$$

arises from (4.1.2) and (4.7.3) when $|a|<1$. The third integral is

$$\int_0^\pi \frac{\mathcal{E}_q(\cos\theta;\alpha)\,\mathcal{E}_q(\cos\theta;\beta)\,\left(e^{2i\theta},e^{-2i\theta};q\right)_\infty}{\left(re^{i\theta+i\varphi},re^{i\theta-i\varphi},re^{i\varphi-i\theta},re^{-i\theta-i\varphi};q\right)_\infty}\,d\theta \tag{4.9.3}$$

$$= 2\pi\sum_{m=0}^{\infty} q^{m^2/4}(\alpha r)^m H_m(\cos\varphi|q)\,\frac{\left(-\alpha\beta q^{(m+1)/2};q\right)_\infty\left(-q^{(1-m)/2}\beta/\alpha;q\right)_m}{\left(q,r^2;q\right)_\infty\left(q\alpha^2,q\beta^2;q^2\right)_\infty(q;q)_m}$$

with $|r|<1$. This follows directly from (4.1.2) and (3.5.4). The limit $r\to 1^-$ gives (4.1.1) again due to Lemma 6.6; see also Ex. 6.13. Introducing new parameters $a=re^{i\varphi}$, $b=e^{-i\varphi}$, one can rewrite (4.9.3) as

$$\int_0^\pi \frac{\mathcal{E}_q(\cos\theta;\alpha)\,\mathcal{E}_q(\cos\theta;\beta)\,\left(e^{2i\theta},e^{-2i\theta};q\right)_\infty}{\left(ae^{i\theta},ae^{-i\theta},be^{i\theta},be^{-i\theta};q\right)_\infty}\,d\theta \tag{4.9.4}$$

$$= 2\pi\sum_{m=0}^{\infty} q^{m^2/4}\alpha^m(ab)^{m/2}\,H_m\left(\left.\frac{a+b}{2\sqrt{ab}}\right|q\right)$$

$$\times\frac{\left(-\alpha\beta q^{(m+1)/2};q\right)_\infty\left(-q^{(1-m)/2}\beta/\alpha;q\right)_m}{(q,ab;q)_\infty\left(q\alpha^2,q\beta^2;q^2\right)_\infty(q;q)_m},$$

when $\max(|a|,|b|)<1$. The limit $b\to 0$ results in (4.9.2).

The next integral is due to Ismail and Stanton [**74**],

$$\int_0^\pi \frac{\mathcal{E}_q(\cos\theta;\alpha)\,\mathcal{E}_q(\cos\theta;\beta)\,\left(e^{2i\theta},e^{-2i\theta};q\right)_\infty}{\left(ae^{2i\theta},ae^{-2i\theta},be^{2i\theta},be^{-2i\theta};q^2\right)_\infty}\,d\theta \tag{4.9.5}$$

$$= 2\pi\sum_{m,n=0}^{\infty}(-1)^n q^{m^2/4}\frac{(\alpha\beta)^m(ab)^{(m+n)/2}}{(q;q)_n\left(q^2;q^2\right)_m}\,H_{m+n}\left(\left.\frac{a+b}{2\sqrt{ab}}\right|q^2\right)$$

$$\times\frac{\left(-q^{1/2}\alpha\beta;q\right)_\infty\left(-q^{1/2}\alpha/\beta,-q^{1/2}\beta/\alpha;q\right)_m}{(q;q)_\infty\left(q\alpha^2,q\beta^2,ab;q^2\right)_\infty\left(-q^{1/2}\alpha\beta;q\right)_m},$$

when $\max(|a|,|b|)<1$. Let $a=re^{i\varphi}$, $b=re^{-i\varphi}$, $|r|<1$ and denote the integral in the left side of (4.9.5) as I. Then, in view of (4.1.1) and (3.5.4),

$$\left(q\alpha^2,q\beta^2,r^2;q^2\right)_\infty\,I \tag{4.9.6}$$

$$=\sum_{n=0}^{\infty}\frac{r^n}{\left(q^2;q^2\right)_n}\,H_n\left(\cos\varphi|q^2\right)$$

$$\times\sum_{m=0}^{\infty} q^{m^2/4}\alpha^m\left(-\alpha\beta q^{(m+1)/2};q\right)_\infty\frac{\left(-q^{(1-m)/2}\beta/\alpha;q\right)_m}{(q;q)_m}$$

$$\times \int_0^\pi H_n\left(\cos 2\theta|q^2\right) H_m\left(\cos\theta|q\right) \left(e^{2i\theta}, e^{-2i\theta}; q\right)_\infty d\theta.$$

But

$$H_n\left(\cos 2\theta|q^2\right) = \sum_{k=0}^{n} (-1)^k \frac{\left(q^2;q^2\right)_n}{(q;q)_k \left(q^2;q^2\right)_{n-k}} H_{2n-2k}\left(\cos\theta|q\right), \tag{4.9.7}$$

see Ex. 4.14, and due to the orthogonality relation (3.5.5) one gets

$$\begin{aligned} &\frac{(q;q)_\infty}{2\pi} \frac{\left(q\alpha^2, q\beta^2, r^2; q^2\right)_\infty}{\left(-q^{1/2}\alpha\beta; q\right)_\infty} I \\ &= \sum_{m,n=0}^{\infty} (-1)^n q^{m^2/4} r^{m+n} \frac{(\alpha\beta)^m}{(q;q)_n \left(q^2;q^2\right)_m} H_{m+n}\left(\cos\varphi|q^2\right) \\ &\quad \times \frac{\left(-q^{1/2}\alpha/\beta, -q^{1/2}\beta/\alpha; q\right)_m}{\left(-q^{1/2}\alpha\beta; q\right)_m}, \end{aligned} \tag{4.9.8}$$

where we interchanged the order of summation, substituted $n = k + m$ and finally replaced k by n. Eq. (4.9.8) is equivalent to (4.9.5).

4.10. Exercises for Chapter 4

(1) Derive the orthogonality relations for the basic trigonometric functions from (4.4.3)–(4.4.5).

(2) Obtain (4.5.2) as the limiting case $\beta \to 0$ of (4.1.7).

(3) Derive the orthogonality relation for the continuous q-Hermite polynomials (3.5.5) as a limiting case $\alpha \to 0$ of (4.2.3).

(4) Obtain the generating function (3.4.15) as the limiting case $\nu \to \infty$ of (4.5.2).

(5) Substitute the $\mathcal{E}$'s from (4.5.2) in (4.1.3) with $\alpha = i\omega$, $\beta = i\varpi$, $m = 0$, $\gamma = q^\nu$ and derive the following expansion [**71**]

$$\begin{aligned} &\sum_{n=0}^{\infty} (-1)^n q^{n^2/2} \frac{\left(1-q^{\nu+n}\right)\left(q^{2\nu};q\right)_n}{\left(1-q^\nu\right)(q;q)_n} J_{\nu+n}^{(2)}(2\omega;q)\ J_{\nu+n}^{(2)}(2\varpi;q) \\ &= (\omega\varpi)^\nu \frac{\left(q^{\nu+1},\ q^{\nu+1},\ \omega\varpi q^{1/2}; q\right)_\infty}{(q,\ q;q)_\infty} \\ &\times {}_2\varphi_2\left(\begin{matrix} -q^{1/2}\omega/\varpi, -q^{1/2}\varpi/\omega \\ q^{\nu+1},\ q^{1/2}\omega\varpi \end{matrix}; q, \omega\varpi q^{\nu+1/2}\right). \end{aligned}$$

Show that if ω and ϖ are replaced by $(1-q)\,\omega$ and $(1-q)\,\varpi$, respectively, and $q \to 1^-$, then the last expansion reduces to the multiplication formula

$$\sum_{n=0}^{\infty} (-1)^n \frac{(\nu+n)\,(2\nu)_n}{\nu\ n!} J_{\nu+n}(2\omega)\ J_{\nu+n}(2\varpi)$$

$$= \left(\frac{\omega \varpi}{\omega + \varpi} \right)^{\nu} \frac{J_\nu \left(2 \left(\omega + \varpi \right) \right)}{\Gamma \left(\nu + 1 \right)},$$

which is the special case $\varphi = 0$ of the Gegenbauer addition theorem; see Eq. (11.4.2) of [**163**].

(6) The Al-Salam and Chihara polynomials given by (4.7.1),

$$p_n(x; a, b) = p_n(x; a, b|q) = p_n(x; a, b, 0, 0|q),$$

have the generating function (4.7.2), which is a two parameter extension of the generation relation for the continuous q-Hermite polynomials (4.7.3). Expanding both sides of the identity

$$\frac{1}{\left(te^{i\theta}, te^{-i\theta}; q\right)_\infty} = \frac{\left(\beta te^{i\varphi}, \beta te^{-i\varphi}; q\right)_\infty}{\left(te^{i\theta}, te^{-i\theta}; q\right)_\infty} \frac{1}{\left(\beta te^{i\varphi}, \beta te^{-i\varphi}; q\right)_\infty}$$

in the powers of t with the help of these generation relations and equating coefficients of the t^n show that

$$\frac{H_n(x|q)}{(q;q)_n} = \sum_{k=0}^{n} \frac{\beta^k}{(q;q)_k \, (q;q)_{n-k}} \, p_{n-k}(x; \beta e^{i\varphi}, \beta e^{-i\varphi}) \; H_k(y|q) \;,$$

where $x = \cos\theta$ and $y = \cos\varphi$. Derive (4.7.4) as a special case $y = 0$ of this relation. Multiply both sides of the last equation by $q^{n^2/4}\alpha^n$, sum over n from zero to infinity and, changing the order of summation, obtain

$$\sum_{n=0}^{\infty} \frac{q^{n^2/4}\alpha^n}{(q;q)_n} H_n\left(x|q\right) = \sum_{n=0}^{\infty} \frac{q^{n^2/4}\alpha^n}{(q;q)_n} p_n(x; \beta e^{i\varphi}, \beta e^{-i\varphi})$$
$$\times \sum_{k=0}^{\infty} \frac{q^{k^2/4}\left(\beta q^{n/2}\right)^k}{(q;q)_k} H_k\left(y|q\right).$$

As a result derive the following expansion formula [**43**], [**69**]

$$\left(q\alpha^2; q^2\right)_\infty \mathcal{E}_q\left(x; \alpha\right) = \sum_{n=0}^{\infty} \frac{q^{n^2/4}\alpha^n}{(q;q)_n} p_n(x; \beta e^{i\varphi}, \beta e^{-i\varphi})$$
$$\times \left(q\beta^2 q^n; q^2\right)_\infty \mathcal{E}_q\left(y; \beta q^{n/2}\right),$$

where $x = \cos\theta$ and $y = \cos\varphi$, with the help of the generating relation (3.4.15).

(7) Replace ω by $\left(1 - q^{1/2}\right)\omega$ and obtain expansion (4.6.20) as a limit $q \to 1^-$ of (4.6.23).

(8) Prove (4.6.11) and (4.6.30).

(9) Prove the following connection coefficient formulas [**73**]:

$$\frac{\left(ae^{i\theta},\; ae^{-i\theta}; q\right)_n}{\left(aq^{-1/4}; q^{1/2}\right)_{2n}}$$

$$= {}_4\varphi_3\left(\begin{matrix} q^{-n/2},\ -q^{-n/2},\ q^{1/4}e^{i\theta},\ q^{1/4}e^{-i\theta} \\ -q^{1/2},\ aq^{-1/4},\ q^{3/4-n}/a \end{matrix};q^{1/2},q^{1/2}\right),$$

$$\frac{\left(q^{1/4}e^{i\theta},\ q^{1/4}e^{-i\theta};q^{1/2}\right)_n}{\left(q^{1/4}a,\ q^{1/4}/a;q^{1/2}\right)_n}$$

$$= {}_3\varphi_2\left(\begin{matrix} q^{-n},\ ae^{i\theta},\ ae^{-i\theta} \\ aq^{(1-2n)/4},\ aq^{(3-2n)/4} \end{matrix};q,q\right),$$

and

$$\left(ae^{i\theta},\ ae^{-i\theta};q\right)_n = \sum_{k=0}^{n}(-a)^k q^{k(k-1)/2}\frac{(q;q)_n\left(-a^2q^k;q^2\right)_{n-k}}{(q;q)_k\,(q;q)_{n-k}}$$
$$\times\frac{1+e^{2i\theta}}{1+q^{-k}e^{2i\theta}}\left(-q^{-k}e^{2i\theta};q^2\right)_k e^{-ik\theta},$$

$$\left(q^{1/4}e^{i\theta},\ q^{1/4}e^{-i\theta};q\right)_n = \sum_{k=0}^{n}(-1)^k q^{k^2/4}\frac{(q;q)_n\left(-q^{1/2};q\right)_{n-k}}{(q;q)_k\,(q;q)_{n-k}}$$
$$\times\frac{1+e^{2i\theta}}{1+q^{-k}e^{2i\theta}}\left(-q^{-k}e^{2i\theta};q^2\right)_k e^{-ik\theta}.$$

(10) Show that [**46**]

$$J_\nu^{(2)}(x;q) = \left(\frac{x}{2}\right)^\nu\frac{\left(q^{\nu+1};q\right)_\infty}{(q;q)_\infty}\left(ix/2;q^{1/2}\right)_\infty$$
$$\times{}_2\varphi_1\left(\begin{matrix} q^{\nu/2+1/4},\ -q^{\nu/2+1/4} \\ q^{\nu+1/2} \end{matrix};q^{1/2},\ \frac{ix}{2}\right).$$

[*Hint*: Consider (4.5.2) as a special case of (4.6.19) with the help of (4.6.11).]

(11) Use the Rodrigues formula (4.8.4), the differentiation formula

$$\frac{d}{dx}P_n^{(\alpha,\beta)}(x) = \frac{1}{2}(\alpha+\beta+n+1)\ P_{n-1}^{(\alpha+1,\beta+1)}(x),$$

and integration by parts, in order to prove the orthogonality relation for the Jacobi polynomials:

$$\int_{-1}^{1}P_n^{(\alpha,\beta)}(x)\,P_m^{(\alpha,\beta)}(x)\ (1-x)^\alpha(1+x)^\beta\ dx$$
$$= 2^{\alpha+\beta+1}\frac{\Gamma(\alpha+n+1)\,\Gamma(\beta+n+1)}{(\alpha+\beta+2n+1)\,n!\Gamma(\alpha+\beta+n+1)}\ \delta_{mn},$$

where $\alpha > -1$ and $\beta > -1$.

(12) Use the Rodrigues-type formula (4.8.10), the difference differentiation formula

$$\frac{\delta}{\delta x} p_n(x(z); a, b, c, d) = -\frac{2q^{(n+1)/2}}{1-q}\left(1-q^{-n}\right)\left(1-abcdq^{n-1}\right)$$
$$\times p_{n-1}\left(x; aq^{1/2}, bq^{1/2}, cq^{1/2}, dq^{1/2}\right),$$

and an analog of integration by parts, to prove the orthogonality relation for the Askey–Wilson polynomials (4.6.2).

(13) An extension of the plane wave expansion (4.5.1) is **[163]**

$$(\sin\theta\sin\varphi)^{-\nu+1/2} J_{\nu-1/2}(r\sin\theta\sin\varphi)\, e^{ir\cos\theta\cos\varphi}$$
$$= \frac{2^{2\nu}}{\sqrt{2\pi r}}\Gamma^2(\nu)\sum_{n=0}^{\infty} i^n \frac{n!\,(\nu+n)}{\Gamma(2\nu+n)} J_{\nu+n}(r)\, C_n^{\nu}(\cos\theta)\, C_n^{\nu}(\cos\varphi).$$

Derive the following q-analog of this formula **[62]**

$$\frac{(q;q)_\infty (2/r)^\nu}{(-r^2/4;q^2)_\infty (q^\nu;q)_\infty}$$
$$\times\sum_{n=0}^{\infty} i^n \frac{(1-q^{\nu+n})(q;q)_n}{(q^{2\nu};q)_n}\, q^{n(n+2\nu)/4}$$
$$\times J_{\nu+n}^{(2)}(r;q)\ C_n(\cos\theta; q^\nu|\, q)\ C_n(\cos\varphi; q^\nu|\, q)$$
$$= \sum_{n=0}^{\infty} \frac{q^{n^2/4}}{(q;q)_n}\left(\frac{r}{2}\right)^n \left(-iq^{(1+\nu-n)/2}e^{i\theta+i\varphi}, -iq^{(1-\nu-n)/2}e^{-i\theta-i\varphi}; q\right)_n$$
$$\times {}_4\varphi_3\left(\begin{matrix} q^{-n}, q^\nu e^{2i\theta}, q^\nu e^{2i\varphi}, q^\nu \\ iq^{(1+\nu-n)/2}e^{i\theta+i\varphi}, -iq^{(1+\nu-n)/2}e^{i\theta+i\varphi}, q^{2\nu} \end{matrix}; q, q\right)$$

and its alternate form

$$\sum_{n=0}^{\infty} i^n \frac{(1-q^{\nu+n})(q;q)_n}{(q^{2\nu};q)_n}\, q^{n(n+2\nu)/4}$$
$$\times J_{\nu+n}^{(2)}(r;q)\ C_n(\cos\theta; q^\nu|\, q)\ C_n(\cos\varphi; q^\nu|\, q)$$
$$= \frac{\left(-r^2/4;q^2\right)_\infty (q^\nu, q^\nu, -q;q)_\infty}{(q, q^{2\nu}, -q^{1+\nu};q)_\infty}$$
$$\times\sum_{n=0}^{\infty} \frac{q^{n(\nu+n/4)}}{(q;q)_n}\left(-iq^{(1-\nu-n)/2}e^{i\theta+i\varphi}, -iq^{(1-\nu-n)/2}e^{-i\theta-i\varphi}; q\right)_n$$
$$\times\left(\frac{r}{2}\right)^{n+\nu} \frac{\left(iq^{(1+3\nu+n)/2}e^{i\theta-i\varphi}, iq^{(1+3\nu+n)/2}e^{i\varphi-i\theta}; q\right)_\infty}{\left(iq^{(1+\nu+n)/2}e^{i\theta-i\varphi}, iq^{(1+\nu+n)/2}e^{i\varphi-i\theta}; q\right)_\infty}$$
$$\times {}_8W_7\left(-q^{-\nu}; q^\nu, iq^{(1+\nu+n)/2}e^{i\theta+i\varphi}, iq^{(1+\nu+n)/2}e^{-i\theta-i\varphi},\right.$$
$$\left. iq^{(1-\nu-n)/2}e^{i\varphi-i\theta}, iq^{(1-\nu-n)/2}e^{i\theta-i\varphi}; q, q^\nu\right).$$

[*Hint*: Use a product formula for the continuous q-ultraspherical polynomials in the integral form (8.4.1) of [**46**] and the Nassrallah–Rahman integral.]

(14) Use the identity

$$\frac{1}{\left(r^2e^{2i\theta}, r^2e^{-2i\theta}; q^2\right)} = \frac{1}{\left(re^{i\theta}, re^{-i\theta}; q\right)} \frac{1}{\left(-re^{i\theta}, -re^{-i\theta}; q\right)}$$

and the generating function (4.7.3) in order to derive the connection relation (4.9.7). [*Hint*: You might want to use the formula

$$\sum_{k=0}^{2n} \frac{(-1)^k}{(q;q)_k\,(q;q)_{2n-k}} = \frac{1}{\left(q^2;q^2\right)_n},$$

which follows from (3.4.16) and (3.4.25).]

Use (4.6.31) and (3.4.15), or (4.6.30) and (4.7.4), in order to derive the connection relation

$$H_n\left(x|q\right) = \sum_{k=0}^{[n/2]} q^k \frac{(q;q)_n}{\left(q^2;q^2\right)_k (q;q)_{n-k}} H_{n-2k}\left(x|\,q^2\right).$$

(15) Use Rogers' connection coefficient formula (4.3.2) with $\gamma = 0$, $\beta \to \infty$ and $\beta \to 0$, $\gamma \to \infty$, respectively, and Ex. 3.11 in order to derive

$$\frac{H_n\left(x|q\right)}{(q;q)_n} = \sum_{k=0}^{[n/2]} \frac{(-1)^k\, q^{k(3k-2n-1)/2}}{(q;q)_k\,(q;q)_{n-2k}} H_{n-2k}\left(x|q^{-1}\right)$$

and

$$\frac{H_n\left(x|q^{-1}\right)}{(q;q)_n} = \sum_{k=0}^{[n/2]} \frac{q^{k^2-nk}}{(q;q)_k\,(q;q)_{n-2k}} H_{n-2k}\left(x|q\right).$$

CHAPTER 5

Introduction of Basic Fourier Series

Our main objective in this chapter is to define basic Fourier series and to establish some elementary facts about them. In our presentation most of the material can be read independently from Chapters 3 and 4, we only assume that the reader is familiar with the basic exponential and basic trigonometric functions covered in Chapter 2.

5.1. Preliminaries

A periodic function with period $2l$,

$$f(x+2l)=f(x), \tag{5.1.1}$$

can be represented as the Fourier series,

$$f(x)=a_0+\sum_{n=1}^{\infty}\left(a_n\cos\frac{\pi n}{l}x+b_n\sin\frac{\pi n}{l}x\right), \tag{5.1.2}$$

where

$$a_0=\frac{1}{2l}\int_{-l}^{l}f(x)\,dx, \tag{5.1.3}$$

$$a_n=\frac{1}{l}\int_{-l}^{l}f(x)\cos\frac{\pi n}{l}x\,dx, \tag{5.1.4}$$

$$b_n=\frac{1}{l}\int_{-l}^{l}f(x)\sin\frac{\pi n}{l}x\,dx. \tag{5.1.5}$$

For convergence conditions of (5.1.2) see, for example, **[3]**, **[20]**, **[156]**, **[164]**, and **[166]**. The formulas (5.1.3)–(5.1.5) for the coefficients of the Fourier series are consequences of the orthogonality relations for trigonometric functions

$$\int_{-l}^{l}\cos\frac{n\pi x}{l}\,\cos\frac{m\pi x}{l}\,dx=0,\qquad m\neq n, \tag{5.1.6}$$

$$\int_{-l}^{l}\sin\frac{n\pi x}{l}\,\sin\frac{m\pi x}{l}\,dx=0,\qquad m\neq n, \tag{5.1.7}$$

$$\int_{-l}^{l}\cos\frac{n\pi x}{l}\,\sin\frac{m\pi x}{l}\,dx=0,\qquad m\neq n. \tag{5.1.8}$$

We shall discuss a q-version of the Fourier series (5.1.2) on the basis of the q-analogs of trigonometric functions introduced in **[17]**, **[76]** and **[140]**;

see Chapter 2. Our first main objective will be to establish analogs of the orthogonality relations (5.1.6)–(5.1.8) for the above basic trigonometric functions on a q-quadratic grid.

There are several ways to prove the orthogonality relations (5.1.6)–(5.1.8) for the trigonometric system. In this chapter we shall extend the classical Sturm–Liouville method using the second order differential equation for harmonic motion,

$$u'' + \omega^2 u = 0, \tag{5.1.9}$$

to the case of the basic trigonometric functions. Consider, for example, two functions $\cos \omega x$ and $\cos \omega' x$, which satisfy (5.1.9) with different eigenvalues ω and ω'. Then

$$\left(\omega^2 - \omega'^2\right) \int_{-l}^{l} \cos \omega x \cos \omega' x \, dx = W\left(\cos \omega x, \cos \omega' x\right) |_{-l}^{l}$$
$$= \left. \begin{vmatrix} \cos \omega x & \cos \omega' x \\ -\omega \sin \omega x & -\omega' \sin \omega' x \end{vmatrix} \right|_{-l}^{l}, \tag{5.1.10}$$

where $W(u,v) = uv' - u'v$ is the Wronskian; see Ex. 2.13. The right side of (5.1.10) vanishes when

$$\sin \omega l = \sin \omega' l = 0, \tag{5.1.11}$$

which gives

$$\omega = \frac{\pi}{l}\, n, \qquad \omega' = \frac{\pi}{l}\, m, \tag{5.1.12}$$

where $n, m = 0, \pm 1, \pm 2, \pm 3, \ldots$. One can prove (5.1.7) in a similar fashion. The last equation (5.1.8) is valid by a symmetry when $x \to -x$. In this chapter we shall extend this consideration to the case of the basic trigonometric functions following the original paper [**26**]. On the other hand, the orthogonality relation for the basic exponential functions (4.4.3) established in the previous chapter gives an independent approach on the basis of a direct integral evaluation; see Ex. 4.1.

The chapter is organized as follows. In Section 5.2 we establish the continuous orthogonality property for the basic trigonometric functions, and then, in Section 5.3, we formally discuss the limit $q \to 1$ of these orthogonality relations. Section 5.4 is devoted to the investigation of some elementary properties of the zeros of the basic trigonometric functions and in Section 5.5 we evaluate the normalization constants. In Section 5.6 we state, once again, the orthogonality relation for the corresponding q-exponential functions for the benefit of the reader who is not familiar with Chapter 4. Finally, we introduce the basic analogs of Fourier series and their modification in Section 5.7, and consider some q-trigonometric identities in Section 5.8. Investigation of convergence of these basic Fourier series, expansions of many elementary and q-functions with respect to the basic trigonometric system, and some applications will be discussed in the next chapters.

5.2. Orthogonality Property for q-Trigonometric Functions

In this section we shall establish the orthogonality relations for the q-trigonometric functions (2.4.6)–(2.4.7), similar to the orthogonality relations for the classical trigonometric system (5.1.6)–(5.1.8), using the Sturm–Liouville type arguments [**26**]. Consider the difference equations for the functions $u(z) = C_q\left(x(z);\omega\right)$ and $v(z) = C_q\left(x(z);\omega'\right)$ in self-adjoint form,

$$\frac{\Delta}{\nabla x_1(z)}\left(\sigma\rho(z)\,\frac{\nabla u(z)}{\nabla x(z)}\right) + \lambda\,\rho(z)u(z) = 0 \tag{5.2.1}$$

and

$$\frac{\Delta}{\nabla x_1(z)}\left(\sigma\rho(z)\,\frac{\nabla v(z)}{\nabla x(z)}\right) + \lambda'\,\rho(z)v(z) = 0, \tag{5.2.2}$$

where function $\rho(z)$ satisfies the Pearson equation [**111**], [**139**],

$$\frac{\rho(z+1)}{\rho(z)} = \frac{\sigma(-z)}{\sigma(z+1)} = 1 = q^{-4z-2}\,\left(q^2\right)^{2z+1}, \tag{5.2.3}$$

see also Ex. 2.21, and the eigenvalues are

$$\lambda = \frac{4q^{1/2}\sigma}{(1-q)^2}\,\omega^2, \qquad \lambda' = \frac{4q^{1/2}\sigma}{(1-q)^2}\,\omega'^2. \tag{5.2.4}$$

One can easily verify that

$$\frac{\rho_0(z+1)}{\rho_0(z)} = q^{-4z-2}, \quad \rho_0(z) = \frac{\left(q^{2z},\, q^{-2z};\, q\right)_\infty}{q^z - q^{-z}} \tag{5.2.5}$$

and

$$\frac{\rho_\alpha(z+1)}{\rho_\alpha(z)} = q^{4z+2}, \quad \rho_\alpha(z) = \left(q^{2\alpha+2z},\, q^{2\alpha-2z},\, q^{2-2\alpha+2z},\, q^{2-2\alpha-2z};\, q^2\right)_\infty^{-1}, \tag{5.2.6}$$

cf. [**63**], [**141**] and [**143**]. Therefore we can choose the following solution of Eq. (5.2.3):

$$\rho(z) = \frac{\left(q^{2z},\, q^{-2z};\, q\right)_\infty\,\left(q^z - q^{-z}\right)^{-1}}{\left(q^{2\alpha+2z},\, q^{2\alpha-2z},\, q^{2-2\alpha+2z},\, q^{2-2\alpha-2z};\, q^2\right)_\infty}, \tag{5.2.7}$$

where α is an arbitrary additional parameter [**26**]. We shall see later that this solution satisfies the correct boundary conditions for the difference analog of the equation for harmonic motion (2.4.16) for certain values of this parameter α.

Let us multiply Eq. (5.2.1) by $v(z)$, Eq. (5.2.2) by $u(z)$ and subtract the second equality from the first one. As a result,

$$\left(\lambda - \lambda'\right)\,u(z)\,v(z)\,\rho(z)\nabla x_1(z) = \Delta\left[\sigma\rho(z)\,W\left(u(z),\, v(z)\right)\right], \tag{5.2.8}$$

where

$$W\left(u(z),\, v(z)\right) \;=\; \begin{vmatrix} u(z) & v(z) \\ \dfrac{\nabla u(z)}{\nabla x(z)} & \dfrac{\nabla v(z)}{\nabla x(z)} \end{vmatrix} \tag{5.2.9}$$

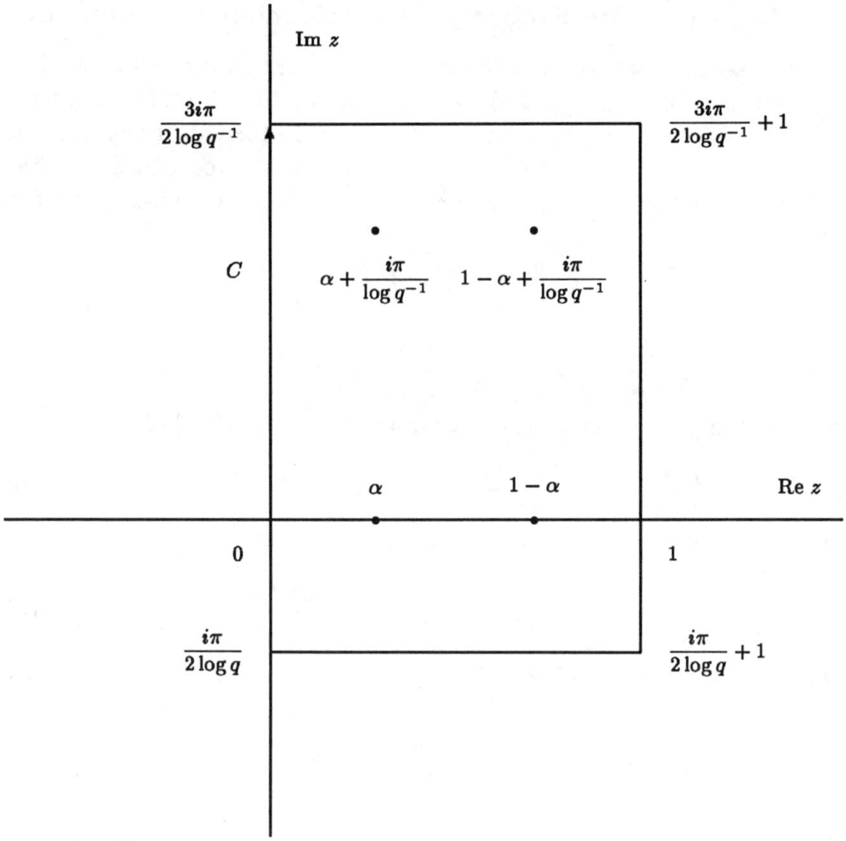

FIGURE 5.1. The contour C: $z = i\theta/\log q^{-1}$ and $-\pi/2 \leq \theta \leq 3\pi/2$ in the complex z-plane.

$$= \quad u(z)\,\frac{\nabla v(z)}{\nabla x(z)} - v(z)\,\frac{\nabla u(z)}{\nabla x(z)}$$

is the difference analog of the Wronskian [**111**]; see also Ex. 2.21.

Integrating (5.2.8) over the contour C indicated in Figure 5.1, where z is such that $z = i\theta/\log q^{-1}$ and $-\pi/2 \leq \theta \leq 3\pi/2$, we obtain

$$(\lambda-\lambda')\int_C u(z)\ v(z)\ \rho(z)\nabla x_1(z)\ dz \qquad (5.2.10)$$
$$= \int_C \Delta\left[\sigma\rho(z)\ W\left(u(z),\ v(z)\right)\right]\ dz.$$

As a function in z, the integrand in the right side of (5.2.10) has a natural purely imaginary period $T = 2\pi i/\log q^{-1}$ when $0 < q < 1$, so this integral is equal to

$$\int_D \sigma\rho(z)\ W\left(u(z),\ v(z)\right)\ dz, \qquad (5.2.11)$$

where D is the boundary of the rectangle on Figure 5.1 oriented counterclockwise.

The basic trigonometric functions $C_q(x;\omega)$ and $S_q(x;\omega)$ are entire functions in the complex z-plane due to (2.4.8)–(2.4.9). Therefore the poles of the integrand in (5.2.11) inside the rectangle in Figure 5.1 are the simple poles of $\rho(z)$ at $z=\alpha$, $z=1-\alpha$ and at $z=\alpha-i\pi/\log q$, $z=1-\alpha-i\pi/\log q$ when $0<\operatorname{Re}\alpha<1/2$. Hence, by the Cauchy theorem,

$$\begin{aligned}\frac{1}{2\pi i}\int_D \rho\, W(u,\ v)\ dz &= \operatorname*{Res}_{z=\alpha} f(z) + \operatorname*{Res}_{z=1-\alpha} f(z) \\ &+ \operatorname*{Res}_{z=\alpha-i\pi/\log q} f(z) + \operatorname*{Res}_{z=1-\alpha-i\pi/\log q} f(z),\end{aligned} \tag{5.2.12}$$

where

$$\begin{aligned} f(z) &= \rho(z)\, W(u(z),\ v(z)) \\ &= \frac{q^{-z}\left(q^{2z},\ q^{1-2z};\ q\right)_\infty\ W(u(z),\ v(z))}{\left(q^{2\alpha+2z},\ q^{2\alpha-2z},\ q^{2-2\alpha+2z},\ q^{2-2\alpha-2z};\ q^2\right)_\infty}.\end{aligned} \tag{5.2.13}$$

Evaluation of the residues at these simple poles gives

$$\begin{aligned}\operatorname*{Res}_{z=\alpha} f(z) &= \lim_{z\to\alpha}\,(z-\alpha)\ f(z) \\ &= -\frac{q^{-\alpha}\left(q^{2\alpha},\ q^{1-2\alpha};\ q\right)_\infty}{2\log q^{-1}\left(q^2,\ q^2,\ q^{4\alpha},\ q^{2-4\alpha};\ q^2\right)_\infty} \\ &\quad\times W(u(z),\ v(z))\big|_{z=\alpha},\end{aligned} \tag{5.2.14}$$

$$\begin{aligned}\operatorname*{Res}_{z=1-\alpha} f(z) &= \lim_{z\to1-\alpha}\,(z-1+\alpha)\ f(z) \\ &= -\frac{q^{-\alpha}\left(q^{2\alpha},\ q^{1-2\alpha};\ q\right)_\infty}{2\log q^{-1}\left(q^2,\ q^2,\ q^{4\alpha},\ q^{2-4\alpha};\ q^2\right)_\infty} \\ &\quad\times W(u(z),\ v(z))\big|_{z=1-\alpha},\end{aligned} \tag{5.2.15}$$

$$\begin{aligned}\operatorname*{Res}_{z=\alpha-i\pi/\log q} f(z) &= \lim_{z\to\alpha-i\pi/\log q}\,(z-\alpha+i\pi/\log q)\ f(z) \\ &= \frac{q^{-\alpha}\left(q^{2\alpha},\ q^{1-2\alpha};\ q\right)_\infty}{2\log q^{-1}\left(q^2,\ q^2,\ q^{4\alpha},\ q^{2-4\alpha};\ q^2\right)_\infty} \\ &\quad\times W(u(z),\ v(z))\big|_{z=\alpha-i\pi/\log q},\end{aligned} \tag{5.2.16}$$

and

$$\begin{aligned}\operatorname*{Res}_{z=1-\alpha-i\pi/\log q} f(z) &= \lim_{z\to1-\alpha}\,(z-1+\alpha+i\pi/\log q)\ f(z) \\ &= \frac{q^{-\alpha}\left(q^{2\alpha},\ q^{1-2\alpha};\ q\right)_\infty}{2\log q^{-1}\left(q^2,\ q^2,\ q^{4\alpha},\ q^{2-4\alpha};\ q^2\right)_\infty} \\ &\quad\times W(u(z),\ v(z))\big|_{z=1-\alpha-i\pi/\log q}.\end{aligned} \tag{5.2.17}$$

However

$$W\left(u(z),\ v(z)\right)=\frac{v(z)u(z-1)-u(z)v(z-1)}{x(z)-x(z-1)} \tag{5.2.18}$$

by (5.2.9) and, therefore,

$$\begin{aligned} &W\left(u(z),\ v(z)\right)\big|_{z=\alpha}=W\left(u(z),\ v(z)\right)\big|_{z=1-\alpha} \\ &=-\ W\left(u(z),\ v(z)\right)\big|_{z=\alpha-i\pi/\log q}=-\ W\left(u(z),\ v(z)\right)\big|_{z=1-\alpha-i\pi/\log q} \end{aligned} \tag{5.2.19}$$

due to the symmetry properties $C_q(x;\omega)=C_q(-x;\omega)$, $x(z)=x(-z)$, and $x(z)=-x(z-i\pi/\log q)$. Thus, all the residues are equal to each other and as a result we get

$$\begin{aligned} &\frac{q^{1/2}}{(1-q)^2}\left(\omega^2-\omega'^2\right)\int_C u(z)\ v(z)\ \rho(z)\nabla x_1(z)\ dz \\ &=-\frac{\pi i\ q^{-\alpha}\left(q^{2\alpha},\ q^{1-2\alpha};\ q\ \right)_\infty}{\log q^{-1}\left(q^2,\ q^2,\ q^{4\alpha},\ q^{2-4\alpha};\ q^2\ \right)_\infty}\ W\left(u(\alpha),\ v(\alpha)\right), \end{aligned} \tag{5.2.20}$$

where $0<\operatorname{Re}\alpha<1/2$.

We have established the main equation (5.2.20) for the case of two basic cosine functions $u(z)=C_q\left(x(z);\omega\right)$ and $v(z)=C_q\left(x(z);\omega'\right)$. In the same way, one can verify that this equation holds for the basic sine functions $u(z)=S_q\left(x(z);\omega\right)$ and $v(z)=S_q\left(x(z);\omega'\right)$. The corresponding analogs of the Wronskians in (5.2.20) can be written as

$$\begin{aligned} &W\left(C_q(x(z);\omega),\ C_q(x(z);\omega')\right) \\ &=\frac{2q^{1/4}}{1-q}\ \left[\omega\ C_q\left(x(z);\omega'\right)S_q\left(x(z-1/2);\omega\right)\right. \\ &\qquad\left.-\ \omega'\ C_q\left(x(z);\omega\right)S_q\left(x(z-1/2);\omega'\right)\right] \end{aligned} \tag{5.2.21}$$

and

$$\begin{aligned} &W\left(S(x(z);\omega),\ S(x(z);\omega')\right) \\ &=\frac{2q^{1/4}}{1-q}\ \left[\omega'\ S\left(x(z);\omega\right)C\left(x(z-1/2);\omega'\right)\right. \\ &\qquad\left.-\ \omega\ S\left(x(z);\omega'\right)C\left(x(z-1/2);\omega\right)\right] \end{aligned} \tag{5.2.22}$$

by (2.4.17) and (2.4.18), respectively. One can see from (5.2.21) and (5.2.22) that the right side of (5.2.20) vanishes in both cases when the eigenvalues ω and ω' are the roots of the following transcendental equation

$$S_q\left(x(1/4);\omega\right)=S_q\left(x(1/4);\omega'\right)=0. \tag{5.2.23}$$

This is the direct analog of (5.1.11) for the basic trigonometric functions.

In the last case under consideration, when $u(z)=C_q\left(x(z);\omega\right)$ and $v(z)=S_q\left(x(z);\omega'\right)$, the left side of (5.2.20) vanishes by the symmetry. It is of interest nonetheless to verify this fact by using the analog of the Sturm–Liouville method as well. Indeed, Eqs. (5.2.1) to (5.2.18) are valid again. But now

$$\begin{aligned} &W\left(u(z),\ v(z)\right)\big|_{z=\alpha} = W\left(u(z),\ v(z)\right)\big|_{z=1-\alpha} \\ &= W\left(u(z),\ v(z)\right)\big|_{z=\alpha-i\pi/\log q} = W\left(u(z),\ v(z)\right)\big|_{z=1-\alpha-i\pi/\log q} \end{aligned} \tag{5.2.24}$$

due to the symmetry properties $C_q(x;\omega) = C_q(-x;\omega)$, $S_q(x;\omega) = -S_q(-x;\omega)$, $x(z) = x(-z)$, and $x(z) = -x(z - i\pi/\log q)$. Therefore

$$\begin{aligned} &\frac{q^{1/2}}{(1-q)^2}\left(\omega^2 - \omega'^2\right)\int_C u(z)\ v(z)\ \rho(z)\nabla x_1(z)\ dz \\ &= -\frac{\pi i\ q^{-\alpha}\left(q^{2\alpha},\ q^{1-2\alpha};\ q\ \right)_\infty}{2\log q^{-1}\left(q^2,\ q^2,\ q^{4\alpha},\ q^{2-4\alpha};\ q^2\ \right)_\infty} \\ &\times\ \left[W\left(u(\alpha),\ v(\alpha)\right) - W\left(u(\alpha),\ v(\alpha)\right)\right] \equiv 0, \end{aligned} \tag{5.2.25}$$

when $0 < \operatorname{Re}\alpha < 1/2$.

Combining all the above cases together, we finally arrive at the *continuous orthogonality relations* for basic trigonometric functions,

$$\begin{aligned} &\int_0^\pi C_q\left(\cos\theta;\omega\right)\ C_q\left(\cos\theta;\omega'\right)\ \left(e^{2i\theta},\ e^{-2i\theta};\ q\right)_{1/2}\ d\theta \\ &= \begin{cases} 0 & \text{if}\quad \omega \neq \omega', \\ \pi\dfrac{\left(q^{1/2};\ q\right)_\infty^2}{\left(q;\ q\right)_\infty^2}\ C_q\left(\eta;\omega\right)\dfrac{\partial}{\partial\omega}S_q\left(\eta;\omega\right) & \text{if}\quad \omega = \omega', \end{cases} \end{aligned} \tag{5.2.26}$$

$$\begin{aligned} &\int_0^\pi S_q\left(\cos\theta;\omega\right)\ S_q\left(\cos\theta;\omega'\right)\ \left(e^{2i\theta},\ e^{-2i\theta};\ q\right)_{1/2}\ d\theta \\ &= \begin{cases} 0 & \text{if}\quad \omega \neq \omega', \\ \pi\dfrac{\left(q^{1/2};\ q\right)_\infty^2}{\left(q;\ q\right)_\infty^2}\ C_q\left(\eta;\omega\right)\dfrac{\partial}{\partial\omega}S_q\left(\eta;\omega\right) & \text{if}\quad \omega = \omega', \end{cases} \end{aligned} \tag{5.2.27}$$

and

$$\int_0^\pi C_q\left(\cos\theta;\omega\right)\ S_q\left(\cos\theta;\omega'\right)\ \left(e^{2i\theta},\ e^{-2i\theta};\ q\right)_{1/2}\ d\theta = 0. \tag{5.2.28}$$

Here $\eta := x(1/4) = \left(q^{1/4} + q^{-1/4}\right)/2$,

$$\left(e^{2i\theta},\ e^{-2i\theta};\ q\right)_{1/2} = \frac{\left(e^{2i\theta},\ e^{-2i\theta};\ q\right)_\infty}{\left(q^{1/2}e^{2i\theta},\ q^{1/2}e^{-2i\theta};\ q\right)_\infty}, \tag{5.2.29}$$

and the eigenvalues ω and ω' are different solutions of the transcendental equation (5.2.23):

$$S_q\left(\frac{1}{2}\left(q^{1/4} + q^{-1/4}\right);\omega\right) = \frac{(-i\omega;q^{1/2})_\infty - (i\omega;q^{1/2})_\infty}{2i(-q\omega^2;q^2)_\infty} = 0 \tag{5.2.30}$$

in view of (2.4.19). A similar consideration shows that the orthogonality relations (5.2.26)–(5.2.28) hold also when ω and ω' are different solutions of the following equation

$$C_q\left(\frac{1}{2}\left(q^{1/4}+q^{-1/4}\right);\omega\right)=\frac{(-i\omega;q^{1/2})_\infty+(i\omega;q^{1/2})_\infty}{2(-q\omega^2;q^2)_\infty}=0. \tag{5.2.31}$$

See also Section 4.4 above and [**144**] for an elementary proof of the orthogonality property of the q-trigonometric system on the basis of the integral (4.1.3) evaluated by Ismail and Stanton [**71**].

For arbitrary eigenvalues $\omega\neq\omega'$ one gets from (5.2.20)–(5.2.22) that

$$\begin{aligned}&\int_0^\pi C_q(\cos\theta;\omega)\ C_q(\cos\theta;\omega')\ \left(e^{2i\theta},\ e^{-2i\theta};\ q\right)_{1/2}\ d\theta \\ &= \frac{2\pi}{\omega^2-\omega'^2}\frac{\left(q^{1/2};q\right)^2_\infty}{(q;q)^2_\infty}\left[\omega\ C_q(\eta;\omega')\,S_q(\eta;\omega)-\omega'\ C_q(\eta;\omega)\,S_q(\eta;\omega')\right]\end{aligned} \tag{5.2.32}$$

and

$$\begin{aligned}&\int_0^\pi S_q(\cos\theta;\omega)\ S_q(\cos\theta;\omega')\ \left(e^{2i\theta},\ e^{-2i\theta};\ q\right)_{1/2}\ d\theta \\ &= \frac{2\pi}{\omega^2-\omega'^2}\frac{\left(q^{1/2};q\right)^2_\infty}{(q;q)^2_\infty}\left[\omega'\ S_q(\eta;\omega)\,C_q(\eta;\omega')-\omega\ S_q(\eta;\omega')\,C_q(\eta;\omega)\right].\end{aligned} \tag{5.2.33}$$

Also, in the limit $\omega\to\omega'$,

$$\begin{aligned}&\int_0^\pi C_q^2(\cos\theta;\omega)\ \left(e^{2i\theta},\ e^{-2i\theta};\ q\right)_{1/2}\ d\theta \\ &\quad= \frac{\pi\left(q^{1/2};\ q\right)^2_\infty}{\omega\,(q;\ q)^2_\infty}\ \bigg[\omega\ C_q(\eta;\omega)\,\frac{\partial}{\partial\omega}S_q(\eta;\omega) \\ &\qquad + C_q(\eta;\omega)\,S_q(\eta;\omega)-\omega\frac{\partial}{\partial\omega}C_q(\eta;\omega)\,S_q(\eta;\omega)\bigg]\end{aligned} \tag{5.2.34}$$

and

$$\begin{aligned}&\int_0^\pi S_q^2(\cos\theta;\omega)\ \left(e^{2i\theta},\ e^{-2i\theta};\ q\right)_{1/2}\ d\theta \\ &\quad= \frac{\pi\left(q^{1/2};\ q\right)^2_\infty}{\omega\,(q;\ q)^2_\infty}\ \bigg[\omega\ C_q(\eta;\omega)\,\frac{\partial}{\partial\omega}S_q(\eta;\omega) \\ &\qquad - C_q(\eta;\omega)\,S_q(\eta;\omega)-\omega\frac{\partial}{\partial\omega}C_q(\eta;\omega)\,S_q(\eta;\omega)\bigg].\end{aligned} \tag{5.2.35}$$

We remind the reader that $\eta=x(1/4)=\left(q^{1/4}+q^{-1/4}\right)/2$. This notation will be widely used throughout the book.

5.3. Formal Limit $q \to 1^-$

In this section we obtain formally the orthogonality property of the trigonometric system as a limiting case of the orthogonality relations (5.2.26)–(5.2.28) for the basic trigonometric functions. According to the limiting relations (2.4.4)–(2.4.5),

$$\lim_{q\to 1^-} C_q\left(x;\omega\left(1-q\right)/2\right) = \cos\omega x, \tag{5.3.1}$$

$$\lim_{q\to 1^-} S_q\left(x;\omega\left(1-q\right)/2\right) = \sin\omega x. \tag{5.3.2}$$

If $\omega \neq \omega'$, Eq. (5.2.26) has the form

$$\int_0^\pi C_q\left(\cos\theta;\omega\right)\, C_q\left(\cos\theta;\omega'\right)\, \left(e^{2i\theta}, e^{-2i\theta};q\right)_{1/2}\, d\theta = 0, \tag{5.3.3}$$

where the eigenvalues ω and ω' satisfy the boundary condition (5.2.23). Using the limit [**46**]

$$\lim_{q\to 1^-} \left(a;q\right)_\alpha = \left(1-a\right)^\alpha, \tag{5.3.4}$$

see also Ex. 1.5, one gets

$$\left(e^{2i\theta}, e^{-2i\theta};q\right)_{1/2} \to 2\sin\theta \tag{5.3.5}$$

as $q \to 1^-$. Therefore, changing ω by $(1-q)\,\omega/2$ in (5.3.3), with the help of (5.3.1) we obtain the orthogonality relation (5.1.6) with $l = 1$. The boundary condition (5.1.11) follows from (5.2.23) in the same limit.

When $\omega = \omega'$, one can rewrite (5.2.26) as

$$\int_0^\pi C_q^2\left(\cos\theta;\omega\right)\, \left(e^{2i\theta}, e^{-2i\theta};q\right)_{1/2}\, d\theta = \frac{\pi\left(1-q\right)}{\Gamma_q^2\left(1/2\right)}\, C_q\left(\eta;\omega\right) \frac{\partial}{\partial\omega} S_q\left(\eta;\omega\right), \tag{5.3.6}$$

where

$$\Gamma_q\left(z\right) = \left(1-q\right)^{1-z}\, \frac{\left(q;q\right)_\infty}{\left(q^z;q\right)_\infty} \tag{5.3.7}$$

is the q-analog of Euler's gamma function $\Gamma\left(z\right)$; see, for example, [**46**] and Ex. 1.6. Changing ω by $(1-q)\,\omega/2$ in (5.3.6) with the help of the limiting relation [**10**], [**46**]

$$\lim_{q\to 1^-} \Gamma_q\left(z\right) = \Gamma\left(z\right), \tag{5.3.8}$$

we get in the limit $q \to 1^-$ that

$$2\int_{-1}^1 \cos^2 \pi n x\; dx = \frac{2\pi}{\Gamma^2\left(1/2\right)}\, \cos^2 \pi n = 2, \tag{5.3.9}$$

where $n = \pm 1, \pm 2, \ldots$.

In a similar manner, one can obtain formally the orthogonality relations (5.1.7) and (5.1.8) from (5.2.27) and (5.2.28), respectively.

5.4. Some Properties of Zeros

In Section 5.2 we have established the orthogonality relations for the basic trigonometric functions (5.2.26)–(5.2.28) under the boundary conditions (5.2.30)–(5.2.31). Here we shall discuss some elementary properties of the ω-zeros for the corresponding basic cosine and sine functions given by

$$S_q(\eta;\omega) = \frac{(-i\omega;q^{1/2})_\infty - (i\omega;q^{1/2})_\infty}{2i(-q\omega^2;q^2)_\infty} = \frac{1}{(-q\omega^2;q^2)_\infty} \sum_{k=0}^{\infty} (-1)^k \frac{q^{k(k+1/2)}}{\left(q^{1/2};q^{1/2}\right)_{2k+1}} \omega^{2k+1} \tag{5.4.1}$$

and

$$C_q(\eta;\omega) = \frac{(-i\omega;q^{1/2})_\infty + (i\omega;q^{1/2})_\infty}{2(-q\omega^2;q^2)_\infty} = \frac{1}{(-q\omega^2;q^2)_\infty} \sum_{k=0}^{\infty} (-1)^k \frac{q^{k(k-1/2)}}{\left(q^{1/2};q^{1/2}\right)_{2k}} \omega^{2k}. \tag{5.4.2}$$

These functions $S_q(\eta;\omega)$ and $C_q(\eta;\omega)$ are special cases of the q-sine $S_q(x;\omega)$ and q-cosine $C_q(x;\omega)$ functions in two independent variables x and ω when $x = \eta = \left(q^{1/4} + q^{-1/4}\right)/2$; see Eqs. (2.4.19)–(2.4.24).

The large ω-asymptotics of $S_q(\eta;\omega)$ and $C_q(\eta;\omega)$ can be investigated with the help of the following expressions

$$\begin{aligned} C_q(\eta;\omega) &= \frac{\left(q^{1/2}\omega^2, q^{3/2}/\omega^2;q^2\right)_\infty}{\left(q^{1/2};q\right)_\infty (q,-q\omega^2,-q/\omega^2;q^2)_\infty} C_q(\eta;\frac{q^{1/2}}{\omega}) \\ &- \omega \frac{\left(q^{3/2}\omega^2, q^{1/2}/\omega^2;q^2\right)_\infty}{\left(q^{1/2};q\right)_\infty (q,-q\omega^2,-q/\omega^2;q^2)_\infty} S_q(\eta;\frac{q^{1/2}}{\omega}) \end{aligned} \tag{5.4.3}$$

and

$$\begin{aligned} S_q(\eta;\omega) &= \omega \frac{\left(q^{3/2}\omega^2, q^{1/2}/\omega^2;q^2\right)_\infty}{\left(q^{1/2};q\right)_\infty (q,-q\omega^2,-q/\omega^2;q^2)_\infty} C_q(\eta;\frac{q^{1/2}}{\omega}) \\ &+ \frac{\left(q^{1/2}\omega^2, q^{3/2}/\omega^2;q^2\right)_\infty}{\left(q^{1/2};q\right)_\infty (q,-q\omega^2,-q/\omega^2;q^2)_\infty} S_q(\eta;\frac{q^{1/2}}{\omega}). \end{aligned} \tag{5.4.4}$$

These formulas follow directly from (2.4.10) and (2.4.11), when we substitute $e^{i\theta} = q^{1/4}$. One can easily see, indeed, that Eqs. (5.4.1)–(5.4.2) and (5.4.3)–(5.4.4) determine the asymptotic behavior of the special basic trigonometric functions $S_q(\eta;\omega)$ and $C_q(\eta;\omega)$ for large values of the variable ω.

The basic sine and cosine functions given by (5.4.1) and (5.4.2) are odd and even functions, respectively, so it is sufficient to study only their nonnegative zeros.

Let us first establish an important result that the basic trigonometric systems under consideration include infinitely many functions.

THEOREM 5.1. *The basic sine $S_q(\eta;\omega)$ and basic cosine $C_q(\eta;\omega)$ functions have an infinity of real ω-zeros when $0<q<1$.*

PROOF. Consider the case of the basic sine function $S_q(\eta;\omega)$. In view of (5.4.4) one needs to show that equation

$$\frac{S_q(\eta;q^{1/2}/\omega)}{C_q(\eta;q^{1/2}/\omega)}=-\omega\frac{\left(q^{3/2}\omega^2,q^{1/2}/\omega^2;q^2\right)_\infty}{\left(q^{1/2}\omega^2,q^{3/2}/\omega^2;q^2\right)_\infty} \tag{5.4.5}$$

has infinitely many solutions. Indeed, in the limit $\omega\to\infty$ the continuous function in the left side of this equation approaches zero, while the right side is changing from $-\infty$ to ∞ on each of the intervals $\left(\pm q^{-n-1/4},\pm q^{-n-5/4}\right)$ when $n=0,1,2,\ldots,\ \infty$. Thus, Eq. (5.4.5) must have at least one solution in each of these intervals. In a similar fashion, with the help of (5.4.3) one can show that the $C_q(\eta;\omega)$ has an infinity of real ω-zeros. □

Let us prove the next result that all the zeros of the q-sine and q-cosine functions under consideration are real.

THEOREM 5.2. *The basic sine $S_q(\eta;\omega)$ and basic cosine $C_q(\eta;\omega)$ functions have only real ω-zeros when $0<q<1$.*

PROOF. Suppose that ω_0 is a zero of the basic sine function $S_q(\eta;\omega)$ which is not real. From (2.4.7) with the help of (III.4) of [**46**] or (A.3.4) we get

$$\begin{aligned} S_q(\eta;\omega) &= \frac{\left(q^{5/2}\omega^2;q^2\right)_\infty}{(-q\omega^2;q^2)_\infty} \\ &\quad\times\frac{\omega}{1-q^{1/2}}\,{}_2\varphi_2\left(\begin{matrix}-q^{3/2},\ -q^{5/2}\\ q^3,\ q^{5/2}\omega^2\end{matrix}\ ;\ q^2,\ q^{3/2}\omega^2\right). \end{aligned} \tag{5.4.6}$$

Now one can see that ω_0 is not purely imaginary, because otherwise our function would be a multiple of a positive function.

Let ω_1 be the complex number conjugate to ω_0, so that ω_1 is also a zero of (5.4.1) because $S_q(\eta;\omega)$ is a real function of ω. Since $\omega_0^2\neq\omega_1^2$ the integral in the orthogonality relation (5.2.26) equals zero, but the integrand on the left is positive, and so we have obtained a contradiction. Hence a complex zero ω_0 cannot exist. One can consider the case of the basic cosine function in a similar fashion. □

THEOREM 5.3. *If $0<q<1$, then the real ω-zeros of the basic sine $S_q(\eta;\omega)$ and basic cosine $C_q(\eta;\omega)$ functions are simple.*

PROOF. This follows directly from the relations (5.2.34) and (5.2.35). Consider, for example, the case of the basic sine function $S_q(\eta;\omega)$. If $\omega=\omega'$, then the integral in the left side of (5.2.34) is positive, which means that $\frac{\partial}{\partial\omega}S_q(\eta;\omega)\neq0$ when $S_q(\eta;\omega)=0$. The same is true for the zeros of the basic cosine function $C_q(\eta;\omega)$. □

Our next goal is to show that the positive zeros of the basic sine function $S_q(\eta;\omega)$ are interlaced with those of the basic cosine function $C_q(\eta;\omega)$.

THEOREM 5.4. *If* $\omega_1, \omega_2, \omega_3, \ldots$ *are the positive zeros of* $S_q(\eta;\omega)$ *arranged in ascending order of magnitude, and* $\varpi_1, \varpi_2, \varpi_3, \ldots$ *are those of* $C_q(\eta;\omega)$*, then*

$$0 = \omega_0 < \varpi_1 < \omega_1 < \varpi_2 < \omega_3 < \varpi_3 < \ldots, \tag{5.4.7}$$

when $0 < q < 1$.

PROOF. Suppose that ω_k and ω_{k+1} are two successive zeros of $S_q(\eta;\omega)$. Then the derivative $\frac{\partial}{\partial\omega} S_q(\eta;\omega)$ has different signs at $\omega = \omega_k$ and $\omega = \omega_{k+1}$. This means, in view of (5.2.35), that $C_q(\eta;\omega)$ changes its sign between ω_k and ω_{k+1} and, therefore, has at least one zero on each interval $(\omega_k,\ \omega_{k+1})$.

To complete the proof of the theorem, we have to show that $C_q(\eta;\omega)$ changes its sign on each interval $(\omega_k,\ \omega_{k+1})$ only once. Suppose that we have $C_q(\eta;\varpi_k) = C_q(\eta;\varpi_{k+1}) = 0$ and $\omega_k < \varpi_k < \varpi_{k+1} < \omega_{k+1}$. Then, by (5.2.34), the function $S_q(\eta;\omega)$ has different signs at $\omega = \varpi_k$ and $\omega = \varpi_{k+1}$ and, therefore, this function has at least one more zero on $(\omega_k,\ \omega_{k+1})$. So, we have obtained a contradiction, and therefore the basic cosine function $C_q(\eta;\omega)$ has exactly one zero between any two successive zeros of the basic sine function $S_q(\eta;\omega)$. □

We shall continue investigation of main properties of the zeros of the basic sine and cosine functions later; asymptotics of these zeros will be found in Sections 6.3 and 8.4. Numerical values of the zeros of $S_q(\eta;\omega)$ and $C_q(\eta;\omega)$ are presented in Appendix C for different values of parameter q.

5.5. Evaluation of Some Constants

In this section we shall find explicitly the normalization constants in the right sides of the orthogonality relations (5.2.26)–(5.2.27) for the basic sine and basic cosine functions. Let us evaluate the following integral

$$\begin{aligned} 2k(\omega) &= \int_0^\pi \left(C_q^2(\cos\theta;\omega) + S_q^2(\cos\theta;\omega)\right) \left(e^{2i\theta},\ e^{-2i\theta};\ q\right)_{1/2} d\theta \\ &= \int_0^\pi C_q(\cos\theta, -\cos\theta;\omega) \left(e^{2i\theta},\ e^{-2i\theta};\ q\right)_{1/2} d\theta, \end{aligned} \tag{5.5.1}$$

where we have used the identity (3.2.12). In view of (2.4.2), when $|\omega| < 1$ one can write

$$\begin{aligned} 2\,\frac{(-q\omega^2;q^2)_\infty}{(-\omega^2;q^2)_\infty}\, k(\omega) &= \sum_{n=0}^{\infty} (-\omega^2)^n \frac{(q^{1/2};q)_n}{(q,-q,-q^{1/2};q)_n} \\ &\times \int_0^\pi \frac{(e^{2i\theta},\ e^{-2i\theta};\ q)_\infty}{(q^{n+1/2}e^{2i\theta},\ q^{n+1/2}e^{-2i\theta};\ q)_\infty}\, d\theta. \end{aligned} \tag{5.5.2}$$

The integral in the right side can be evaluated as a special case of the Askey–Wilson integral [**15**]; see also (A.4.1):

$$\int_0^\pi \frac{\left(e^{2i\theta},\ e^{-2i\theta};\ q\right)_\infty}{\left(q^{n+1/2}e^{2i\theta},\ q^{n+1/2}e^{-2i\theta};\ q\right)_\infty}\,d\theta \tag{5.5.3}$$
$$= \frac{2\pi\left(q^{2n+2};q\right)_\infty}{\left(q,-q^{n+1/2},q^{n+1},-q^{n+1},q^{n+1},-q^{n+1},-q^{n+3/2};q\right)_\infty}.$$

Therefore

$$2\frac{\left(-q\omega^2;q^2\right)_\infty}{\left(-\omega^2;q^2\right)_\infty}\,k(\omega) = \frac{2\pi\left(q^{1/2};q\right)_\infty}{\left(q,q,-q,-q^{1/2};q\right)_\infty}\sum_{n=0}^\infty \frac{\left(-\omega^2\right)^n}{1-q^{n+1/2}}, \tag{5.5.4}$$

where we have used the identity

$$\left(q^{2n+2};q\right)_\infty = \left(q^{n+1},-q^{n+1},q^{n+3/2},-q^{n+3/2};q\right)_\infty. \tag{5.5.5}$$

However,

$$\sum_{n=0}^\infty \frac{\left(-\omega^2\right)^n}{1-q^{n+1/2}} = \sum_{k=0}^\infty \frac{q^{k/2}}{1+\omega^2q^k}, \tag{5.5.6}$$

which provides an analytic continuation in the complex ω-plane.

As a result,

$$k(\omega) = \frac{1}{2}\int_0^\pi \left(C_q^2(\cos\theta;\omega) + S_q^2(\cos\theta;\omega)\right)\left(e^{2i\theta},\ e^{-2i\theta};\ q\right)_{1/2}\,d\theta$$
$$= \pi\frac{\left(q^{1/2};q\right)_\infty^2}{(q;q)_\infty^2}\,\frac{\left(-\omega^2;q^2\right)_\infty}{\left(-q\omega^2;q^2\right)_\infty}\sum_{k=0}^\infty \frac{q^{k/2}}{1+\omega^2q^k}. \tag{5.5.7}$$

It is convenient for further consideration to rewrite (5.5.7) in a compact form

$$k(\omega) = \pi\frac{\left(q^{1/2};q\right)_\infty^2}{(q;q)_\infty^2}\,\frac{\left(-\omega^2;q^2\right)_\infty}{\left(-q\omega^2;q^2\right)_\infty}\,\kappa(\omega), \tag{5.5.8}$$

where by the definition

$$\kappa(\omega) := \sum_{k=0}^\infty \frac{q^{k/2}}{1+\omega^2q^k}. \tag{5.5.9}$$

With the help of (5.5.7) one can rewrite (5.2.34) and (5.2.35) in the following explicit forms

$$\int_0^\pi C_q^2(\cos\theta;\omega)\ \left(e^{2i\theta},\ e^{-2i\theta};\ q\right)_{1/2}\,d\theta \tag{5.5.10}$$
$$= k(\omega) + \frac{\pi\left(q^{1/2};q\right)_\infty^2}{\omega\,(q;q)_\infty^2}\ C_q(\eta;\omega)\,S_q(\eta;\omega)$$

and

$$\int_0^\pi S_q^2(\cos\theta;\omega)\ \left(e^{2i\theta},\ e^{-2i\theta};\ q\right)_{1/2}\,d\theta \tag{5.5.11}$$

$$= k(\omega) - \frac{\pi \left(q^{1/2};q\right)_\infty^2}{\omega \left(q;q\right)_\infty^2} \ C_q(\eta;\omega)\, S_q(\eta;\omega).$$

These basic integrals are obviously q-extensions of the following elementary integrals

$$\int_{-1}^{1} \cos^2 \omega x \, dx = 1 + \frac{1}{\omega} \ \sin\omega \cos\omega \tag{5.5.12}$$

and

$$\int_{-1}^{1} \sin^2 \omega x \, dx = 1 - \frac{1}{\omega} \ \sin\omega \cos\omega, \tag{5.5.13}$$

respectively.

When ω satisfies the boundary conditions (5.2.30) or (5.2.31), the last terms in the right sides of (5.2.10) and (5.2.11) vanish and we obtain the values of the $\mathcal{L}^2$-norm in the orthogonality relations (5.2.26)–(5.2.27) in terms of the function $k(\omega)$ given by (5.5.7).

On the other hand, Eqs. (5.2.34)–(5.2.35) and (5.5.7) imply that

$$k(\omega) = \pi \frac{\left(q^{1/2};q\right)_\infty^2}{\left(q;q\right)_\infty^2} \ W\left(C_q(\eta;\omega),\ S_q(\eta;\omega)\right), \tag{5.5.14}$$

where

$$W\left(C_q(\eta;\omega),\ S_q(\eta;\omega)\right) = \begin{vmatrix} C_q(\eta;\omega) & S_q(\eta;\omega) \\ \dfrac{\partial}{\partial\omega} C_q(\eta;\omega) & \dfrac{\partial}{\partial\omega} S_q(\eta;\omega) \end{vmatrix}$$

$$= C_q(\eta;\omega) \frac{\partial}{\partial\omega} S_q(\eta;\omega) - S_q(\eta;\omega) \frac{\partial}{\partial\omega} C_q(\eta;\omega) \tag{5.5.15}$$

is the Wronskian.

We shall also establish the asymptotic behavior of $k(\omega)$ as $|\omega| \to \infty$. It is more convenient to use (5.5.9) instead of (5.5.8) and to work with the function (5.5.9). One can easily see that

$$\sum_{k=-\infty}^{\infty} \frac{q^{k/2}}{1+\omega^2 q^k} = \kappa(\omega) + \frac{q^{1/2}}{\omega^2} \ \kappa\left(\frac{q^{1/2}}{\omega}\right) \tag{5.5.16}$$

$$= \frac{1}{1+\omega^2} \ {}_1\psi_1 \left(\begin{matrix} -\omega^2 \\ -q\omega^2 \end{matrix} ; q, q^{1/2} \right) = \frac{\left(q, q, -q^{1/2}\omega^2, -q^{1/2}/\omega^2; q\right)_\infty}{\left(q^{1/2}, q^{1/2}, -\omega^2, -q/\omega^2; q\right)_\infty}$$

by the Ramanujan summation theorem (A.2.5). Therefore

$$\kappa(\omega) = \frac{\left(q, q, -q^{1/2}\omega^2, -q^{1/2}/\omega^2; q\right)_\infty}{\left(q^{1/2}, q^{1/2}, -\omega^2, -q/\omega^2; q\right)_\infty} - \frac{q^{1/2}}{\omega^2} \ \kappa\left(\frac{q^{1/2}}{\omega}\right), \tag{5.5.17}$$

which gives a complete asymptotic expansion of the constant $k(\omega)$ in (5.5.7) and (5.5.8)–(5.5.9) in term of q/ω^2 :

$$k(\omega) = \pi \frac{(-\omega^2; q^2)_\infty}{(-q\omega^2; q^2)_\infty} \frac{(-q^{1/2}\omega^2; q)_\infty}{(-\omega^2; q)_\infty} \sum_{k=0}^{\infty} \frac{(q^{-1/2}; q)_k}{(q; q)_k} \left(-\frac{q}{\omega^2}\right)^k \tag{5.5.18}$$

$$- \pi \frac{(q^{1/2}; q)^2_\infty}{(q; q)^2_\infty} \frac{(-\omega^2; q^2)_\infty}{(-q\omega^2; q^2)_\infty} \frac{q^{1/2}}{\omega^2} \sum_{k=0}^{\infty} \frac{1}{1 - q^{k+1/2}} \left(-\frac{q}{\omega^2}\right)^k .$$

This expansion confirms the leading term (6.8) of [**26**].

5.6. Orthogonality Relations for q-Exponential Functions

Euler's formula,

$$e^{i\omega x} = \cos\omega x + i \sin\omega x, \tag{5.6.1}$$

allows us to rewrite the orthogonality relations for the trigonometric functions (5.1.6)–(5.1.8) in a complex form,

$$\frac{1}{2l} \int_{-l}^{l} \exp\left(i \frac{\pi m}{l} x\right) \exp\left(-i \frac{\pi n}{l} x\right) dx = \delta_{mn}, \tag{5.6.2}$$

where

$$\delta_{mn} = \begin{cases} 1 & \text{if} \quad m = n, \\ 0 & \text{if} \quad m \neq n. \end{cases} \tag{5.6.3}$$

The q-analog of Euler's formula is

$$\mathcal{E}_q(x; i\omega) = C_q(x; \omega) + iS_q(x; \omega), \tag{5.6.4}$$

where the $\mathcal{E}_q(x; \alpha)$ with $\alpha = i\omega$ is the q-exponential function and the $C_q(x; \omega)$ and $S_q(x; \omega)$ are the basic cosine and sine functions. The orthogonality relations for the basic trigonometric functions (5.2.26)–(5.2.28) result in the *orthogonality property* for the q-exponential function

$$\int_0^\pi \mathcal{E}_q(\cos\theta; i\omega_m)\, \mathcal{E}_q(\cos\theta; -i\omega_n) \left(e^{2i\theta}, e^{-2i\theta}; q\right)_{1/2} d\theta$$
$$= 2k(\omega_n)\, \delta_{mn}, \tag{5.6.5}$$

where $\omega_m, \omega_n = 0, \pm\omega_1, \pm\omega_2, \pm\omega_3, \ldots$ and $\omega_0 = 0, \omega_1, \omega_2, \omega_3, \ldots$, are nonnegative zeros of the basic sine function $S_q(\eta; \omega)$ arranged in ascending order of magnitude; the normalization constants $k(\omega_n)$ are defined by (5.5.7). An independent proof of this relation was given in Section 4.4 by a direct evaluation of the integral.

The basic analog of

$$e^{i\omega(x+y)} = \cos\omega(x+y) + i \sin\omega(x+y) \tag{5.6.6}$$

is

$$\mathcal{E}_q(x, y; i\omega) = C_q(x, y; \omega) + iS_q(x, y; \omega), \tag{5.6.7}$$

see (2.4.1), so that the exponential function on a q-quadratic grid $\mathcal{E}_q(x, y; i\omega)$ has the following orthogonality property.

THEOREM 5.5.

$$\int_0^\pi \mathcal{E}_q(\cos\theta,\cos\varphi;i\omega_m)\,\mathcal{E}_q(\cos\theta,\cos\varphi';-i\omega_n)\left(e^{2i\theta},\,e^{-2i\theta};\,q\right)_{1/2}d\theta$$
$$=2k(\omega_n)\;\mathcal{E}_q(\cos\varphi;i\omega_n)\;\mathcal{E}_q(\cos\varphi';-i\omega_n)\;\delta_{mn}, \tag{5.6.8}$$

where $\omega_m,\omega_n=0,\pm\omega_1,\pm\omega_2,\pm\omega_3,\ldots$ *and* $\omega_0=0,\omega_1,\omega_2,\omega_3,\ldots,$ *are nonnegative zeros of the basic sine function* $S(\eta;\omega)$ *arranged in ascending order of magnitude; the normalization constants* $k(\omega_n)$ *are defined by* (5.5.7).

PROOF. Use the addition theorem for basic exponential functions **[140]**,

$$\mathcal{E}_q(x,y;i\omega)=\mathcal{E}_q(x;i\omega)\;\mathcal{E}_q(y;i\omega), \tag{5.6.9}$$

and the orthogonality relation (5.6.5) to get (5.6.8). □

In a similar fashion, one can establish the following results.

THEOREM 5.6. *We have*

$$\int_0^\pi C_q(\cos\theta,\cos\varphi;\omega_m)\;C_q(\cos\theta,\cos\varphi';\omega_n) \tag{5.6.10}$$
$$\times\left(e^{2i\theta},\,e^{-2i\theta};\,q\right)_{1/2}d\theta$$
$$=\begin{cases}0 & \text{if } m\neq n,\\ k(\omega_n)\;C_q(\cos\varphi,-\cos\varphi';\omega_n) & \text{if } m=n;\end{cases}$$

$$\int_0^\pi S_q(\cos\theta,\cos\varphi;\omega_m)\;S_q(\cos\theta,\cos\varphi';\omega_n) \tag{5.6.11}$$
$$\times\left(e^{2i\theta},\,e^{-2i\theta};\,q\right)_{1/2}d\theta$$
$$=\begin{cases}0 & \text{if } m\neq n,\\ k(\omega_n)\;C_q(\cos\varphi,-\cos\varphi';\omega_n) & \text{if } m=n;\end{cases}$$

and

$$\int_0^\pi C_q(\cos\theta,\cos\varphi;\omega_m)\;S_q(\cos\theta,\cos\varphi';\omega_n) \tag{5.6.12}$$
$$\times\left(e^{2i\theta},\,e^{-2i\theta};\,q\right)_{1/2}d\theta$$
$$=\begin{cases}0 & \text{if } m\neq n,\\ k(\omega_n)\;S_q(\cos\varphi,-\cos\varphi';\omega_n) & \text{if } m=n;\end{cases}$$

where $\omega_m,\omega_n=\omega_1,\omega_2,\omega_3,\ldots,$ *are positive zeros of the basic sine function* $S(\eta;\omega)$ *arranged in ascending order of magnitude; the normalization constants* $k(\omega_n)$ *are defined by* (5.5.7).

PROOF. Use the addition theorem for the basic trigonometric functions (3.2.7)–(3.2.8) and the orthogonality relations (5.2.26)–(5.2.28). □

5.7. Basic Fourier Series

In a complete analogy with (5.1.2), we define a *q-version of the Fourier series* as

$$f(\cos\theta) = a_0 + \sum_{n=1}^{\infty} \left(a_n C_q(\cos\theta;\omega_n) + b_n S_q(\cos\theta;\omega_n)\right), \tag{5.7.1}$$

where $\omega_0 = 0, \omega_1, \omega_2, \omega_3, \ldots$ are nonnegative zeros of the basic sine function $S_q(\eta;\omega)$ given by (5.2.30), arranged in ascending order of magnitude, and the q-Fourier coefficients are

$$a_0 = \frac{1}{2k(0)} \int_0^{\pi} f(\cos\theta) \left(e^{2i\theta}, e^{-2i\theta}; q\right)_{1/2} d\theta, \tag{5.7.2}$$

$$a_n = \frac{1}{k(\omega_n)} \int_0^{\pi} f(\cos\theta)\, C_q(\cos\theta;\omega_n) \left(e^{2i\theta}, e^{-2i\theta}; q\right)_{1/2} d\theta, \tag{5.7.3}$$

$$b_n = \frac{1}{k(\omega_n)} \int_0^{\pi} f(\cos\theta)\, S_q(\cos\theta;\omega_n) \left(e^{2i\theta}, e^{-2i\theta}; q\right)_{1/2} d\theta. \tag{5.7.4}$$

The *complex form* of the basic Fourier series (5.7.1) is

$$f(\cos\theta) = \sum_{n=-\infty}^{\infty} c_n\, \mathcal{E}_q(\cos\theta; i\omega_n) \tag{5.7.5}$$

with

$$c_n = \frac{1}{2k(\omega_n)} \int_0^{\pi} f(\cos\theta)\ \mathcal{E}_q(\cos\theta; -i\omega_n) \left(e^{2i\theta}, e^{-2i\theta}; q\right)_{1/2} d\theta, \tag{5.7.6}$$

where $\omega_n = 0, \pm\omega_1, \pm\omega_2, \pm\omega_3, \ldots$ and $\omega_0 = 0 < \omega_1 < \omega_2 < \omega_3 < \ldots$ are nonnegative zeros of the basic sine function $S_q(\eta;\omega)$; the normalization constants $k(\omega_n)$ are defined by (5.5.7). These expressions, of course, merely indicate how the coefficients of our basic Fourier series are to be determined on the hypothesis that the expansion exists and is uniformly convergent. We shall study the convergence of the series (5.7.1) and (5.7.5) in the next chapter. As usual, due to (5.6.4) the q-Fourier coeffitients in (5.7.2)–(5.7.4) and (5.7.6) are related as

$$c_0 = a_0, \tag{5.7.7}$$

$$c_n = \frac{1}{2}(a_n - ib_n), \qquad c_{-n} = \frac{1}{2}(a_n + ib_n), \tag{5.7.8}$$

where $n = 1, 2, 3, \ldots$.

The q-Fourier series of f in either of the forms (5.7.1) and (5.7.5) will be denoted in a usual manner by $\mathbf{S}_q[f]$.

We shall also consider expansions of certain functions with respect to the basic trigonometric system $\{\mathcal{E}_q(x; i\varpi_n)\}_{n=-\infty}^{\infty}$, where $\varpi_n = \pm\varpi_1, \pm\varpi_2, \pm\varpi_3, \ldots$ and $0 < \varpi_1 < \varpi_2 < \varpi_3 < \ldots$ are nonnegative zeros of $C_q(\eta;\omega)$;

see (5.2.31). The corresponding *"modified" basic Fourier series* are

$$f(\cos\theta) = \sum_{n=1}^{\infty} \left(a_n C_q(\cos\theta; \varpi_n) + b_n S_q(\cos\theta; \varpi_n)\right), \tag{5.7.9}$$

where

$$a_n = \frac{1}{k(\varpi_n)} \int_0^{\pi} f(\cos\theta)\ C_q(\cos\theta; \varpi_n) \left(e^{2i\theta},\ e^{-2i\theta};\ q\right)_{1/2}\ d\theta, \tag{5.7.10}$$

$$b_n = \frac{1}{k(\varpi_n)} \int_0^{\pi} f(\cos\theta)\ S_q(\cos\theta; \varpi_n) \left(e^{2i\theta},\ e^{-2i\theta};\ q\right)_{1/2}\ d\theta, \tag{5.7.11}$$

and the complex form is

$$f(\cos\theta) = \sum_{n=-\infty}^{\infty} c_n\ \mathcal{E}_q(\cos\theta; i\varpi_n) \tag{5.7.12}$$

with

$$c_n = \frac{1}{2k(\varpi_n)} \int_0^{\pi} f(\cos\theta)\ \mathcal{E}_q(\cos\theta; -i\varpi_n)\ \left(e^{2i\theta},\ e^{-2i\theta};\ q\right)_{1/2}\ d\theta. \tag{5.7.13}$$

The normalization constants $k(\varpi_n)$ are defined by (5.5.7). We shall study the convergence of the series (5.7.9) and (5.7.12) in the next chapter.

5.8. Some Basic Trigonometric Identities

One of the most important formulas for the trigonometric functions is the main trigonometric identity,

$$\cos^2 \omega x + \sin^2 \omega x = 1. \tag{5.8.1}$$

It follows from the Pythagorean Theorem or from the addition formulas for the trigonometric functions, but one can also prove this identity on the basis of the differential equation for harmonic motion. The functions $\cos \omega x$ and $\sin \omega x$ are two solutions of (5.1.9) corresponding to the same eigenvalue ω. Therefore

$$\frac{d}{dx}\left[W(\cos\omega x,\ \sin\omega x)\right] = 0, \tag{5.8.2}$$

or

$$\cos^2 \omega x + \sin^2 \omega x = \text{constant}. \tag{5.8.3}$$

Substituting $x = 0$, we obtain (5.8.1).

In this section we extend this consideration to the case of the basic trigonometric functions. Consider Eq. (5.2.8) with $u(z) = C_q(x(z); \omega)$, $v(z) = S_q(x(z); \omega)$ and $\rho(z) = 1$, namely,

$$\Delta\left[W(u(z),\ v(z))\right] = 0, \tag{5.8.4}$$

where

$$\begin{aligned} W(u,\ v) &= W\left(C_q(x; \omega),\ S_q(x; \omega)\right) \\ &= \frac{2q^{1/4}\omega}{1-q}\left[C_q(x(z); \omega)\, C_q(x(z-1/2); \omega)\right. \end{aligned} \tag{5.8.5}$$

$$+ S_q(x(z);\omega)\, S_q(x(z-1/2);\omega)]$$

is the analog of the Wronskian (5.2.9) and we have used the difference-differentiation formulas (2.4.17)–(2.4.18) with $y=0$. One can easily see that the $W(u,\ v)$ here is a doubly periodic function in z without poles in the rectangle in Figure 5.1. Therefore this function is a constant by Liouville's theorem:

$$C_q(x(z);\omega)\, C_q(x(z-1/2);\omega) + S_q(x(z);\omega)\, S_q(x(z-1/2;\omega)) = C.$$

The value of this constant C can be found by choosing $x=0$, which gives

$$C = \frac{(-\omega^2;q^2)_\infty^2}{(-q\omega^2;q^2)_\infty^2}\, {}_2\varphi_1\left(\begin{matrix} q,\ q \\ q \end{matrix};q^2,-\omega^2\right) {}_2\varphi_1\left(\begin{matrix} 1,\ q \\ q \end{matrix};q^2,-\omega^2\right)$$
$$= \frac{(-\omega^2;q^2)_\infty^2}{(-q\omega^2;q^2)_\infty^2}\, {}_1\varphi_0\left(\begin{matrix} q \\ - \end{matrix};q^2,-\omega^2\right) = \frac{(-\omega^2;q^2)_\infty}{(-q\omega^2;q^2)_\infty}$$

by the q-binomial theorem; see (A.2.3). As a result, one gets

$$C_q(\cos\theta;\omega)\, C_q(\cos(\theta+i\log q/2);\omega) \tag{5.8.6}$$
$$+ S_q(\cos\theta;\omega)\, S_q(\cos(\theta+i\log q/2);\omega) = \frac{(-\omega^2;q^2)_\infty}{(-q\omega^2;q^2)_\infty}$$

as a q-extension of the main trigonometric identity (5.8.1). The special case of (5.8.6) when $z=1/4$ and $\eta = x(1/4)$ has the simplest form

$$C_q^2(\eta;\omega) + S_q^2(\eta;\omega) = \frac{(-\omega^2;q^2)_\infty}{(-q\omega^2;q^2)_\infty}. \tag{5.8.7}$$

The identity (5.8.6) can also be derived as a special case of the addition theorem for the basic trigonometric functions.

In a similar fashion, we can find an analog of the identity

$$\cos^2\omega(x+y) + \sin^2\omega(x+y) = 1 \tag{5.8.8}$$

considering more general basic sine and cosine functions, $C_q(x,y;\omega)$ and $S_q(x,y;\omega)$, as two linearly independent solutions of equation (2.4.16). The result is

$$C_q(\cos\theta,\cos\varphi;\omega)\, C_q(\cos(\theta+i\log q/2),\cos\varphi;\omega) \tag{5.8.9}$$
$$+ S_q(\cos\theta,\cos\varphi;\omega)\, S_q(\cos(\theta+i\log q/2),\cos\varphi;\omega)$$
$$= \frac{(-\omega^2;q^2)_\infty}{(-q\omega^2;q^2)_\infty}\left[C_q^2(\cos\varphi;\omega) + S_q^2(\cos\varphi;\omega)\right]$$
$$= \frac{(-\omega^2;q^2)_\infty}{(-q\omega^2;q^2)_\infty}\, C_q(\cos\varphi,-\cos\varphi;\omega).$$

We have used (3.2.12) here. This identity can also be verified with the aid of the addition theorems for the basic trigonometric functions.

Identity (5.8.7) gives the values of the basic cosine function $C_q(\eta;\omega)$ at the zeros of the basic sine function $S_q(\eta;\omega)$,

$$C_q(\eta;\omega_n) = (-1)^n \sqrt{\frac{(-\omega_n^2;q^2)_\infty}{(-q\omega_n^2;q^2)_\infty}}, \tag{5.8.10}$$

and vice versa,

$$S_q(\eta;\varpi_n) = (-1)^{n-1} \sqrt{\frac{(-\varpi_n^2;q^2)_\infty}{(-q\varpi_n^2;q^2)_\infty}}, \quad n \geq 1 \tag{5.8.11}$$

with the help of Theorem 5.4. From here

$$\mathcal{E}_q(\eta;i\omega_n) = C_q(\eta;\omega_n) = (-1)^n \sqrt{\frac{(-\omega_n^2;q^2)_\infty}{(-q\omega_n^2;q^2)_\infty}} \tag{5.8.12}$$

and

$$\mathcal{E}_q(\eta;i\varpi_n) = iS_q(\eta;\varpi_n) = i(-1)^{n-1} \sqrt{\frac{(-\varpi_n^2;q^2)_\infty}{(-q\varpi_n^2;q^2)_\infty}}, \quad n \geq 1 \tag{5.8.13}$$

due to the q-Euler formula (5.6.4).

Eq. (5.8.6) shows that the basic cosine $C_q(x;\omega)$ and basic sine $S_q(x;\omega)$ functions do not have common x-zeros for the same real value of ω.

5.9. Exercises for Chapter 5

(1) Derive the following product formulas by direct series manipulations [**122**]

$$C_q^2(\cos\theta;\omega) = \frac{1}{2}\left(-\omega^2;q^2\right)_{1/2}$$
$$\times\left[{}_3\varphi_2\left(\begin{matrix} q^{1/2},\ q^{1/2}e^{2i\theta},\ q^{1/2}e^{-2i\theta} \\ -q,\ -q^{1/2} \end{matrix};\ q, -\omega^2\right)\right.$$
$$\left. + {}_3\varphi_2\left(\begin{matrix} -q^{1/2},\ -q^{1/2}e^{2i\theta},\ -q^{1/2}e^{-2i\theta} \\ -q,\ q^{1/2} \end{matrix};\ q, -\omega^2\right)\right],$$

$$S_q^2(\cos\theta;\omega) = \frac{1}{2}\left(-\omega^2;q^2\right)_{1/2}$$
$$\times\left[{}_3\varphi_2\left(\begin{matrix} q^{1/2},\ q^{1/2}e^{2i\theta},\ q^{1/2}e^{-2i\theta} \\ -q,\ -q^{1/2} \end{matrix};\ q, -\omega^2\right)\right.$$
$$\left. - {}_3\varphi_2\left(\begin{matrix} -q^{1/2},\ -q^{1/2}e^{2i\theta},\ -q^{1/2}e^{-2i\theta} \\ -q,\ q^{1/2} \end{matrix};\ q, -\omega^2\right)\right],$$

and show that

$$\int_0^\pi C_q^2(\cos\theta;\omega)\left(e^{2i\theta},\ e^{-2i\theta};\ q\right)_{1/2}\ d\theta$$

$$= \frac{\pi}{1-q^{1/2}} \frac{\left(q^{1/2};q\right)_\infty^2}{(q;q)_\infty^2} \frac{\left(-\omega^2;q^2\right)_\infty}{\left(-q\omega^2;q^2\right)_\infty}$$

$$\times \left[{}_2\varphi_1\left(\begin{matrix} q^{1/2},\ q \\ q^{3/2}\end{matrix}\ q,-\omega^2\right) + {}_2\varphi_1\left(\begin{matrix} -q^{1/2},-q \\ q^{3/2}\end{matrix}\ ;q,-\omega^2\right)\right],$$

$$\int_0^\pi S_q^2(\cos\theta;\omega)\left(e^{2i\theta},\ e^{-2i\theta};\ q\right)_{1/2}\ d\theta$$

$$= \frac{\pi}{1-q^{1/2}} \frac{\left(q^{1/2};q\right)_\infty^2}{(q;q)_\infty^2} \frac{\left(-\omega^2;q^2\right)_\infty}{\left(-q\omega^2;q^2\right)_\infty}$$

$$\times \left[{}_2\varphi_1\left(\begin{matrix} q^{1/2},\ q \\ q^{3/2}\end{matrix}\ ;q,-\omega^2\right) - {}_2\varphi_1\left(\begin{matrix} -q^{1/2},-q \\ q^{3/2}\end{matrix}\ ;q,-\omega^2\right)\right].$$

The last two formulas give alternate expressions for the right sides of (5.5.10) and (5.5.11).
Verify that the product formulas are consequences of Eqs. (2.4.2), (3.2.10) and (3.2.12) and the above integrals follow from Eqs. (5.5.7) and (5.5.10)–(5.5.11) with the help of Eqs. (2.4.3) and (3.2.11).
Show that both the ${}_2\varphi_1$ series above can be summed by a consequence of the q-binomial theorem, see Ex. 1.4,

$${}_2\varphi_1\left(\begin{matrix} -q^{1/2},-q \\ q^{3/2}\end{matrix}\ ;q,-\omega^2\right) = \frac{1-q^{1/2}}{2i\omega}\sum_{n=0}^\infty \frac{\left(-1;q^{1/2}\right)_{2n+1}}{\left(q^{1/2};q^{1/2}\right)_{2n+1}}(i\omega)^{2n+1}$$

$$= \frac{1-q^{1/2}}{4i\omega}\ \frac{(-i\omega;q^{1/2})_\infty^2 - (i\omega;q^{1/2})_\infty^2}{(-\omega^2;q)_\infty}$$

and

$${}_2\varphi_1\left(\begin{matrix} q^{1/2},\ q \\ q^{3/2}\end{matrix}\ ;q,-\omega^2\right) = \lim_{\epsilon\to 1}\ {}_2\varphi_1\left(\begin{matrix} \epsilon q^{1/2},\ \epsilon q \\ q^{3/2}\end{matrix}\ ;q,-\omega^2\right)$$

$$= \frac{1-q^{1/2}}{2i\omega}\lim_{\epsilon\to 1}\frac{1}{1-\epsilon}\left[\frac{(i\epsilon\omega;q^{1/2})_\infty}{(i\omega;q^{1/2})_\infty} - \frac{(-i\epsilon\omega;q^{1/2})_\infty}{(-i\omega;q^{1/2})_\infty}\right]$$

$$= \frac{1-q^{1/2}}{2i}\frac{d}{d\omega}\log\frac{(-i\omega;q^{1/2})_\infty}{(i\omega;q^{1/2})_\infty}.$$

Derive as a result that

$$k(\omega) = \frac{\pi\left(q^{1/2};q\right)_\infty^2\left(-\omega^2;q^2\right)_\infty}{i\,(q;q)_\infty^2\left(-q\omega^2;q^2\right)_\infty}\frac{d}{d\omega}\log\frac{(-i\omega;q^{1/2})_\infty}{(i\omega;q^{1/2})_\infty}$$

and obtain (5.5.7) from here.

(2) Prove the orthogonality relation for the Askey–Wilson polynomials (4.6.2) using the Sturm–Liouville method for the difference equation of hypergeometric type [**18**], [**111**]. [*Hint:* See Ex. 2.22.]

(3) *Analogs of the Dirichlet and Fejér kernels for basic Fourier series.* The *partial sum* Sf of order m of the q-Fourier series is defined by $(S_0 f)(x) = c_0 = a_0$ if $m = 0$ and

$$(S_m f)(x) := \sum_{n=-m}^{m} c_n \, \mathcal{E}_q(x; i\omega_n)$$
$$= a_0 + \sum_{n=1}^{m} \left(a_n C_q(x; \omega_n) + b_n S_q(x; \omega_n) \right)$$

if $m = 1, 2, 3, \ldots$. Use the partial sums in order to introduce *q-analog of the Dirichlet kernels* as follows

$$D_0(x, -y; q) = \frac{1}{2k(0)},$$
$$D_m(x, -y; q) = \sum_{n=-m}^{m} \frac{\mathcal{E}_q(x, -y; i\omega_n)}{2k(\omega_n)}$$
$$= \frac{1}{2k(0)} + \sum_{n=1}^{m} \frac{C_q(x, -y; i\omega_n)}{k(\omega_n)},$$

when $m = 1, 2, 3, \ldots$, and show that

$$(S_m f)(x) = \int_{-1}^{1} D_m(x, -y; q) \; f(y) \; \rho(y) \, dy,$$

where $\rho(x)$ is the weight function in the orthogonality relation (5.6.5).

The *Cesàro mean* σf of order m of the q-Fourier series is defined by $(\sigma_0 f)(x) = (S_0 f)(x)$ if $m = 0$ and

$$(\sigma_m f)(x) := \frac{(S_0 f)(x) + \ldots + (S_m f)(x)}{m+1}$$

if $m = 1, 2, 3, \ldots$. Use the Cesàro means in order to introduce *q-analog of the Fejér kernels* as follows

$$K_0(x, -y; q) = \frac{1}{2k(0)},$$
$$K_m(x, -y; q) = \frac{D_0(x, -y; q) + \ldots + D_m(x, -y; q)}{m+1}$$
$$= \frac{1}{2k(0)} + \sum_{n=1}^{m} \left(1 - \frac{n}{m+1}\right) \frac{C_q(x, -y; i\omega_n)}{k(\omega_n)},$$

when $m = 1, 2, 3, \ldots$ and show that

$$(\sigma_m f)(x) = \int_{-1}^{1} K_m(x, -y; q) \; f(y) \; \rho(y) \, dy,$$

where $\rho(x)$ is the weight function in the orthogonality relation (5.6.5).

(4) *The third Jackson q-Bessel function.* The following q-analog of the Bessel function

$$J_{\nu}^{(3)}(z;q) = \frac{\left(q^{\nu+1};q\right)_{\infty}}{(q;q)_{\infty}} \sum_{n=0}^{\infty} \frac{(-1)^n q^{n(n+1)/2} z^{\nu+2n}}{(q;q)_n \left(q^{\nu+1};q\right)_n}$$

was introduced by Jackson [**77**], [**78**], [**79**]. This function was also studied by Hahn and Exton and sometimes appears in the literature as the Hahn–Exton q-Bessel function. The current notation is due to Ismail [**57**].

Prove that $J_{\nu}^{(3)}(z;q)$ satisfies the orthogonality relation [**92**], [**150**]

$$\begin{aligned} &\int_0^1 x\, J_{\nu}^{(3)}\left(j_{n,\nu}qx;q^2\right) J_{\nu}^{(3)}\left(j_{m,\nu}qx;q^2\right) d_qx \\ &\quad = \frac{1-q}{2q^2} J_{\nu+1}^{(3)}\left(j_{n,\nu};q^2\right) \frac{d}{dx} J_{\nu+1}^{(3)}\left(j_{n,\nu};q^2\right) \delta_{mn}, \end{aligned}$$

where $j_{1,\nu} < j_{2,\nu} < j_{3,\nu} < \dots$ are the positive zeros of $J_{\nu}^{(3)}\left(z;q^2\right)$ arranged in ascending order and the q-integral on the interval $(0,a)$ is defined by

$$\int_0^a f(x)\, d_qx = (1-q)\, a \sum_{n=0}^{\infty} f(aq^n)\, q^n.$$

Show that

$$c_q(x) = q^{-3/8} \frac{\left(q^2;q^2\right)_{\infty}}{\left(q;q^2\right)_{\infty}} x^{1/2} J_{-1/2}^{(3)}\left(q^{-3/4}x;q^2\right),$$

$$s_q(x) = q^{1/8} \frac{\left(q^2;q^2\right)_{\infty}}{\left(q;q^2\right)_{\infty}} x^{1/2} J_{1/2}^{(3)}\left(q^{-1/4}x;q^2\right),$$

where the basic cosine $c_q(x)$ and basic sine $s_q(x)$ functions on a q-linear grid are given by (2.5.7) and (2.5.8), respectively.

(5) *The orthogonality property of basic trigonometric functions on a q-linear grid.* Consider the basic cosine $c_q(x)$ and basic sine $s_q(x)$ functions on a q-linear grid given by (2.5.7) and (2.5.8), respectively. Let ω and ω' be roots of $s_q(x)$. Prove the following orthogonality relations [**25**]

$$\int_{-1}^1 c_q\left(q^{1/2}\omega x\right) c_q\left(q^{1/2}\omega' x\right) d_qx = \begin{cases} 0, & \text{if } \omega \neq \omega' \\ 2, & \text{if } \omega = \omega' = 0 \\ k(\omega) & \text{if } \omega = \omega' \neq 0 \end{cases}$$

$$\begin{aligned} &\int_{-1}^1 s_q(q\omega x)\, s_q(q\omega' x)\, d_qx \\ &\quad = \begin{cases} 0, & \text{if } \omega \neq \omega' \text{ or } \omega = \omega' = 0 \\ q^{-1/2} k(\omega) & \text{if } \omega = \omega' \neq 0 \end{cases} \end{aligned}$$

and

$$\int_{-1}^{1} c_q\left(\omega x\right)\ s_q\left(\omega' x\right)\ d_q x = 0,$$

where

$$k\left(\omega\right) = \left(1-q\right) c_q\left(q^{1/2}\omega\right) \frac{d}{d\omega} s_q\left(\omega\right)$$

and the q-integral on the interval (a,b) is defined by

$$\int_a^b f\left(x\right)\ d_q x = \int_0^b f\left(x\right)\ d_q x - \int_0^a f\left(x\right)\ d_q x.$$

Use these orthogonality relations in order to introduce basic Fourier series on a q-linear grid [**25**].

(6) *Analog of Bessel functions on a q-quadratic grid.* Consider the ${}_2\varphi_1$ function [**63**],

$$J_\nu(x;r) = \left(\frac{r}{2}\right)^\nu \frac{\left(q^{\nu+1}, -r^2/4; q\right)_\infty}{(q;q)_\infty} \times {}_2\varphi_1\left(\begin{matrix} q^{(\nu+1)/2}e^{i\theta},\ q^{(\nu+1)/2}e^{-i\theta} \\ q^{\nu+1} \end{matrix}\ ;\ q,\ -\frac{r^2}{4}\right)$$

with $x = \cos\theta$, as a q-analog of the Bessel function [**163**],

$$J_\nu(x) = \left(\frac{x}{2}\right)^\nu \sum_{n=0}^{\infty} \frac{\left(-x^2/4\right)^n}{n!\,\Gamma(\nu+n+1)}.$$

Prove the following orthogonality property for these q-Bessel functions,

$$\int_0^\pi \frac{J_\nu(\cos\theta;r)\,J_\nu(\cos\theta;r')\,\left(e^{2i\theta},\ e^{-2i\theta};q\right)_\infty}{\left(q^\alpha e^{i\theta}, q^\alpha e^{-i\theta}, q^{1-\alpha}e^{i\theta}, q^{1-\alpha}e^{-i\theta};q\right)_\infty} \times \left(q^{(\nu+1)/2}e^{i\theta}, q^{(\nu+1)/2}e^{-i\theta}, q^{(\nu+1)/2}e^{i\theta}, q^{(\nu+1)/2}e^{-i\theta};q\right)_\infty^{-1} d\theta = 0$$

if $r \neq r'$ and

$$\int_0^\pi \frac{\left(J_\nu(\cos\theta;r)\right)^2\left(e^{2i\theta},\ e^{-2i\theta};q\right)_\infty}{\left(q^\alpha e^{i\theta}, q^\alpha e^{-i\theta}, q^{1-\alpha}e^{i\theta}, q^{1-\alpha}e^{-i\theta};q\right)_\infty} \times \left(q^{(\nu+1)/2}e^{i\theta}, q^{(\nu+1)/2}e^{-i\theta}, q^{(\nu+1)/2}e^{i\theta}, q^{(\nu+1)/2}e^{-i\theta};q\right)_\infty^{-1} d\theta$$
$$= \frac{-4\pi(1-q)q^{-(\nu+1)/2}}{\left(q, q^{(\nu+1)/2+\alpha}, q^{(\nu+1)/2-\alpha+1};q\right)_\infty^2} \frac{\partial J_\nu(x(\alpha);r)}{\partial r^2} \frac{\nabla J_\nu(x(\alpha);r)}{\nabla x(\alpha)}$$

if $r = r'$. Here r and r' are two roots of the equation

$$J_\nu(x(\alpha);r) = J_\nu(x(\alpha);r') = 0$$

with $x(z) = (q^z + q^{-z})/2$ and $\operatorname{Re}\nu > -1$, $0 < \operatorname{Re}\alpha < 1$. [*Note:* This is a q-version of the orthogonality relation for the classical Bessel function

$$\int_0^1 x\, J_\nu(rx)\, J_\nu(r'x)\, dx = \begin{cases} 0 & \text{if } r \neq r', \\ \dfrac{1}{2}\,(J_{\nu+1}(r))^2 & \text{if } r = r', \end{cases}$$

under the conditions $J_\nu(r) = J_\nu(r') = 0$; see [**163**].]
The classical Bessel functions have the differentiation formulas

$$\frac{d}{dx}\left(x^{\pm\nu}\, J_\nu(x)\right) = \pm x^{\pm\nu}\, J_{\nu\mp 1}(x).$$

Introduce the following functions

$$J_\nu^{(+)}(z,r) := J_\nu(x(z);r),$$

$$J_\nu^{(-)}(z,r) := \frac{\left(q^{(1-\nu)/2+z},\ q^{(1-\nu)/2-z};q\right)_\infty}{\left(q^{(1+\nu)/2+z},\ q^{(1+\nu)/2-z};q\right)_\infty}\, J_\nu(x(z);r)$$

and show that the corresponding difference-differentiation formulas have the form

$$\frac{\delta}{\delta x}\, J_\nu^{(\pm)}(x,r) = \pm\frac{rq^{(1\pm\nu)/2}}{1-q}\, J_{\nu\pm1}^{(\pm)}(x,r).$$

The Lommel formula for the classical Bessel functions [**163**],

$$J_\nu(x)\, J_{1-\nu}(x) + J_{-\nu}(x)\, J_{\nu-1}(x) = \frac{2\sin\pi\nu}{\pi x},$$

gives the value of the Wronskian of $J_\nu(x)$ and $J_{-\nu}(x)$, so that these two functions are linearly independent for the noninteger ν.
Derive the corresponding q-Lommel formula

$$\left(q^z - q^{(1-\nu)/2}\right)\left(q^{-z} - q^{(\nu-1)/2}\right) J_\nu(x\,(z)\,;r)\, J_{1-\nu}\left(x\left(z-\frac{1}{2}\right);r\right)$$
$$+ J_{-\nu}(x\,(z)\,;r)\, J_{\nu-1}\left(x\left(z-\frac{1}{2}\right);r\right)$$
$$= \frac{2(q^\nu,\, q^{1-\nu},\, -r^2/4,\, -r^2/4;q)_\infty}{r(q,\, q;q)_\infty},$$

or

$$J_\nu^{(+)}(x\,(z)\,;r)\, J_{-\nu-1}^{(-)}\left(x\left(z-\frac{1}{2}\right);r\right)$$
$$+ J_{-\nu}^{(-)}(x\,(z)\,;r)\, J_{\nu+1}^{(+)}\left(x\left(z-\frac{1}{2}\right);r\right)$$
$$= \frac{2\left(q^{-\nu},\, q^{1+\nu},\, -r^2/4,\, -r^2/4;q\right)_\infty}{r\,(q,q;q)_\infty\left(q^{z-(1+\nu)/2},\, q^{-z+(1-\nu)/2};q\right)_{\nu+1}}.$$

Rewrite this relation in terms of the difference analog of the Wronskian,

$$W\left(J_\nu^{(+)}(z,r),\ J_{-\nu}^{(-)}(z,r)\right) = \begin{vmatrix} J_\nu^{(+)}(z,r) & J_{-\nu}^{(-)}(z,r) \\ \dfrac{\nabla J_\nu^{(+)}(z,r)}{\nabla x(z)} & \dfrac{\nabla J_{-\nu}^{(-)}(z,r)}{\nabla x(z)} \end{vmatrix}$$

$$= -\frac{2q^{(\nu+1)/2}}{1-q} \frac{\left(q^{-\nu},\ q^{1+\nu},\ -r^2/4,\ -r^2/4; q\right)_\infty}{(q,q;q)_\infty \left(q^{z-(1+\nu)/2},\ q^{-z+(1-\nu)/2}; q\right)_{\nu+1}},$$

and show that two functions $J_\nu^{(+)}$ and $J_{-\nu}^{(-)}$ are linearly dependent when ν is an integer.

Introduce corresponding analog of the Fourier–Bessel series [**63**].

(7) Extend the results of the previous exercise to the following analog of the Bessel function

$$J_\nu^\mu(x;r) = \left(\frac{r}{2}\right)^\nu \frac{\left(q^{\nu+1},\ -r^2/4; q\right)_\infty}{(q;q)_\infty} \times {}_2\varphi_1\left(\begin{matrix} q^{(\mu+\nu+1)/2}e^{i\theta},\ q^{(\mu+\nu+1)/2}e^{-i\theta} \\ q^{\nu+1} \end{matrix}\ ;\ q,\ -\frac{r^2}{4}\right),$$

where μ is an extra free parameter and $\left|r^2/4\right| < 1$.

[*Note*: Similar q-Bessel functions appeared in Rahman's addition formula for the q-Bessel functions $J_\nu^{(1)}$ and $J_\nu^{(2)}$ of Jackson [**117**] and in the papers [**17**], [**63**], [**81**], [**87**], [**88**], [**89**], and [**107**]. Koelink has obtained the ${}_2\varphi_1$-Bessel functions as generalized spherical elements for irreducible unitary corepresentations of the quantized function algebra for the group of plane motions [**87**], [**88**], and [**89**]. These q-Bessel functions have appeared also as matrix elements of irreducible unitary corepresentations of the quantized function algebra for $SU(1,1)$ [**107**], [**81**], and [**80**]. In [**17**], the ${}_2\varphi_1$ functions have appeared as solutions of a difference equation of hypergeometric type on a q-quadratic grid. In [**63**], similar q-Bessel functions have arisen as special cases of the bilinear generating function for the continuous q-ultraspherical polynomials. The orthogonality properties of the q-Bessel function in Exs. 5.3 and 5.4 are different from the orthogonality relations in [**87**], [**88**], [**89**], [**107**], [**81**], and [**80**]. Another type of continuous orthogonality of the ${}_2\varphi_1$-Bessel functions was found by Askey, Rahman, and Suslov [**14**].]

(8) *Orthogonal very-well poised* ${}_8\varphi_7$ *functions.* Consider the *difference equation of hypergeometric type*

$$\sigma(z)\,\frac{\Delta}{\nabla x_1(z)}\left(\frac{\nabla u(z)}{\nabla x(z)}\right) + \tau(z)\,\frac{\Delta u(z)}{\Delta x(z)} + \lambda\, u(z) = 0$$

on a q-quadratic grid $x(z) = \left(q^z + q^{-z}\right)/2$ with $x_1(z) = x\left(z + 1/2\right)$ and

$$\Delta f(z) = \nabla f(z+1) = f(z+1) - f(z),$$

where

$$\sigma(z) = q^{-2z}\left(q^z - a\right)\left(q^z - b\right)\left(q^z - c\right)\left(q^z - d\right),$$

$$\tau(z) = \frac{\sigma(-z) - \sigma(z)}{\nabla x_1(z)}$$

$$= \frac{2q^{1/2}}{1-q}\left(abc + abd + acd + bcd - a - b - c - d + 2(1 - abcd)\, x\right),$$

$$\lambda = \lambda_\nu = \frac{4q^{3/2}}{(1-q)^2}\left(1 - q^{-\nu}\right)\left(1 - abcd\, q^{\nu-1}\right),$$

see Ex. 2.20 and **[139]**, **[141]**, **[143]**.

Show that *particular solutions* of this equation have the form **[66]**, **[119]**, **[120]**, **[121]**, **[139]**, **[141]**, **[143]**

$$u_1 = u_\nu(x; a, b, c; d), \qquad u_2 = u_\nu(x; a, b, d; c),$$

where

$$u_\nu(x; a, b, c; d) = \frac{\left(qa/d, bcq^\nu, q^{1-\nu+z}/d, q^{1-\nu-z}/d; q\right)_\infty}{\left(q^{1-\nu}a/d, bc, q^{1+z}/d, q^{1-z}/d; q\right)_\infty}$$

$$\times\ {}_8W_7\left(\frac{aq^{-\nu}}{d}; q\sqrt{\frac{aq^{-\nu}}{d}}, -q\sqrt{\frac{aq^{-\nu}}{d}}, q^{-\nu}, \frac{q^{1-\nu}}{bd}, \frac{q^{1-\nu}}{cd}, aq^z, aq^{-z}; q, bcq^\nu\right)$$

$$= \frac{\left(bcq^\nu, q^{1-\nu}/ad; q\right)_\infty}{\left(bc, q/ad; q\right)_\infty}\ {}_4\varphi_3\left(\begin{matrix} q^{-\nu}, abcdq^{\nu-1}, aq^z, aq^{-z} \\ ab, ac, ad \end{matrix}; q, q\right)$$

$$+ \frac{\left(q^{-\nu}, abcdq^{\nu-1}, qb/d, qc/d; q\right)_\infty}{\left(ab, ac, bc, ad/q; q\right)_\infty}$$

$$\times \frac{\left(aq^z, aq^{-z}; q\right)_\infty}{\left(q^{1+z}/d, q^{1-z}/d; q\right)_\infty}\ {}_4\varphi_3\left(\begin{matrix} q^{1-\nu}/ad, bcq^\nu, q^{1+z}/d, q^{1-z}/d \\ qb/d, qc/d, q^2/ad \end{matrix}; q, q\right).$$

Show that the alternate forms are **[143]**

$$u_\nu(x; a, b, c; d) = \frac{\left(abcq^{\nu+z}, q^{1-\nu+z}/d; q\right)_\infty}{\left(abcq^z, q^{1+z}/d; q\right)_\infty}$$

$$\times\ {}_8W_7\left(abcq^{z-1}; q^{-\nu}, abcdq^{\nu-1}, aq^z, bq^z, cq^z;\ q,\ \frac{q^{1-z}}{d}\right)$$

$$= \frac{\left(abq^\nu, acq^\nu, bcq^\nu, abcdq^{\nu-1}, q^{1-\nu+z}/d, q^{1-\nu-z}/d; q\right)_\infty}{\left(ab, ac, bc, abcdq^{2\nu-1}, q^{1+z}/d, q^{1-z}/d; q\right)_\infty}$$

$$\times\ {}_4\varphi_3\left(\begin{matrix} q^{-\nu}, q^{1-\nu}/ad, q^{1-\nu}/bd, q^{1-\nu}/cd \\ q^{2-2\nu}/abcd, q^{1-\nu+z}/d, q^{1-\nu-z}/d \end{matrix}; q, q\right)$$

$$+\frac{\left(q^{-\nu},q^{1-\nu}/ad,q^{1-\nu}/bd,q^{1-\nu}/cd,abcq^{\nu+z},abcq^{\nu-z};q\right)_\infty}{(ab,ac,bc,q^{1-2\nu}/abcd,q^{1+z}/d,q^{1-z}/d;q)_\infty}$$

$$\times\ {}_4\varphi_3\left(\begin{matrix}abcdq^{\nu-1},abq^\nu,acq^\nu,bcq^\nu\\ abcdq^{2\nu},abcq^{\nu+z},abcq^{\nu-z}\end{matrix};\ q,\ q\right)$$

$$=\frac{\left(aq^{-z},bq^{-z},cq^{-z},abcq^{\nu+z},q^{1-\nu+z}/d;q\right)_\infty}{(ab,ac,bc,q^{-2z},q^{1+z}/d;q)_\infty}$$

$$\times\ {}_4\varphi_3\left(\begin{matrix}aq^z,bq^z,cq^z,q^{1+z}/d\\ q^{1+2z},abcq^{\nu+z},q^{1-\nu+z}\end{matrix};\ q,\ q\right)$$

$$+\frac{\left(aq^{z},bq^{z},cq^{z},abcq^{\nu-z},q^{1-\nu-z}/d;q\right)_\infty}{(ab,ac,bc,q^{2z},q^{1-z}/d;q)_\infty}$$

$$\times\ {}_4\varphi_3\left(\begin{matrix}aq^{-z},bq^{-z},cq^{-z},q^{1-z}/d\\ q^{1-2z},abcq^{\nu-z},q^{1-\nu-z}\end{matrix};\ q,\ q\right)$$

(these representations reveal that $u_\nu(x;a,b,c;d)$ is symmetric in a, b, and c) and [**119**], [**143**]

$$u_\nu(x;a,b,c;d)=\frac{\left(bq^{1+\nu+z},cq^{1+\nu+z},dq^{\nu-z},abcq^{\nu+z},bcdq^{\nu+z};q\right)_\infty}{(q^{1+\nu},bdq^\nu,cdq^\nu,bcq^{1+\nu+2z};q)_\infty}$$

$$\times\frac{\left(aq^{-z},bq^{-z},cq^{-z},q^{1-\nu+z}/d;q\right)_\infty}{(ab,ac,bc,q^{-2z},q^{1+z}/d;q)_\infty}$$

$$\times\ {}_8W_7\left(bcq^{\nu+2z};bcq^\nu,bq^z,cq^z,q^{1+z}/a,q^{1+z}/d;q,\ adq^\nu\right)$$

$$+\frac{\left(bq^{1+\nu-z},cq^{1+\nu-z},dq^{\nu+z},abcq^{\nu-z},bcdq^{\nu-z};q\right)_\infty}{(q^{1+\nu},bdq^\nu,cdq^\nu,bcq^{1+\nu-2z};q)_\infty}$$

$$\times\frac{\left(aq^{z},bq^{z},cq^{z},q^{1-\nu-z}/d;q\right)_\infty}{(ab,ac,bc,q^{2z},q^{1-z}/d;q)_\infty}$$

$$\times\ {}_8W_7\left(bcq^{\nu-2z};bcq^\nu,bq^{-z},cq^{-z},q^{1-z}/a,q^{1-z}/d;q,\ adq^\nu\right).$$

Verify that [**120**]

$$\lim_{q\to1^-}u_\nu(x;q^{\alpha+1/2},q^{\alpha+3/4},-q^{\beta+1/2};-q^{\beta+3/4})$$

$$={}_2F_1\left(-\nu,\alpha+\beta+\nu+1;\alpha+1;\frac{1-x}{2}\right).$$

Derive the following *difference-differentiation formula* [**143**]

$$\frac{\delta}{\delta x(z)}u_\nu(x(z);a,b,c;d)=\frac{2q}{(1-q)d}\,\frac{\left(1-q^{-\nu}\right)\left(1-abcdq^{\nu-1}\right)}{(1-ab)(1-ac)(1-bc)}$$

$$\times\ u_{\nu-1}\left(x(z);aq^{1/2},bq^{1/2},cq^{1/2};dq^{1/2}\right),$$

where $\delta f(z)=f(z+1/2)-f(z-1/2)$.

The *Pearson-type equation* can be rewritten in the form

$$\frac{\rho(z+1)}{\rho(z)}=\frac{\sigma(-z)}{\sigma(z+1)}=q^{-4z-2}\,q^{2z+1}\times\frac{(1-aq^z)(1-bq^z)(1-cq^z)(1-q^{-z}/d)}{(1-aq^{-z-1})(1-bq^{-z-1})(1-cq^{-z-1})(1-q^{z+1}/d)},$$

see Ex. 2.21 and [**141**], [**143**]. Verify that

$$\frac{\rho_0(z+1)}{\rho_0(z)}=q^{-4z-2}\quad\text{if}\quad \rho_0(z)=\frac{(q^{2z},q^{-2z};q)_\infty}{q^z-q^{-z}},$$

$$\frac{\rho_\alpha(z+1)}{\rho_\alpha(z)}=q^{-2z-1}\quad\text{if}\quad \rho_\alpha(z)=(\alpha q^z,\alpha q^{-z},q^{1+z}/\alpha,q^{1-z}/\alpha;q)_\infty,$$

$$\frac{\rho_a(z+1)}{\rho_a(z)}=\frac{1-aq^{-z-1}}{1-aq^z}\quad\text{if}\quad \rho_a(z)=(aq^z,aq^{-z};q)_\infty$$

and obtain the following solution

$$\rho(z)=\frac{(q^z-q^{-z})^{-1}(q^{2z},q^{-2z},q^{1+z}/d,q^{1-z}/d;q)_\infty}{(\alpha q^z,\alpha q^{-z},q^{1+z}/\alpha,q^{1-z}/\alpha,aq^z,aq^{-z},bq^z,bq^{-z},cq^z,cq^{-z};q)_\infty},$$

where α is an arbitrary parameter.
Rewrite the Pearson equation as

$$\frac{\rho(z+1)}{\rho(z)}=\frac{(1-aq^z)(1-bq^z)(1-q^{-z}/c)(1-q^{-z}/d)}{(1-aq^{-z-1})(1-bq^{-z-1})(1-q^{z+1}/c)(1-q^{z+1}/d)}$$

and find the corresponding solution

$$\rho(z)=\frac{\left(q^{1+z}/c,q^{1-z}/c,q^{1+z}/d,q^{1-z}/d;q\right)_\infty}{(aq^z,aq^{-z},bq^z,bq^{-z};q)_\infty}.$$

Find the following multiparameter extension

$$\rho(z)=\frac{(q^z-q^{-z})^{-1}(q^{2z},q^{-2z},q^{1+z}/d,q^{1-z}/d;q)_\infty}{(\alpha q^z,\alpha q^{-z},q^{1+z}/\alpha,q^{1-z}/\alpha,aq^z,aq^{-z},bq^z,bq^{-z},cq^z,cq^{-z};q)_\infty}\times\prod_{k=1}^{n}\frac{(\beta_kq^z,\beta_kq^{-z},q^{1+z}/\beta_k,q^{1-z}/\beta_k;q)_\infty}{(\alpha_kq^z,\alpha_kq^{-z},q^{1+z}/\alpha_k,q^{1-z}/\alpha_k;q)_\infty},$$

where $\alpha_i\neq\beta_k$. [**143**]
Introduce

$$v_\nu(x(z);a,b,c;d):=\left(q^{1+z}/d,q^{1-z}/d;q\right)_\infty\,u_\nu(x(z);a,b,c;d)$$

and prove that $v_\nu(x(z);a,b,c;d)$ is an entire function in the complex z-plane.
Prove that [**143**]

$$\int_0^\pi\frac{v_\mu(\cos\theta;a,b,c;d)\;v_\nu(\cos\theta;a,b,c;d)}{(ae^{i\theta},ae^{-i\theta},be^{i\theta},be^{-i\theta},ce^{i\theta},ce^{-i\theta},qe^{i\theta}/d,qe^{-i\theta}/d;q)_\infty}$$

$$\times \frac{\left(e^{2i\theta},\, e^{-2i\theta}; q\right)_\infty}{\left(\alpha e^{i\theta}, \alpha e^{-i\theta}, qe^{i\theta}/\alpha, qe^{-i\theta}/\alpha; q\right)_\infty}\, d\theta$$

$$= \left(q, q, \alpha a, qa/\alpha, \alpha b, qb/\alpha, \alpha c, qc/\alpha, q\alpha/d, q^2/\alpha d; q\right)_\infty^{-1}$$

$$\times \frac{-4\pi q^{1/2} d}{1-q}\ \frac{W\left(v_\mu\left(\eta; a, b, c; d\right),\ v_\nu\left(\eta; a, b, c; d\right)\right)}{\lambda_\mu - \lambda_\nu}$$

and

$$\int_0^\pi \frac{\left(v_\nu(\cos\theta; a, b, c; d)\right)^2}{\left(ae^{i\theta}, ae^{-i\theta}, be^{i\theta}, be^{-i\theta}, ce^{i\theta}, ce^{-i\theta}, qe^{i\theta}/d, qe^{-i\theta}/d; q\right)_\infty}$$

$$\times \frac{\left(e^{2i\theta},\, e^{-2i\theta}; q\right)_\infty}{\left(\alpha e^{i\theta}, \alpha e^{-i\theta}, qe^{i\theta}/\alpha, qe^{-i\theta}/\alpha; q\right)_\infty}\, d\theta$$

$$= \left(q, q, \alpha a, qa/\alpha, \alpha b, qb/\alpha, \alpha c, qc/\alpha, q\alpha/d, q^2/\alpha d; q\right)_\infty^{-1}$$

$$\times \frac{-4\pi q^{1/2} d}{1-q} \left[\frac{\partial}{\partial\lambda_\nu} v_\nu(\eta; a, b, c; d) \left(\frac{\nabla}{\nabla x} v_\nu(x; a, b, c; d) \right) \Bigg|_{x=\eta} \right.$$

$$\left. - v_\nu(\eta; a, b, c; d) \frac{\partial}{\partial\lambda_\nu} \left(\frac{\nabla}{\nabla x} v_\nu(x; a, b, c; d) \right) \Bigg|_{x=\eta} \right].$$

Choose eigenvalues μ and ν as ε-solutions of the equation

$$v_\varepsilon\left(\frac{1}{2}\left(\alpha + \alpha^{-1}\right); a, b, c; d\right) = 0$$

and derive the *continuous orthogonality relation* of the ${}_8\varphi_7$ functions under consideration [**143**],

$$\int_0^\pi \frac{v_\mu(\cos\theta; a, b, c; d)\ v_\nu(\cos\theta; a, b, c; d)}{\left(ae^{i\theta}, ae^{-i\theta}, be^{i\theta}, be^{-i\theta}, ce^{i\theta}, ce^{-i\theta}, qe^{i\theta}/d, qe^{-i\theta}/d; q\right)_\infty}$$

$$\times \frac{\left(e^{2i\theta},\, e^{-2i\theta}; q\right)_\infty}{\left(\alpha e^{i\theta}, \alpha e^{-i\theta}, qe^{i\theta}/\alpha, qe^{-i\theta}/\alpha; q\right)_\infty}\, d\theta = 0$$

if $\mu \neq \nu$, and

$$\int_0^\pi \frac{\left(v_\nu(\cos\theta; a, b, c; d)\right)^2}{\left(ae^{i\theta}, ae^{-i\theta}, be^{i\theta}, be^{-i\theta}, ce^{i\theta}, ce^{-i\theta}, qe^{i\theta}/d, qe^{-i\theta}/d; q\right)_\infty}$$

$$\times \frac{\left(e^{2i\theta},\, e^{-2i\theta}; q\right)_\infty}{\left(\alpha e^{i\theta}, \alpha e^{-i\theta}, qe^{i\theta}/\alpha, qe^{-i\theta}/\alpha; q\right)_\infty}\, d\theta$$

$$= \left(q, q, \alpha a, qa/\alpha, \alpha b, qb/\alpha, \alpha c, qc/\alpha, q\alpha/d, q^2/\alpha d; q\right)_\infty^{-1}$$

$$\times \frac{-4\pi q^{1/2} d}{1-q} \frac{\partial}{\partial\lambda_\nu} v_\nu(\eta; a, b, c; d) \left(\frac{\nabla}{\nabla x} v_\nu(x; a, b, c; d) \right) \Bigg|_{x=\eta}$$

if $\mu = \nu$, respectively. Here $\max\left(|a|, |b|, |c|, |q/d|\right) < 1$.
Verify that the $v_\varepsilon(x; a, b, c; d)$ with $\varepsilon = \mu, \nu$ are real-valued functions of x for real ε which are orthogonal with respect to a positive weight

function when all parameters a, b, c, d, and α are real, or when any two of a, b, c are complex conjugate and all other parameters are real.

Show that

$$
\begin{aligned}
& d_\nu^2 \\
&= \int_0^\pi \frac{v_\nu^2(\cos\theta; a, b, c; d)}{(ae^{i\theta}, ae^{-i\theta}, be^{i\theta}, be^{-i\theta}, ce^{i\theta}, ce^{-i\theta}, qe^{i\theta}/d, qe^{-i\theta}/d; q)_\infty} \\
&\quad \times \frac{(e^{2i\theta}, e^{-2i\theta}; q)_\infty}{(\alpha e^{i\theta}, \alpha e^{-i\theta}, qe^{i\theta}/\alpha, qe^{-i\theta}/\alpha; q)_\infty} \, d\theta \\
&= \frac{2\pi}{(1-ac)(q, ab; q)^2} \frac{(qa/d, abq^\nu, bcq^\nu, q^{1-\nu}/ad, q^{1-\nu}/cd; q)_\infty}{(bc, b/a, q/cd, \alpha a, qa/\alpha, \alpha c, qc/\alpha; q)_\infty} \\
&\quad \times \sum_{n=0}^\infty q^n \frac{(q^{-\nu}, abcdq^{\nu-1}, \alpha c, qc/\alpha; q)_n}{(q, \alpha cq, bc, cd; q)_n} \\
&\quad \times {}_5\varphi_4\left(\begin{matrix} acq^n, acq^\nu, q^{1-\nu}/bd, \alpha a, qa/\alpha \\ acq^{n+1}, ac, qa/b, qa/d \end{matrix} ; q, q \right) \\
&+ \frac{2\pi}{(1-bc)(q, ab; q)^2} \frac{(qb/d, abq^\nu, acq^\nu, q^{1-\nu}/bd, q^{1-\nu}/cd; q)_\infty}{(ac, a/b, q/cd, \alpha b, qb/\alpha, \alpha c, qc/\alpha; q)_\infty} \\
&\quad \times \sum_{n=0}^\infty q^n \frac{(q^{-\nu}, abcdq^{\nu-1}, \alpha c, qc/\alpha; q)_n}{(q, q\alpha c, ac, cd; q)_n} \\
&\quad \times {}_5\varphi_4\left(\begin{matrix} bcq^n, bcq^\nu, q^{1-\nu}/ad, \alpha b, qb/\alpha \\ bcq^{n+1}, bc, qb/a, qb/d \end{matrix} ; q, q \right) \\
&+ \frac{2\pi\, (qa/d; q)_\infty^2}{(1-qa/d)(q, ab, bc; q)^2} \frac{(q^{-\nu}, abcdq^{\nu-1}, q^{1-\nu}/ad, bcq^\nu, qb/d; q)_\infty}{(ac, b/a, cd/q, \alpha a, qa/\alpha, q\alpha/d, q^2/\alpha d; q)_\infty} \\
&\quad \times \sum_{n=0}^\infty q^n \frac{(abq^\nu, q^{1-\nu}/cd, q\alpha/d, q^2/\alpha d; q)_n}{(q, qb/d, q^2a/d, q^2/cd; q)_n} \\
&\quad \times {}_5\varphi_4\left(\begin{matrix} aq^{n+1}/d, acq^\nu, q^{1-\nu}/bd, \alpha a, qa/\alpha \\ aq^{n+2}/d, ac, qa/b, qa/d \end{matrix} ; q, q \right) \\
&+ \frac{2\pi\, (qb/d; q)_\infty^2}{(1-qb/d)(q, ab, ac; q)^2} \frac{(q^{-\nu}, abcdq^{\nu-1}, q^{1-\nu}/bd, acq^\nu, qa/d; q)_\infty}{(bc, a/b, cd/q, \alpha b, qb/\alpha, q\alpha/d, q^2/\alpha d; q)_\infty} \\
&\quad \times \sum_{n=0}^\infty q^n \frac{(abq^\nu, q^{1-\nu}/cd, q\alpha/d, q^2/\alpha d; q)_n}{(q, qa/d, q^2b/d, q^2/cd; q)_n} \\
&\quad \times {}_5\varphi_4\left(\begin{matrix} bq^{n+1}/d, bcq^\nu, q^{1-\nu}/ad, \alpha b, qb/\alpha \\ bq^{n+2}/d, bc, qb/a, qb/d \end{matrix} ; q, q \right),
\end{aligned}
$$

where the second and the fourth terms are equal to the first and the third ones, respectively, with a and b interchanged. [**143**]

Show that the *difference analog of the Wronskian*, see Ex. 2.21 for the definition, of the particular solutions u_1 and u_2 is

$$\begin{aligned} &W\left(u_1(z),\ u_2(z)\right) \\ &\quad = \frac{2q\left(c/d, qd/c, q^{-\nu}, abcdq^{\nu-1}, abq^{\nu}, q^{1-\nu}/cd; q\right)_\infty}{(1-q)c\left(ab, ab, ac, ad, bc, bd; q\right)_\infty} \\ &\qquad \times \frac{\left(aq^z, aq^{1-z}, bq^z, bq^{1-z}; q\right)_\infty}{\left(q^z/c, q^{1-z}/c, q^z/d, q^{1-z}/d; q\right)_\infty} \end{aligned}$$

and derive the following identity

$$\begin{aligned} &\frac{d\left(ac, adq; q\right)_\infty}{\left(q^{2-\nu}a/c, q^{1-\nu}a/d; q\right)_\infty} \\ &\quad \times \frac{\left(q^{1-\nu+z}/c, q^{2-\nu-z}/c, q^{1-\nu+z}/d, q^{1-\nu-z}/d; q\right)_\infty}{\left(q^z/c, q^{1-z}/c, q^{1+z}/d, q^{1-z}/d; q\right)_\infty} \\ &\quad \times {}_8W_7\left(q^{1-\nu}a/c; q^{1-\nu}, q^{1-\nu}/bc, q^{1-\nu}/cd, aq^z, aq^{1-z}; q,\ q\right) \\ &\quad \times {}_8W_7\left(q^{-\nu}a/d; q^{-\nu}, q^{1-\nu}/bd, q^{1-\nu}/cd, aq^z, aq^{-z}; q,\ q\right) \\ &- \frac{c\left(ad, acq; q\right)_\infty}{\left(q^{2-\nu}a/d, q^{1-\nu}a/c; q\right)_\infty} \\ &\quad \times \frac{\left(q^{1-\nu+z}/d, q^{2-\nu-z}/d, q^{1-\nu+z}/c, q^{1-\nu-z}/c; q\right)_\infty}{\left(q^z/d, q^{1-z}/d, q^{1+z}/c, q^{1-z}/c; q\right)_\infty} \\ &\quad \times {}_8W_7\left(q^{1-\nu}a/d; q^{1-\nu}, q^{1-\nu}/bd, q^{1-\nu}/cd, aq^z, aq^{1-z}; q,\ q\right) \\ &\quad \times {}_8W_7\left(q^{-\nu}a/c; q^{-\nu}, q^{1-\nu}/bc, q^{1-\nu}/cd, aq^z, aq^{-z}; q,\ q\right) \\ &= \frac{\left(c/d, qd/c, q^{1-\nu}, abcdq^{\nu}, abq^{\nu}, q^{1-\nu}/cd; q\right)_\infty}{c\left(ab, abq, qa/c, qa/d, bcq^{\nu}, bdq^{\nu}; q\right)_\infty} \\ &\quad \times \frac{d\left(aq^z, aq^{1-z}, bq^z, bq^{1-z}; q\right)_\infty}{\left(q^z/c, q^{1-z}/c, q^z/d, q^{1-z}/d; q\right)_\infty}, \end{aligned}$$

where the second term on the left side is the same as the first one with c and d interchanged. [**143**]

Show that solutions $u_1(z)$ and $u_2(z)$ are linearly dependent when parameter ν is an integer and that in this case both solutions are just the Askey–Wilson polynomials (4.6.1) up to a factor, which are related by Sears' transformation (A.3.12)–(A.3.13).
On the other hand, use the analog of the Wronskian above in order to prove that there is no analog of Sears's transformation at the level of nonterminating very-well-poised ${}_8\varphi_7$ functions.

[*Note*: Another orthogonality property of very-well-poised ${}_8\varphi_7$ functions was established by Askey, Rahman, and Suslov [**14**]. Recently

Koelink and Stokman [90], [91] introduced the Askey–Wilson function transform on the basis of yet another orthogonality property of the very-well-poised ${}_8\varphi_7$ functions under consideration when the measure has an absolutely continuous part and infinite series of mass points.]

(9) *The $\mathcal{A}\Delta$ wave equation* [39]. Consider analytic difference wave equation, or q-wave equation, of the form

$$\frac{\delta^2 u(x,y)}{\delta x(t)^2} = c^2 \frac{\delta^2 u(x,y)}{\delta y(s)^2},$$

where c is a constant, $x(t) = \left(q^t + q^{-t}\right)/2$ is the "time lattice" and $y(s) = \left(q^s + q^{-s}\right)/2$ is the "space lattice", subject to the following "boundary",

$$u(x,0) = u(x,\eta) = 0, \qquad \eta = \left(q^{1/4} + q^{-1/4}\right)/2,$$

and "initial",

$$\left.\frac{\delta u(x,y)}{\delta x(t)}\right|_{x=0} = \frac{2q^{1/4}}{1-q}\, g(y),$$

conditions.

Use separation of the variables in order to find the following solution of the above "initial value" problem

$$u(x,y) = \sum_{n=1}^{\infty} \left(C_n\, C_q(x;c\omega_n) + D_n\, S_q(x;c\omega_n)\right)\, S_q(y;\omega_n),$$

where $0 = \omega_0 < \omega_1 < \omega_2 < \ldots$ are the positive zeros of $S_q(\eta;\omega)$, C_n and D_n are the q-Fourier coefficients in the expansions

$$\begin{aligned} f(y) &= \sum_{n=1}^{\infty} C_n\, C_q(y;\omega_n), \\ g(y) &= c\sum_{n=1}^{\infty} D_n\, \omega_n C_q(y;\omega_n). \end{aligned}$$

Show that this solution can be written in the form

$$\begin{aligned} &u(x,y) \\ &= \sum_{n=1}^{\infty} \left(\frac{C_n}{2}\, S_q(y,x;\omega_n,c\omega_n) - \frac{D_n}{2}\, C_q(y,x;\omega_n,c\omega_n)\right) \\ &+ \sum_{n=1}^{\infty} \left(\frac{C_n}{2}\, S_q(y,-x;\omega_n,c\omega_n) + \frac{D_n}{2}\, C_q(y,-x;\omega_n,c\omega_n)\right), \end{aligned}$$

which resembles the d'Alembert solution

$$u(x,t) = f_1(x+ct) + f_2(x-ct)$$

of the classical wave equation

$$\frac{\partial^2 u}{\partial t^2} = c^2 \frac{\partial^2 u}{\partial x^2}.$$

(10) *The $\mathcal{A}\Delta$ heat equation* [**39**]. Consider analytic difference heat equation, or q-heat equation, of the form

$$\frac{\delta u(x,y)}{\delta x(t)} = c^2 \frac{\delta^2 u(x,y)}{\delta y(s)^2},$$

where c is a constant, $x(t) = \left(q^t + q^{-t}\right)/2$ is the "time lattice" and $y(s) = \left(q^s + q^{-s}\right)/2$ is the "space lattice", subject to the following "boundary",

$$\left.\frac{\delta u(x,y)}{\delta y(s)}\right|_{y=0} = \left.\frac{\delta u(x,y)}{\delta y(s)}\right|_{y=\eta} = 0, \qquad \eta = \left(q^{1/4} + q^{-1/4}\right)/2,$$

and "initial",

$$u(0,y) = f(y),$$

conditions.

Use separation of the variables to find the following solution of the above "initial value" problem

$$u(x,y) = D_0 + \sum_{n=1}^{\infty} D_n\ C_q(y;\omega_n)\ \mathcal{E}_q(x;-\lambda_n),$$

where $0 = \omega_0 < \omega_1 < \omega_2 < \ldots$ are the positive zeros of $S_q(\eta;\omega)$,

$$\lambda_n = \frac{2q^{1/4}}{1-q}\, c^2 \omega_n^2,$$

and D_n are the q-Fourier coefficients in the expansion

$$f(y) = D_0 + \sum_{n=1}^{\infty} D_n\ C_q(y;\omega_n).$$

[*Remark*: See [**39**] for more details on similar solutions of other basic analogs of equations of mathematical physics.]

CHAPTER 6

Investigation of Basic Fourier Series

In this chapter we shall present elementary results on the convergence of the basic Fourier series, including a proof of completeness of the basic trigonometric systems, and will establish several convenient tools, such as asymptotics of zeros, which are important for practical investigation of these series in the next chapters. Methods of summation and a few explicit examples of q-Fourier series will be also discussed among other things.

6.1. Uniform Bounds

In order to establish completeness of the basic trigonometric systems as a first step to investigation of convergence of the corresponding q-Fourier series, one needs to provide convenient uniform upper bounds for the q-exponential function. We shall use a connection between the basic trigonometric functions and the continuous q-Hermite polynomials given by

$$H_n(\cos\theta|q) = \sum_{k=0}^{n} \frac{(q;q)_n}{(q;q)_k (q;q)_{n-k}} \, e^{i(n-2k)\theta}. \tag{6.1.1}$$

These polynomials have two generating functions

$$\sum_{n=0}^{\infty} \frac{r^n}{(q;q)_n} H_n(\cos\theta|q) = \frac{1}{(re^{i\theta}, re^{-i\theta};q)_\infty}, \qquad |r|<1 \tag{6.1.2}$$

and

$$\sum_{n=0}^{\infty} \frac{q^{n^2/4}}{(q;q)_n} \alpha^n H_n(\cos\theta|q) = e(\cos\theta, \alpha), \tag{6.1.3}$$

where by the definition

$$e(\cos\theta, \alpha) := \left(q\alpha^2;q^2\right)_\infty \mathcal{E}_q(\cos\theta;\alpha), \tag{6.1.4}$$

see, for example, [**46**], [**76**], and [**140**]; the proof of (6.1.3) is given in Section 3.4. See also generating relation (4.6.31).

The following simple inequalities hold [**26**].

LEMMA 6.1. *Let* $-\cosh\tau \le -1 \le x \le 1 \le \cosh\tau$, *where* $x = \cos\theta$, $0 \le \theta \le \pi$, *and* $\tau \ge 0$. *Then*

$$|e(\cos\theta;\alpha)| \le e(\cosh\tau;|\alpha|) \tag{6.1.5}$$

and

$$e(\cosh\tau;|\alpha|) \le e(\cosh\tau_1;|\alpha|) \tag{6.1.6}$$

if $\tau < \tau_1$.

PROOF. Let us rewrite (6.1.1) as

$$H_n(\cos\theta|q) = 2\sum_{k=0}^{[n/2]} \frac{(q;q)_n}{(q;q)_k\,(q;q)_{n-k}} \cos(n-2k)\,\theta. \tag{6.1.7}$$

Then

$$\begin{aligned}|H_n(\cos\theta|q)| &\le 2\sum_{k=0}^{[n/2]} \frac{(q;q)_n}{(q;q)_k\,(q;q)_{n-k}} \cosh(n-2k)\,\tau \\ &= H_n(\cosh\tau|q)\end{aligned} \tag{6.1.8}$$

and

$$\left|\left(q\alpha^2;q^2\right)_\infty \mathcal{E}_q(\cos\theta;\alpha)\right| \le \sum_{n=0}^{\infty} \frac{q^{n^2/4}}{(q;q)_n}\,|\alpha|^n\,|H_n(\cos\theta|q)| \tag{6.1.9}$$

$$\le \sum_{n=0}^{\infty} \frac{q^{n^2/4}}{(q;q)_n}\,|\alpha|^n\,H_n(\cosh\tau|q) = \left(q\,|\alpha|^2;q^2\right)_\infty \mathcal{E}_q(\cosh\tau;|\alpha|)$$

by (6.1.3)–(6.1.4) and (6.1.8). This completes the proof of (6.1.5). The monotonicity property (6.1.6) follows from the monotonicity of the hyperbolic cosine function. □

When $\tau = 0$ this lemma provides a natural uniform bound for the basic exponential function $e(\cos\theta,\alpha)$ by its value at the end point of the orthogonality interval $[-1,1]$:

$$|e(\cos\theta,\alpha)| \le e(1,|\alpha|), \qquad 0 \le \theta \le \pi \tag{6.1.10}$$

for all α. Unfortunately, there is a removable singularity in the expression of $e(\cos\theta,\alpha)$ at $x=1$ for $|\alpha|>1$; see Eqs. (5.6.4) and (2.4.10)–(2.4.11). One way to resolve this problem is go a bit further in the variable $x=\cos\theta$ and to establish the following convenient uniform bounds outside the interval of orthogonality.

LEMMA 6.2. *Let* $-x(\varepsilon) < -1 \le x = \cos\theta \le 1 < x(\varepsilon)$, *where* $x(\varepsilon) = (q^\varepsilon + q^{-\varepsilon})/2$, $0<q<1$ *and* $0<\varepsilon<1/2$. *The following inequalities hold*

$$\begin{aligned}|e(\cos\theta,i\omega)| &= \left(-q\omega^2;q^2\right)_\infty |\mathcal{E}_q(\cos\theta;i\omega)| \\ &< \frac{\left(-q^{1/4-\varepsilon}\,|\omega|;q^{1/2}\right)_\infty}{\left(q,q^{2\varepsilon},q^{1-2\varepsilon};q\right)_\infty}\end{aligned} \tag{6.1.11}$$

and

$$\begin{aligned}|e(\cos\theta,i\omega)| &= \left(-q\omega^2;q^2\right)_\infty |\mathcal{E}_q(\cos\theta;i\omega)| \\ &< \frac{\left(q^{1/2},-q^{1/4-\varepsilon}\,|\omega|,-q^{1/4+\varepsilon}/\,|\omega|;q^{1/2}\right)_\infty}{\left(q,q^{2\varepsilon},q^{1-2\varepsilon};q\right)_\infty}\end{aligned} \tag{6.1.12}$$

for all real values of the ω.

PROOF. By (6.1.8)

$$H_n(\cos\theta|q) \le H_n(x(\varepsilon)|q), \tag{6.1.13}$$

where $x(\varepsilon) = (q^{\varepsilon} + q^{-\varepsilon})/2$ and $q^{\varepsilon} = e^{-\tau}$. Using (7.4.5) of [**46**] with $e^{i\theta} = q^{\varepsilon}$, see also Ex. 6.1, one gets

$$\begin{aligned} H_n(x(\varepsilon)|q) &= \frac{q^{-n\varepsilon}}{(q^{2\varepsilon};q)_\infty} \sum_{k=0}^{\infty} \frac{q^{(1+n)k}}{(q,q^{1-2\varepsilon};q)_k} \\ &\quad - \frac{q^{(n+2)\varepsilon}}{(1-q^{2\varepsilon})(q^{1-2\varepsilon};q)_\infty} \sum_{k=0}^{\infty} \frac{q^{(1+n)k}}{(q,q^{1+2\varepsilon};q)_k} \\ &< \frac{q^{-n\varepsilon}}{(q^{2\varepsilon};q)_\infty} \sum_{k=0}^{\infty} \frac{q^{(1+n)k}}{(q,q^{1-2\varepsilon};q)_k} \end{aligned} \tag{6.1.14}$$

when $0 < \varepsilon < 1/2$. But $(a;q)_k > (a;q)_\infty$ for $0 < a, q < 1$, and

$$\sum_{k=0}^{\infty} \frac{q^{(1+n)k}}{(q,q^{1-2\varepsilon};q)_k} < \frac{1}{(q^{1-2\varepsilon};q)_\infty} \sum_{k=0}^{\infty} \frac{q^{(1+n)k}}{(q;q)_k} = \frac{(q;q)_n}{(q,q^{1-2\varepsilon};q)_\infty} \tag{6.1.15}$$

by (A.2.1). Therefore

$$|H_n(\cos\theta|q)| < q^{-n\varepsilon} \frac{(q;q)_n}{(q,q^{2\varepsilon},q^{1-2\varepsilon};q)_\infty}. \tag{6.1.16}$$

A similar inequality holds for the continuous q-ultraspherical polynomials; see Ex. 6.1. An extension to a region in the complex x-plane is given by (6.9.13).

The uniform bounds (6.1.11) and (6.1.12) for the $\mathcal{E}_q(\cos\theta; i\omega)$ can be obtained now by means of the generating function (6.1.3). Indeed, for all real ω and $0 < q < 1$ the following inequalities hold

$$\left(-q\omega^2;q^2\right)_\infty |\mathcal{E}_q(x;i\omega)| \le \sum_{k=0}^{\infty} \frac{q^{k^2/4}|\omega|^k}{(q;q)_k} |H_k(x|q)| \tag{6.1.17}$$

$$\begin{aligned} &< \frac{1}{(q,q^{2\varepsilon},q^{1-2\varepsilon};q)_\infty} \sum_{k=0}^{\infty} q^{k^2/4-k\varepsilon} |\omega|^k \\ &< \frac{1}{(q,q^{2\varepsilon},q^{1-2\varepsilon};q)_\infty} \sum_{k=0}^{\infty} \frac{q^{(k^2-k)/4}\left(q^{1/4-\varepsilon}|\omega|\right)^k}{\left(q^{1/2};q^{1/2}\right)_k} \\ &= \frac{\left(-q^{1/4-\varepsilon}|\omega|;q^{1/2}\right)_\infty}{(q,q^{2\varepsilon},q^{1-2\varepsilon};q)_\infty}, \end{aligned}$$

which proves (6.1.11). In order to derive (6.1.12), one can rewrite the second line here as

$$\sum_{k=0}^{\infty} q^{k^2/4}\left(q^{-\varepsilon}|\omega|\right)^k < \sum_{k=-\infty}^{\infty} q^{k^2/4}\left(q^{-\varepsilon}|\omega|\right)^k \tag{6.1.18}$$

$$= \left(q^{1/2}, -q^{1/4-\varepsilon}\,|\omega|\,, -q^{1/4+\varepsilon}/\,|\omega|\,; q^{1/2}\right)_\infty$$

by the Jacobi triple product identity (A.2.4). □

We shall see later that the uniform bounds (6.1.11)–(6.1.12) are convenient tools for investigation of convergence of the basic Fourier series. Extensions of the bounds (6.1.11)–(6.1.12) to a region in the complex x-plane will be discussed in Section 6.9; see (6.9.7)–(6.9.8).

6.2. Completeness of Basic Trigonometric System

Completeness of the trigonometric system $\left\{e^{i\pi nx}\right\}_{n=-\infty}^{\infty}$ on the interval $(-1,1)$ is one of the fundamental facts in the theory of trigonometric series; see, for example, **[3]**, **[20]**, **[93]**, **[99]**, **[101]**, **[108]**, **[156]**, and **[166]**. In this section we shall follow the original paper **[26]** with somewhat different details in order to establish a similar property for the system of the basic trigonometric function $\{\mathcal{E}_q(x; i\omega_n)\}$, where $\omega_n = 0, \pm\omega_1, \pm\omega_2, \pm\omega_3, ...$ and $\omega_0 = 0 < \omega_1 < \omega_2 < \omega_3 < ...$ are nonnegative zeros of the basic sine function $S_q(\eta;\omega)$. Further investigation of the completeness of basic trigonometric systems in certain $\mathcal{L}^p$ spaces will be given in the next chapter.

6.2.1. Preliminaries. We first establish elementary results on the analyticity of the basic exponential and basic trigonometric functions on a q-quadratic grid.

LEMMA 6.3. *The following functions*

$$e(x,\alpha) = \left(q\alpha^2; q^2\right)_\infty \mathcal{E}_q(x;\alpha)\,, \tag{6.2.1}$$

$$s(x,\omega) = \left(-q\omega^2; q^2\right) S_q(x;\omega) = \frac{1}{2i}\left(e(x,i\omega) - e(x,-i\omega)\right), \tag{6.2.2}$$

$$c(x,\omega) = \left(-q\omega^2; q^2\right) C_q(x;\omega) = \frac{1}{2}\left(e(x,i\omega) + e(x,-i\omega)\right) \tag{6.2.3}$$

are entire functions of order zero in α and ω, respectively, for all finite values of x.

PROOF. The generating function (6.1.3) provides a power series expansion in α for the function (6.2.1),

$$e(x,\alpha) = \sum_{n=0}^{\infty} h_n\, \alpha^n \tag{6.2.4}$$

with the coefficients

$$h_n = h_n(x) = \frac{q^{n^2/4}}{(q;q)_n}\, H_n(x|q)\,. \tag{6.2.5}$$

The radius of convergence of this series is infinity in view of

$$\frac{1}{R} = \lim_{n\to\infty}\left(|h_n|\right)^{1/n} = \lim_{n\to\infty}\left(\left|\frac{q^{n^2/4}}{(q;q)_n}\, H_n(x|q)\right|\right)^{1/n} = 0. \tag{6.2.6}$$

Indeed, due to (6.1.2) for any finite x there is a real r such that

$$\lim_{n\to\infty} \left| \frac{r^n}{(q;q)_n} \; H_n(x|q) \right| = 0 \tag{6.2.7}$$

and

$$\lim_{n\to\infty} \left(r^{-n} q^{n^2/4} \right)^{1/n} = 0, \qquad 0 < q < 1. \tag{6.2.8}$$

Thus, $e(x,\alpha)$ is an entire function in α. The order of this entire function is given by **[99]**

$$\lim_{n\to\infty} \left(\frac{n \log n}{-\log |h_n|} \right) = \lim_{n\to\infty} \left(\frac{n \log n}{-\log \left| q^{n^2/4} \; H_n(x|q) \, / \, (q;q)_n \right|} \right) = 0, \tag{6.2.9}$$

see also (B.1.10). Functions (6.2.2) and (6.2.3) are just a sum or difference of two functions of type (6.2.1), so they are also entire functions of order zero. This proves the lemma. □

6.2.2. Proof of Completeness. Now we have all the necessary tools to prove the completeness property of the basic trigonometric series established in [**26**]. Clearly the system $\{\mathcal{E}_q(x; i\omega_n)\}_{n=-\infty}^{\infty}$ is complete if the equivalent system $\{e(x, i\omega_n)\}_{n=-\infty}^{\infty}$ is closed; see, for example, [**3**], [**93**], [**108**].

Suppose that the system $\{e(x, i\omega_n)\}_{n=-\infty}^{\infty}$ is not closed on $(-1,1)$. This means that there exists at least one function, say $\phi(x)$, not identically zero, such that

$$\int_{-1}^{1} \phi(x) \; e(x, i\omega_n) \; \rho(x) \; dx = 0, \qquad n = 0, \pm 1, \pm 2, \ldots , \tag{6.2.10}$$

where $\rho(x)$ is the weight function in the orthogonality relation (5.6.5). Then the following function

$$f(\omega) = \int_{-1}^{1} \phi(x) \; e(x, i\omega) \; \rho(x) \; dx \tag{6.2.11}$$

is an entire function of order zero and $f(\omega_n) = 0$ for all $n = 0, \pm 1, \pm 2, \ldots$. Thus the study of closure amounts to the study of zeros of a certain entire function.

Indeed, suppose that $\phi(x)$ is integrable on $(-1,1)$,

$$\int_{-1}^{1} |\phi(x)| \; \rho(x) \; dx = A < \infty, \tag{6.2.12}$$

or $\phi \in \mathcal{L}^1_\rho(-1,1)$. Then

$$\begin{aligned} |f(\omega)| &\leq \int_{-1}^{1} |\phi(x) \; e(x, i\omega)| \; \rho(x) \; dx \\ &\leq e(x(\varepsilon), |\omega|) \int_{-1}^{1} |\phi(x)| \; \rho(x) \; dx \\ &= A \; e(x(\varepsilon), |\omega|) \end{aligned} \tag{6.2.13}$$

with $x(\varepsilon) = (q^{\varepsilon}+q^{-\varepsilon})/2$, $0 \le \varepsilon < 1/2$ by (6.1.5) and (6.2.12). This inequality shows that $f(\omega)$ is an entire function of order zero in view of Lemma 6.3.

Our next step is to show that function $f(\omega)$ is identically zero and the $\phi(x)$ does not exist. Consider the quotient

$$g(\omega) = \frac{f(\omega)}{s(\omega)} \tag{6.2.14}$$

of two entire functions, $f(\omega)$ and $s(\omega) = s(\eta,\omega)$ defined by (6.2.11) and (6.2.2) with $x = \eta = \left(q^{1/4}+q^{-1/4}\right)/2$, respectively. The functions $f(\omega)$ and $s(\omega)$ have the same zeros and the ω-zeros of the $s(\omega)$ are simple, so the $g(\omega)$ is an entire function. The order of this entire function is zero because both the $f(\omega)$ and the $s(\omega)$ are of order zero; see [**99**], Corollary of Theorem 12 on p. 24 and Appendix B. Moreover, this function $g(\omega)$ is bounded on the imaginary axis. Indeed, by (6.2.13)

$$|g(\omega)| = \left|\frac{f(\omega)}{s(\omega)}\right| \le A\left|\frac{e(x(\varepsilon),|\omega|)}{s(\eta,\omega)}\right| \tag{6.2.15}$$

and

$$\lim_{|\omega|\to\infty}\left|\frac{e(x(\varepsilon),|\omega|)}{s(\eta,\omega)}\right| = 0, \qquad 0<\varepsilon<1/4. \tag{6.2.16}$$

This limiting relation can be verified in the following manner. By (5.4.4)

$$\begin{aligned} s(\eta,\omega) &= \left(-q\omega^2;q^2\right)_\infty\ S_q(\eta;\omega) \\ &= \omega\frac{\left(q^{3/2}\omega^2,q^{1/2}/\omega^2;q^2\right)_\infty}{\left(q^{1/2};q\right)_\infty\left(q,-q/\omega^2;q^2\right)_\infty}\ C_q(\eta;\frac{q^{1/2}}{\omega}) \\ &+ \frac{\left(q^{1/2}\omega^2,q^{3/2}/\omega^2;q^2\right)_\infty}{\left(q^{1/2};q\right)_\infty\left(q,-q/\omega^2;q^2\right)_\infty}\ S_q(\eta;\frac{q^{1/2}}{\omega}) \end{aligned}$$

and on the imaginary axis for sufficiently large $|\omega|$:

$$|s(\eta,\omega)| \ge \left|\frac{\omega\left(q^{3/2}\omega^2,q^{1/2}/\omega^2;q^2\right)_\infty}{\left(q^{1/2};q\right)_\infty\left(q,-q/\omega^2;q^2\right)_\infty}\ C_q(\eta;\frac{q^{1/2}}{\omega})\right|. \tag{6.2.17}$$

In view of (6.1.11) and (6.2.17)

$$\begin{aligned} \left|\frac{e(x(\varepsilon),|\omega|)}{s(\eta,\omega)}\right| &< \frac{\left(-q^{1/4-\varepsilon}|\omega|;q^{1/2}\right)_\infty}{\left(q,q^{2\varepsilon},q^{1-2\varepsilon};q\right)_\infty\ |s(\eta,\omega)|} \\ &< \frac{\left(q^{1/2};q\right)_\infty\left|\left(q,-q/\omega^2;q^2\right)_\infty\right|\left(-q^{1/4-\varepsilon}|\omega|;q^{1/2}\right)_\infty}{\left(q,q^{2\varepsilon},q^{1-2\varepsilon};q\right)_\infty\ \left|\omega\left(q^{3/2}\omega^2,q^{1/2}/\omega^2;q^2\right)_\infty\ C_q(\eta;q^{1/2}/\omega)\right|}. \end{aligned}$$

But

$$\lim_{\omega\to\infty} C_q(\eta;q^{1/2}/\omega) = \lim_{\omega\to\infty}\left(-q/\omega^2;q^2\right)_\infty = \lim_{\omega\to\infty}\left(q^{1/2}/\omega^2;q^2\right)_\infty = 1$$

and as a result

$$\left| \frac{e\left(x\left(\varepsilon\right), |\omega|\right)}{s(\eta, \omega)} \right| < B \; \frac{\left(-q^{1/4-\varepsilon} \, |\omega| \, ; q^{1/2}\right)_\infty}{\left|\omega \left(q^{3/2}\omega^2; q^2\right)_\infty\right|}, \tag{6.2.18}$$

where B is some constant.

Now let us consider an arbitrary sequence $\{\omega_k\}$ of purely imaginary numbers such that $|\omega_k| \to \infty$. For sufficiently large k one can represent these numbers in an "exponential" form

$$\omega_k = \pm i\alpha_k q^{-n_k}, \qquad 1 \leq \alpha_k < q^{-1} \tag{6.2.19}$$

and by (I.7) of [**46**] or (1.4.1)

$$\begin{aligned} &\left(-q^{1/4-\varepsilon} \, |\omega_k| \, ; q^{1/2}\right)_\infty \\ &\quad = \left(-\alpha_k q^{1/4-\varepsilon}; q^{1/2}\right)_\infty \left(-q^{1/4+\varepsilon}/\alpha_k; q^{1/2}\right)_{2n_k} \; \alpha_k^{2n_k} q^{-n_k^2-2\varepsilon n_k} \\ &\quad < \left(-q^{-3/4-\varepsilon}, -q^{1/4+\varepsilon}; q^{1/2}\right)_\infty \; \alpha_k^{2n_k} q^{-n_k^2-2\varepsilon n_k} \end{aligned} \tag{6.2.20}$$

and

$$\begin{aligned} &|\omega_k| \left(q^{3/2}\omega_k^2; q^2\right)_\infty \\ &\quad = \left(-\alpha_k^2 q^{3/2}; q^2\right)_\infty \left(-q^{1/2}/\alpha_k^2; q^2\right)_{n_k} \; \alpha_k^{2n_k+1} q^{-n_k^2-n_k/2} \\ &\quad > \left(-q^{3/2}; q^2\right)_\infty \left(-q^{5/2}; q^2\right)_{n_k} \; \alpha_k^{2n_k+1} q^{-n_k^2-n_k/2}. \end{aligned} \tag{6.2.21}$$

Thus, for an arbitrary sequence $\omega_k \to \infty$ on the imaginary axis one gets

$$\begin{aligned} &\frac{\left(-q^{1/4-\varepsilon} \, |\omega_k| \, ; q^{1/2}\right)_\infty}{|\omega_k| \left(q^{3/2}\omega_k^2; q^2\right)_\infty} \\ &\quad < \frac{\left(-q^{-3/4-\varepsilon}, -q^{1/4+\varepsilon}; q^{1/2}\right)_\infty}{\alpha_k \left(-q^{3/2}; q^2\right)_\infty \left(-q^{5/2}; q^2\right)_{n_k}} \; q^{n_k(1-4\varepsilon)/2} \to 0 \end{aligned} \tag{6.2.22}$$

as $k \to \infty$ when $0 < \varepsilon < 1/4$. The limiting relation (6.2.16) results in from (6.2.18) and (6.2.22).

From (6.2.15) and (6.2.16), it follows that the entire function $g(\omega)$ is bounded on the imaginary axis. But an entire function of order zero bounded on a line must be a constant; see Theorems 21–22 and Corollary on pp. 49–51 of [**99**] or Appendix B, Theorem B.5. Then

$$f(\omega) = C \; s(\omega) \tag{6.2.23}$$

and, therefore,

$$\begin{aligned} |C| &= \left| \int_{-1}^{1} \phi(x) \; \frac{e\left(x, i\omega\right)}{s(\omega)} \; \rho(x) \; dx \right| \\ &\leq \int_{-1}^{1} \left| \phi(x) \; \frac{e\left(x, i\omega\right)}{s(\omega)} \right| \; \rho(x) \; dx \end{aligned} \tag{6.2.24}$$

$$\leq A \left| \frac{e(x(\varepsilon), |\omega|)}{s(\eta, \omega)} \right| \to 0$$

as $|\omega| \to \infty$ and $0 < \varepsilon < 1/4$. Thus, $f(\omega)$ is identically zero and

$$f(\omega) = \int_{-1}^{1} \phi(x)\ e(x, i\omega)\ \rho(x)\ dx \tag{6.2.25}$$

$$= \sum_{n=0}^{\infty} (i\omega)^n \frac{q^{n^2/4}}{(q;q)_n} \int_{-1}^{1} \phi(x)\ H_n(x|q)\ \rho(x)\ dx \equiv 0,$$

or

$$\int_{-1}^{1} \phi(x)\ H_n(x|q)\ \rho(x)\ dx = 0, \qquad n = 0, 1, 2, \dots . \tag{6.2.26}$$

Therefore, an integrable function $\phi \in \mathcal{L}^1_\rho(-1,1)$, which is not identical zero almost everywhere, does not exist due to the completeness of the system of the continuous q-Hermite polynomials.

Thus, we have established the following theorem.

THEOREM 6.1. *The system of basic trigonometric functions* $\{\mathcal{E}_q(x; i\omega_n)\}$, *where* $n = 0, \pm1, \pm2, \dots$, $\omega_{-n} = -\omega_n$, *and* $\omega_0 = 0, \omega_1, \omega_2, \omega_3, \dots$ *are nonnegative zeros of the basic sine function* $S_q(\eta; \omega)$ *arranged in ascending order of magnitude, is complete on* $\mathcal{L}^1_\rho(-1,1)$, *where* $\rho(x)$ *is the weight function in the orthogonality relation* (5.6.5).

We shall continue investigation of completeness of the basic trigonometric systems in Chapter 7.

6.2.3. Some Consequences. Consider formal basic Fourier series generated by function $f(x)$:

$$f(x) \sim \sum_{n=-\infty}^{\infty} c_n\ \mathcal{E}_q(x; i\omega_n) = \mathbf{S}_q[f], \tag{6.2.27}$$

where the coefficients are given by (5.7.6).

LEMMA 6.4. *For every integrable function* $f \in \mathcal{L}^1_\rho(-1,1)$, *where* $\rho(x)$ *is the weight function in the orthogonality relation* (5.6.5), *the* $\mathbf{S}_q[f]$ *does exist.*

PROOF. Indeed, by (5.7.6) and (6.1.10)

$$2k(\omega_n)|c_n| \leq \int_{-1}^{1} |f(x)\ \mathcal{E}_q(x; i\omega_n)|\ \rho(x)\, dx$$

$$\leq \frac{(q\omega_n^2; q^2)_\infty\ \mathcal{E}_q(1; |\omega_n|)}{(-q\omega_n^2; q^2)_\infty} \int_{-1}^{1} |f(x)|\ \rho(x)\, dx < \infty.$$

□

As corollaries of Theorem 6.1 we have the following extensions of the classical results [**20**], [**156**], [**166**].

THEOREM 6.2. *If integrable functions* $f, g \in \mathcal{L}^1_\rho(-1,1)$ *have the same q-Fourier series, then* $f(x) \equiv g(x)$.

PROOF. The q-Fourier coefficients of the difference $f - g$ all vanish:

$$\int_{-1}^{1} (f(x) - g(x))\ \mathcal{E}_q(x; i\omega_n)\ \rho(x)\, dx = 0, \qquad n = 0, \pm 1, \pm 2, \dots ,$$

so that $f - g \equiv 0$ due to the closure of the q-trigonometric system in $\mathcal{L}^1_\rho(-1,1)$. □

THEOREM 6.3. *If* $f(x)$ *is continuous and* $\mathbf{S}_q[f]$, *the q-Fourier series of function* f, *converges uniformly, then its sum is* $f(x)$.

PROOF. Let $g(x)$ denote the sum of $\mathbf{S}_q[f]$, the q-Fourier series in the right side of (6.2.27). Then the coefficients of the $\mathbf{S}_q[f]$ are the q-Fourier coefficients of the g because a uniformly convergent series can be integrated termwise. Hence, $\mathbf{S}_q[f] = \mathbf{S}_q[g]$, so that $f \equiv g$ by the previous theorem and, the f and g being continuous, $f(x) \equiv g(x)$. □

Near the end of this section we shall consider functions $f \in \mathcal{L}^2_\rho(-1,1)$, which means that $|f(x)|^2$ is integrable on $(-1,1)$ with respect to the weight function $\rho(x)$ in the orthogonality relation (5.6.5):

$$\int_{-1}^{1} |f(x)|^2\ \rho(x)\ dx < \infty.$$

By the Cauchy–Schwarz inequality,

$$\int_{-1}^{1} |f(x)|\ \rho(x)\ dx \le \left(\int_{-1}^{1} |f(x)|^2\ \rho(x)\ dx \right)^{1/2} \left(\int_{-1}^{1} \rho(x)\ dx \right)^{1/2} < \infty,$$

these functions are automatically integrable, $f \in \mathcal{L}^1_\rho(-1,1)$, and due to Theorem 6.1 the basic trigonometric system $\{\mathcal{E}_q(x; i\omega_n)\}_{n=-\infty}^{\infty}$, where $\omega_{-n} = -\omega_n$ and $\omega_0 = 0, \omega_1, \omega_2, \omega_3, \dots$ are nonnegative zeros of the basic sine function $S_q(\eta; \omega)$ arranged in ascending order of magnitude, is also complete in $\mathcal{L}^2_\rho(-1,1)$. An independent proof of the completeness in $\mathcal{L}^2$ will be given in Section 7.1.

The system $\{\mathcal{E}_q(x; i\omega_n)\}_{n=-\infty}^{\infty}$ forms an orthogonal basis in $\mathcal{L}^2_\rho(-1,1)$ and the q-Fourier series (5.7.5) are expansions of functions $f \in \mathcal{L}^2_\rho(-1,1)$ with respect to this basis, which converge in the mean square to f. The *Bessel inequality* takes the form

$$\sum_{n=-m}^{m} 2k(\omega_n)\ |c_n|^2 \le \int_{-1}^{1} |f(x)|^2\ \rho(x)\ dx, \tag{6.2.28}$$

where c_n are the q-Fourier coefficients of the $f(x)$ defined by (5.7.6). When $m \to \infty$ we get *Parseval's formula*

$$2k(0)\ a_0^2 + \sum_{n=1}^{\infty} k(\omega_n)\ \left(a_n^2 + b_n^2\right) \tag{6.2.29}$$

$$= \sum_{n=-\infty}^{\infty} 2k(\omega_n)\ |c_n|^2 = \int_{-1}^{1} |f(x)|^2\ \rho(x)\ dx$$

due to the completeness of the q-trigonometric system $\{\mathcal{E}_q(x; i\omega_n)\}_{n=-\infty}^{\infty}$ and the space $\mathcal{L}_\rho^2(-1,1)$ [**3**], [**93**], [**108**]; see also Ex. 6.3.

Since the terms of a convergent series tend to zero, it follows from Bessel's inequality that $k(\omega_n)\ |c_n|^2$ converge to zero if $f \in \mathcal{L}_\rho^2(-1,1)$. This is an analog of the classical Riemann–Lebesgue Lemma for the Fourier series; see Ex. 6.4 for an extension of this result.

6.3. Asymptotics of Zeros

Our next step is to establish asymptotics of zeros of basic sine and cosine functions arising in the theory of q-Fourier series.

6.3.1. Preliminaries. In this section we shall continue the study of the zeros of the basic sine $S_q(\eta;\omega)$ and cosine $C_q(\eta;\omega)$ functions given by (5.4.1) and (5.4.2), respectively. The ω-zeros of the $S_q(\eta;\omega)$ are the eigenvalues related to the basic Fourier series on a q-quadratic grid [**26**]. Their asymptotics are very important for investigation of the convergence of these series [**146**]. The main properties of these zeros were discussed in [**55**], [**56**], [**26**] and in Section 5.4 here, from different viewpoints. We remind the reader that when $0 < q < 1$ all zeros of $S_q(\eta;\omega)$ and $C_q(\eta;\omega)$ are real. Also these zeros are simple, the positive zeros of the basic sine function $S_q(\eta;\omega)$ are interlaced with those of the basic cosine function $C_q(\eta;\omega)$; see Theorems 5.1–5.4. Asymptotic behavior of the large zeros of these q-trigonometric functions has been discussed in [**26**] and [**147**]; see also [**48**] for numerical investigation of these zeros.

Let $0 = \omega_0 < \omega_1 < \omega_2 < \omega_3 < \ldots$ be positive zeros of $S_q(\eta;\omega)$ and let $\varpi_1 < \varpi_2 < \varpi_3 < \ldots$ be positive zeros of $C_q(\eta;\omega)$. Gosper and Suslov [**48**] have found numerically the following asymptotic formulas[1]

$$\omega_n = q^{1/4-n} - c_1(q) + \mathrm{o}(1) \tag{6.3.1}$$

and

$$\varpi_n = q^{3/4-n} - c_1(q) + \mathrm{o}(1) \tag{6.3.2}$$

as $n \to \infty$. Here,

$$c_1(q) = \frac{q^{1/4}}{2\left(1-q^{1/2}\right)} \frac{\left(q;q^2\right)_\infty^2}{\left(q^2;q^2\right)_\infty^2} = \frac{q^{1/4}\left(1+q^{1/2}\right)(1+q)}{2\Gamma_{q^2}^2(1/2)} \tag{6.3.3}$$

and $\Gamma_{q^2}(z)$ is the q-gamma function. The function $c_1(q)$ is nonnegative and increasing on $[0,1]$; see [**48**] or Figure 6.1 below for the graph of this

[1]We use the symbol o in a usual sense; see Appendix B.4 for the definition.

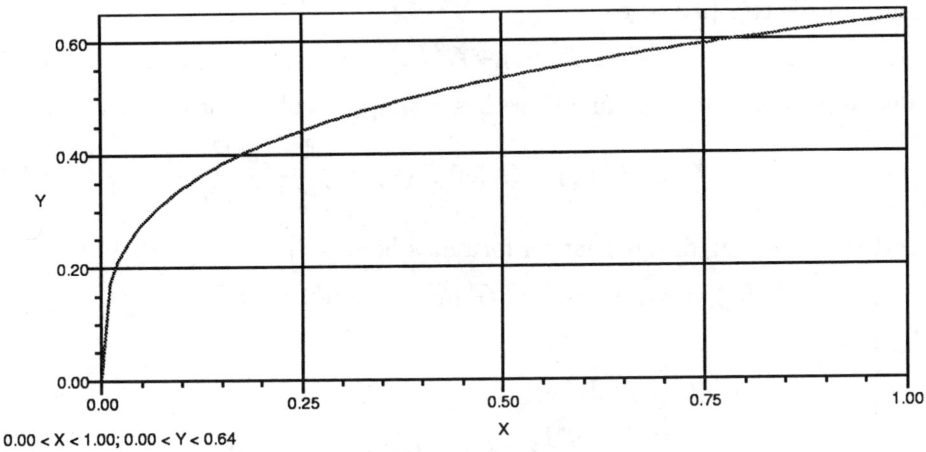

FIGURE 6.1. Function $c_1(q)$, $0 \le q \le 1$.

function and Section 8.6 for the proof of the monotonicity. The maximum value of this function on $[0, 1]$ is

$$\lim_{q \to 1^-} c_1(q) = 2/\pi \simeq 0.63661977236758. \tag{6.3.4}$$

Numerical analysis in [**48**] has shown that asymptotic formulas (6.3.1)–(6.3.2) are pretty accurate.

Our main objective in this section is to present a rigorous proof of these formulas [**147**]. The next terms in these asymptotic expansions will be found in Chapter 8.

6.3.2. Some Properties of q-Sine and Cosine Functions. We first remind the reader that Eqs. (5.4.3)–(5.4.4) determine the large ω-asymptotics of $S_q(\eta;\omega)$ and $C_q(\eta;\omega)$.

It has been shown in [**48**] that the graphs of $S_q(\eta;\omega)$ and $C_q(\eta;\omega)$ look much more elegant if we choose a different normalization for these functions. An analog of the main trigonometric identity for the basic trigonometric functions is

$$C_q^2(\eta;\omega) + S_q^2(\eta;\omega) = \frac{(-\omega^2;q^2)_\infty}{(-q\omega^2;q^2)_\infty} \tag{6.3.5}$$

and introducing functions

$$F(\omega) = \sqrt{\frac{(-q\omega^2;q^2)_\infty}{(-\omega^2;q^2)_\infty}}\, S_q(\eta;\omega), \tag{6.3.6}$$

$$G(\omega) = \sqrt{\frac{(-q\omega^2;q^2)_\infty}{(-\omega^2;q^2)_\infty}}\, C_q(\eta;\omega), \tag{6.3.7}$$

one can rewrite (6.3.5) as

$$F^2(\omega) + G^2(\omega) = 1. \tag{6.3.8}$$

The Wronskian of these functions has a simple explicit form

$$\kappa(\omega) := G(\omega) F'(\omega) - G'(\omega) F(\omega) = \sum_{k=0}^{\infty} \frac{q^{k/2}}{1 + \omega^2 q^k} \tag{6.3.9}$$

and the following differentiation formulas hold [**48**]

$$F'(\omega) = \kappa(\omega) G(\omega), \quad G'(\omega) = -\kappa(\omega) F(\omega). \tag{6.3.10}$$

Indeed,

$$\begin{aligned} &W(G(\omega),\ F(\omega)) &\quad (6.3.11)\\ &\quad = \frac{(-q\omega^2; q^2)_\infty}{(-\omega^2; q^2)_\infty}\, W(C_q(\eta;\omega),\ S_q(\eta;\omega)) = \kappa(\omega) \end{aligned}$$

by (5.5.14)–(5.5.15) and (5.5.8)–(5.5.9). The differentiation formulas (6.3.10) follow from (6.3.9) and the relation

$$F(\omega) F'(\omega) + G(\omega) G'(\omega) = 0, \tag{6.3.12}$$

which is a consequence of (6.3.8).

The functions $F(\omega)$ and $G(\omega)$ have the same real zeros as the original functions, $S_q(\eta;\omega)$ and $C_q(\eta;\omega)$, but they obey nice properties similar to those for the classical trigonometric functions [**48**]. For example, the functions $F(\omega)$ and $G(\omega)$ are bounded for all real values of ω and change from -1 to 1. Moreover, all extrema of function $F(\omega)$ $(G(\omega))$ are located at zeros of $G(\omega)$ $(F(\omega))$; the function $F(\omega)$ $(G(\omega))$ is monotone between any two successive zeros of $G(\omega)$ $(F(\omega))$. These properties are direct consequences of (6.3.8)–(6.3.10). See [**48**] and Chapter 11 for the graphs of $F(\omega)$ and $G(\omega)$ for different values of parameter q. It is worth mentioning that both functions, $F(\omega)$ and $G(\omega)$, satisfy the following differential equation

$$u'' + \kappa^2 u = (\log \kappa)' u'. \tag{6.3.13}$$

We shall use these properties of the functions $F(\omega)$ and $G(\omega)$ in order to prove the asymptotic formulas (6.3.1)–(6.3.2).

6.3.3. Proof of Asymptotic Formulas. In this section we give a rigorous proof of (6.3.1)–(6.3.2) by means of, essentially, elementary calculus tools only [**147**]. These formulas have been conjectured in [**48**]. Let us reformulate the main result in the form of a theorem.

THEOREM 6.4. *Let* $0 = \omega_0 < \omega_1 < \omega_2 < \omega_3 < \ldots$ *be positive zeros of* $S_q(\eta;\omega)$ *and let* $\varpi_1 < \varpi_2 < \varpi_3 < \ldots$ *be positive zeros of* $C_q(\eta;\omega)$ *for* $0 < q < 1$. *Then*

$$\omega_n = q^{1/4-n} - c_1(q) + o(1), \tag{6.3.14}$$

$$\varpi_n = q^{3/4-n} - c_1(q) + o(1) \tag{6.3.15}$$

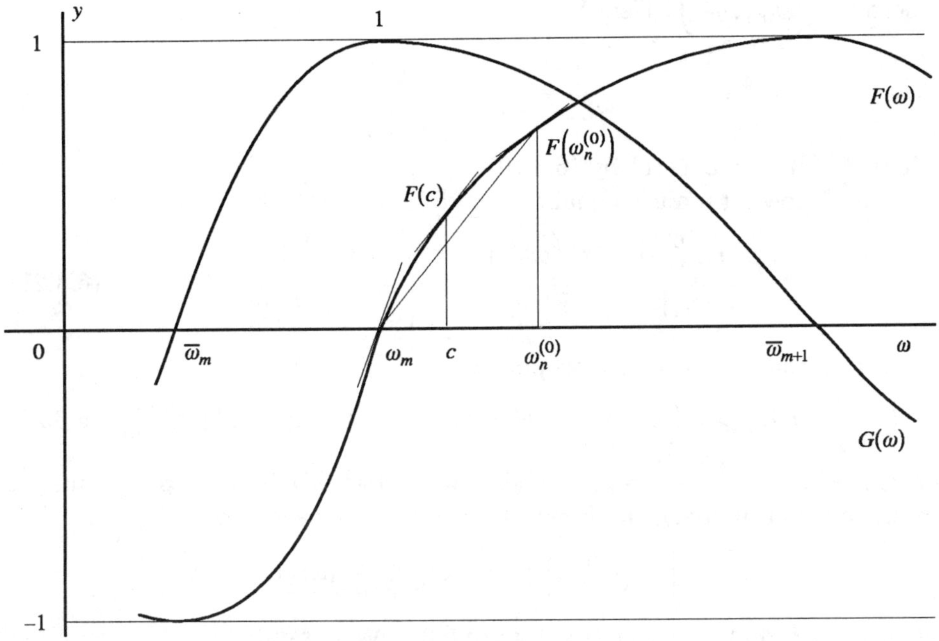

FIGURE 6.2. The Mean Value Theorem and Concavity of $F(\omega)$ on the interval $\left[\omega_m, \omega_n^{(0)}\right]$, $m = 2l$.

as $n \to \infty$, *where*

$$c_1(q) = \frac{q^{1/4}}{2\left(1 - q^{1/2}\right)} \frac{\left(q; q^2\right)_\infty^2}{\left(q^2; q^2\right)_\infty^2}. \tag{6.3.16}$$

PROOF. Denote

$$\omega_n^{(0)} = q^{1/4-n}. \tag{6.3.17}$$

In view of (5.4.3)–(5.4.4) and (6.3.6)–(6.3.7),

$$\frac{F\left(\omega_n^{(0)}\right)}{G\left(\omega_n^{(0)}\right)} = \frac{S_q(\eta; \omega_n^{(0)})}{C_q(\eta; \omega_n^{(0)})} = \frac{S_q(\eta; q^{1/2}/\omega_n^{(0)})}{C_q(\eta; q^{1/2}/\omega_n^{(0)})} > 0 \tag{6.3.18}$$

for all $q^{1/2}/\omega_n^{(0)} < \varpi_1$ or $q^{1/4+n} < \varpi_1$. The last inequality holds for all sufficiently large values of n for any $0 < q < 1$. This means that for sufficiently large n we always have $\omega_m < \omega_n^{(0)} < \varpi_{m+1}$, where m, generally speaking, may be different from n (we shall show later that $m = n$ for sufficiently large n). By the Mean Value Theorem for the interval $\left[\omega_m, \omega_n^{(0)}\right]$ one can write (see Figure 6.2)

$$F\left(\omega_n^{(0)}\right) = F'(c)\left(\omega_n^{(0)} - \omega_m\right), \tag{6.3.19}$$

where $c \in \left(\omega_m, \omega_n^{(0)}\right)$. Hence,

$$\omega_n^{(0)} - \omega_m = \frac{F\left(\omega_n^{(0)}\right)}{F'(c)}, \tag{6.3.20}$$

where $F'(c) = \kappa(c)\, G(c)$ by (6.3.10).

The following main inequalities hold

$$0 < \frac{\left|F\left(\omega_n^{(0)}\right)\right|}{\kappa(\omega_m)} < \frac{F\left(\omega_n^{(0)}\right)}{F'(c)} < \frac{F\left(\omega_n^{(0)}\right)}{\kappa\left(\omega_n^{(0)}\right) G\left(\omega_n^{(0)}\right)} \tag{6.3.21}$$

in view of the monotonicity properties

$$\kappa(\omega_m) > \kappa(c) > \kappa\left(\omega_n^{(0)}\right), \quad |G(\omega_m)| = 1 > |G(c)| > \left|G\left(\omega_n^{(0)}\right)\right| \tag{6.3.22}$$

on $\omega_m < c < \omega_n^{(0)} < \varpi_{m+1}$. These properties admit a simple geometric interpretation, namely, the function $|F(\omega)|$ is concave and

$$\left|F'\left(\omega_n^{(0)}\right)\right| < |F'(c)| < |F'(\omega_m)|$$

on $\omega_m < c < \omega_n^{(0)} < \varpi_{m+1}$ (see Figure 6.2). As a result

$$0 < \frac{\left|F\left(\omega_n^{(0)}\right)\right|}{\kappa(\omega_m)} < \omega_n^{(0)} - \omega_m < \frac{F\left(\omega_n^{(0)}\right)}{\kappa\left(\omega_n^{(0)}\right) G\left(\omega_n^{(0)}\right)} \tag{6.3.23}$$

by (6.3.20) and (6.3.21).

Our next step is to show that

$$\lim_{n\to\infty} \frac{F\left(\omega_n^{(0)}\right)}{\kappa\left(\omega_n^{(0)}\right) G\left(\omega_n^{(0)}\right)} = \frac{q^{1/4}}{2\left(1-q^{1/2}\right)} \frac{\left(q;q^2\right)_\infty^2}{\left(q^2;q^2\right)_\infty^2} = c_1(q). \tag{6.3.24}$$

Indeed, by (5.4.1)

$$\lim_{n\to\infty} q^{-n} S_q(\eta; q^{1/4+n}) = \frac{q^{1/4}}{1-q^{1/2}} \tag{6.3.25}$$

and by (5.5.17)

$$\begin{aligned}
&\lim_{n\to\infty} q^{-n} \kappa(q^{1/4-n}) \qquad (6.3.26)\\
&= \frac{(q;q)_\infty^2}{\left(q^{1/2};q\right)_\infty^2} \lim_{n\to\infty} q^{-n} \frac{\left(-q^{1-2n};q\right)_\infty}{\left(-q^{1/2-2n};q\right)_\infty}\\
&= \frac{(q;q)_\infty^2}{\left(q^{1/2};q\right)_\infty^2} \lim_{n\to\infty} q^{-n} \frac{\left(-q^{1-2n};q\right)_{2n} (-q;q)_\infty}{\left(-q^{1/2-2n};q\right)_{2n} \left(-q^{1/2};q\right)_\infty}\\
&= \frac{(q;q)_\infty^2}{\left(q^{1/2};q\right)_\infty^2} \frac{(-q;q)_\infty}{\left(-q^{1/2};q\right)_\infty} \lim_{n\to\infty} \frac{(-1;q)_{2n}}{\left(-q^{1/2};q\right)_{2n}}
\end{aligned}$$

$$= 2\frac{(q,-q;q)_\infty^2}{\left(q^{1/2},-q^{1/2};q\right)_\infty^2} = 2\frac{\left(q^2;q^2\right)_\infty^2}{\left(q;q^2\right)_\infty^2}.$$

We have used (I.9) of [46] or (1.4.4) in the third line here. Thus, from (6.3.18) and (6.3.25)–(6.3.26),

$$\lim_{n\to\infty}\frac{F\left(\omega_n^{(0)}\right)}{\kappa\left(\omega_n^{(0)}\right)G\left(\omega_n^{(0)}\right)} = \lim_{n\to\infty}\frac{S_q\left(\eta;q^{1/2}/\omega_n^{(0)}\right)}{\kappa\left(\omega_n^{(0)}\right)C_q\left(\eta;q^{1/2}/\omega_n^{(0)}\right)}$$
$$= \lim_{n\to\infty}\frac{q^{-n}S_q\left(\eta;q^{1/4+n}\right)}{q^{-n}\kappa\left(q^{1/4-n}\right)} = c_1(q).$$

Let us rewrite (6.3.23) as

$$0 < 1 - \frac{\omega_m}{\omega_n^{(0)}} < \frac{1}{\omega_n^{(0)}}\,\frac{F\left(\omega_n^{(0)}\right)}{\kappa\left(\omega_n^{(0)}\right)G\left(\omega_n^{(0)}\right)}. \tag{6.3.27}$$

Taking the limit $n\to\infty$ one gets by (6.3.24) and the Squeeze Theorem

$$\lim_{n\to\infty}\frac{\omega_m}{\omega_n^{(0)}} = 1, \quad \text{or} \quad \lim_{n\to\infty} q^n\omega_m = q^{1/4}. \tag{6.3.28}$$

This justifies the leading term in (6.3.14) (cf. Theorem 5 of [**26**]), if one can show that $m=n$. This can be done on the basis of Jensen's Theorem and it is the only "non-elementary" part of our proof.

Jensen's Theorem [**99**] states that if $f(z)$ is holomorphic in a circle of radius R with the center at the origin, and $f(0)\neq 0$, then

$$\int_0^R \frac{n_f(r)}{r}\,dr = \frac{1}{2\pi}\int_0^{2\pi}\log\left|f\left(Re^{i\theta}\right)\right|\,d\theta - \log|f(0)|, \tag{6.3.29}$$

where $n_f(r)$ is the number of zeros of $f(z)$ in the circle $|z|<r$; see Theorem B.2 in Appendix B.

In view of (6.3.28), one can write $\omega_{m+1} < \omega_{n+1}^{(0)} < \varpi_{m+2}$. Indeed, if $\omega_m < \omega_n^{(0)} < \omega_{n+1}^{(0)} < \varpi_{m+1}$, then in a similar fashion,

$$1 = \lim_{n\to\infty}\frac{\omega_m}{\omega_{n+1}^{(0)}} = \lim_{n\to\infty}\left(\frac{\omega_m}{\omega_n^{(0)}}\,\frac{\omega_n^{(0)}}{\omega_{n+1}^{(0)}}\right) = q,$$

which is a contradiction. By the definition of $n_f(r)$ and (6.3.28)

$$\int_{\omega_n^{(0)}}^{\omega_{n+1}^{(0)}}\frac{n_f(r)}{r}\,dr = \int_{\omega_n^{(0)}}^{\omega_{m+1}}\frac{n_f(r)}{r}\,dr + \int_{\omega_{m+1}}^{\omega_{n+1}^{(0)}}\frac{n_f(r)}{r}\,dr$$
$$= 2m\log\frac{\omega_{m+1}}{\omega_n^{(0)}} + 2(m+1)\log\frac{\omega_{n+1}^{(0)}}{\omega_{m+1}}$$

$$= 2m \log \frac{\omega_{n+1}^{(0)}}{\omega_n^{(0)}} + 2 \log \frac{\omega_{n+1}^{(0)}}{\omega_{m+1}}$$
$$= 2m \log q^{-1} + \mathrm{o}\left(1\right), \qquad n \to \infty, \tag{6.3.30}$$

where $f(\omega)$ is an entire function with the simple zeros at $\omega = \pm\omega_1, \pm\omega_2, \pm\omega_3, \ldots$ defined by

$$f(\omega) = \left(-q\omega^2; q^2\right) \frac{S_q(\eta; \omega)}{\omega}$$
$$= \frac{\left(q^{3/2}\omega^2, q^{1/2}/\omega^2; q^2\right)}{\left(q^{1/2}; q\right)_\infty \left(q, -q/\omega^2; q^2\right)_\infty} C_q(\eta; \frac{q^{1/2}}{\omega})$$
$$+ \frac{\left(q^{1/2}\omega^2, q^{3/2}/\omega^2; q^2\right)}{\left(q^{1/2}; q\right)_\infty \left(q, -q/\omega^2; q^2\right)_\infty} S_q(\eta; \frac{q^{1/2}}{\omega})$$
$$= \frac{\left(q^2; q^2\right)_\infty}{\left(q^{1/2}; q^{1/2}\right)_\infty} \left(q^{3/2}\omega^2; q^2\right) \left(1 + \mathrm{o}\left(1\right)\right) \tag{6.3.31}$$

by (5.4.1)–(5.4.2) as $\omega = \beta q^{-n} \to \infty$. By (6.3.29)

$$\int_{\omega_n^{(0)}}^{\omega_{n+1}^{(0)}} \frac{n_f(r)}{r}\, dr = \frac{1}{2\pi} \int_0^{2\pi} \log \left| \frac{f\left(\omega_{n+1}^{(0)} e^{i\theta}\right)}{f\left(\omega_n^{(0)} e^{i\theta}\right)} \right| d\theta \tag{6.3.32}$$

and in view of (6.3.31)

$$\frac{f\left(\omega_{n+1}^{(0)} e^{i\theta}\right)}{f\left(\omega_n^{(0)} e^{i\theta}\right)} = \left(1 - q^{-2n} e^{2i\theta}\right) \left(1 + \mathrm{o}\left(1\right)\right)$$
$$= -q^{-2n} e^{2i\theta} \left(1 + \mathrm{o}\left(1\right)\right), \quad n \to \infty.$$

Thus

$$\frac{1}{2\pi} \int_0^{2\pi} \log \left| \frac{f\left(\omega_{n+1}^{(0)} e^{i\theta}\right)}{f\left(\omega_n^{(0)} e^{i\theta}\right)} \right| d\theta = 2n \log q^{-1} + \mathrm{o}\left(1\right), \quad n \to \infty \tag{6.3.33}$$

and, finally, by (6.3.30) and (6.3.32)–(6.3.33) we obtain

$$m = n + \mathrm{o}\left(1\right), \quad n \to \infty, \tag{6.3.34}$$

which implies that $m = n$ for sufficiently large n because m and n are both integers.

In order to complete the proof of the theorem one can now rewrite (6.3.23) with $m = n$ as

$$\left| G\left(\omega_n^{(0)}\right) \right| \frac{\kappa\left(\omega_n^{(0)}\right)}{\kappa\left(\omega_n\right)} \frac{F\left(\omega_n^{(0)}\right)}{\kappa\left(\omega_n^{(0)}\right) G\left(\omega_n^{(0)}\right)} \tag{6.3.35}$$

$$< \omega_n^{(0)} - \omega_n < \frac{F\left(\omega_n^{(0)}\right)}{\kappa\left(\omega_n^{(0)}\right) G\left(\omega_n^{(0)}\right)}.$$

Due to (6.3.28) where $m = n$

$$\lim_{n\to\infty} \left|G\left(\omega_n^{(0)}\right)\right| = |G(\omega_n)| = 1 \tag{6.3.36}$$

and

$$\lim_{n\to\infty} \kappa\left(\omega_n^{(0)}\right) / \kappa(\omega_n) = 1. \tag{6.3.37}$$

Indeed, by (5.4.3) and (6.3.7)

$$G\left(\omega_n^{(0)}\right) = \sqrt{\frac{\left(-q^{3/2-2n};q^2\right)_\infty}{\left(-q^{1/2-2n};q^2\right)_\infty}} \frac{1}{\left(q^{1/2};q\right)_\infty} \times \frac{\left(q^{1-2n}, q^{1+2n};q^2\right)_\infty}{\left(q, -q^{3/2-2n}, -q^{1/2+2n};q^2\right)_\infty} C_q(\eta; q^{1/4+n}).$$

Using (I.9) of [**46**] or (1.4.4)

$$\begin{aligned} &\lim_{n\to\infty} \left|G\left(\omega_n^{(0)}\right)\right| \\ &= \lim_{n\to\infty} \left(q^{n/2} \sqrt{\frac{\left(-q^{1/2};q^2\right)_n \left(-q^{3/2};q^2\right)_\infty}{\left(-q^{3/2};q^2\right)_n \left(-q^{1/2};q^2\right)_\infty}} \frac{1}{\left(q^{1/2};q\right)_\infty} \right. \\ &\left. \times\, q^{-n/2} \frac{\left(q;q^2\right)_n \left(q^{1+2n};q^2\right)_\infty}{\left(-q^{1/2};q^2\right)_n \left(-q^{1/2+2n}, -q^{3/2};q^2\right)_\infty} C_q(\eta; q^{1/4+n}) \right) \\ &= \frac{\left(q;q^2\right)_\infty}{\left(q^{1/2};q\right)_\infty \left(-q^{1/2}, -q^{3/2};q^2\right)_\infty} = \frac{\left(q;q^2\right)_\infty}{\left(q;q^2\right)_\infty} = 1. \end{aligned}$$

In a similar fashion, in view of (5.5.17) and (6.3.28) with $m = n$

$$\begin{aligned} \lim_{n\to\infty} \frac{\kappa\left(\omega_n^{(0)}\right)}{\kappa(\omega_n)} &= \lim_{n\to\infty} \frac{\left(-q^{1-2n}, -\omega_n^2;q\right)_\infty}{\left(-q^{1/2-2n}, -q^{1/2}\omega_n^2;q\right)_\infty} \\ &= \frac{(-q;q)_\infty}{\left(-q^{1/2};q\right)_\infty} \lim_{n\to\infty} \frac{\left(-1, -q/\left(\omega_n^2 q^{2n}\right);q\right)_{2n}}{\left(-q^{1/2}, -q^{1/2}/\left(\omega_n^2 q^{2n}\right);q\right)_{2n}} \\ &\times \lim_{n\to\infty} \frac{\left(-\left(\omega_n^2 q^{2n}\right);q\right)_\infty}{\left(-q^{1/2}\left(\omega_n^2 q^{2n}\right);q\right)_\infty} \\ &= \frac{\left(-q, -1, -q^{1/2}, -q^{1/2};q\right)_\infty}{\left(-q^{1/2}, -q^{1/2}, -1, -q;q\right)_\infty} = 1. \end{aligned}$$

Our final step is to take the limit $n \to \infty$ in (6.3.35). By the Squeeze Theorem

$$\lim_{n\to\infty} \left(\omega_n^{(0)} - \omega_n\right) = c_1(q) \tag{6.3.38}$$

due to (6.3.24), (6.3.36) and (6.3.37). This proves (6.3.14). The asymptotic formula (6.3.15) can be justified in a similar fashion. We leave the details to the reader. □

The asymptotic formulas (6.3.14)–(6.3.15) can be modified in the following manner to give a better approximation for small zeros.

THEOREM 6.5. *Let* $0 = \omega_0 < \omega_1 < \omega_2 < \omega_3 < \dots$ *be positive zeros of* $S_q(\eta;\omega)$ *and let* $\varpi_1 < \varpi_2 < \varpi_3 < \dots$ *be positive zeros of* $C_q(\eta;\omega)$ *for* $0 < q < 1$. *Then*

$$\omega_n = \omega_n^{(0)} \sqrt{1 - 2c_1(q) / \omega_n^{(0)}} + o(1), \tag{6.3.39}$$

$$\varpi_n = \varpi_n^{(0)} \sqrt{1 - 2c_1(q) / \varpi_n^{(0)}} + o(1) \tag{6.3.40}$$

as $n \to \infty$. *Here* $\omega_n^{(0)} = q^{1/4-n}$, $\varpi_n^{(0)} = q^{3/4-n}$, *and* $c_1(q)$ *is defined by* (6.3.16).

PROOF. Let us consider the case of the q-sine function. Introduce

$$\omega_n^{(1)} = \omega_n^{(0)} \sqrt{1 - 2c_1(q) / \omega_n^{(0)}} \tag{6.3.41}$$

and rewrite (6.3.23) where $m = n$ in the form

$$\omega_n^{(1)} - \omega_n^{(0)} + \frac{\left|F\left(\omega_n^{(0)}\right)\right|}{\kappa(\omega_n)} < \omega_n^{(1)} - \omega_n < \omega_n^{(1)} - \omega_n^{(0)} + \frac{F\left(\omega_n^{(0)}\right)}{F'\left(\omega_n^{(0)}\right)}. \tag{6.3.42}$$

Taking the limit $n \to \infty$ with the help of the same arguments as in Theorem 6.4 one gets

$$\lim_{n\to\infty} \left(\omega_n^{(1)} - \omega_n\right) = 0, \tag{6.3.43}$$

which proves (6.3.39). The proof of (6.3.40) is similar. □

Numerical analysis shows that asymptotics (6.3.39)–(6.3.40) are pretty accurate. It is of interest, nonetheless, to find next terms in the asymptotic expansions (6.3.14)–(6.3.15). Numerical analysis similar to [**48**] strongly indicates that the following asymptotics hold.

THEOREM 6.6. *Let* $0 = \omega_0 < \omega_1 < \omega_2 < \omega_3 < \dots$ *be positive zeros of* $S_q(\eta;\omega)$ *and let* $\varpi_1 < \varpi_2 < \varpi_3 < \dots$ *be positive zeros of* $C_q(\eta;\omega)$ *for* $0 < q < 1$. *Then*

$$\omega_n = \omega_n^{(0)} - c_1(q) - c_1^2(q) / \left(2\omega_n^{(0)}\right) + O\left(1/\left(\omega_n^{(0)}\right)^2\right), \tag{6.3.44}$$

$$\varpi_n = \varpi_n^{(0)} - c_1(q) - c_1^2(q) / \left(2\varpi_n^{(0)}\right) + O\left(1/\left(\varpi_n^{(0)}\right)^2\right) \tag{6.3.45}$$

as $n \to \infty$. *Here* $\omega_n^{(0)} = q^{1/4-n}$, $\varpi_n^{(0)} = q^{3/4-n}$, *and* $c_1(q)$ *is defined by* (6.3.16).

These asymptotics formally appear also if one expands the first terms in (6.3.39)–(6.3.40). This observation proves our next result.

THEOREM 6.7. *The symbols* $o(1)$ *in* (6.3.39)–(6.3.40) *should be replaced by* $o\left(1/\omega_n^{(0)}\right)$ *and* $o\left(1/\varpi_n^{(0)}\right)$, *respectively.*

It does not look that there are simple proofs of these theorems by the methods of elementary calculus. We shall derive these and other improved asymptotics in Chapter 8 by a completely different method.

6.4. Pointwise Asymptotics of Basis

Let us discuss the large ω-asymptotics of the basic sine $S_q(x;\omega)$ and basic cosine $C_q(x;\omega)$ functions when the independent variable $x = \cos\theta$ belongs to the interval of orthogonality $-1 < x < 1$. By (2.4.10) and (2.4.11)

$$C_q(\cos\theta;\omega) \tag{6.4.1}$$
$$= \frac{\left(q\omega^2 e^{2i\theta},\ e^{-2i\theta}/\omega^2;q^2\right)_\infty}{\left(e^{-2i\theta};q\right)_\infty \left(q,-q\omega^2,-q/\omega^2;q^2\right)_\infty}\ {}_2\varphi_1\left(\begin{matrix}-e^{2i\theta},-qe^{2i\theta}\\ q^2e^{4i\theta}\end{matrix};q^2,-\frac{q^2}{\omega^2}\right)$$
$$+\frac{\left(q\omega^2 e^{-2i\theta},\ e^{2i\theta}/\omega^2;q^2\right)_\infty}{\left(e^{2i\theta};q\right)_\infty \left(q,-q\omega^2,-q/\omega^2;q^2\right)_\infty}\ {}_2\varphi_1\left(\begin{matrix}-e^{-2i\theta},-qe^{-2i\theta}\\ q^2e^{-4i\theta}\end{matrix};q^2,-\frac{q^2}{\omega^2}\right)$$

and

$$S_q(\cos\theta;\omega) \tag{6.4.2}$$
$$= \frac{q^{1/4}\omega e^{i\theta}\left(q^2\omega^2 e^{2i\theta},e^{-2i\theta}/\omega^2;q^2\right)_\infty}{\left(e^{-2i\theta};q\right)_\infty \left(q,-q\omega^2,-q/\omega^2;q^2\right)_\infty}\ {}_2\varphi_1\left(\begin{matrix}-e^{2i\theta},-qe^{2i\theta}\\ q^2e^{4i\theta}\end{matrix};q^2,-\frac{q^2}{\omega^2}\right)$$
$$+\frac{q^{1/4}\omega e^{-i\theta}\left(q^2\omega^2 e^{-2i\theta},e^{2i\theta}/\omega^2;q^2\right)_\infty}{\left(e^{2i\theta};q\right)_\infty \left(q,-q\omega^2,-q/\omega^2;q^2\right)_\infty}\ {}_2\varphi_1\left(\begin{matrix}-e^{-2i\theta},-qe^{-2i\theta}\\ q^2e^{-4i\theta}\end{matrix};q^2,-\frac{q^2}{\omega^2}\right).$$

It follows from these expressions that for $|x|<1$, $|q|<1$ the leading terms in the asymptotic expansions of $C_q(\cos\theta;\omega)$ and $S_q(\cos\theta;\omega)$ are given by[2]

$$C_q(\cos\theta;\omega) \tag{6.4.3}$$
$$\sim \frac{\left(q\omega^2 e^{2i\theta};q^2\right)_\infty}{\left(e^{-2i\theta};q\right)_\infty\left(q,-q\omega^2;q^2\right)_\infty} + \frac{\left(q\omega^2 e^{-2i\theta};q^2\right)_\infty}{\left(e^{2i\theta};q\right)_\infty\left(q,-q\omega^2;q^2\right)_\infty}$$

and

$$S_q(\cos\theta;\omega) \tag{6.4.4}$$
$$\sim \frac{q^{1/4}\omega e^{i\theta}\left(q^2\omega^2 e^{2i\theta};q^2\right)_\infty}{\left(e^{-2i\theta};q\right)_\infty\left(q,-q\omega;q^2\right)_\infty} + \frac{q^{1/4}\omega e^{-i\theta}\left(q^2\omega^2 e^{-2i\theta};q^2\right)_\infty}{\left(e^{2i\theta};q\right)_\infty\left(q,-q\omega^2;q^2\right)_\infty}$$

as $|\omega|\to\infty$. In particular, when $\omega=\omega_n$ are large zeros of the basic sine function $S_q(\eta;\omega)$, one can estimate

$$C_q(\cos\theta;\omega_n) \sim C_q\left(\cos\theta;q^{1/4-n}\right), \tag{6.4.5}$$

[2]We use the symbol $\sim$ in a usual sense; see Appendix B.4 for the definition.

$$S_q(\cos\theta;\omega_n) \sim S_q\left(\cos\theta; q^{1/4-n}\right) \tag{6.4.6}$$

as $n \to \infty$, due to the asymptotics of the zeros (6.3.14). Eqs. (6.4.3)–(6.4.4) lead to the following theorem.

THEOREM 6.8. *For $-1 < x = \cos\theta < 1$ and $|q| < 1$ the leading term in the asymptotic expansion of $C_q(\cos\theta;\omega_n)$ as $n \to \infty$ is given by*

$$C_q\left(\cos\theta; q^{1/4-n}\right) \sim 2\frac{\left(q^{1/2};q\right)_\infty}{\left(q;q^2\right)_\infty^2} \left|A\left(e^{i\theta}\right)\right| \cos\left(\left(2\theta+\pi\right)n - \chi\right) \tag{6.4.7}$$

where

$$A\left(e^{i\theta}\right) = \left(1 - q^{1/2}e^{2i\theta}\right) \frac{\left(q^{3/2}e^{-2i\theta}, q^{5/2}e^{2i\theta}; q^2\right)_\infty}{\left(e^{2i\theta};q\right)_\infty}, \tag{6.4.8}$$

$$\left|A\left(e^{i\theta}\right)\right|^{-2} = \frac{\left(e^{2i\theta},\ e^{-2i\theta};q\right)_\infty}{\left(q^{1/2}e^{2i\theta},\ q^{1/2}e^{-2i\theta};q\right)_\infty}, \tag{6.4.9}$$

and

$$\chi = \arg A\left(e^{i\theta}\right). \tag{6.4.10}$$

For $-1 < x = \cos\theta < 1$ and $|q| < 1$ the leading term in the asymptotic expansion of $S_q(\cos\theta;\omega_n)$ as $n \to \infty$ is given by

$$S_q\left(\cos\theta; q^{1/4-n}\right) \sim 2\frac{\left(q^{1/2};q\right)_\infty}{\left(q;q^2\right)_\infty^2} \left|B\left(e^{i\theta}\right)\right| \cos\left(\left(2\theta+\pi\right)(n-1) - \psi\right) \tag{6.4.11}$$

where

$$B\left(e^{i\theta}\right) = e^{i\theta} \frac{\left(q^{1/2}e^{-2i\theta}, q^{3/2}e^{2i\theta}; q^2\right)_\infty}{\left(e^{2i\theta};q\right)_\infty}, \tag{6.4.12}$$

$$\left|B\left(e^{i\theta}\right)\right|^{-2} = \frac{\left(e^{2i\theta},\ e^{-2i\theta};q\right)_\infty}{\left(q^{1/2}e^{2i\theta},\ q^{1/2}e^{-2i\theta};q\right)_\infty}, \tag{6.4.13}$$

and

$$\psi = \arg B\left(e^{i\theta}\right). \tag{6.4.14}$$

From (6.4.8) and (6.4.12),

$$A\left(e^{i\theta}\right) = e^{i\theta} \frac{\left(e^{-2i\theta};q\right)_\infty}{\left(e^{2i\theta};q\right)_\infty} B\left(e^{-i\theta}\right). \tag{6.4.15}$$

It is worth noting that the factor $\left|A\left(e^{i\theta}\right)\right|^{-2} = \left|B\left(e^{i\theta}\right)\right|^{-2}$ coincides with the weight function in the orthogonality relations (5.2.26)–(5.2.28) for the basic trigonometric functions.

In a similar fashion, Eqs. (2.4.12)–(2.4.13) may be considered as "asymptotic expansions" of the basic sine and cosine functions for the large values of ω, which are not in terms of the usual asymptotic sequence $\left\{(x\omega)^{-n}\right\}_{n=0}^{\infty}$, but are sums of two asymptotic expansions in terms of the "inverse generalized powers" $\left(q^2\omega^2 e^{\pm 2i\theta};q^2\right)_n^{-1}$; see **[54]**.

6.5. Bilinear Generating Function

The Poisson kernel for the trigonometric functions is

$$\sum_{n=-\infty}^{\infty} r^{|n|} e^{in(x-y)} = \frac{1-r^2}{1-2r\cos(x-y)+r^2}, \qquad 0 \leq r < 1, \tag{6.5.1}$$

see Ex. 6.7. In this section we shall derive the following bilinear generating relation

$$\sum_{n=-\infty}^{\infty} \frac{\left(-qr^2\omega_n^2; q^2\right)_\infty}{\left(-q\omega_n^2; q^2\right)_\infty k(\omega_n)} \mathcal{E}_q(\cos\theta; i\omega_n)\, \mathcal{E}_q(\cos\varphi; -ir\omega_n)$$
$$= \frac{\left(q, r^2, q^{1/2}e^{2i\theta},\ q^{1/2}e^{-2i\theta};\ q\right)_\infty}{\pi\left(re^{i\theta+i\varphi}, re^{i\theta-i\varphi}, re^{i\varphi-i\theta}, re^{-i\theta-i\varphi}; q\right)_\infty}, \qquad |r|<1 \tag{6.5.2}$$

for the basic exponential functions as an analog of (6.5.1). Here as before $\omega_n = 0, \pm\omega_1, \pm\omega_2, \pm\omega_3, \ldots$ and $\omega_0 = 0 < \omega_1 < \omega_2 < \omega_3 < \ldots$ are non-negative zeros of the basic sine function $S_q(\eta;\omega)$. An independent proof of (6.5.2) will be given in Section 9.3.3. We shall use this generating function for investigation of summability of the basic Fourier series (5.7.5) in the subsequent section.

Let us start from the connecting relation (3.5.3):

$$\frac{\left(q\alpha^2r^2; q^2\right)_\infty}{\left(q\alpha^2; q^2\right)_\infty} \mathcal{E}_q(\cos\theta; \alpha r) \tag{6.5.3}$$
$$= \frac{1}{2\pi}\int_0^\pi \frac{\left(q, r^2, e^{2i\varphi},\ e^{-2i\varphi};\ q\right)_\infty}{\left(re^{i\theta+i\varphi}, re^{i\theta-i\varphi}, re^{i\varphi-i\theta}, re^{-i\theta-i\varphi}; q\right)_\infty} \mathcal{E}_q(\cos\varphi; \alpha)\, d\varphi$$

where $|r| < 1$. One can easily see that if we can prove uniform convergence in the variable $x = \cos\theta$ of the series in the left side of (6.5.2), then the integral in (6.5.3) gives the correct values of the basic Fourier coefficients; see (5.7.5)–(5.7.6); which verifies the generating relation (6.5.2) by Theorem 6.3.

Uniform convergence of the series in (6.5.2) can be justified with the help of the upper bound (6.1.11). Indeed, the n-th term in this series can be estimated as

$$\left|\frac{\left(-qr^2\omega_n^2; q^2\right)_\infty\ \mathcal{E}_q(\cos\theta; i\omega_n)\, \mathcal{E}_q(\cos\varphi; -ir\omega_n)}{\left(-q\omega_n^2; q^2\right)_\infty\ k(\omega_n)}\right|$$
$$< \frac{\left(-q^{1/4-\varepsilon}\left|\omega_n\right|, -q^{1/4-\varepsilon}r\left|\omega_n\right|; q^{1/2}\right)_\infty}{\left(q, q^{2\varepsilon}, q^{1-2\varepsilon}; q\right)_\infty^2 \left(-q\omega_n^2; q^2\right)_\infty^2\ k(\omega_n)}$$
$$= A\, a_n, \qquad A = \left(q, q^{2\varepsilon}, q^{1-2\varepsilon}; q\right)_\infty^{-2} \tag{6.5.4}$$

for $-\left(q^\varepsilon + q^{-\varepsilon}\right)/2 < x, y < \left(q^\varepsilon + q^{-\varepsilon}\right)/2$ and $0 < \varepsilon < 1/2$. Suppose $0 < r < 1$ and $n \geq 0$, the case of negative n is similar. Consider

$$b_n = B\ \left(rq^{-2\varepsilon}\right)^{2n}, \qquad B = \frac{1}{\pi} \tag{6.5.5}$$

with $rq^{-2\varepsilon}<1$ and $\sum_{n=0}^{\infty} b_n<\infty$. In view of (5.5.18),

$$\begin{aligned}
\lim_{n\to\infty}\frac{a_n}{b_n} &= \lim_{n\to\infty}\frac{\left(-q^{1/4-\varepsilon}\omega_n,-q^{1/4-\varepsilon}r\omega_n;q^{1/2}\right)_\infty}{\left(-q^{1/2}\omega_n^2;q\right)_\infty\left(rq^{-2\varepsilon}\right)^{2n}}\\
&= \lim_{n\to\infty}\left[\frac{\left(-q^{1/4-\varepsilon}\omega_n,-q^{1/4-\varepsilon}r\omega_n;q^{1/2}\right)_{2n}}{\left(-q^{1/2}\omega_n^2;q\right)_{2n}\left(rq^{-2\varepsilon}\right)^{2n}}\right.\\
&\quad\times\left.\frac{\left(-q^{1/4-\varepsilon+n}\omega_n,-q^{1/4-\varepsilon+n}r\omega_n;q^{1/2}\right)_\infty}{\left(-q^{1/2+2n}\omega_n^2;q\right)_\infty}\right]\\
&= \frac{\left(-q^{1/2-\varepsilon},-rq^{1/2-\varepsilon};q^{1/2}\right)_\infty}{(-q;q)_\infty}\\
&\quad\times\lim_{n\to\infty}\frac{\left(-q^{1/4-\varepsilon}\omega_n,-q^{1/4-\varepsilon}r\omega_n;q^{1/2}\right)_{2n}}{\left(-q^{1/2}\omega_n^2;q\right)_{2n}\left(rq^{-2\varepsilon}\right)^{2n}}
\end{aligned}\tag{6.5.6}$$

because $\lim_{n\to\infty} q^n\omega_n=q^{1/4}$ by (6.3.14). Using (I.7) of [**46**] or (1.4.1)

$$\begin{aligned}
\left(-q^{1/4-\varepsilon}r\omega_n;q^{1/2}\right)_{2n} &= \left(-q^{1/4+\varepsilon}/rq^n\omega_n;q^{1/2}\right)_{2n}(r\omega_n)^{2n}q^{n^2-2n\varepsilon},\\
\left(-q^{1/2}\omega_n^2;q\right)_{2n} &= \left(-q^{1/2}/q^{2n}\omega_n^2;q\right)_{2n}\omega_n^{4n}q^{2n^2},
\end{aligned}$$

and, therefore,

$$\begin{aligned}
&\lim_{n\to\infty}\frac{\left(-q^{1/4-\varepsilon}\omega_n,-q^{1/4-\varepsilon}r\omega_n;q^{1/2}\right)_{2n}}{\left(-q^{1/2}\omega_n^2;q\right)_{2n}\left(rq^{-2\varepsilon}\right)^{2n}}\\
&\quad= \lim_{n\to\infty}\frac{\left(-q^{1/4+\varepsilon}/q^n\omega_n,-q^{1/4+\varepsilon}/rq^n\omega_n;q^{1/2}\right)_{2n}}{\left(-q^{1/2}/q^{2n}\omega_n^2;q\right)_{2n}}\\
&\quad= \lim_{n\to\infty}\frac{\left(-q^{1/4+\varepsilon}/q^n\omega_n,-q^{1/4+\varepsilon}/rq^n\omega_n;q^{1/2}\right)_\infty}{\left(-q^{1/2}/q^{2n}\omega_n^2;q\right)_\infty}\\
&\quad\times\lim_{n\to\infty}\frac{\left(-q^{1/2}/\omega_n^2;q\right)_\infty}{\left(-q^{1/4+\varepsilon}/\omega_n,-q^{1/4+\varepsilon}/r\omega_n;q^{1/2}\right)_\infty}\\
&\quad= \frac{\left(-q^{\varepsilon},-q^{\varepsilon}/r;q^{1/2}\right)_\infty}{(-1;q)_\infty}.
\end{aligned}\tag{6.5.7}$$

As a result from (6.5.6) and (6.5.7) one gets

$$0<\lim_{n\to\infty}\frac{a_n}{b_n}<\infty \tag{6.5.8}$$

and the series (6.5.2) converges absolutely and uniformly on $[-1,1]$ when $0<r<1$ due to the Weierstrass M-Test and the Limit Comparison Test. This completes the proof of the pointwise convergence by Theorem 6.3. By analytic continuation the generating relation (6.5.2) holds in an open disk $|r|<1$.

It is worth mentioning a few special cases of (6.5.2). When $r = 0$ we obtain the following generating relation

$$\sum_{n=-\infty}^{\infty} \frac{\pi}{(-q\omega_n^2;q^2)_\infty \; k(\omega_n)} \; \mathcal{E}_q(\cos\theta; i\omega_n) \tag{6.5.9}$$
$$= \left(q, q^{1/2}e^{2i\theta}, \; q^{1/2}e^{-2i\theta}; \; q\right)_\infty$$

for the basic trigonometric system $\{\mathcal{E}_q(x; i\omega_n)\}_{n=-\infty}^{\infty}$. If $\varphi = \pi/2$, one gets another q-Fourier series expansion,

$$\sum_{n=-\infty}^{\infty} \frac{(-qr^2\omega_n^2;q^2)_\infty}{(-q\omega_n^2;q^2)_\infty} \, k^{-1}(\omega_n) \; \mathcal{E}_q(\cos\theta; i\omega_n) \tag{6.5.10}$$
$$= \frac{\left(q, r^2, q^{1/2}e^{2i\theta}, \; q^{1/2}e^{-2i\theta}; \; q\right)_\infty}{\pi\left(-r^2e^{2i\theta}, -r^2e^{-2i\theta}; q^2\right)_\infty}.$$

A terminating case of this generating relation occurs when $r^2 = -1/q\omega_m^2$ for an integer $m \neq 0$,

$$\sum_{n=-|m|}^{|m|} \frac{(\omega_n^2/\omega_m^2;q^2)_\infty}{(-q\omega_n^2;q^2)_\infty} \, k^{-1}(\omega_n) \; \mathcal{E}_q(\cos\theta; i\omega_n) \tag{6.5.11}$$
$$= \frac{\left(q, \; -1/q\omega_m^2, \; q^{1/2}e^{2i\theta}, \; q^{1/2}e^{-2i\theta}; \; q\right)_\infty}{\pi\left(e^{2i\theta}/q\omega_m^2, \; e^{-2i\theta}/q\omega_m^2; q^2\right)_\infty}.$$

Here $m = \pm1, \pm2, \pm3, \ldots$. In the limit $|m| \to \infty$ one obtains formally expansion (6.5.9).

6.6. Methods of Summation of Basic Fourier Series

According to Theorem 6.3, for a function $f(x)$ the basic Fourier series $\mathbf{S}_q[f]$ converges to $f(x)$ if it converges uniformly. In this section we shall discuss other methods of summation of basic Fourier series, which are similar to Abel's summation method for the classical Fourier series; see, for example, [**156**], [**165**] and Ex. 6.7–6.10.

Let $f \in \mathcal{L}_\rho^1(-1,1)$, which means that $|f(x)|$ is integrable on $(-1,1)$ with respect to the weight function $\rho(x)$ in the orthogonality relation (5.6.5), and let $\mathbf{S}_q[f]$ be its q-Fourier series defined by the right side of (5.7.5). Following [**26**] let us replace this series by

$$\mathbf{S}_q[f;r] = \sum_{n=-\infty}^{\infty} c_n(r) \; \mathcal{E}_q(\cos\theta; i\omega_n) \tag{6.6.1}$$

where by the definition the new q-Fourier coefficients are given by

$$c_n(r) = \frac{(-qr^2\omega_n^2;q^2)_\infty}{(-q\omega_n^2;q^2)_\infty} \tag{6.6.2}$$

$$\times \frac{1}{2k(\omega_n)} \int_0^\pi f(\cos\theta)\ \mathcal{E}_q(\cos\theta; -ir\omega_n) \left(e^{2i\theta}, e^{-2i\theta}; q\right)_{1/2}\ d\theta$$

provided that $0 < r < 1$. Comparing (6.6.2) and (5.7.6) one can easily see that

$$\lim_{r\to 1^-} c_n(r) = c_n, \tag{6.6.3}$$

where c_n are the regular q-Fourier coefficients of $f(x)$. Indeed, the interchange of limit and integral can be justified by the Lebesgue Dominated Convergence Theorem; in a complete analogy with (6.1.9) one gets

$$\left|\left(-qr^2\omega_n^2; q^2\right)_\infty \mathcal{E}_q(\cos\theta; -ir\omega_n)\right| \le \sum_{n=0}^\infty \frac{q^{n^2/4}}{(q;q)_n}\ r\,|\omega_n|^n\, |H_n(\cos\theta|q)|$$

$$\le \sum_{n=0}^\infty \frac{q^{n^2/4}}{(q;q)_n}\ |\omega_n|^n\, H_n(\cosh\tau|q) = \left(q\omega_n^2; q^2\right)_\infty \mathcal{E}_q(\cosh\tau; |\omega_n|)$$

when $0 < r < 1$ and

$$|c_n(r)| \le \frac{\left(q\omega_n^2; q^2\right)_\infty \mathcal{E}_q(1; |\omega_n|)}{\left(-q\omega_n^2; q^2\right)_\infty} \int_{-1}^1 |f(x)|\ \rho(x)\,dx < \infty. \tag{6.6.4}$$

The modified q-Fourier series (6.6.1) has the advantage that they are uniformly convergent for integrable functions.

LEMMA 6.5. *If $f \in \mathcal{L}^1_\rho(-1,1)$, then q-Fourier series* (6.6.1) *converge absolutely and uniformly on $[-1,1]$ when $0 \le r < 1$.*

PROOF. The same estimate as in the previous section for the n-th term gives

$$\begin{aligned} &|c_n(r)\ \mathcal{E}_q(\cos\theta; i\omega_n)| \\ &\quad< \frac{\left(-q^{1/4-\varepsilon}|\omega_n|, -q^{1/4-\varepsilon} r\,|\omega_n|; q^{1/2}\right)_\infty}{\left(q, q^{2\varepsilon}, q^{1-2\varepsilon}; q\right)^2_\infty \left(-q\omega_n^2; q^2\right)^2_\infty\ k(\omega_n)} \\ &\quad\times \int_0^\pi |f(\cos\varphi)| \left(e^{2i\varphi}, e^{-2i\varphi}; q\right)_{1/2}\ d\varphi \sim C\left(rq^{-2\varepsilon}\right)^{2n} \end{aligned} \tag{6.6.5}$$

as $n \to \infty$ for $-\left(q^\varepsilon + q^{-\varepsilon}\right)/2 < x, y < \left(q^\varepsilon + q^{-\varepsilon}\right)/2$, $0 < \varepsilon < 1/2$. In view of (6.5.4)–(6.5.8) the series (6.6.1) converge absolutely and uniformly on $[-1,1]$ when $0 < r < 1$ due to the Weierstrass M-Test and the Limit Comparison Test. By analytic continuation these series represent analytic functions in the open disk $|r| < 1$. □

The study of convergence of the series (6.6.1) will relay on the following theorem [**165**].

THEOREM 6.9. *Let $K_r(x,y)$ be nonnegative. Let*

$$\lim_{r\to 1^-} \int_a^b K_r(x,y)\ dy = 1 \tag{6.6.6}$$

uniformly in y *for* $a < x < b$, *and if* $\varepsilon > 0$, *let*

$$\lim_{r \to 1^-} K_r(x, y) = 0 \tag{6.6.7}$$

uniformly for all x *and* y *in* (a, b) *for which* $|x - y| > \varepsilon$. *Let* $f(x)$ *be bounded over* (a, b). *Then*

$$\lim_{r \to 1^-} \int_a^b K_r(x, y) f(y)\ dy = f(x) \tag{6.6.8}$$

boundedly over the set of points interior to (a, b) *at which* $f(x)$ *is continuous.*

PROOF. This proof is due to N. Wiener [**165**]; p. 41, Proposition X_{55}.
To prove this, let us put $a + \varepsilon < x < b - \varepsilon$, and let us note that

$$\left| f(x) - \int_a^b K_r(x, y) f(y)\ dy \right|, \tag{6.6.9}$$

by (6.6.6), is asymptotically equal to

$$\left| \int_a^b K_r(x, y) (f(x) - f(y))\ dy \right|. \tag{6.6.10}$$

This does not exceed

$$2 \limsup |f(x)| \left[\int_a^{x-\varepsilon} + \int_{x+\varepsilon}^b \right] K_r(x, y)\ dy \tag{6.6.11}$$
$$+ \limsup_{|x-y|<\varepsilon} |f(x) - f(y)| \int_a^b K_r(x, y)\ dy.$$

At any point of continuity of $f(x)$, we may choose ε so small that the second term of (6.6.11) is less than δ in view of (6.6.6). We may also by (6.6.7) choose r so near 1 that the first term of (6.6.11) is less than δ. It follows that for r near enough to 1, (6.6.9) does not exceed 2δ, and hence that it tends to 0 as $r \to 1$. The boundedness of (6.6.9) results from the fact that (6.6.10) does not exceed

$$2 \limsup |f(x)| \int_a^b K_r(x, y)\ dy$$

and the proof is complete. □

The argument of the last theorem has further proved the following result [**165**]; p. 42, Proposition X_{56}.

COROLLARY 6.1. *On the hypothesis of Theorem* 6.9, *Eq.* (6.6.8) *holds uniformly over any interval interior to* (a, b) *and to one in which* $f(x)$ *is continuous.*

PROOF. We merely need bring in the fact that a function is uniformly continuous over any closed interval of continuity. □

Theorem 6.9 allows us to establish the next result [**14**].

LEMMA 6.6. *The following limiting relation holds*

$$\lim_{r\to 1^-}\frac{1}{2\pi}\int_0^\pi \frac{\left(q,r^2,e^{2i\varphi},\ e^{-2i\varphi};\ q\right)_\infty\ f(\cos\varphi)}{\left(re^{i\theta+i\varphi},re^{i\theta-i\varphi},re^{i\varphi-i\theta},re^{-i\theta-i\varphi};q\right)_\infty}\,d\varphi=f(\cos\theta) \quad (6.6.12)$$

for every bounded function $f(\cos\theta)$ *that is continuous on* $0<\theta<\pi$.

PROOF. We need to verify the hypothesis of Theorem 6.9 for the kernel

$$K_r(\theta,\varphi)=\frac{\left(q,r^2,e^{2i\varphi},\ e^{-2i\varphi};\ q\right)_\infty}{2\pi\left(re^{i\theta+i\varphi},re^{i\theta-i\varphi},re^{i\varphi-i\theta},re^{-i\theta-i\varphi};q\right)_\infty}, \quad (6.6.13)$$

which coincides with the Poisson kernel for the continuous q-Hermite polynomials (3.5.4) up to a constant. Indeed, if we set $a=0$ and $b=\pi$, then

$$\int_0^\pi K_r(\theta,\varphi)\ d\varphi=\sum_{n=0}^\infty \frac{r^n}{(q;q)_n}H_n(\cos\theta|q) \quad (6.6.14)$$
$$\times\frac{(q;q)_\infty}{2\pi}\int_0^\pi H_n(\cos\varphi|q)\left(e^{2i\varphi},\ e^{-2i\varphi};\ q\right)_\infty\ d\varphi=1$$

due to uniform convergence of the series in (3.5.4) for $|r|<1$ and the orthogonality of the continuous q-Hermite polynomials (3.5.5). Thus the property (6.6.6) is established.

Again, if $0<|\theta-\varphi|\le\theta+\varphi<2\pi-\epsilon$, $\epsilon<\theta<\pi-\epsilon$ and $0<\varphi<\pi$, then

$$K_r(\theta,\varphi)<\frac{\left(q,r^2,e^{2i\varphi},\ e^{-2i\varphi};q\right)_\infty}{2\pi\left(re^{i\epsilon},re^{-i\epsilon};q\right)_\infty^2}<C\frac{\left(r^2;q\right)_\infty}{2\pi\left(re^{i\epsilon},re^{-i\epsilon};q\right)_\infty^2}, \quad (6.6.15)$$

where C is a constant, since

$$\left(re^{i(\theta\pm\varphi)},re^{i(\theta\pm\varphi)};q\right)_\infty=\prod_{k=0}^\infty\left(1-2r\cos(\theta\pm\varphi)\ q^k+r^2q^{2k}\right)$$
$$>\prod_{k=0}^\infty\left(1-2r\cos\epsilon\ q^k+r^2q^{2k}\right)=\left(re^{i\epsilon},re^{-i\epsilon};q\right)_\infty.$$

In the limit $r\to 1^-$ the right side of (6.6.15) approaches to zero and the validity of (6.6.7) is thus assured. The positivity of $K_r(\theta,\varphi)$ is obvious when $0<r<1$. Therefore, by (6.6.8)

$$\lim_{r\to 1^-}\int_0^\pi K_r(\theta,\varphi)\,g(\varphi)\ d\varphi=g(\theta) \quad (6.6.16)$$

for every bounded function g that is continuous on $(0,\pi)$, which is equivalent to (6.6.12). □

We can now show that

$$\lim_{r\to 1^-}\mathbf{S}_q[f;r]=f(x) \quad (6.6.17)$$

under the hypothesis of Lemma 6.6. From (6.6.1)–(6.6.2) one gets

$$\mathbf{S}_q[f;r] = \sum_{n=-\infty}^{\infty} \frac{\left(-qr^2\omega_n^2;q^2\right)_\infty}{\left(-q\omega_n^2;q^2\right)_\infty} \mathcal{E}_q(\cos\theta; i\omega_n) \tag{6.6.18}$$

$$\times \frac{1}{2k(\omega_n)} \int_0^\pi f(\cos\varphi)\ \mathcal{E}_q(\cos\varphi; -ir\omega_n)\left(e^{2i\varphi}, e^{-2i\varphi};q\right)_{1/2}\, d\varphi$$

$$= \frac{1}{2}\int_0^\pi \left(\sum_{n=-\infty}^{\infty} \frac{\left(-qr^2\omega_n^2;q^2\right)_\infty}{\left(-q\omega_n^2;q^2\right)_\infty k(\omega_n)} \mathcal{E}_q(\cos\theta; i\omega_n)\,\mathcal{E}_q(\cos\varphi; -ir\omega_n)\right)$$

$$\times f(\cos\varphi)\ \left(e^{2i\varphi}, e^{-2i\varphi};q\right)_{1/2}\, d\varphi,$$

where the series can be summed by the bilinear generating function (6.5.2) and the interchange of summation and integration can be justified by the uniform convergence of the series when $|r|<1$. As a result we have written $\mathbf{S}_q[f;r]$ as an analog of the Poisson integral in Lemma 6.6:

$$\mathbf{S}_q[f;r] = \frac{\left(q^{1/2}e^{2i\theta}, q^{1/2}e^{-2i\theta};q\right)_\infty}{2\pi} \tag{6.6.19}$$

$$\times \int_0^\pi \frac{\left(q, r^2, e^{2i\varphi}, e^{-2i\varphi};q\right)_\infty}{\left(re^{i\theta+i\varphi}, re^{i\theta-i\varphi}, re^{i\varphi-i\theta}, re^{-i\theta-i\varphi};q\right)_\infty} \frac{f(\cos\varphi)}{\left(q^{1/2}e^{2i\varphi}, q^{1/2}e^{-2i\varphi};q\right)_\infty} d\varphi$$

and one can evaluate the limit as $r\to 1^-$. Thus, we have proved the following summability theorem for the q-Fourier series.

THEOREM 6.10. *Let $f(x)$ be a bounded function that is continuous on $(-1,1)$ and let $f(x,r) = \mathbf{S}_q[f;r]$ be a function defined by the series* (6.6.1)–(6.6.2) *with $0\le r<1$. Then $\lim_{r\to1^-} f(x,r) = f(x)$.*

The bilinear generating function (6.5.2) is not symmetric with respect to interchanging θ and φ. This gives a possibility to introduce another version of Abel's summation method for the q-Fourier series. Consider

$$\widetilde{\mathbf{S}}_q[f;r] = \sum_{n=-\infty}^{\infty} c_n \frac{\left(-qr^2\omega_n^2;q^2\right)_\infty}{\left(-q\omega_n^2;q^2\right)_\infty} \mathcal{E}_q(\cos\theta; ir\omega_n) \tag{6.6.20}$$

with $0<r<1$, where the regular q-Fourier coefficients are

$$c_n = \frac{1}{2k(\omega_n)}\int_0^\pi f(\cos\theta)\ \mathcal{E}_q(\cos\theta; -i\omega_n)\left(e^{2i\theta}, e^{-2i\theta};q\right)_{1/2} d\theta. \tag{6.6.21}$$

An analog of the Poisson integral is

$$\widetilde{\mathbf{S}}_q[f;r] = \int_0^\pi \frac{\left(q, r^2, e^{2i\varphi}, e^{-2i\varphi};q\right)_\infty f(\cos\varphi)}{\left(re^{i\theta+i\varphi}, re^{i\theta-i\varphi}, re^{i\varphi-i\theta}, re^{-i\theta-i\varphi};q\right)_\infty} d\varphi \tag{6.6.22}$$

and the following summability theorem holds.

THEOREM 6.11. *Let $f(x)$ be a bounded function that is continuous on $(-1,1)$ and let $\widetilde{f}(x,r) = \widetilde{\mathbf{S}}_q[f;r]$ be a function defined by the series* (6.6.20)–(6.6.21) *with $0\le r<1$. Then $\lim_{r\to1^-} \widetilde{f}(x,r) = f(x)$.*

6.7. Basic Trigonometric System and q-Legendre Polynomials

The trigonometric system $\{e^{i\pi nx}\}_{n=-\infty}^{\infty}$ and the system of the Legendre polynomials $\{P_m(x)\}_{m=0}^{\infty}$ are two complete systems in $\mathcal{L}^2(-1,1)$. The corresponding unitary transformation between these two orthogonal bases and its inverse are

$$e^{i\pi nx} = \left(\frac{2}{\pi n}\right)^{1/2} \sum_{m=0}^{\infty} i^m (m+1/2)\ J_{m+1/2}(\pi n)\ P_m(x) \tag{6.7.1}$$

and

$$P_m(x) = \sum_{n=-\infty}^{\infty} (-i)^m \left(\frac{1}{2\pi n}\right)^{1/2} J_{m+1/2}(\pi n)\ e^{i\pi nx}, \tag{6.7.2}$$

respectively. Relation (6.7.1) is a special case of a more general expansion

$$e^{irx} = \left(\frac{2}{r}\right)^{\nu} \Gamma(\nu) \sum_{m=0}^{\infty} i^m (\nu+m)\ J_{\nu+m}(r)\ C_m^{\nu}(x) \tag{6.7.3}$$

where $C_m^{\nu}(x)$ are the ultraspherical polynomials and $J_{\nu+m}(r)$ are the Bessel functions [**163**]. Expansion (6.7.2) is the Fourier series of the Legendre polynomials on $(-1,1)$. The orthogonality properties of the trigonometric system and Legendre polynomials lead to the orthogonality relations

$$\sum_{m=0}^{\infty} \frac{m+1/2}{\pi n}\ J_{m+1/2}(\pi n)\ J_{m+1/2}(\pi l) = \delta_{nl}, \tag{6.7.4}$$

$$\sum_{n=-\infty}^{\infty} \frac{m+1/2}{\pi n}\ J_{m+1/2}(\pi n)\ J_{p+1/2}(\pi n) = \delta_{mp} \tag{6.7.5}$$

for the corresponding Bessel functions.

The basic trigonometric system $\{\mathcal{E}_q(x;i\omega_n)\}_{n=-\infty}^{\infty}$ and the system of the continuous q-ultraspherical polynomials $\{C_m(x;\beta|\,q)\}_{m=0}^{\infty}$ with $\beta = q^{1/2}$, which are the basic analogs of the Legendre polynomials, are two complete orthogonal systems in $\mathcal{L}_{\rho}^2(-1,1)$ where ρ is the weight function in the orthogonality relation (5.6.5). Therefore there exists a q-version of the unitary transformation (6.7.1)–(6.7.2).

Ismail and Zhang [**76**] have found a q-analog of (6.7.3) as

$$\mathcal{E}_q(x;i\omega) = \frac{(q;q)_{\infty}\,\omega^{-\nu}}{(-q\omega^2;q^2)_{\infty}\,(q^{\nu};q)_{\infty}} \tag{6.7.6}$$

$$\times \sum_{m=0}^{\infty} i^m \left(1-q^{\nu+m}\right)\ q^{m^2/4}\ J_{\nu+m}^{(2)}(2\omega;q)\ C_m(x;q^{\nu}|\,q),$$

see (4.5.2), where $J_{\nu+m}^{(2)}(2\omega;q)$ is Jackson's q-Bessel function given by (4.5.3). The special case $\nu = 1/2$ gives the basic analog of the expansion (6.7.1) as

$$\mathcal{E}_q(x;i\omega_n) = \frac{(q;q)_\infty \, \omega_n^{-1/2}}{(-q\omega_n^2;q^2)_\infty \, (q^{1/2};q)_\infty} \times \sum_{m=0}^{\infty} i^m \left(1-q^{m+1/2}\right) q^{m^2/4} \, J_{m+1/2}^{(2)}(2\omega_n;q) \, C_m\left(x;q^{1/2}\middle| q\right), \tag{6.7.7}$$

where $\omega_{-n} = -\omega_n$ and $\omega_0 = 0 < \omega_1 < \omega_2 < \omega_3 < \ldots$ are nonnegative zeros of the basic sine function $S_q(\eta;\omega)$.

On the other hand, the continuous q-ultraspherical polynomials of the form $C_m\left(x;q^{1/2}\middle| q\right)$ can be expanded in the q-Fourier series as

$$C_m\left(x;q^{1/2}\middle| q\right) = \pi \frac{(q^{1/2};q)_\infty}{(q;q)_\infty} \times \sum_{n=-\infty}^{\infty} (-i)^m q^{m^2/4} \frac{\omega_n^{-1/2}}{k(\omega_n)(-q\omega_n^2;q^2)_\infty} J_{m+1/2}^{(2)}(2\omega_n;q) \, \mathcal{E}_q(x;i\omega_n). \tag{6.7.8}$$

Indeed, by (5.7.5)–(5.7.6) we have

$$C_m\left(x;q^{1/2}\middle| q\right) = \sum_{n=-\infty}^{\infty} c_n \, \mathcal{E}_q(x;i\omega_n) \tag{6.7.9}$$

where

$$c_n = \frac{1}{2k(\omega_n)} \int_0^\pi C_m\left(\cos\theta;q^{1/2}\middle| q\right) \mathcal{E}_q(\cos\theta;-i\omega_n) \times \left(e^{2i\theta}, e^{-2i\theta};q\right)_{1/2} d\theta. \tag{6.7.10}$$

Using (6.7.7), where the series on the right converge uniformly in x for any ω, and the orthogonality relation,

$$\int_0^\pi C_m\left(\cos\theta;q^{1/2}\middle| q\right) C_p\left(\cos\theta;q^{1/2}\middle| q\right) \left(e^{2i\theta}, e^{-2i\theta};q\right)_{1/2} d\theta = 2\pi \frac{(q^{1/2};q)_\infty^2}{(q;q)_\infty^2} \left(1-q^{m+1/2}\right)^{-1} \delta_{mp}, \tag{6.7.11}$$

see, for example, [**15**], [**46**] and (4.3.5), one gets

$$c_n = \pi \frac{(q^{1/2};q)_\infty}{(q;q)_\infty} (-i)^m q^{m^2/4} \times \frac{\omega_n^{-1/2}}{k(\omega_n)(-q\omega_n^2;q^2)_\infty} J_{m+1/2}^{(2)}(2\omega_n;q). \tag{6.7.12}$$

Thus, we have established expansion (6.7.8) formally and now the uniform convergence of this series needs to be verified.

Taking into account (2.4.27) with $\nu = m + 1/2$,

$$
\begin{aligned}
c_n &= \pi \frac{\left(q^{1/2};q\right)_\infty^2}{(q;q)_\infty^2} \frac{(-i)^m q^{m^2/4}}{\left(q^{1/2};q\right)_{m+1}} \frac{\omega_n^m \left(-\omega_n^2;q^2\right)_\infty}{k(\omega_n)\left(-q\omega_n^2;q^2\right)_\infty} \\
&\quad \times {}_2\varphi_1\left(\begin{matrix} -q^{m+3/2}, \ -q^{m+5/2} \\ q^{2m+3} \end{matrix} ; \ q^2, \ -\omega_n^2 \right) \qquad (6.7.13) \\
&= \pi \frac{\left(q^{1/2};q\right)_\infty^2}{(q;q)_\infty^2} \frac{(-i)^m q^{m^2/4} \omega_n^m}{\left(q^{1/2};q\right)_{m+1} k(\omega_n)} \\
&\quad \times \left[\frac{\left(-q^{m+3/2};q\right)_\infty}{\left(q, q^{2m+3};q^2\right)_\infty} \frac{\left(\omega_n^2 q^{m+3/2}, q^{1/2-m}/\omega_n^2;q^2\right)_\infty}{\left(-q\omega_n^2, -q^2/\omega_n^2;q^2\right)_\infty} \right. \\
&\quad \times {}_2\varphi_1\left(\begin{matrix} -q^{m+3/2}, \ -q^{1/2-m} \\ q \end{matrix} ; \ q^2, \ -\frac{q}{\omega_n^2} \right) \\
&\quad + \frac{\left(-q^{m+1/2};q\right)_\infty}{\left(q^{-1}, q^{2m+3};q^2\right)_\infty} \frac{\left(\omega_n^2 q^{m+5/2}, q^{-1/2-m}/\omega_n^2;q^2\right)_\infty}{\left(-q\omega_n^2, -q^2/\omega_n^2;q^2\right)_\infty} \\
&\quad \left. \times {}_2\varphi_1\left(\begin{matrix} -q^{m+5/2}, \ -q^{3/2-m} \\ q^3 \end{matrix} ; \ q^2, \ -\frac{q}{\omega_n^2} \right) \right] \qquad (6.7.14)
\end{aligned}
$$

by (III.32) of [**46**] or (A.3.8). The last equation gives the large asymptotics of the basic Fourier coefficients as follows

$$
k(\omega_n)\, c_n \sim \begin{cases} (-1)^n A \dfrac{\left(q^{-m}, q^{2+m};q^2\right)_\infty}{\left(-q^{1/2};q\right)_\infty} q^{n/2+m/4}, & m = 2k+1 \\ (-1)^n B \dfrac{\left(q^{-m-1}, q^{3+m};q^2\right)_\infty}{\left(-q^{1/2};q\right)_\infty} q^{3n/2+m/4}, & m = 2k \end{cases} \qquad (6.7.15)
$$

as $n \to \infty$; the values of the constants A and B here can be easily found explicitly but they are not essential for our consideration.

Indeed, by (6.7.14) when $m = 2k + 1$ the leading term is

$$
k(\omega_n)\, c_n \sim A\, \omega_n^m \frac{\left(q^{3/2+m}\omega_n^2;q^2\right)_\infty}{\left(-q\omega_n^2;q^2\right)_\infty} \qquad (6.7.16)
$$

as $n \to \infty$ and by (I.7) of [**46**] or (1.4.2) one gets

$$
\begin{aligned}
\frac{\left(q^{3/2+m}\omega_n^2;q^2\right)_\infty}{\left(-q\omega_n^2;q^2\right)_\infty} &= \frac{\left(q^{3/2+m}\omega_n^2;q^2\right)_n}{\left(-q\omega_n^2;q^2\right)_n} \frac{\left(q^{3/2+m+2n}\omega_n^2;q^2\right)_\infty}{\left(-q^{1+2n}\omega_n^2;q^2\right)_\infty} \\
&= (-1)^n q^{(m+1/2)n} \frac{\left(q^{1/2-m-2n}/\omega_n^2;q^2\right)_n}{\left(-q^{1-2n}/\omega_n^2;q^2\right)_n} \frac{\left(q^{3/2+m+2n}\omega_n^2;q^2\right)_\infty}{\left(-q^{1+2n}\omega_n^2;q^2\right)_\infty} \\
&\sim (-1)^n q^{(m+1/2)n} \frac{\left(q^{-m}, q^{2+m};q^2\right)_\infty}{\left(-q^{1/2};q\right)_\infty}, \quad n \to \infty, \qquad (6.7.17)
\end{aligned}
$$

because

$$\lim_{n\to\infty} \frac{\left(q^{3/2+m+2n}\omega_n^2; q^2\right)_\infty}{\left(-q^{1+2n}\omega_n^2; q^2\right)_\infty} = \frac{\left(q^{2+m}; q^2\right)_\infty}{\left(-q^{3/2}; q^2\right)_\infty} \tag{6.7.18}$$

and

$$\begin{aligned} \lim_{n\to\infty} \frac{\left(q^{1/2-m-2n}/\omega_n^2; q^2\right)_n}{\left(-q^{1-2n}/\omega_n^2; q^2\right)_n} &= \lim_{n\to\infty} \frac{\left(q^{1/2-m-2n}/\omega_n^2; q^2\right)_\infty}{\left(-q^{1-2n}/\omega_n^2; q^2\right)_\infty} \\ \times \lim_{n\to\infty} \frac{\left(-q/\omega_n^2; q^2\right)_\infty}{\left(q^{1/2-m}/\omega_n^2; q^2\right)_\infty} &= \frac{\left(q^{-m}, q^{2+m}; q^2\right)_\infty}{\left(-q^{1/2}; q\right)_\infty} \end{aligned} \tag{6.7.19}$$

by (6.3.14). As a result the first asymptotic in (6.7.15) follows from (6.7.16)–(6.7.19). The proof in the case $m = 2k$ is similar.

Our next step is to establish the following estimates which are convenient tools for investigation of convergence of the q-Fourier series here and in the subsequent chapters.

LEMMA 6.7. *Let $-x(\varepsilon) < -1 \le x = \cos\theta \le 1 < x(\varepsilon)$, where $x(\varepsilon) = (q^\varepsilon + q^{-\varepsilon})/2$, $0 < q < 1$ and $0 < \varepsilon < 1/2$. If $0 = \omega_0 < \omega_1 < \omega_2 < \omega_3 < \dots$ be nonnegative zeros of $S_q(\eta;\omega)$ and $\omega_{-n} = -\omega_n$, and if $\varpi_1 < \varpi_2 < \varpi_3 < \dots$ be positive zeros of $C_q(\eta;\omega)$ and $\varpi_{-n} = -\varpi_n$, then the following asymptotic estimates hold*

$$\begin{aligned} \left|\mathcal{E}_q(\cos\theta; i\omega_n)\right| &< \frac{\left(-q^{1/4-\varepsilon}|\omega_n|; q^{1/2}\right)_\infty}{\left(q, q^{2\varepsilon}, q^{1-2\varepsilon}; q\right)_\infty \left(-q\omega_n^2; q^2\right)_\infty} \\ &\sim \frac{q^{-2n\varepsilon}}{\left(-q^{1/2}, q; q\right)_\infty \left(q^\varepsilon, q^{1/2-\varepsilon}; q^{1/2}\right)_\infty} \end{aligned} \tag{6.7.20}$$

and

$$\begin{aligned} \left|\mathcal{E}_q(\cos\theta; i\varpi_n)\right| &< \frac{\left(-q^{1/4-\varepsilon}|\varpi_n|; q^{1/2}\right)_\infty}{\left(q, q^{2\varepsilon}, q^{1-2\varepsilon}; q\right)_\infty \left(-q\varpi_n^2; q^2\right)_\infty} \\ &\sim \frac{q^{(1-2n)\varepsilon}}{\left(-q^{1/2}, q; q\right)_\infty \left(q^\varepsilon, q^{1/2-\varepsilon}; q^{1/2}\right)_\infty} \end{aligned} \tag{6.7.21}$$

as $n \to \infty$. Also

$$k(\omega_n) \sim 2\pi \frac{(-q;q)_\infty^2}{\left(-q^{1/2};q\right)_\infty^2}, \qquad k(\varpi_n) \sim 2\pi \frac{(-q;q)_\infty^2}{\left(-q^{1/2};q\right)_\infty^2} \tag{6.7.22}$$

and

$$\kappa(\omega_n) \sim 2\frac{\left(q^2;q^2\right)_\infty^2}{\left(q;q^2\right)_\infty^2}\, q^n, \qquad \kappa(\varpi_n) \sim 2\frac{\left(q^2;q^2\right)_\infty^2}{\left(q;q^2\right)_\infty^2}\, q^{n-1/2} \tag{6.7.23}$$

as $n \to \infty$. Moreover,

$$\frac{\left(-q\omega_n^2;q^2\right)_\infty}{\left(-\omega_n^2;q^2\right)_\infty} \sim q^n, \qquad \frac{\left(-q\varpi_n^2;q^2\right)_\infty}{\left(-\varpi_n^2;q^2\right)_\infty} \sim q^{n-1/2} \tag{6.7.24}$$

as $n \to \infty$.

PROOF. The uniform bounds are given by (6.1.11). In view of (I.7) of [46] or (1.4.1), we obtain

$$\frac{\left(-q^{1/4-\varepsilon}\left|\omega_n\right|;q^{1/2}\right)_\infty}{\left(-q\omega_n^2;q^2\right)_\infty} = \frac{\left(-q^{1/4-\varepsilon}\left|\omega_n\right|;q^{1/2}\right)_{2n}}{\left(-q\omega_n^2;q^2\right)_n}\frac{\left(-q^{1/4-\varepsilon+n}\left|\omega_n\right|;q^{1/2}\right)_\infty}{\left(-q^{1+2n}\omega_n^2;q^2\right)_\infty}$$
$$= q^{-2n\varepsilon}\frac{\left(-q^{1/4-n+\varepsilon}/\left|\omega_n\right|;q^{1/2}\right)_{2n}}{\left(-q^{1-2n}\omega_n^2;q^2\right)_n}\frac{\left(-q^{1/4-\varepsilon+n}\left|\omega_n\right|;q^{1/2}\right)_\infty}{\left(-q^{1+2n}\omega_n^2;q^2\right)_\infty}$$
$$\sim q^{-2n\varepsilon}\frac{\left(-q^{\varepsilon},-q^{1/2-\varepsilon};q^{1/2}\right)_\infty}{\left(-q^{1/2};q\right)_\infty}, \qquad n\to\infty.$$

Indeed,

$$\lim_{n\to\infty}\frac{\left(-q^{1/4-\varepsilon+n}\left|\omega_n\right|;q^{1/2}\right)_\infty}{\left(-q^{1+2n}\omega_n^2;q^2\right)_\infty} = \frac{\left(-q^{1/2-\varepsilon};q^{1/2}\right)_\infty}{\left(-q^{3/2};q\right)_\infty}$$

and

$$\lim_{n\to\infty}\frac{\left(-q^{1/4-n+\varepsilon}/\left|\omega_n\right|;q^{1/2}\right)_{2n}}{\left(-q^{1-2n}/\omega_n^2;q^2\right)_n} = \lim_{n\to\infty}\frac{\left(-q^{1/4+\varepsilon}/q^n\left|\omega_n\right|;q^{1/2}\right)_\infty}{\left(-q/q^{2n}\omega_n^2;q^2\right)_\infty}$$
$$\times\lim_{n\to\infty}\frac{\left(-q/\omega_n^2;q^2\right)_\infty}{\left(-q^{1/4+\varepsilon}/\left|\omega_n\right|;q^{1/2}\right)_\infty} = \frac{\left(-q^{\varepsilon};q^{1/2}\right)_\infty}{\left(-q^{1/2};q\right)_\infty}$$

by (6.3.14). This proves (6.7.20). The case of (6.7.21) is similar.

In a similar fashion, from (5.5.17) one gets

$$\kappa\left(\omega_n\right) \sim \frac{(q;q)_\infty^2}{\left(q^{1/2};q\right)_\infty^2}\frac{\left(-q^{1/2}\omega_n^2;q\right)_\infty}{\left(-\omega_n^2;q\right)_\infty}$$
$$= \frac{(q;q)_\infty^2}{\left(q^{1/2};q\right)_\infty^2}\frac{\left(-q^{1/2}\omega_n^2;q\right)_{2n}}{\left(-\omega_n^2;q\right)_{2n}}\frac{\left(-q^{1/2+2n}\omega_n^2;q\right)_\infty}{\left(-q^{2n}\omega_n^2;q\right)_\infty}$$
$$= \frac{(q;q)_\infty^2}{\left(q^{1/2};q\right)_\infty^2}q^n\frac{\left(-q^{1/2-2n}/\omega_n^2;q\right)_{2n}}{\left(-q^{1-2n}/\omega_n^2;q\right)_{2n}}\frac{\left(-q^{1/2+2n}\omega_n^2;q\right)_\infty}{\left(-q^{2n}\omega_n^2;q\right)_\infty}$$
$$\sim 2\frac{\left(q^2;q^2\right)_\infty^2}{\left(q;q^2\right)_\infty^2}\,q^n, \qquad n\to\infty$$

by (I.7) of [46] or (1.4.2) and (6.3.14), which is the first asymptotic estimate in (6.7.23); the second case is similar. Eqs. (6.7.22) follow from (5.5.8) and (6.7.23). □

As a result from (6.7.15), (6.7.20) and (6.7.22) we obtain

$$\left|c_n\mathcal{E}_q\left(\cos\theta;i\omega_n\right)\right| < a_n \sim \begin{cases} Cq^{(1-4\varepsilon)n/2}, & m=2k+1 \\ Dq^{(3-4\varepsilon)n/2}, & m=2k \end{cases} \tag{6.7.25}$$

as $n\to\infty$, where C and D are some positive constants. Therefore the series on the right side of (6.7.8) converge absolutely and uniformly on every closed subinterval of $(-\eta,\eta)$ and $(-\eta_1,\eta_1)$ with $\eta = \left(q^{1/4}+q^{-1/4}\right)/2$ and $\eta_1 =$

$(q^{1/2}+q^{-1/2})/2$ for $m=2k+1$ and $m=2k$, respectively, by the Weierstrass M-test and the Limit Comparison Test. Thus, we have established the expansion of the q-Legendre polynomials $C_m\left(x;q^{1/2}\middle|\,q\right)$ in terms of the basic trigonometric functions $\mathcal{E}_q(x;i\omega_n)$ due to Theorem 6.3. We shall continue the study the q-Fourier expansions (6.7.8) in Chapter 9 where important connections with the q-Lommel polynomials will be established and analytic continuation will be discussed; see Section 9.1.2.

Relations (6.7.7)–(6.7.8) define a unitary operator acting in $\mathcal{L}^2_\rho(-1,1)$ [**5**]. Orthogonality relations of the matrix of this operator lead to the following orthogonality properties

$$\sum_{m=0}^{\infty}\frac{\pi\left(1-q^{m+1/2}\right)}{\omega_n\;k(\omega_n)\;\left(-q\omega_n^2;q^2\right)^2_\infty}\;q^{m^2/2}\;J^{(2)}_{m+1/2}(2\omega_n;q)\;J^{(2)}_{m+1/2}(2\omega_l;q)=\delta_{nl} \tag{6.7.26}$$

and

$$\sum_{n=-\infty}^{\infty}\frac{\pi\left(1-q^{m+1/2}\right)}{\omega_n\;k(\omega_n)\;\left(-q\omega_n^2;q^2\right)^2_\infty}\;q^{m^2/2}\;J^{(2)}_{m+1/2}(2\omega_n;q)\;J^{(2)}_{p+1/2}(2\omega_n;q)=\delta_{mp} \tag{6.7.27}$$

for the corresponding Jackson q-Bessel function. These relations are clearly q-analogs of (6.7.4)–(6.7.5). Connections with the q-Lommel polynomials will be discussed in Section 9.1.2; see also Ex. 7.6.

6.8. Examples

Let us consider a periodic function $p_1(x)$ which is defined in the interval $(-1,1)$ by $p_1(x)=x$. Its Fourier coefficients are

$$c_0=0;$$
$$c_n=\frac{1}{2}\int_{-1}^{1}xe^{-i\pi nx}\,dx=\frac{(-1)^{n-1}}{i\pi n},\qquad n\neq 0.$$

Therefore

$$\begin{aligned} x &= \sum_{n=-\infty}^{\infty}{}'\;\frac{(-1)^{n-1}}{i\pi n}\;e^{i\pi nx} \\ &= 2\sum_{n=1}^{\infty}(-1)^{n-1}\;\frac{\sin\pi nx}{\pi n}. \end{aligned} \tag{6.8.1}$$

The special case $m=1$ of (6.7.8),

$$C_1\left(x;q^{1/2}\middle|\,q\right)=-i\pi\frac{\left(q^{1/2};q\right)_\infty}{(q;q)_\infty}\;q^{1/4} \tag{6.8.2}$$
$$\times\sum_{n=-\infty}^{\infty}\frac{\omega_n^{-1/2}}{k(\omega_n)\left(-q\omega_n^2;q^2\right)_\infty}\;J^{(2)}_{3/2}(2\omega_n;q)\;\mathcal{E}_q(x;i\omega_n),$$

gives us a possibility to establish a q-analog of (6.8.1). Let us first simplify the right side of (6.8.2). Using the three-term recurrence relation for the q-Bessel functions,

$$q^{\nu}\, J_{\nu+1}^{(2)}(r;q)=\frac{2\left(1-q^{\nu}\right)}{r}\, J_{\nu}^{(2)}(r;q)-J_{\nu-1}^{(2)}(r;q), \tag{6.8.3}$$

see [**56**] or Ex. 1.25 of [**46**], and connections with the basic trigonometric functions (2.4.25)–(2.4.26), one gets

$$\begin{aligned} J_{3/2}^{(2)}(2\omega_n;q) &= -q^{-1/2} J_{-1/2}^{(2)}(2\omega_n;q) \\ &= -\frac{\left(q^{1/2};q\right)_\infty}{(q;q)_\infty}\, \frac{\left(-q\omega_n^2;q^2\right)_\infty}{\left(q\omega_n\right)^{1/2}}\, C_q(\eta;\omega_n). \end{aligned} \tag{6.8.4}$$

On the other hand,

$$C_1\left(x;q^{1/2}\middle|\, q\right)=\frac{2}{1+q^{1/2}}\,x, \tag{6.8.5}$$

see Ex. 6.1. Combining (6.8.2)–(6.8.5) and (5.8.10), we finally obtain

$$\begin{aligned} x &= \left(q^{1/4}+q^{-1/4}\right) \\ &\quad \times {\sum_{n=-\infty}^{\infty}}' \frac{(-1)^{n-1}}{2i\,\kappa(\omega_n)\,\omega_n} \sqrt{\frac{\left(-q\omega_n^2;q^2\right)_\infty}{\left(-\omega_n^2;q^2\right)_\infty}}\, \mathcal{E}_q(x;i\omega_n) \\ &= \left(q^{1/4}+q^{-1/4}\right) \\ &\quad \times \sum_{n=1}^{\infty} \frac{(-1)^{n-1}}{\kappa(\omega_n)\,\omega_n} \sqrt{\frac{\left(-q\omega_n^2;q^2\right)_\infty}{\left(-\omega_n^2;q^2\right)_\infty}}\, S_q(x;\omega_n). \end{aligned} \tag{6.8.6}$$

These equations are clearly q-analogs of (6.8.1). These series converge absolutely and uniformly on every closed subinterval of the open interval $|x|<\eta=\left(q^{1/4}+q^{-1/4}\right)/2$; analytic continuation to the larger domain of the complex x-plane will be discussed in the next section.

In a similar fashion, the basic Fourier series for the quadratic function $p_2(x)=x^2$ can be found as the special case $m=2$ of Eq. (6.7.8):

$$\begin{aligned} x^2 &= \frac{\left(1+q^{1/2}\right)^2}{4\left(1+q^{1/2}+q\right)}+\frac{\left(1+q^{1/2}\right)\left(1-q^2\right)}{4q} \\ &\quad \times {\sum_{n=-\infty}^{\infty}}' \frac{(-1)^{n}}{\kappa(\omega_n)\,\omega_n^2} \sqrt{\frac{\left(-q\omega_n^2;q^2\right)_\infty}{\left(-\omega_n^2;q^2\right)_\infty}}\, \mathcal{E}_q(x;i\omega_n) \\ &= \frac{\left(1+q^{1/2}\right)^2}{4\left(1+q^{1/2}+q\right)}+\frac{\left(1+q^{1/2}\right)\left(1-q^2\right)}{2q} \\ &\quad \times \sum_{n=1}^{\infty} \frac{(-1)^{n}}{\kappa(\omega_n)\,\omega_n^2} \sqrt{\frac{\left(-q\omega_n^2;q^2\right)_\infty}{\left(-\omega_n^2;q^2\right)_\infty}}\, C_q(x;\omega_n), \end{aligned} \tag{6.8.7}$$

which is clearly a q-analog of

$$x^2 = \frac{1}{3} + 4\sum_{n=1}^{\infty} (-1)^n \frac{\cos \pi n x}{(\pi n)^2}. \tag{6.8.8}$$

The series (6.8.7) converge absolutely and uniformly on every closed subinterval of the open interval $|x| < \eta_1 = \left(q^{1/2} + q^{-1/2}\right)/2$. We leave the details to the reader; see Ex. 6.18. Analytic continuation to the larger domain of the complex x-plane will be discussed in the next section.

More explicit q-Fourier expansions of elementary and q-functions will be found in Chapters 9 and 10.

6.9. Analytic Continuation of Basic Fourier Series

So far we have only discussed convergence of the q-Fourier series on the intervals of the real line containing the interval of orthogonality $[-1, 1]$ of the basic trigonometric system. At the same time, these series are infinite sums of analytic functions and by the Weierstrass theorem the uniform convergence guarantees analyticity of the limit function is some region of the complex x-plane. This raises an interesting question about the maximum domain of analyticity of these series. It is surprising that the problem has a partial solution. In this section we shall derive an analog of Cauchy–Hadamard's formula for the basic Fourier series. This gives an independent test for convergence of these series in the complex plane.

THEOREM 6.12. *Let E_ε be an ellipse in the complex x-plane*

$$\left(\frac{\text{Re } x}{a_\varepsilon}\right)^2 + \left(\frac{\text{Im } x}{b_\varepsilon}\right)^2 = 1 \tag{6.9.1}$$

with the semiaxes given by

$$a_\varepsilon = \frac{1}{2}\left(q^\varepsilon + q^{-\varepsilon}\right), \qquad b_\varepsilon = \frac{1}{2}\left|q^\varepsilon - q^{-\varepsilon}\right| \tag{6.9.2}$$

and the focal points at ± 1 and let I_ε be the interior of the ellipse E_ε. The basic Fourier series

$$f(x) = \sum_{n=-\infty}^{\infty} c_n\, \mathcal{E}_q(x; i\omega_n) \tag{6.9.3}$$

converges absolutely for every $x \in I_\varepsilon$, in the interior of the ellipse E_ε :

$$\left(\frac{\text{Re } x}{a_\varepsilon}\right)^2 + \left(\frac{\text{Im } x}{b_\varepsilon}\right)^2 < 1 \tag{6.9.4}$$

with

$$q^{2\varepsilon} = \limsup_{|n|\to\infty} \left(|c_n|\right)^{1/|n|}, \tag{6.9.5}$$

when $\varepsilon > 0$. If $0 < \varepsilon' < \varepsilon$ the convergence is uniform in the closure of the interior of the subellipse $\overline{I}_{\varepsilon'} = I_{\varepsilon'} \cup E_{\varepsilon'}$:

$$\left(\frac{\text{Re } x}{a_{\varepsilon'}}\right)^2 + \left(\frac{\text{Im } x}{b_{\varepsilon'}}\right)^2 \leq 1. \tag{6.9.6}$$

For every $x \in I_\varepsilon$, in the interior of the ellipse E_ε, the sum of the series is an analytic function.

We shall call the ellipse E_ε given by (6.9.1)–(6.9.2) and (6.9.5) the *ellipse of convergence.* Nothing is claimed about convergence on the ellipse. When $\varepsilon < 0$ the series (6.9.3) does not converge uniformly on $[-1, 1]$ because its q-Fourier coefficients do not satisfy the analog of the Riemann–Lebesgue lemma, see Ex. 6.4; when $\varepsilon = 0$ our test is inconclusive.

The proof of this theorem uses the following extensions of the uniform bounds for the q-exponential function (6.1.11)–(6.1.12) to a region in the complex plane.

LEMMA 6.8. *The following inequalities hold*

$$|\mathcal{E}_q (x; i\omega)| < \frac{\left(-q^{1/4-\varepsilon} |\omega| ; q^{1/2}\right)_\infty}{(1 - q^{2\varepsilon}) (q; q)_\infty^2 (-q\omega^2; q^2)_\infty} \tag{6.9.7}$$

and

$$|\mathcal{E}_q (x; i\omega)| < \frac{\left(q^{1/2}, -q^{1/4-\varepsilon} |\omega| , -q^{1/4+\varepsilon} / |\omega| ; q^{1/2}\right)_\infty}{(1 - q^{2\varepsilon}) (q; q)_\infty^2 (-q\omega^2; q^2)_\infty} \tag{6.9.8}$$

for all real values of ω and every $\varepsilon > 0$ when $x \in I_\varepsilon \cup E_\varepsilon$ is in the interior I_ε or on the ellipse E_ε given by (6.9.1)–(6.9.2).

PROOF. As in the proof of Lemma 6.2 we use the relation between the basic exponential functions and the continuous q-Hermite polynomials given by

$$H_n (x|q) = \sum_{k=0}^{n} \frac{(q; q)_n}{(q; q)_k (q; q)_{n-k}} q^{(n-2k)z}, \qquad x = \frac{1}{2} \left(q^z + q^{-z}\right). \tag{6.9.9}$$

Let $z = \text{Re } z + i \,\text{Im } z$ with $\text{Re } z = \varepsilon'$, $q^{i \text{Im } z} = e^{i\theta}$, or $q^z = e^{i\theta} q^{\varepsilon'}$, where $0 \leq \varepsilon' \leq \varepsilon$. Then

$$x = \frac{1}{2} \left(q^{\varepsilon'} + q^{-\varepsilon'}\right) \cos\theta + i\frac{1}{2} \left(q^{\varepsilon'} - q^{-\varepsilon'}\right) \sin\theta \tag{6.9.10}$$

and the image of the vertical line segment given by $\text{Re}\, z = \varepsilon'$, $\text{Im}\, z = \theta/\log q$, $-\pi \leq \theta \leq \pi$ is the ellipse $E_{\varepsilon'}$ in the complex x-plane; see Figure 6.3. The case $-\varepsilon \leq -\varepsilon' \leq 0$ corresponds to the opposite orientation of the ellipse $E_{\varepsilon'}$ and we will skip the details. Every x inside the largest ellipse E_ε belongs to some subellipse $E_{\varepsilon'}$ and, therefore, the following inequalities hold

$$|H_n (x|q)| \leq \sum_{k=0}^{n} \frac{(q; q)_n}{(q; q)_k (q; q)_{n-k}} q^{(n-2k)\varepsilon'}$$

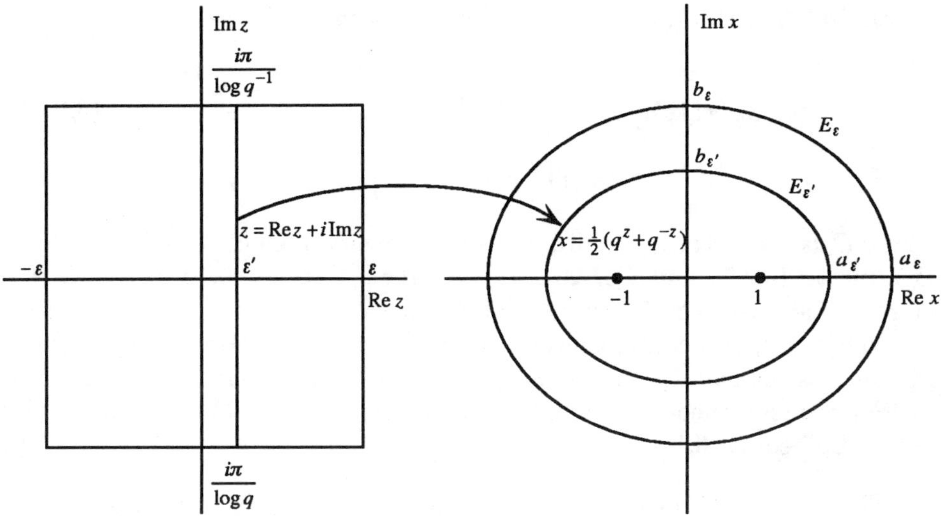

FIGURE 6.3. Mapping by $x = (q^z + q^{-z})/2$.

$$= \quad H_n\left(x\left(\varepsilon'\right)|q\right) \leq H_n\left(x\left(\varepsilon\right)|q\right). \tag{6.9.11}$$

One can also use the principal of maximum of the modulus for analytic functions and use the fact that the maxima of $|H_n(x|q)|$ on the ellipse E_ε occur at the points $\pm x = a_\varepsilon = x(\varepsilon)$. From (6.9.9) we obtain

$$H_n\left(x\left(\varepsilon\right)|q\right) = \sum_{k=0}^{n} \frac{(q;q)_n}{(q;q)_k\,(q;q)_{n-k}}\, q^{(n-2k)\varepsilon} \tag{6.9.12}$$

$$< \frac{(q;q)_n}{(q;q)_\infty^2} \sum_{k=0}^{n} q^{(n-2k)\varepsilon} = \frac{(q;q)_n}{(q;q)_\infty^2}\, q^{-n\varepsilon} \frac{1-q^{2(n+1)\varepsilon}}{1-q^{2\varepsilon}}$$

because $(q;q)_\infty < (q;q)_k$ for $0 < q < 1$. As a result, for every x inside or on the ellipse E_ε with $\varepsilon > 0$ the following inequalities hold

$$\frac{|H_n(x|q)|}{(q;q)_n} < \frac{q^{-n\varepsilon}}{(q;q)_\infty^2} \frac{1-q^{2(n+1)\varepsilon}}{1-q^{2\varepsilon}} < \frac{q^{-n\varepsilon}}{\left(1-q^{2\varepsilon}\right)(q;q)_\infty^2} \tag{6.9.13}$$

as extensions of the upper bound (6.1.16) to the larger domain in the complex x-plane. The same consideration as at the end of the proof of Lemma 6.2 results in (6.9.7) and (6.9.8). This completes the proof of Lemma 6.8. □

Now we can prove Theorem 6.12.

PROOF. Let E_ε be the ellipse of convergence given by (6.9.1)–(6.9.2) and (6.9.5). Let $0 < \delta < \varepsilon$ and $q^{2\varepsilon} < q^{2(\varepsilon-\delta)}$. Then by the definition of limes superior there exists an $n_0 > 0$ such that

$$\left(|c_n|\right)^{1/|n|} < q^{2(\varepsilon-\delta)}, \qquad \text{or} \qquad |c_n| < q^{2(\varepsilon-\delta)|n|} \tag{6.9.14}$$

for all $|n| > n_0$. If $0 < \varepsilon' < \varepsilon$, the following estimate holds

$$\begin{aligned} |c_n \, \mathcal{E}_q(x; i\omega)| &< C\, q^{2(\varepsilon-\delta)|n|} \frac{\left(-q^{1/4-\varepsilon'}\,|\omega_n|\,; q^{1/2}\right)_\infty}{\left(-q\omega_n^2; q^2\right)_\infty} \\ &= \mathrm{O}\left(q^{2(\varepsilon-\delta-\varepsilon')|n|}\right), \qquad |n| \to \infty, \end{aligned} \tag{6.9.15}$$

where C is a constant, for all $|n| > n_0$ and every x inside or on the ellipse $E_{\varepsilon'}$ due to (6.9.7) and (6.7.20). Thus, when $\varepsilon' < \varepsilon - \delta$, $0 < \delta < \varepsilon$, for large n the basic Fourier series (6.9.3) has a convergent geometric series as a majorant, and consequently converges absolutely. For every subellipse $E_{\varepsilon'}$ and its interior $I_{\varepsilon'}$ the convergence is uniform by the M-test and due to Weierstrass's theorem the limit function is analytic. This completes the proof of Theorem 6.12. □

Theorem 6.12 provides an effective tool for investigation of the convergence of the basic Fourier series. For example, in the case of the linear function (6.8.6) one gets

$$|c_n| = \frac{1}{2\kappa(\omega_n)\,\omega_n} \sqrt{\frac{(-q\omega_n^2; q^2)_\infty}{(-\omega_n^2; q^2)_\infty}} \sim C\, q^{|n|/2}, \qquad |n| \to \infty \tag{6.9.16}$$

by (6.3.1), (6.7.23)–(6.7.24) and

$$\limsup_{|n|\to\infty} \left(|c_n|\right)^{1/|n|} = q^{1/2} = q^{2\varepsilon}. \tag{6.9.17}$$

Thus, the series (6.8.6) converges to the linear function x inside the ellipse of convergence $E_{1/4}$ in the complex x-plane.

In the similar fashion, for the quadratic function (6.8.7) we have

$$|c_n| = \frac{1}{\kappa(\omega_n)\,\omega_n^2} \sqrt{\frac{(-q\omega_n^2; q^2)_\infty}{(-\omega_n^2; q^2)_\infty}} \sim D\, q^{3|n|/2}, \qquad |n| \to \infty, \tag{6.9.18}$$

$$\limsup_{|n|\to\infty} \left(|c_n|\right)^{1/|n|} = q^{3/2} = q^{2\varepsilon} \tag{6.9.19}$$

and the series (6.8.7) converges to x^2 inside the ellipse $E_{3/4}$.

Expansion (6.5.9) gives an example of a q-Fourier series that converges in the whole complex x-plane. Indeed, by (1.4.1), (6.3.1) and (6.7.22) one gets

$$\begin{aligned} |c_n| &= \frac{\pi}{(-q\omega_n^2; q^2)_\infty \; k(\omega_n)} \\ &= \frac{\pi}{\left(-q^{1-2|n|}/\omega_n^2; q^2\right)_{|n|} \; \omega_n^{2|n|} q^{n^2} \; \left(-q^{1+2|n|}\omega_n^2; q^2\right)_\infty \; k(\omega_n)} \\ &\sim C\, q^{n^2}, \qquad |n| \to \infty \end{aligned} \tag{6.9.20}$$

and therefore

$$\limsup_{|n|\to\infty} \left(|c_n|\right)^{1/|n|} = 0, \tag{6.9.21}$$

which corresponds to the ellipse of convergence E_∞.

More examples of analytic continuation of the basic Fourier series will be discussed in Chapters 9 and 10.

6.10. Miscellaneous Results

By the Hadamard theorem an entire function $f(z)$ of order zero which does not have zero at the origin can be represented as an infinite product of the form

$$f(z) = f(0) \prod_{n=1}^{\infty} \left(1 - \frac{z}{a_n}\right), \tag{6.10.1}$$

where $a_1,\ a_2,\ a_3, \ldots$ are all zeros of the function $f(z)$. See, for example, **[21]**; **[23]**; **[106]**; **[99]**, p. 24 or Theorem B.3 from Appendix B. Consider the entire function of order zero

$$f(\omega) = \left(-q\omega^2; q^2\right)_\infty \frac{S_q(\eta;\omega)}{\omega}, \tag{6.10.2}$$

which has only simple real zeros at $\omega = \pm\omega_n$ with $n = 1, 2, 3, \ldots$ by Theorems 5.1–5.3. In this case

$$f(0) = \frac{1}{1-q^{1/2}}$$

by (5.4.1) and therefore

$$f(\omega) = \frac{1}{1-q^{1/2}} \prod_{n=1}^{\infty} \left(1 - \frac{\omega}{\omega_n}\right)\left(1 + \frac{\omega}{\omega_n}\right).$$

As a result we arrive at the infinite product representation for the basic sine function

$$\begin{aligned} S_q(\eta;\omega) &= \frac{1}{1-q^{1/2}} \frac{\omega}{\left(-q\omega^2;q^2\right)_\infty} \prod_{n=1}^{\infty} \left(1 - \frac{\omega}{\omega_n}\right)\left(1 + \frac{\omega}{\omega_n}\right) \\ &= \frac{1}{1-q^{1/2}} \frac{\omega}{\left(-q\omega^2;q^2\right)_\infty} \prod_{n=1}^{\infty} \left(1 - \frac{\omega^2}{\omega_n^2}\right). \end{aligned} \tag{6.10.3}$$

In a similar manner, one can obtain an infinite product representation for the basic cosine function

$$\begin{aligned} C_q(\eta;\omega) &= \frac{1}{\left(-q\omega^2;q^2\right)_\infty} \prod_{n=1}^{\infty} \left(1 - \frac{\omega}{\varpi_n}\right)\left(1 + \frac{\omega}{\varpi_n}\right) \\ &= \frac{1}{\left(-q\omega^2;q^2\right)_\infty} \prod_{n=1}^{\infty} \left(1 - \frac{\omega^2}{\varpi_n^2}\right). \end{aligned} \tag{6.10.4}$$

Equations (5.8.10)–(5.8.11) and (6.10.3)–(6.10.4) result in the following relations

$$(-1)^n \sqrt{(-\omega_m^2;q)_\infty} = \prod_{n=1}^{\infty}\left(1-\frac{\omega_m^2}{\varpi_n^2}\right) \tag{6.10.5}$$

and

$$(-1)^{n-1} \sqrt{(-\varpi_m^2;q)_\infty} = \frac{\varpi_m}{1-q^{1/2}}\prod_{n=1}^{\infty}\left(1-\frac{\varpi_m^2}{\omega_n^2}\right) \tag{6.10.6}$$

between the zeros of the basic sine $S_q(\eta;\omega)$ and basic cosine $C_q(\eta;\omega)$ functions.

6.11. Exercises for Chapter 6

(1) The continuous q-ultraspherical polynomials are

$$C_n(\cos\theta;\beta|q) = \sum_{k=0}^{n} \frac{(\beta;q)_k\,(\beta;q)_{n-k}}{(q;q)_k\,(q;q)_{n-k}}\, e^{i(n-2k)\theta}.$$

Show that they are generated by the three-term recurrence relation

$$2xC_n(x;\beta|q) = \frac{1-q^{n+1}}{1-\beta q^n} C_{n+1}(x;\beta|q) + \frac{1-\beta^2 q^{n-1}}{1-\beta q^n} C_{n-1}(x;\beta|q)$$

$n \geq 0$ with $C_{-1}(x;\beta|q) = 0$, $C_0(x;\beta|q) = 1$ and verify that

$$C_1(x;\beta|q) = \frac{1-\beta}{1-q}(2x),$$

$$C_2(x;\beta|q) = \frac{1-\beta}{1-q^2}\left[\frac{1-\beta q}{1-q}(2x)^2 - 1 - \beta\right],$$

$$C_3(x;\beta|q) = \frac{(1-\beta)(1-\beta q)}{(1-q^2)(1-q^3)}$$

$$\times (2x)\left[\frac{1-\beta q^2}{1-q}(2x)^2 - 2 - q - \beta - 2\beta q\right],$$

$$C_4(x;\beta|q) = \frac{(1-\beta)(1-\beta q)}{(1-q^2)(1-q^4)}\left[\frac{(1-\beta q^2)(1-\beta q^3)}{(1-q)(1-q^3)}(2x)^4\right.$$

$$\left. - \frac{(1-\beta q^2)(3+2q+q^2+\beta+2\beta q+3\beta q^2)}{1-q^3}(2x)^2 + (1+\beta)(1+\beta q)\right].$$

Show that

$$|C_n(\cos\theta;\beta|q)| \leq C_n(1;\beta|q)$$

if $-1 < q, \beta < 1$. Use the following representation [**46**]

$$C_n(\cos\theta;\beta|q) = \frac{(\beta;q)_\infty\,(\beta^2;q)_n}{(\beta^2;q)_\infty\,(q;q)_n}$$

$$\times\left[\frac{(\beta e^{-2i\theta};q)_\infty}{(e^{-2i\theta};q)_\infty}e^{in\theta}\ {}_2\varphi_1\left(\begin{matrix}\beta,\ \beta e^{2i\theta}\\ qe^{2i\theta}\end{matrix};q,\ q^{n+1}\right)\right.$$
$$\left.+\frac{(\beta e^{2i\theta};q)_\infty}{(e^{2i\theta};q)_\infty}e^{-in\theta}\ {}_2\varphi_1\left(\begin{matrix}\beta,\ \beta e^{-2i\theta}\\ qe^{-2i\theta}\end{matrix};q,\ q^{n+1}\right)\right]$$

in order to show that

$$|C_n(\cos\theta;\beta|q)|\le C_n(1;\beta|q)<C_n(x(\varepsilon);\beta|q)$$
$$<q^{-n\varepsilon}\frac{(\beta^2;q)_n\,(\beta,\ \beta q^{2\varepsilon},\ \beta q^{1-2\varepsilon};q)_\infty}{(\beta q^{1-2\varepsilon};q)_n\,(q,\ q^{2\varepsilon},\ q^{1-2\varepsilon},\ \beta^2;q)_\infty}$$

if $x(\varepsilon)=(q^\varepsilon+q^{-\varepsilon})/2$, $0<\varepsilon<1/2$; $-1<\beta<q^{2\varepsilon}$ and $0<q<1$.

(2) Prove that

$$\left|\frac{H_n(x,y|q)}{(q;q)_n}\right|<Cq^{-n^2/4+n/2}\frac{(q^{1/2},-q^{\varepsilon+3/4},-q^{-\varepsilon-1/4};q^{1/2})_\infty}{(q,q,q^{2\varepsilon},q^{1-2\varepsilon};q)_\infty},$$

where $H_n(x,y|q)$ are the continuous q-Hermite polynomials in two variables given by (3.7.1)–(3.7.2) or (3.7.5), $-1\le x,y\le 1$, $0<\varepsilon<1/2$, and the constant C is defined in Ex. 3.4. [*Hint:* Use (3.7.12), (6.1.16) and Ex. 3.4 in order to obtain

$$\left|\frac{H_n(x,y|q)}{(q;q)_n}\right|\le\sum_{k=0}^{n}\frac{q^{(k-n)(k+n-2)/4}}{(q;q)_k\,(q;q)_{n-k}}\,|H_k(x|q)|\ |\phi_{n-k}(y;q)|$$
$$<\frac{Cq^{-n(n-2)/4}}{(q,q^{2\varepsilon},q^{1-2\varepsilon};q)_\infty}\sum_{k=0}^{n}\frac{q^{k^2/4-(\varepsilon+1/2)k}}{(q;q)_{n-k}}$$
$$<\frac{Cq^{-n(n-2)/4}}{(q,q,q^{2\varepsilon},q^{1-2\varepsilon};q)_\infty}\sum_{k=-\infty}^{\infty}q^{k^2/4}\left(q^{-\varepsilon-1/2}\right)^k$$

and then use the Jacobi triple product identity (A.2.4).]

(3) *Parseval's formula.* Let $f,g\in\mathcal{L}^2_\rho(-1,1)$, where $\rho(x)$ is the weight function in the orthogonality relation (5.6.5). Show that

$$\int_{-1}^{1}f(x)\,\overline{g(x)}\ \rho(x)\ dx=\sum_{n=-\infty}^{\infty}2k(\omega_n)\ c_n\ \overline{d_n},$$

where c_n and d_n are the q-Fourier coefficients of the functions f and g, respectively.

(4) *Analog of the Riemann–Lebesgue Lemma.* If $f\in\mathcal{L}^2_\rho(-1,1)$, prove that

$$\lim_{n\to\infty}\int_{-1}^{1}f(x)\ \mathcal{E}_q(x;-i\omega_n)\ \rho(x)\ dx=0,$$

where $\rho(x)$ is the weight function in the orthogonality relation (5.6.5). [*Hint*: Use Parseval's identity (6.2.29) and asymptotic (6.7.22).]

(5) Consider two differentiable functions $f(x)$ and $g(x)$ such that

$$f^2(x) + g^2(x) = 1$$

$$W(f,g) = f(x)\,g'(x) - f'(x)\,g(x) = h(x).$$

Show that

$$f'(x) = h(x)\,g(x), \qquad g'(x) = -h(x)\,f(x)$$

and

$$f(x) = \sin\left(\int_{x_0}^{x} h(t)\ dt\right),$$

$$g(x) = \cos\left(\int_{x_0}^{x} h(t)\ dt\right).$$

Prove that both functions $f(x)$ and $g(x)$ are bounded for all real values of x and change from -1 to 1. Show that all extrema of function $f(x)$ $(g(x))$ are located at zeros of $g(x)$ $(f(x))$ and that function $f(x)$ $(g(x))$ is monotone between any two successive zeros of $g(x)$ $(f(x))$.

(6) Use Jensen's theorem in order to prove that there is only one positive zero of the basic sine (cosine) function $S_q(\eta;\omega)$ $(C_q(\eta;\omega))$ located between the points $\omega = \gamma q^{-n}$ and $\omega = \gamma q^{-n-1}$ for sufficiently large values of n. [**26**]

(7) *The Poisson kernel.* Show that

$$1 + 2\sum_{n=1}^{\infty} r^n \cos n\theta = \frac{1-r^2}{1-2r\cos\theta + r^2},$$

$$\sum_{n=1}^{\infty} r^n \sin n\theta = \frac{r\sin\theta}{1-2r\cos\theta + r^2},$$

when $0 \le r < 1$.

(8) *Abel's summation method.* Consider the series

$$\sum_{n=0}^{\infty} u_n = u_0 + u_1 + u_2 + \ldots \tag{6.11.1}$$

and

$$\sum_{n=0}^{\infty} u_n r^n = u_0 + u_1 r + u_2 r^2 + \ldots, \tag{6.11.2}$$

where we assume that the series (6.11.2) converges for $0 < r < 1$. (This will always be the case if the terms of the series (6.11.1) are bounded.) If $\sigma(r)$ is the sum of (6.11.2), we say that the series (6.11.1) is summable by Abel's method to the value σ, when

$$\lim_{r\to 1^-} \sigma(r) = \sigma.$$

Prove that if the series (6.11.1) converges to its sum σ, then this series is summable by Abel's method to the same number σ.

(9) *The Poisson integral.* If $f \in \mathcal{L}^1(-\pi,\pi)$ and hence if $f \in \mathcal{L}^2(-\pi,\pi)$, use the Poisson kernel from Ex. 6.4 to show that

$$\sum_{n=-\infty}^{\infty} r^{|n|} e^{inx} \frac{1}{2\pi} \int_{-\pi}^{\pi} f(y)\, e^{-iny}\, dy$$
$$= \frac{1}{2\pi} \int_{-\pi}^{\pi} f(y) \frac{1-r^2}{1-2r\cos(x-y)+r^2}\, dy,$$

when $0 \le r < 1$.

(10) *Abel's method for the classical Fourier series.* Let $f(x)$ be an absolutely integrable periodic function of period 2π and

$$f(x) \sim \sum_{n=-\infty}^{\infty} c_n e^{inx}$$

be its Fourier series. Prove that the Fourier series is summable by Abel's method to the value $f(x)$ at every point of continuity of $f(x)$ and to $(f(x+0)+f(x-0))/2$ at every point of jump discontinuity of $f(x)$.

(11) Introduce an analog of the Poisson kernel as

$$K_r(x,y) := \sum_{n=-\infty}^{\infty} \frac{\mathcal{E}_q(x;i\omega_n)\ \mathcal{E}_q(y;-i\omega_n)}{2k(\omega_n)}\, r^{|n|}$$
$$= \sum_{n=-\infty}^{\infty} \frac{\mathcal{E}_q(x,-y;i\omega_n)}{2k(\omega_n)}\, r^{|n|}, \qquad 0 \le r < 1$$

and establish the following analog of the Poisson integral

$$\sum_{n=-\infty}^{\infty} r^{|n|}\ \mathcal{E}_q(x;i\omega_n)\ \frac{1}{2k(\omega_n)} \int_{-1}^{1} f(y)\ \mathcal{E}_q(y;-i\omega_n)\ \rho(y)\, dy$$
$$= \int_{-1}^{1} K_r(x,y)\ f(y)\ \rho(y)\, dy, \qquad 0 \le r < 1,$$

where $\rho(x)$ is the weight function in the orthogonality relation (5.6.5). Invesigate convergense when $r \to 1^-$.
Show that

$$\int_{-1}^{1} K_{r_1}(x,y)\ K_{r_2}(y,z)\ \rho(y)\, dy = K_{r_1 r_2}(x,z),$$

where $\max(|r_1|,|r_2|) < 1$.

(12) *The Poisson integral for expansions in continuous q-Hermite polynomials.* Let $f \in \mathcal{L}^1_\rho(-1,1)$, where $\rho(x)$ is the weight function in the orthogonality relation (3.5.5). Use the Poisson kernel (3.5.4) to show that

$$\sum_{n=0}^{\infty} \frac{(q;q)_\infty}{(q;q)_n} r^n \; H_n(\cos\theta|q)$$
$$\times \frac{1}{2\pi}\int_0^\pi f(\cos\varphi)\; H_n(\cos\varphi|q)\; \left(e^{2i\varphi}, e^{-2i\varphi}; q\right)_\infty \, d\varphi$$
$$= \frac{1}{2\pi}\int_0^\pi \frac{\left(q, r^2, e^{2i\varphi},\; e^{-2i\varphi};\; q\right)_\infty \; f(\cos\varphi)}{\left(re^{i\theta+i\varphi}, re^{i\theta-i\varphi}, re^{i\varphi-i\theta}, re^{-i\theta-i\varphi}; q\right)_\infty}\, d\varphi,$$

when $|r| < 1$.

(13) *Abel's method for expansions in continuous q-Hermite polynomials.* Let $f \in \mathcal{L}^1_\rho(-1,1)$, where $\rho(x)$ is the weight function in the orthogonality relation (3.5.5). Consider the following expansions

$$\sigma_q(x) = \sum_{n=0}^{\infty} c_n \; H_n(x|q),$$

and

$$\sigma_q(x;r) = \sum_{n=0}^{\infty} r^n c_n \; H_n(x|q), \quad 0 \le r < 1,$$

where

$$c_n = \frac{(q;q)_\infty}{(q;q)_n}\frac{1}{2\pi}\int_0^\pi f(\cos\theta)\; H_n(\cos\theta|q)\; \left(e^{2i\theta}, e^{-2i\theta}; q\right)_\infty \, d\theta.$$

Show that the series $\sigma_q(x;r)$ converges absolutely and uniformly when $f \in \mathcal{L}^1_\rho(-1,1)$ and $|r| < 1$.
Let $f(x)$ be a bounded function that is continuous on $(-1,1)$. Use the Poisson integral from Ex. 6.8 and Lemma 6.6 to show that $\lim_{r\to 1^-} \sigma_q(x;r) = f(x)$.

(14) *Poisson's kernel and the Askey–Wilson integral.* The Poisson kernel for the continuous q-Hermite polynomials is

$$K_r(x,y) = \sum_{n=0}^{\infty} \frac{r^n}{(q;q)_n} H_n(x|q)\; H_n(y|q)$$
$$= \frac{\left(r^2;\; q\right)_\infty}{\left(re^{i\theta+i\varphi}, re^{i\theta-i\varphi}, re^{i\varphi-i\theta}, re^{-i\theta-i\varphi}; q\right)_\infty},$$

where $x = \cos\theta$, $y = \cos\varphi$ and $|r| < 1$. Show that

$$\int_{-1}^{1} K_{r_1}(x,y)\; K_{r_2}(y,z)\; \rho(y)\; dy = \frac{2\pi}{(q;q)_\infty} K_{r_1 r_2}(x,z),$$

where $\max(|r_1|, |r_2|) < 1$ and $\rho(x)$ is the weight function in the orthogonality relation (3.5.5) for the continuous q-Hermite polynomials. Evaluate the Askey–Wilson integral (A.4.1) on the basis of this integral relation.

(15) *Classical Fourier transform.* Prove that

$$f(x) = F[g](x) = \frac{1}{\sqrt{2\pi}} \int_{-\infty}^{\infty} e^{ixy} g(y)\ dy,$$

$$g(y) = F[f]^{-1}(y) = \frac{1}{\sqrt{2\pi}} \int_{-\infty}^{\infty} e^{-ixy} f(x)\ dx.$$

[*Remark*: This kernel is the special case $t = i$ of the Poisson kernel for the Hermite polynomials $H_n(x)$:

$$K_t(x,y) = \sum_{n=0}^{\infty} t^n \Psi_n(x)\, \Psi_n(y)$$
$$= \frac{1}{\sqrt{\pi(1-t^2)}} \exp\left[\frac{4txy - (1+t^2)(x^2+y^2)}{2(1-t^2)}\right],$$

where

$$\Psi_n(x) = \frac{1}{\sqrt{2^n n! \sqrt{\pi}}}\ H_n(x)\ e^{-x^2/2}$$

is the normalized wave function for the harmonic oscillator in Quantum Mechanics [**97**]. See, for example, [**154**], [**165**] for an extensive theory of the Fourier transform.]

(16) *An analog of Fourier transform.* Let $\mathcal{A}$ be a space of analytic functions for which the expansions in the continuous q-Hermite polynomials,

$$f(x) = \sum_{n=0}^{\infty} c_n\ H_n(x|q),$$

converge absolutely and uniformly in the closure of the interior of the ellipse E_ε given by (6.9.1) and (6.9.2) with $\varepsilon > 0$; see also Ex. 6.21. If $f, g \in \mathcal{A}$, prove that

$$f(x) = F_q[g](x) = \lim_{r \to 1^-} \int_{-1}^{1} T_{ir}(x,y)\ g(y)\ dy,$$

$$g(y) = F_q^{-1}[f](y) = \lim_{r \to 1^-} \int_{-1}^{1} T_{-ir}(x,y)\ f(x)\ dx,$$

where $\rho(x)$ is the weight function in the orthogonality relation (3.5.5) for the continuous q-Hermite polynomials and

$$T_t(x,y) = \frac{(q;q)_\infty}{2\pi} \sum_{n=0}^{\infty} \frac{t^n}{(q;q)_n}\ H_n(x|q)\ H_n(y|q)$$

$$= \frac{(q, r^2;\ q)_\infty}{2\pi \left(re^{i\theta+i\varphi}, re^{i\theta-i\varphi}, re^{i\varphi-i\theta}, re^{-i\theta-i\varphi}; q\right)_\infty}, \qquad |t| < 1$$

with $x = \cos\theta$ and $y = \cos\varphi$. **[13]**, **[14]**, **[125]**.
Show that if $g \in \mathcal{A}$, $0 \le r < 1$ and $0 < \tau < \pi$, then

$$\begin{aligned} f(x) &= \lim_{r \to 1^-} \int_{-1}^{1} T_{re^{i\tau}}(x, y)\ g(y)\ dy \\ &= Pv \int_{-1}^{1} T_{re^{i\tau}}(x, y)\ g(y)\ dy \\ &\quad + \frac{1}{2} \frac{\left(e^{-2i\tau-2i\theta}; q\right)_\infty}{\left(e^{-2i\theta}; q\right)_\infty}\ g(\cos(\tau+\theta)) \\ &\quad + \frac{1}{2} \frac{\left(e^{-2i\tau+2i\theta}; q\right)_\infty}{\left(e^{2i\theta}; q\right)_\infty}\ g(\cos(\tau-\theta)), \end{aligned}$$

where Pv is the Cauchy principal value integral.
Show that

$$F_q\left[\mathcal{E}_q(y; \alpha)\right] = \frac{\left(-q\alpha^2; q^2\right)_\infty}{\left(q\alpha^2; q^2\right)_\infty}\ \mathcal{E}_q(x; i\alpha).$$

[*Hint*: Use the connection relation (3.5.3).]
Show that

$$\begin{aligned} F_q\left[\mathcal{E}_q(y; \alpha)\, \mathcal{E}_q(y; \beta)\right] &= \sum_{m=0}^{\infty} q^{m^2/4} (i\alpha)^m H_m(x|q) \\ &\times \frac{\left(-\alpha\beta q^{(m+1)/2}; q\right)_\infty \left(-q^{(1-m)/2}\beta/\alpha; q\right)_m}{\left(q\alpha^2, q\beta^2; q^2\right)_\infty (q; q)_m}. \end{aligned}$$

[*Hint*: Use (4.9.3).]

(17) Prove that

$$\begin{aligned} J_\nu^{(2)}(r; q) &= \frac{\left(q^{1/2}; q\right)_\infty}{2\,(q; q)_\infty} \left(\frac{r}{2}\right)^\nu \Bigg[\left(i\ \frac{r}{2}\ q^{(\nu+1/2)/2}; q^{1/2}\right)_\infty \\ &\times \sum_{n=0}^{\infty} q^{n/2} \frac{\left(q^{\nu+1/2}; q\right)_n}{(q; q)_n} \left(i\ \frac{r}{2}\ q^{(\nu+1/2)/2}; q^{1/2}\right)_n^{-1} \\ &+ \left(-i\ \frac{r}{2}\ q^{(\nu+1/2)/2}; q^{1/2}\right)_\infty \\ &\times \sum_{n=0}^{\infty} q^{n/2} \frac{\left(q^{\nu+1/2}; q\right)_n}{(q; q)_n} \left(-i\ \frac{r}{2}\ q^{(\nu+1/2)/2}; q^{1/2}\right)_n^{-1} \Bigg]. \end{aligned}$$

[*Remark*: Chen, Ismail, and Muttalib **[54]** call this representation a "complete asymptotic expansion" of $J_\nu^{(2)}(r; q)$ for the large argument. This follows also from Exs. 3.15 and 3.8 of **[46]**.]

Show that the basic sine $S_q(\eta;\omega)$ and cosine $C_q(\eta;\omega)$ functions given by (2.4.19) and (2.4.20) are multiples of $J^{(2)}_{1/2}(2\omega;q)$ and $J^{(2)}_{-1/2}(2\omega;q)$; see Eqs. (2.4.25)–(2.4.26).

(18) Derive the q-Fourier series (6.8.7) as the special case $m = 2$ of (6.7.8).

(19) Suppose that $x = u + iv \in E$:

$$\left(\frac{\text{Re } x}{a}\right)^2 + \left(\frac{\text{Im } x}{b}\right)^2 = 1, \qquad a^2 - b^2 = 1.$$

Show that

$$a^2 = \frac{1}{2}\left(u^2 + v^2 + 1 + \sqrt{\left((u+1)^2 + v^2\right)\left((u-1)^2 + v^2\right)}\right),$$

$$b^2 = \frac{1}{2}\left(u^2 + v^2 - 1 + \sqrt{\left((u+1)^2 + v^2\right)\left((u-1)^2 + v^2\right)}\right),$$

where $u = \text{Re } x$ and $v = \text{Im } x$.

(20) *Analog of Abel's theorem.* Abel's theorem states that if the power series

$$\sum_{n=0}^{\infty} c_n\, z^n$$

converges at some point $z_0 \neq 0$, then it converges absolutely in open disk $|z| < |z_0|$ and it converges absolutely and uniformly in every closed disk $|z| \leq r < |z_0|$. **[136]**
Show that if the series

$$\sum_{n=-\infty}^{\infty} c_n\, \mathcal{E}_q(x; |\omega_n|)$$

converges absolutely at some point $x = a > 1$, then the q-Fourier series (6.9.3) converges absolutely and uniformly in the every compact subset inside the ellipse E defined in Ex. 6.19. [*Hint:* Use the Weierstrass M-test of uniform convergence.]

(21) *Analog of the Cauchy–Hadamard formula for expansions in the continuous q-Hermite polynomials.* Prove that the series

$$\sum_{n=0}^{\infty} c_n\, H_n(x|q)$$

converges absolutely for every point x in the interior of the ellipse E_ε given by (6.9.1)–(6.9.2) with

$$q^{\varepsilon} = \limsup_{n\to\infty} \left(|c_n|\right)^{1/n}$$

when $\varepsilon > 0$.

Show that if $0 < \varepsilon' < \varepsilon$ the convergence is uniform in the closure of the interior of the ellipse $E_{\varepsilon'}$ and that for every x in the interior of the ellipse E_ε the sum of the series is an analytic function. [*Hint*: Use estimate (6.9.12) and follow the proof of Theorem 6.12.]

(22) Use the analog of Cauchy–Hadamard formula from the previous exercise in order to show that the series in (4.1.1) and (6.1.3) are entire functions.
Verify that the ellipse of convergence E_ε of the series (6.1.2) is given by (6.9.1)–(6.9.2) with $q^\varepsilon = r$.

CHAPTER 7

Completeness of Basic Trigonometric Systems

In this chapter we continue the study of completeness of the basic trigonometric system initiated in Section 6.2. Here we first present an independent proof of this important property in $\mathcal{L}^2$ and then extend it to certain $\mathcal{L}^p$-spaces.

7.1. Completeness in $\mathcal{L}^2$ and q-Lommel Polynomials

In this section we shall give an independent proof of the completeness property for q-trigonometric systems. Bustoz and Suslov **[26]** proved that the basic trigonometric system $\{\mathcal{E}_q(x; i\omega_n)\}_{n=-\infty}^{\infty}$ is complete in the weighted $\mathcal{L}^2_{\rho_1}(-1,1)$ space where $\rho_1(x)$ is the weight function for the continuous q-ultraspherical polynomials $C_m(x;\beta|q)$ with $\beta = q^{1/2}$ when $\omega_n = 0, \pm\omega_1, \pm\omega_2, \pm\omega_3, \ldots$ and $\omega_0 = 0 < \omega_1 < \omega_2 < \omega_3 < \ldots$ are nonnegative zeros of the basic sine function $S_q(\eta;\omega)$, which can be viewed as an analog of $\sin \pi\omega$; see Section 6.2. The basic sine $S_q(\eta;\omega)$ and cosine $C_q(\eta;\omega)$ functions are multiples of special Jackson's q-Bessel functions $J^{(2)}_{1/2}(2\omega;q)$ and $J^{(2)}_{-1/2}(2\omega;q)$:

$$S_q(\eta;\omega) = \frac{(q;q)_\infty}{(q^{1/2};q)_\infty} \frac{\omega^{1/2}}{(-q\omega^2;q^2)_\infty} J^{(2)}_{1/2}(2\omega;q), \tag{7.1.1}$$

$$C_q(\eta;\omega) = \frac{(q;q)_\infty}{(q^{1/2};q)_\infty} \frac{\omega^{1/2}}{(-q\omega^2;q^2)_\infty} J^{(2)}_{-1/2}(2\omega;q). \tag{7.1.2}$$

It raises a natural question about completeness of the basic trigonometric systems $\{\mathcal{E}_q(x;i\omega_n)\}_{n=-\infty}^{\infty}$, where ω_n are related to the zeros of $J^{(2)}_\nu(2\omega;q)$ of a general form. The following theorem holds.

THEOREM 7.1. *Let $j_{\mu,k}(q)$ be the zeros of Jackson's q-Bessel function $J^{(2)}_\mu(x;q)$. The q-trigonometric system $\{\mathcal{E}_q(x;i\omega_k)\}_{k=-\infty}^{\infty}$ is complete in $\mathcal{L}^p_\rho(-1,1)$, where $1 \le p < \infty$ and $\rho(x)$ is an integrable function, positive almost everywhere on $(-1,1)$, for $0 < \nu \le 1/2$, if:*

(a) $\omega_k = \frac{1}{2} j_{\nu-1,k}(q)$, $k = \pm 1, \pm 2, \ldots$;

(b) $\omega_k = \frac{1}{2} j_{\nu,k}(q)$, $k = 0, \pm 1, \pm 2, \ldots$.

In the special case when $p = 2$, $0 < \nu < 1/2$ and $\rho(x)$ is the weight function for the continuous q-ultraspherical polynomials, this result has been established in [**144**] on the basis of a completely different method using the Ismail and Zhang formula and relations with the q-Lommel polynomials. Originally, this approach did not cover the most important practical cases $\nu = 1/2$ when the corresponding systems are orthogonal. In the next paper [**148**] this approach has been extended to the larger interval $0 < \nu \leq 1/2$ in $\mathcal{L}^2_\rho(-1,1)$. This gives, finally, an independent proof of the completeness of the basic trigonometric systems in the orthogonal cases. An extension of this result to the larger interval $0 < \nu < 1$ and $1 \leq p < \infty$ is given in [**145**] by using methods of the theory of analytic functions; see also Section 7.5 below.

In order to prove Theorem 7.1 we shall use the Ismail and Zhang [**76**] expansion formula

$$\mathcal{E}_q(x; i\omega) = \frac{(q;q)_\infty \, \omega^{-\nu}}{(q^\nu;q)_\infty \left(-q\omega^2;q^2\right)_\infty} \times \sum_{m=0}^{\infty} i^m \left(1-q^{\nu+m}\right) q^{m^2/4} \, J^{(2)}_{\nu+m}(2\omega;q) \; C_m\left(x;q^\nu \,|\, q\right), \tag{7.1.3}$$

where Jackson's q-Bessel function $J^{(2)}_{\nu+m}(2\omega;q)$ are given by (4.5.3) and $C_m(x;\beta|q)$ are continuous q-ultraspherical polynomials; see Section 4.5 for the proof of (7.1.3) and more details. Relations with the q-Lommel polynomials, that are crucial for further consideration, were noticed for the first time in [**144**] and [**146**]; see also [**60**].

In this section we shall give the proof of Theorem 7.1 only when $p = 2$, $0 < \nu \leq 1/2$ and $\rho(x)$ is the weight function for the continuous q-ultraspherical polynomials. An extension will be discussed in the next sections. Consider the first case: $\omega_k = \frac{1}{2} j_{\nu-1,k}(q)$, $k = \pm1, \pm2, \ldots$. The q-Lommel polynomials introduced in [**56**] are generated by the recurrence relation

$$q^{n\nu+n(n-1)/2} J^{(2)}_{\nu+n}(x;q) = h^\nu_n\left(\frac{1}{x};q\right) J^{(2)}_\nu(x;q) - h^{\nu+1}_{n-1}\left(\frac{1}{x};q\right) J^{(2)}_{\nu-1}(x;q). \tag{7.1.4}$$

We shall use the notation $h^\nu_n(s;q) := h_{n,\nu}(s;q)$ instead of the original one in [**56**]. Letting $x = 2\omega_k = j_{\nu-1,k}(q)$ in (7.1.4), one gets

$$q^{n\nu+n(n-1)/2} J^{(2)}_{\nu+n}\left(j_{\nu-1,k}(q);q\right) = h^\nu_n\left(\frac{1}{j_{\nu-1,k}(q)};q\right) J^{(2)}_\nu\left(j_{\nu-1,k}(q);q\right) \tag{7.1.5}$$

and (7.1.3) takes a simpler form

$$\mathcal{E}_q(x;i\omega_k)\left(-q\omega_k^2;q^2\right)_\infty \omega_k^\nu = \frac{(q;q)_\infty}{(q^\nu;q)_\infty} J^{(2)}_\nu\left(j_{\nu-1,k}(q);q\right) \tag{7.1.6}$$

$$\times \sum_{n=0}^{\infty} i^n \left(1-q^{\nu+n}\right) q^{n(2-4\nu-n)/4}\, h_n^{\nu}\left(\frac{1}{j_{\nu-1,k}(q)};q\right) C_n\left(x;q^{\nu}|\,q\right).$$

The orthogonality relation for the q-Lommel polynomials established in [**56**] has the form

$$\left(1+(-1)^{m+n}\right)\sum_{k=1}^{\infty} \frac{A_k(\nu)}{j_{\nu-1,k}^2(q)}\, h_n^{\nu}\left(\frac{1}{j_{\nu-1,k}(q)};q\right) h_m^{\nu}\left(\frac{1}{j_{\nu-1,k}(q)};q\right)$$
$$= \frac{q^{n\nu+n(n-1)/2}}{1-q^{\nu+n}}\,\delta_{mn}, \qquad \nu>0, \tag{7.1.7}$$

where the jumps of the step function $A_k(\nu)$ can be found from the partial fraction decomposition

$$\sum_{k=1}^{\infty} A_k(\nu)\,\frac{z}{j_{\nu-1,k}^2(q)-z^2} = \frac{J_{\nu}^{(2)}(z;q)}{J_{\nu-1}^{(2)}(z;q)} \tag{7.1.8}$$

as

$$A_k(\nu) = -2\frac{J_{\nu}^{(2)}\left(j_{\nu-1,k}(q);q\right)}{\frac{d}{dz}J_{\nu-1}^{(2)}\left(j_{\nu-1,k}(q);q\right)} = 2q^{1-\nu}\frac{J_{\nu-2}^{(2)}\left(j_{\nu-1,k}(q);q\right)}{\frac{d}{dz}J_{\nu-1}^{(2)}\left(j_{\nu-1,k}(q);q\right)}, \tag{7.1.9}$$

the second equality follows from (6.8.3); see also Ex. 7.5 for the dual orthogonality relation of the q-Lommel polynomials. We shall also write the orthogonality relation (7.1.7) in terms of the Lebesgue integral with respect to the corresponding purely discrete measure $\mu=\mu(s)$, $s=s_k=1/j_{\nu-1,k}(q)$ with the only cluster point of its bounded support $Supp\,\mu$ at the origin $s=0$.

The q-trigonometric system $\{\mathcal{E}_q(x;i\omega_k)\}_{k=-\infty}^{\infty}$ is complete on $(-1,1)$ if it is closed [**3**]. Suppose that this system is not closed. Thus, there exists at least one function $\chi\in\mathcal{L}_{\rho}^2(-1,1)$, not identically zero, that

$$\int_{-1}^{1}\chi(x)\;\mathcal{E}_q(x;i\omega_k)\;\rho(x)\;dx=0, \qquad k=\pm1,\pm2,\dots\,, \tag{7.1.10}$$

where $\rho(x)$ is the weight function in the orthogonality relation for the continuous q-ultraspherical polynomials (4.3.5). It can be shown that the series in (7.1.3) and (7.1.6) are uniformly convergent in x on compact sets. Substituting (7.1.5) in (7.1.10) one gets

$$\sum_{n=0}^{\infty} i^n\left(1-q^{\nu+n}\right) q^{n(2-4\nu-n)/4}\, h_n^{\nu}\left(\frac{1}{j_{\nu-1,k}(q)};q\right) \tag{7.1.11}$$
$$\times\int_{-1}^{1}\chi(x)\;C_n\left(x;q^{\nu}|\,q\right)\;\rho(x)\;dx=0$$

and our goal is to show that

$$\int_{-1}^{1}\chi(x)\;C_n\left(x;q^{\nu}|\,q\right)\;\rho(x)\;dx=0 \tag{7.1.12}$$

for all $n = 0, 1, 2, \dots$, which means that $\chi(x) = 0$ almost everywhere due to the completeness of the system of the continuous q-ultraspherical polynomials $C_n(x; q^\nu | q)$ in $\mathcal{L}^2_\rho(-1, 1)$. In [**144**] this had been done with the help of the orthogonality relation of the q-Lommel polynomials (7.1.7); see also Remark 7.1 below. This method requires the change of the order of summation in the corresponding double series which can be justified for $0 < \nu < 1/2$ only. So it does not cover the orthogonal case when $\nu = 1/2$; see [**144**] for the details. This gap has been fulfilled in [**148**] and we shall follow this approach here in order to give an independent proof of completeness of the basic trigonometric system.

Let us rewrite (7.1.11) as

$$\sum_{n=0}^{\infty} c_n \, \varphi_n(s) = 0, \tag{7.1.13}$$

where

$$\varphi_n(s) = \left(1 - q^{\nu+n}\right)^{1/2} q^{-n\nu/2 - n(n-1)/4} \, h_n^\nu(s; q), \tag{7.1.14}$$

$$s = \frac{1}{j_{\nu-1,k}(q)} \in Supp \ \mu$$

and

$$c_n = i^n \left(1 - q^{\nu+n}\right)^{1/2} q^{(1-2\nu)n/4} \times \int_{-1}^{1} \chi(x) \ C_n(x; q^\nu | q) \ \rho(x) \ dx. \tag{7.1.15}$$

Here $\{\varphi_n(s)\}_{n=0}^\infty$ is an orthonormal system with respect to the purely discrete measure μ in the orthogonality relation (7.1.7) for the q-Lommel polynomials.

Parseval's identity for the continuous q-ultraspherical polynomials,

$$\sum_{n=0}^{\infty} \frac{(q;q)_n \ (1 - q^{\nu+n})}{(q^{2\nu};q)_n \ (1 - q^\nu)} \left| \int_{-1}^{1} \chi(x) \ C_n(x; q^\nu | q) \ \rho(x) \ dx \right|^2 = 2\pi \frac{(q^\nu, q^{\nu+1}; q)_\infty}{(q, q^{2\nu}; q)_\infty} \int_{-1}^{1} |\chi(x)|^2 \ \rho(x) \ dx, \tag{7.1.16}$$

and the Limit Comparison Tests show that

$$\sum_{n=0}^{\infty} |c_n|^2 < \infty \tag{7.1.17}$$

when $0 < \nu \leq 1/2$ and our goal is to prove that $c_n = 0$ for all $n = 0, 1, 2, \dots$. Also, by the Cauchy–Schwarz inequality

$$\sum_{n=0}^{\infty} |c_n| \leq \left(\sum_{n=0}^{\infty} \left(1 - q^{\nu+n}\right) q^{(1-2\nu)n/2} \right)^{1/2} \tag{7.1.18}$$

$$\times \left(\sum_{n=0}^{\infty} \left| \int_{-1}^{1} \chi(x) \; C_n\left(x; q^{\nu} | q\right) \; \rho(x) \; dx \right|^2 \right)^{1/2}$$

and the both series in the right hand side converge when $0 < \nu < 1/2$. As a result

$$\sum_{n=0}^{\infty} |c_n| < \infty \tag{7.1.19}$$

when $0 < \nu < 1/2$ and we arrive at the following lemma.

LEMMA 7.1. *Sequence* $\{c_n\}_{n=0}^{\infty} \in \ell^1$ *for* $0 < \nu < 1/2$ *and* $\{c_n\}_{n=0}^{\infty} \in \ell^2$ *for* $0 < \nu \leq 1/2$.

Consider the partial sums

$$f_m(s) = \sum_{n=0}^{m} c_n \; \varphi_n(s), \tag{7.1.20}$$

$$g_m(s) = \sum_{n=0}^{m} |c_n| \; |\varphi_n(s)|. \tag{7.1.21}$$

In view of (7.1.13),

$$\lim_{m \to \infty} f_m(s) = 0 \tag{7.1.22}$$

pointwise for all $s \in E = Supp \; \mu$. Moreover, a simple estimate using the Cauchy–Schwarz inequality,

$$\begin{aligned} g_m(s) &= \sum_{n=0}^{m} |c_n| \; |\varphi_n(s)| \\ &\leq \left(\sum_{n=0}^{m} |c_n|^2 \right)^{1/2} \left(\sum_{n=0}^{m} |\varphi_n(s)|^2 \right)^{1/2} \\ &\leq \left(\sum_{n=0}^{\infty} |c_n|^2 \right)^{1/2} \left(\sum_{n=0}^{\infty} |\varphi_n(s)|^2 \right)^{1/2}, \end{aligned} \tag{7.1.23}$$

shows that the sequence $\{g_m(s)\}_{m=0}^{\infty}$ is bounded for all $s \in E = Supp \; \mu$. The series

$$\sum_{n=0}^{\infty} |\varphi_n(s)|^2 = \frac{1}{\mu(s)} \tag{7.1.24}$$

here can be summed by the dual orthogonality relation for the q-Lommel polynomials [**60**] for all $s \in E = Supp \; \mu$; see also Ex. 7.5. The monotonicity property

$$0 \leq g_m(s) \leq g_{m+1}(s), \qquad m = 0, 1, 2, ... \tag{7.1.25}$$

and (7.1.23) show that this sequence converges

$$\lim_{m \to \infty} g_m(s) = g(s) < \infty \tag{7.1.26}$$

pointwise for all $s \in E = Supp \; \mu$.

Our next step is to prove that this limiting function $g(s)$ is integrable in the sense $g(s) \in \mathcal{L}^1(\mu)$ for $0 < \nu \leq 1/2$. Indeed, when $0 < \nu < 1/2$ one can write

$$\int_E g_m(s)\ d\mu = \sum_{n=0}^{m} |c_n| \int_E |\varphi_n(s)|\ d\mu \tag{7.1.27}$$
$$\leq \sum_{n=0}^{m} |c_n| \left(\int_E |\varphi_n(s)|^2\ d\mu \right)^{1/2} \left(\int_E d\mu \right)^{1/2} \leq \sum_{n=0}^{\infty} |c_n| < \infty,$$

see (7.1.19), and by the Lebesgue Monotone Convergence Theorem [**108**], [**128**]

$$\int_E g(s)\ d\mu = \int_E \lim_{m\to\infty} g_m(s)\ d\mu = \lim_{m\to\infty} \int_E g_m(s)\ d\mu < \infty. \tag{7.1.28}$$

We could also use Beppo Levi's theorem [**93**], [**127**] here.

On the other hand, inequality (7.1.23) together with (7.1.24) give

$$\int_E g_m(s)\ d\mu \leq \left(\sum_{n=0}^{\infty} |c_n|^2 \right)^{1/2} \int_E \frac{d\mu}{\mu^{1/2}(s)}. \tag{7.1.29}$$

In general it does not look easy to give an estimate for the last integral, but luckily in the orthogonal case $\nu = 1/2$ the measure of the corresponding q-Lommel polynomials $h_n^{1/2}(s;q)$ can be found explicitly [**146**], see also Ex. 7.6 and Section 9.1.4 below, as[1]

$$\int_E d\mu = \left(1 - q^{1/2}\right) \sum_{k=-\infty}^{\infty} \frac{1}{\kappa(\omega_k)\ \omega_k^2} \tag{7.1.30}$$
$$= 2\left(1 - q^{1/2}\right) \sum_{k=1}^{\infty} \frac{1}{\kappa(\omega_k)\ \omega_k^2} = 1,$$

where

$$\kappa(\omega) = \sum_{n=0}^{\infty} \frac{q^{n/2}}{1 + \omega^2 q^n} \tag{7.1.31}$$

and therefore

$$\int_E \frac{d\mu}{\mu^{1/2}(s)} = 2\left(1 - q^{1/2}\right)^{1/2} \sum_{k=1}^{\infty} \frac{1}{\kappa^{1/2}(\omega_k)\ |\omega_k|} < \infty \tag{7.1.32}$$

in view of the asymptotic relations $|\omega_k|\,\kappa(\omega_k) \sim$constant and $|\omega_k| \sim q^{3/4-|k|}$ as $|k| \to \infty$; see (6.7.23) and (6.3.15), respectively. Thus, (7.1.28) holds also when $\nu = 1/2$ and we have proven the following result.

LEMMA 7.2. *Function $g(s) \in \mathcal{L}^1(\mu)$ for $0 < \nu \leq 1/2$.*

---

[1] We used the notation ϖ_n for the zeros of the basic sine function $C_q(\eta;\omega)$ before.

By the Riesz–Fisher theorem **[93]**, **[127]**, **[128]** the series (7.1.20) converges in mean square to some element $f \in \mathcal{L}^2(\mu)$,

$$\lim_{m\to\infty} \int_E |f_m(s) - f(s)|^2 \, d\mu = 0 \tag{7.1.33}$$

and

$$c_n = \int_E f(s) \; \varphi_n(s) \; d\mu, \tag{7.1.34}$$

$$\sum_{n=0}^{\infty} |c_n|^2 = \int_E |f(s)|^2 \; d\mu \tag{7.1.35}$$

where $E = \mathit{Supp}\ \mu$. In the case of the finite measure under consideration, convergence in the mean square in (7.1.33) implies convergence in the mean,

$$\lim_{m\to\infty} \int_E |f_m(s) - f(s)| \; d\mu = 0, \tag{7.1.36}$$

and our final step is to prove that $f(s) \equiv 0$ for all $s \in E = \mathit{Supp}\ \mu$. In principle, this can be done using the fact that convergence in mean implies convergence in measure, which, in turn, implies the existence of a convergent subsequence **[93]**, **[108]**

$$f_{m_k}(s) \to f(s), \tag{7.1.37}$$

which converges to zero due to (7.1.13), but these arguments involve convergence almost everywhere, and in our degenerate case of a purely discrete measure this condition should be replaced by convergence everywhere on the support of the measure.

On second thought, using the facts established above, we can give a direct proof interchanging the limit and integral in (7.1.36). The following inequalities hold

$$\begin{aligned} |f_m(s) - f(s)| &\leq |f_m(s)| + |f(s)| \\ &\leq g_m(s) + |f(s)| \leq g(s) + |f(s)| \end{aligned} \tag{7.1.38}$$

for all $m = 0, 1, 2, ...$ due to the monotonicity of $g_m(s)$. Since $g(s) \in \mathcal{L}^1(\mu)$, $f(s) \in \mathcal{L}^2(\mu)$ (and, therefore, $f(s) \in \mathcal{L}^1(\mu)$ in the case of the finite measure under consideration), the hypotheses of the Lebesgue Dominated Convergence Theorem **[93]**, **[108]**, **[128]** are satisfied and

$$\begin{aligned} 0 &= \lim_{m\to\infty} \int_E |f_m(s) - f(s)| \; d\mu = \int_E \lim_{m\to\infty} |f_m(s) - f(s)| \; d\mu \\ &= \int_E \left| \lim_{m\to\infty} f_m(s) - f(s) \right| \; d\mu = \int_E |f(s)| \; d\mu. \end{aligned} \tag{7.1.39}$$

But the measure $\mu(s)$ is purely discrete. Therefore $f(s) \equiv 0$ for all $s \in E = \mathit{Supp}\ \mu$ and $c_n = 0$ for all $n = 0, 1, 2, ...$ in view of (7.1.34) or (7.1.35). This completes the proof in the first case.

The second case, when $\omega_k = \frac{1}{2} j_{\nu,k}(q)$, $k = 0, \pm 1, \pm 2, \dots$, can be considered in a similar manner. The analog of (7.1.13) is

$$\sum_{n=0}^{\infty} c_n \, \phi_{n-1}(s) = 0, \tag{7.1.40}$$

where

$$\phi_{n-1}(s) = \varphi_n(s)|_{n \to n-1,\ \nu \to \nu+1} \tag{7.1.41}$$

and c_n are given by (7.1.15). For $0 < \nu \le 1/2$ the same arguments show that $c_n = 0$, for all $n = 1, 2, 3, \dots$. Moreover, $c_0 = 0$ because condition (7.1.10) holds for $\omega_0 = 0$ in this case. The measure of the q-Lommel polynomials $h_n^{3/2}(s;q)$ for the corresponding orthogonal case $\nu = 1/2$ has been found explicitly in **[146]**; see also Ex. 7.6 and Section 9.1.2. This completes the proof.

REMARK 7.1. In view of (7.1.13) and (7.1.20) one can also write formally

$$0 = \int_E \left(\lim_{m \to \infty} f_m(s) \right) \varphi_n(s) \; d\mu = \lim_{m \to \infty} \int_E f_m(s) \; \varphi_n(s) \; d\mu = c_n.$$

Interchanging of the limit and the integral here can be justified by the Lebesgue Dominated Convergence Theorem when $0 < \nu < 1/2$ due to

$$\int_E |f_m(s) \; \varphi_n(s)| \; d\mu \le \int_E g_m(s) \; |\varphi_n(s)| \; d\mu$$

$$= \sum_{k=0}^{m} |c_k| \int_E |\varphi_k(s) \; \varphi_n(s)| \; d\mu \le \sum_{k=0}^{m} |c_k| \le \sum_{k=0}^{\infty} |c_k| < \infty,$$

see (7.1.19). This consideration is equivalent to justification of interchanging of the order of summation in **[144]**.

REMARK 7.2. The q-Lommel polynomials $\{h_n^{\nu}(s;q)\}_{n=0}^{\infty}$ form a complete set in $\mathcal{L}^2(\mu)$ when $\nu > 0$. Our method shows that the corresponding dual system $\{h_n^{\nu}(s_k;q)\}_{k=-\infty}^{\infty}$, where $s_k = 1/j_{\nu-1,k}(q) \in Supp\ \mu$, is closed in ℓ^1 for $\nu > 0$. Moreover, in view of (7.1.29) this system is closed in the dual ℓ^2 if

$$\int_E \frac{d\mu}{\mu^{1/2}(s)} < \infty.$$

We were able to verify the last condition for $\nu = 1/2, 3/2$ when the corresponding measures are known explicitly. This result can be generalized to other systems of discrete orthogonal polynomials.

7.2. Completeness in $\mathcal{L}^p$: General Results

In this section we shall study completeness of the basic trigonometric systems in certain $\mathcal{L}^p$-spaces. The set $\mathcal{L}_\rho^p(a,b)$, $1 \le p < \infty$ consists of all complex-valued functions satisfying

$$\int_a^b |f|^p \, \rho \; dx < \infty, \tag{7.2.1}$$

where ρ is an integrable function positive almost everywhere on (a,b). We identify two functions in $\mathcal{L}_\rho^p(a,b)$ whenever they are equal almost everywhere and define the $\mathcal{L}^p$-norm as

$$\|f\|_p = \left(\int_a^b |f|^p \rho\, dx\right)^{1/p}, \qquad 1 \le p < \infty. \tag{7.2.2}$$

Important special cases are $p = 1$ and $p = 2$. The set $\mathcal{L}^\infty(a,b)$ consists of all complex-valued functions with $|f|$ bounded almost everywhere on (a,b). The $\mathcal{L}^p$-spaces when $p \ge 1$ are sometimes referred to in the literature as *classical Banach spaces.* For an introduction to their rich theory, see, for example, **[3]**, **[93]**, **[108]**, **[127]**, and **[128]**.

7.2.1. Preliminaries. In this section we shall establish several results on completeness of the basic trigonometric system $\{\mathcal{E}_q(x; i\omega_n)\}_{n=-\infty}^{\infty}$ in the weighted $\mathcal{L}_\rho^p$-spaces, $1 \le p < \infty$, using methods of the theory of entire functions **[23]**, **[26]**, **[99]**, **[100]** and **[101]**.

The completeness of the trigonometric system $\{e^{inx}\}_{n=-\infty}^{\infty}$ on the interval $(-\pi,\pi)$ admits interesting generalizations. Levinson **[101]** and Levin **[99]** had established the completeness of systems of functions $\{e^{i\lambda_n x}\}$ on the interval $(-\pi,\pi)$ where $\{\lambda_n\}$ is a suitable sequence of complex numbers. We shall mention only two classical results here; see **[101]**, **[99]** and **[100]** for more details.

THEOREM 7.2. *If $|\lambda_n| \le |n| + 1/2p$, $n = 0, \pm1, \pm2, \ldots$, then the system $\{e^{i\lambda_n x}\}$ is complete in $\mathcal{L}^p(-\pi,\pi)$, $1 < p < \infty$.*

THEOREM 7.3. *If $\{\lambda_n\}$ is the set of zeros of an entire function $h(\lambda)$ of exponential type and*

$$\lim_{|\gamma|\to\infty} |h(i\gamma)|\, e^{-\pi|\gamma|} > 0, \tag{7.2.3}$$

then the system $\{e^{i\lambda_n x}\}$ is complete in $\mathcal{L}^p(-\pi,\pi)$, $1 \le p < \infty$.

Let us consider

$$e(x,\alpha) := \left(q\alpha^2; q^2\right)_\infty \mathcal{E}_q(x;\alpha) = \sum_{n=0}^{\infty} \frac{q^{n^2/4}}{(q;q)_n}\, \alpha^n H_n(x|q), \tag{7.2.4}$$

where $H_n(x|q)$ are the continuous q-Hermite polynomials given by (6.1.1). The $e(x,\alpha)$ is an entire function of order zero in α for all finite values of x by Lemma 6.1. Here we shall establish theorems similar to Theorems 7.2 and 7.3 for the basic trigonometric systems $\{e(x, i\omega_n)\}$ following **[145]**.

7.2.2. Completeness Theorems. A set $\mathcal{B}$ of vectors of the linear normed space V is said to be *complete* in V if every vector $\chi \in V$ can be approximated to any degree of accuracy by means of an expression of the form

$$\alpha_1 e_1 + \alpha_2 e_2 + \ldots + \alpha_n e_n, \tag{7.2.5}$$

where $e_k \in \mathcal{B}$. We shall use the Criterion for the Completeness of a Set of Vectors in Linear Normed Spaces **[3]**.

CRITERION 7.1. *The necessary and sufficient condition that a set $\mathcal{B}$ in V be a complete set is, that every linear functional φ in V which vanishes for any vector $e_k \in \mathcal{B}$ be identically zero.*

The completeness of the basic trigonometric system, when the corresponding eigenvalues are the zeros of Jackson's q-Bessel function $J_\nu^{(2)}(x;q)$, is stated in Theorem 7.1. The proof of a special case of this theorem in the previous section uses Ismail and Zhang's basic analog of the expansion formula of the plane wave in terms of the spherical harmonics [**76**] and the q-Lommel polynomials introduced in [**56**]. Here we establish more general results on the basis of the theory of entire functions.

The study of completeness of the classical and basic trigonometric systems amounts to the study of zeros of certain entire function. We shall first extend the following result by Levinson [**101**], [**99**].

THEOREM 7.4. *For the system $\{e^{i\lambda_n x}\}$, in which the λ_n, $n = 1, 2, 3, \ldots$ are complex numbers, to be incomplete in $\mathcal{L}^p(-\pi, \pi)$, $1 \le p < \infty$, it is necessary and sufficient that there exist an entire function $f(\lambda)$, zero at all λ_n, $n = 1, 2, 3, \ldots$, that can be represented in the form*

$$f(\lambda) = \int_{-\pi}^{\pi} e^{i\lambda x} \phi(x)\, dx, \tag{7.2.6}$$

where $\phi(x) \in \mathcal{L}^r(-\pi, \pi)$ and $1/p + 1/r = 1$ (for $p = 1$ the $\phi(x)$ is a bounded measurable function).

Let $\rho(x)$ be an integrable function,

$$\int_{-1}^{1} \rho(x)\, dx < \infty, \tag{7.2.7}$$

positive almost everywhere, $\rho(x) > 0$, on $(-1, 1)$. We shall assume that this condition holds throughout this chapter and shall consider the $\mathcal{L}^p_\rho(-1, 1)$ spaces with respect to this weight function if it is not stated otherwise. The case of a more general finite measure can be discussed in a similar fashion. For the basic trigonometric system a somewhat similar theorem holds [**145**].

THEOREM 7.5. *The basic trigonometric system $\{e(x, i\omega_n)\}$, where $\{\omega_n\}$ is an infinite sequence of complex numbers, is incomplete in the weighted space $\mathcal{L}^p_\rho(-1, 1)$, $1 \le p < \infty$ if and only if there exist an entire function $f(\omega)$ of order zero, such that $f(\omega_n) = 0$ for all ω_n, that can be represented in the form*

$$f(\omega) = \int_{-1}^{1} e(x, i\omega)\ \phi(x)\ \rho(x)\, dx, \tag{7.2.8}$$

where $\phi(x) \in \mathcal{L}^r_\rho(-1, 1)$ not equivalent to zero and $1/p + 1/r = 1$ (for $p = 1$ the $\phi(x)$ is a bounded measurable function).

PROOF. We can use arguments similar to the proof of Theorem 7.4 in [**101**] and [**99**]. If the system $\{e(x, i\omega_n)\}$ is incomplete in $\mathcal{L}^p_\rho(-1, 1)$, $1 \le p < \infty$, then by the Criterion for the Completeness of a Set of Vectors in

Linear Normalized Spaces [3] there is a linear functional φ in $\mathcal{L}^p_\rho(-1,1)$, not identically zero, which vanishes for all the function $e(x, i\omega_n) : \varphi(e(x, i\omega_n)) = 0$. By the Riesz Representation Theorem [3], [108], [127] the general form of a linear functional in $\mathcal{L}^p_\rho(-1,1)$ is

$$\varphi(\chi) = \int_{-1}^{1} \chi(x)\ \phi(x)\ \rho(x)\ dx, \tag{7.2.9}$$

where $\phi(x) \in \mathcal{L}^r_\rho(-1,1)$ and $1/p + 1/r = 1$ (for $p = 1$ the $\phi(x)$ is a bounded measurable function). Thus,

$$\varphi(e(x, i\omega_n)) = \int_{-1}^{1} e(x, i\omega_n)\ \phi(x)\ \rho(x)\ dx = 0 \tag{7.2.10}$$

and function $f(\omega)$ given by (7.2.8) is zero at all the points ω_n.

Therefore the study of completeness of the basic trigonometric systems amounts to the study of zeros of function $f(\omega)$. By the Hölder inequality

$$\int_{-1}^{1} |\phi(x)|\ \rho(x)\ dx \le \left(\int_{-1}^{1} |\phi(x)|^r\ \rho(x)\ dx\right)^{1/r} \left(\int_{-1}^{1} \rho(x)\ dx\right)^{1/p},$$

thus, $\phi(x)$ is integrable on $(-1,1)$ for $1 \le p < \infty$,

$$\int_{-1}^{1} |\phi(x)|\ \rho(x)\ dx = A < \infty. \tag{7.2.11}$$

Let us show that the $f(\omega)$ given by (7.2.8) is an entire function. The series in (7.2.4) converges absolutely and uniformly in x on $[-1,1]$ and integrating with respect to the finite measure $d\mu(x) = |\phi(x)|\, \rho(x) dx$ we obtain by the Lebesgue Dominated Convergence Theorem

$$\begin{aligned} f(\omega) &= \int_{-1}^{1} \phi(x) \left(\sum_{k=0}^{\infty} \frac{q^{k^2/4}}{(q;q)_k}\ (i\omega)^k\, H_k(x|q) \right) \rho(x)\, dx \\ &= \sum_{k=0}^{\infty} \frac{q^{k^2/4}}{(q;q)_k}\ (i\omega)^k \int_{-1}^{1} \phi(x)\, H_k(x|q)\, \rho(x)\, dx, \end{aligned} \tag{7.2.12}$$

or

$$f(\omega) = \sum_{k=0}^{\infty} c_k\ \omega^k \tag{7.2.13}$$

with

$$c_k = i^k \frac{q^{k^2/4}}{(q;q)_k} \int_{-1}^{1} \phi(x)\, H_k(x|q)\, \rho(x)\, dx. \tag{7.2.14}$$

Using the uniform upper bound for the continuous q-Hermite polynomials

$$|H_k(x|q)| < q^{-k\varepsilon} \frac{(q;q)_k}{(q, q^{2\varepsilon}, q^{1-2\varepsilon}; q)_\infty}, \tag{7.2.15}$$

when $|x| < x(\varepsilon) = (q^\varepsilon + q^{-\varepsilon})/2$ and $0 < \varepsilon < 1/2$; see (6.1.16); one gets

$$|c_k| \le B q^{k^2/4 - \varepsilon k}, \tag{7.2.16}$$

where B is a positive constant. The radius of convergence of series (7.2.12) is

$$\frac{1}{R} = \lim_{k\to\infty} \left(|c_k|\right)^{1/k} \leq \lim_{k\to\infty} \left(B^{1/k} q^{k/4-\varepsilon}\right) = 0. \tag{7.2.17}$$

Therefore $f(\omega)$ is an entire function. The order of this entire function is [99]

$$\begin{aligned}\lim_{k\to\infty} \left(\frac{k\log k}{-\log|c_k|}\right) &\leq \lim_{k\to\infty} \left(\frac{k\log k}{-\log\left(Bq^{k^2/4-\varepsilon k}\right)}\right) \\ &= \lim_{k\to\infty} \left(\frac{k\log k}{-\left(k^2/4-\varepsilon k\right)\log q - \log B}\right) = 0,\end{aligned} \tag{7.2.18}$$

see also (B.1.10). This completes the proof. □

REMARK 7.3. As in the classical case [99], Theorem 7.6 holds also for $p=\infty$. The corresponding entire function of order zero can be represented as

$$f(\omega) = \int_{-1}^{1} e(x, i\omega)\ d\sigma(x), \tag{7.2.19}$$

where $\sigma(x)$ is a function of bounded variation on the interval $(-1,1)$. This result follows directly from the general form of the linear functional in $C(-1,1)$ [3], [93], [108], [127].

We can now prove the following analog of Theorem 7.3.

THEOREM 7.6. *Let $\{\omega_n\}$ be the set of zeros of an entire function $h(\omega)$ of order zero and suppose that all these zeros ω_n are simple. If on the imaginary axis*

$$\lim_{|\omega|\to\infty} \frac{e(x(\varepsilon), |\omega|)}{|h(\omega)|} = 0, \tag{7.2.20}$$

where $x(\varepsilon) = \left(q^{\varepsilon}+q^{-\varepsilon}\right)/2$ and $0 \leq \varepsilon < 1/2$, then the system $\{e(x, i\omega_n)\}$ is complete in $\mathcal{L}_\rho^p(-1,1)$, $1 \leq p < \infty$.

PROOF. In view of Theorem 7.6 the q-trigonometric system $\{e(x, i\omega_n)\}$ is incomplete in $\mathcal{L}_\rho^p(-1,1)$ if there is an entire function of order zero $f(\omega)$ given by (7.2.8) where $\phi(x) \in \mathcal{L}_\rho^r(-1,1)$ is not zero almost everywhere on $[-1,1]$. In view of (7.2.11),

$$\begin{aligned}|f(\omega)| &\leq \int_{-1}^{1} |\phi(x)\ e(x, i\omega)|\ \rho(x)\ dx \\ &\leq e(x(\varepsilon), |\omega|) \int_{-1}^{1} |\phi(x)|\ \rho(x)\ dx \\ &= A\ e(x(\varepsilon), |\omega|),\end{aligned} \tag{7.2.21}$$

where we have used the uniform bound from Lemma 6.1,

$$|e(\cos\theta, i\omega)| \leq e(x(\varepsilon), |\omega|), \quad \varepsilon > 0. \tag{7.2.22}$$

Our next step is to show that under the hypotheses of the theorem function $f(\omega)$ is identically zero. Consider the quotient

$$g(\omega) = \frac{f(\omega)}{h(\omega)} \tag{7.2.23}$$

of two entire functions. The functions $f(\omega)$ and $h(\omega)$ have the same zeros and all zeros of $h(\omega)$ are simple, so $g(\omega)$ is an entire function. The order of this entire function is zero because both $f(\omega)$ and $h(\omega)$ are of order zero; see [**99**], Corollary of Theorem 12 on p. 24. Moreover, this function $g(\omega)$ is bounded on the imaginary axis. Indeed, by (7.2.21)

$$|g(\omega)| \leq \frac{Ae\left(x\left(\varepsilon\right), |\omega|\right)}{|h(\omega)|} \tag{7.2.24}$$

and the limit when $|\omega| \to \infty$ of the right hand side exists by (7.2.20). But an entire function of order zero bounded on a line must be a constant; see Theorems 21–22 and Corollary on pp. 49–51 of [**99**] or Appendix B, Theorem B.5. Therefore

$$f(\omega) = C\, h(\omega) \tag{7.2.25}$$

and

$$|C| \leq \frac{Ae\left(x\left(\varepsilon\right), |\omega|\right)}{|h(\omega)|} \to 0 \tag{7.2.26}$$

as $|\omega| \to \infty$ by (7.2.20). Thus, $f(\omega)$ is identically zero and by (7.2.12)

$$\int_{-1}^{1} H_k\left(x|q\right) \phi\left(x\right) \rho\left(x\right) dx = 0 \tag{7.2.27}$$

for all $k = 0, 1, 2, \ldots$. But the system $\{H_k\left(x|q\right)\}_{k=0}^{\infty}$ is complete in $\mathcal{L}_\rho^p\left(-1,1\right)$, $1 \leq r < \infty$ (see, for example, [**3**], [**151**] or [**51**]) and, therefore, $\phi\left(x\right) = 0$ almost everywhere on $(-1,1)$ due to the positivity of the $\rho\left(x\right)$. This contradiction proves that the system $\{e\left(x, i\omega_n\right)\}$ is complete in $\mathcal{L}_\rho^p\left(-1,1\right)$ with $1 \leq p < \infty$. □

In order to verify the limiting condition (7.2.20) one can use the uniform upper bounds for the q-exponential function from Lemmas 6.2 and 6.8.

The proof of Theorem 7.6 is based on the upper bound (7.2.22) for the entire function $f\left(\omega\right)$. In the case of the $\mathcal{L}^2$-space the Cauchy–Schwarz inequality leads to another estimate for this function.

LEMMA 7.3. *Let $f(\omega)$ be defined by* (7.2.8) *where $\phi\left(x\right)$ is in $\mathcal{L}_\rho^2\left(-1,1\right)$ with $\rho\left(x\right) = \rho_1\left(x\right) w\left(x\right)$, $w\left(x\right) > 0$ is a continuous function on $[-1,1]$ and $\rho_1\left(x\right)$ is the weight function of the continuous q-ultraspherical polynomials $C_m\left(x; q^{1/2}|q\right)$. Then*

$$|f(\omega)| \leq \sqrt{\pi}\frac{\left(q^{1/2};q\right)_\infty}{\left(q;q\right)_\infty} \left\|\phi\left(x\right)\sqrt{w\left(x\right)}\right\|_\rho \tag{7.2.28}$$

$$\times \left(\frac{\left|\left(i\omega;q^{1/2}\right)_\infty\right|^2 - \left|\left(-i\omega;q^{1/2}\right)_\infty\right|^2}{2\,\mathrm{Im}\ \omega}\right)^{1/2},$$

where

$$\left\|\phi(x)\sqrt{w(x)}\right\|_{\rho} = \left(\int_{-1}^{1}\left|\phi(x)\sqrt{w(x)}\right|^{2}\rho(x)\,dx\right)^{1/2}. \tag{7.2.29}$$

PROOF. By the Cauchy–Schwarz inequality

$$\begin{aligned} |f(\omega)| &\le \int_{-1}^{1} |e(x, i\omega)|\ |\phi(x)|\ \rho(x)\,dx \qquad (7.2.30)\\ &\le \left(\int_{-1}^{1} |e(x,i\omega)|^{2}\ \rho_1(x)\,dx\right)^{1/2} \\ &\times \left(\int_{-1}^{1}\left|\phi(x)\sqrt{w(x)}\right|^{2}\ \rho(x)\,dx\right)^{1/2}. \end{aligned}$$

The integral involving the q-exponential functions in the right hand side can be evaluated as the following special case

$$\begin{aligned} &\int_{0}^{\pi} e(\cos\theta;\alpha)\ e(\cos\theta;\beta)\ \left(e^{2i\theta},\ e^{-2i\theta};\ q\right)_{1/2}\ d\theta \qquad (7.2.31)\\ &= \pi\frac{\left(q^{1/2};q\right)^{2}_{\infty}}{(q;q)^{2}_{\infty}}\ \frac{\left(-\alpha,\ -\beta;q^{1/2}\right)_{\infty}-\left(\alpha,\ \beta;q^{1/2}\right)_{\infty}}{(\alpha+\beta)} \end{aligned}$$

of an integral evaluated by Ismail and Stanton [**71**]; see (4.4.6). By the hypothesis of the lemma function $w(x)$ is bounded on $[-1,1]$ and the integral in (7.2.29) converges. This completes the proof. □

The last inequality leads to the following theorem.

THEOREM 7.7. *Let $\{\omega_n\}$ be the set of zeros of an entire function $h(\omega)$ of order zero and suppose that all these zeros ω_n are simple. If*

$$\lim_{|\gamma|\to\infty}\frac{\left(-\gamma;q^{1/2}\right)^{2}_{\infty}-\left(\gamma;q^{1/2}\right)^{2}_{\infty}}{2\gamma\,|h(i\gamma)|^{2}} = 0, \tag{7.2.32}$$

then the system $\{e(x,i\omega_n)\}$ is complete in $\mathcal{L}^2_\rho(-1,1)$, if $\rho(x)=\rho_1(x)\,w(x)$, $w(x)>0$ is a continuous function on $[-1,1]$ and $\rho_1(x)$ is the weight function for the continuous q-ultraspherical polynomials $C_m\left(x;q^{1/2}|q\right)$.

PROOF. Suppose that the system $\{e(x,i\omega_n)\}$ is incomplete in $\mathcal{L}^2_\rho(-1,1)$. Then by Theorem 7.5 there is an entire function of order zero $f(\omega)$ given by (7.2.8), where the corresponding function $\phi(x)$, not equivalent to zero, is in $\mathcal{L}^2_\rho(-1,1)$, such that $f(\omega_n)=0$ at all the points ω_n. Moreover, inequality (7.2.28) holds. In the same manner as in the proof of Theorem 7.6, we can see that function $g(\omega)$ given by (7.2.23) is an entire function of order zero and on the imaginary axis

$$|g(\omega)| = \left|\frac{f(\omega)}{h(\omega)}\right| \le B\left(\frac{\left(-\gamma;q^{1/2}\right)^{2}_{\infty}-\left(\gamma;q^{1/2}\right)^{2}_{\infty}}{2\gamma\,|h(i\gamma)|^{2}}\right)^{1/2} \to 0 \tag{7.2.33}$$

as $|\gamma| \to \infty$ by (7.2.32) (B is a constant). Thus, $g(\omega)$ is bounded on the imaginary axis and, therefore, must be a constant. The value of this constant is zero due to (7.2.33). Thus, $f(\omega)$ is identically zero and by (7.2.12) equations (7.2.27) hold for all $k = 0, 1, 2, \ldots$. But the system $\{H_k(x|q)\}_{k=0}^{\infty}$ is closed and $\phi(x) = 0$ almost everywhere on $(-1, 1)$. Therefore the system $\{e(x, i\omega_n)\}$ is complete in $\mathcal{L}_\rho^2(-1, 1)$. □

REMARK 7.4. In view of

$$\left(-|\gamma|; q^{1/2}\right)_\infty^2 - \left(|\gamma|; q^{1/2}\right)_\infty^2 \le \left(-|\gamma|; q^{1/2}\right)_\infty^2, \tag{7.2.34}$$

relation (7.2.32) can be replaced by a weaker condition

$$\lim_{|\gamma|\to\infty} \frac{\left(-|\gamma|; q^{1/2}\right)_\infty}{|\gamma|^{1/2} |h(i\gamma)|} = 0. \tag{7.2.35}$$

In the next three sections we shall discuss several examples of complete basic trigonometric systems $\{e(x, i\omega_n)\}$ using Theorems 7.6 and 7.7.

7.3. Example: Some Infinite Products

Consider the entire function of order zero

$$h(\omega) = \left(\omega^2/\alpha^2; q^2\right)_\infty = \prod_{k=0}^{\infty} \left(1 - \frac{\omega^2}{\alpha^2} q^{2k}\right), \tag{7.3.1}$$

where the zeros are given explicitly as

$$\omega_n = \alpha q^{-n}, \qquad \omega_{-n} = -\omega_n, \qquad n = 0, 1, 2, \ldots . \tag{7.3.2}$$

By Theorem 7.6 the system $\{e(x, i\omega_n)\}_{n=-\infty}^{\infty}$ is complete in $\mathcal{L}_\rho^p(-1, 1)$ if the condition (7.2.20) is satisfied. In view of (6.1.11),

$$\left|\frac{e(x, i\omega)}{h(\omega)}\right| < C \frac{\left(-q^{1/4-\varepsilon}|\omega|; q^{1/2}\right)_\infty}{\left(-|\omega|^2/\alpha^2; q^2\right)_\infty} \tag{7.3.3}$$

on the imaginary axis $\omega = i\gamma$. Consider an arbitrary sequence $\{\gamma_k\}$ of real numbers such that $|\gamma_k| \to \infty$ and present these numbers in an "exponential form",

$$\gamma_k = \beta_k q^{-n_k}, \qquad 1 \le |\beta_k| < q^{-1}. \tag{7.3.4}$$

Then, by (I.7) of [46] or (1.4.1)

$$\begin{aligned} &\frac{\left(-q^{1/4-\varepsilon}|\gamma_k|; q^{1/2}\right)_\infty}{\left(-\gamma_k^2/\alpha^2; q^2\right)_\infty} \\ &= \frac{\left(-q^{1/4-\varepsilon}|\gamma_k|; q^{1/2}\right)_{2n_k}}{\left(-\gamma_k^2/\alpha^2; q^2\right)_{n_k}} \frac{\left(-q^{1/4-\varepsilon+n_k}|\gamma_k|; q^{1/2}\right)_\infty}{\left(-q^{2n_k}\gamma_k^2/\alpha^2; q^2\right)_\infty} \\ &= \left(q^{1-2\varepsilon}\alpha^2\right)^{n_k} \frac{\left(-q^{1/4+\varepsilon-n_k}/|\gamma_k|; q^{1/2}\right)_{2n_k}}{\left(-q^{2-2n_k}\alpha^2/\gamma_k^2; q^2\right)_{n_k}} \end{aligned} \tag{7.3.5}$$

$$\times \frac{\left(-q^{1/4-\varepsilon+n_k}\,|\gamma_k|\,;q^{1/2}\right)_\infty}{\left(-q^{2n_k}\gamma_k^2/\alpha^2;q^2\right)_\infty}$$

$$< \left(q^{1-2\varepsilon}\alpha^2\right)^{n_k} \frac{\left(-q^{1/4+\varepsilon};q^{1/2}\right)_{2n_k}}{\left(-q^4\alpha^2;q^2\right)_{n_k}} \frac{\left(-q^{-3/4-\varepsilon};q^{1/2}\right)_\infty}{\left(-1/\alpha^2;q^2\right)_\infty} \to 0$$

as $k \to \infty$ when $q^{1-2\varepsilon}\alpha^2 < 1$. Thus, the system $\{e\,(x, \pm i\alpha q^{-n})\}_{n=0}^{\infty}$ is complete in $\mathcal{L}_\rho^p\,(-1,1)$, $1 \le p < \infty$, provided that $0 < \alpha^2 < q^{-1}$.

Choosing the function

$$h\,(\omega) = \omega\,\left(\omega^2/\alpha^2;q^2\right)_\infty = \omega \prod_{k=0}^{\infty}\left(1 - \frac{\omega^2}{\alpha^2}\,q^{2k}\right), \tag{7.3.6}$$

one gets that the system $\{1, e\,(x, \pm i\alpha q^{-n})\}_{n=0}^{\infty}$ is complete in $\mathcal{L}_\rho^p\,(-1,1)$, $1 \le p < \infty$ if $0 < \alpha^2 < q^{-2}$. Condition (7.2.35) and Theorem 7.7 give the same results but in $\mathcal{L}_\rho^2\,(-1,1)$.

In a similar fashion, we can establish the following result somewhat similar to Theorem 7.2.

THEOREM 7.8. *The system $\{e\,(x, \pm i\omega_n)\}_{n=1}^{\infty}$, where $\omega_n = \alpha_n q^{1-n}$, is complete in $\mathcal{L}_\rho^p\,(-1,1)$, $1 \le p < \infty$, if $q < \alpha_n < 1$. The similar system $\{e\,(x, i\omega_n)\}_{n=-\infty}^{\infty}$, where $\omega_n = \alpha_n q^{1-n}$ and $\omega_{-n} = -\omega_n$, is complete in $\mathcal{L}_\rho^p\,(-1,1)$, $1 \le p < \infty$, if $1 < \alpha_n < q^{-1}$.*

PROOF. Consider the first case. By definition the convergence exponent of the sequence $\{\omega_n\}$ is the greatest lower bound of numbers λ for which the series

$$\sum_{n=1}^{\infty} \frac{1}{|\omega_n|^\lambda} \tag{7.3.7}$$

converges. For the given sequence $\{\omega_n\}$ this series converges for every $\lambda > 0$ and the convergence exponent is zero. By Borel's Theorem [**99**] the order of the corresponding canonical product does not exceed the convergence exponent of the sequence of zeros; see also Theorem B.4 from Appendix B. Thus, the infinite product

$$h\,(\omega) = \prod_{n=1}^{\infty}\left(1 - \frac{\omega^2}{\alpha_n^2}\,q^{2n-2}\right) \tag{7.3.8}$$

is an entire function of order zero. One can repeat the previous consideration to verify the validity of the limiting condition (7.2.20). Indeed, for a sequence of real numbers $\{\gamma_k\}$ of the form (7.3.4) we can write

$$h\,(i\gamma_k) = \prod_{n=1}^{n_k}\left(1 + \frac{\gamma_k^2}{\alpha_n^2}\,q^{2n-2}\right) \prod_{n=n_k+1}^{\infty}\left(1 + \frac{\gamma_k^2}{\alpha_n^2}\,q^{2n-2}\right) \tag{7.3.9}$$

$$= \frac{\gamma_k^{2n_k} q^{n_k(n_k-1)}}{\prod_{n=1}^{n_k}\alpha_n^2} \prod_{n=1}^{n_k}\left(1 + \frac{\alpha_n^2}{\gamma_k^2}\,q^{2-2n}\right) \prod_{n=n_k+1}^{\infty}\left(1 + \frac{\gamma_k^2}{\alpha_n^2}\,q^{2n-2}\right)$$

and

$$\frac{\left(-q^{1/4-\varepsilon}\left|\gamma_k\right|;q^{1/2}\right)_\infty}{\left|h\left(i\gamma_k\right)\right|} \tag{7.3.10}$$

$$= \left(q^{1-2\varepsilon}\right)^{n_k}\left(\prod_{n=1}^{n_k}\alpha_n^2\right)\frac{\left(-q^{1/4+\varepsilon-n_k}/\left|\gamma_k\right|;q^{1/2}\right)_{2n_k}}{\prod_{n=1}^{n_k}\left(1+q^{2-2n}\alpha_n^2/\gamma_k^2\right)}$$

$$\times\frac{\left(-q^{1/4-\varepsilon+n_k}\left|\gamma_k\right|;q^{1/2}\right)_\infty}{\prod_{n=n_k+1}^{\infty}\left(1+q^{2n-2}\gamma_k^2/\alpha_n^2\right)}$$

$$< \left(q^{1-2\varepsilon}\right)^{n_k}\frac{\left(-q^{1/4+\varepsilon};q^{1/2}\right)_{2n_k}}{\left(-q^2;q^2\right)_{n_k}}\frac{\left(-q^{-3/4-\varepsilon};q^{1/2}\right)_\infty}{\left(-1;q^2\right)_\infty}\to 0$$

as $k\to\infty$ when $0<\varepsilon<1/2$. Thus, the system $\{e(x,\pm i\alpha_n q^{-n})\}_{n=0}^{\infty}$ is complete in $\mathcal{L}_\rho^p(-1,1)$, $1\le p<\infty$.

In the second case we repeat the same consideration for the infinite product of the form

$$h(\omega)=\omega\prod_{n=1}^{\infty}\left(1-\frac{\omega^2}{\alpha_n^2}\,q^{2n-2}\right). \tag{7.3.11}$$

This completes the proof. □

7.4. Example: Basic Sine and Cosine Functions

The q-sine $S_q(\eta;\omega)$ and cosine $C_q(\eta;\omega)$ functions can be introduced by (5.4.1) and (5.4.2), respectively; their large ω-asymptotics can be investigated with the help of Eqs. (5.4.3)–(5.4.4).

Let $0=\omega_0<\omega_1<\omega_2<\omega_3<\ldots$ be nonnegative zeros of $S_q(\eta;\omega)$ and let $\varpi_1<\varpi_2<\varpi_3<\ldots$ be positive zeros of $C_q(\eta;\omega)$. We remind the reader that when $0<q<1$ all zeros of $S_q(\eta;\omega)$ and $C_q(\eta;\omega)$ are real. Also these zeros are simple, the positive zeros of the basic sine function $S_q(\eta;\omega)$ are interlaced with those of the basic cosine function $C_q(\eta;\omega)$; see Theorems 5.1–5.4. Asymptotic behavior of the large zeros of these q-trigonometric functions has been discussed in Theorem 6.4.

Completeness of the basic trigonometric system $\{\mathcal{E}_q(x;i\omega_n)\}_{n=-\infty}^{\infty}$ has been established in Theorem 6.1; see also [**26**]. Here we shall extend this result to $\mathcal{L}_\rho^p(-1,1)$, $1\le p<\infty$.

THEOREM 7.9. *The following systems of the basic trigonometric function* $\{\mathcal{E}_q(x;i\omega_n)\}_{n=-\infty}^{\infty}$, *where* $0=\omega_0<\omega_1<\omega_2<\omega_3<\ldots$ *are the nonnegative zeros of* $S_q(\eta;\omega)$, $\omega_{-n}=-\omega_n$, *and* $\{\mathcal{E}_q(x;i\varpi_n)\}_{n=-\infty}^{\infty}$, *where* $\varpi_1<\varpi_2<\varpi_3<\ldots$ *are the positive zeros of* $C_q(\eta;\omega)$, $\varpi_{-n}=-\varpi_n$, *are complete in* $\mathcal{L}_\rho^p(-1,1)$, $1\le p<\infty$.

PROOF. In the case of the q-sine function consider the entire function of order zero

$$h(\omega)=(-q\omega^2;q^2)_\infty\,S_q(\eta;\omega). \tag{7.4.1}$$

We need to verify the limiting condition (7.2.20), say, using the sequences of the real numbers $\{\gamma_k\}$ of the form (7.3.4). The large asymptotic of this function follows from (5.4.4). The leading term is $\omega\left(q^{3/2}\omega^2;q^2\right)_\infty$, which corresponds to $\alpha^2 = q^{-3/2} < q^{-2}$ in (7.3.6) and condition (7.2.20) is satisfied. The case of the basic cosine function can be considered in a similar manner. We leave the details to the reader. □

The systems of the basic trigonometric functions in Theorem 7.9 are orthogonal.

7.5. Example: Jackson's q-Bessel Functions

Let us consider an extension of Theorem 7.1. The Jackson q-Bessel functions given by (4.5.3) can be transformed with the help of a quadratic transformation formula (2.4.27) as

$$J_\nu^{(2)}(x;q) = \frac{\left(q^{\nu+1};q\right)_\infty}{(q;q)_\infty}\left(\frac{x}{2}\right)^\nu\left(-x^2/4;q^2\right)_\infty \tag{7.5.1}$$

$$\times\ {}_2\varphi_1\left(\begin{matrix}-q^{\nu+1},\ -q^{\nu+2} \\ q^{2\nu+2}\end{matrix}\ ;\ q^2,\ -x^2/4\right)$$

$$= \frac{\left(q^{\nu+1};q\right)_\infty}{(q;q)_\infty}\left(\frac{x}{2}\right)^\nu \tag{7.5.2}$$

$$\times\ \frac{\left(-q^{\nu+1},-q^{\nu+2},x^2q^{\nu+1}/4,4q^{1-\nu}/x^2;q^2\right)_\infty}{\left(q,q^{2\nu+2},-4q^2/x^2;q^2\right)_\infty}$$

$$\times\ {}_2\varphi_1\left(\begin{matrix}-q^{1+\nu},\ -q^{1-\nu} \\ q\end{matrix}\ ;\ q^2,\ -\frac{4q}{x^2}\right)$$

$$+\ \frac{\left(q^{\nu+1};q\right)_\infty}{(q;q)_\infty}\left(\frac{x}{2}\right)^\nu$$

$$\times\ \frac{\left(-q^{\nu},-q^{\nu+1},x^2q^{\nu+2}/4,4q^{-\nu}/x^2;q^2\right)_\infty}{\left(q^{-1},q^{2\nu+2},-4q^2/x^2;q^2\right)_\infty}$$

$$\times\ {}_2\varphi_1\left(\begin{matrix}-q^{2+\nu},\ -q^{2-\nu} \\ q^3\end{matrix}\ ;\ q^2,\ -\frac{4q}{x^2}\right),$$

where we have used (III.32) of [**46**] or (A.3.8) in order to transform (7.5.1) to (7.5.2). Let us consider the entire function

$$h(\omega) = \omega^{1-\nu}\ J_{\nu-1}^{(2)}(2\omega;q). \tag{7.5.3}$$

Expression (7.5.2) shows that the leading term of the $h(\omega)$ as $|\omega|\to\infty$ is $\left(q^\nu\omega^2;q^2\right)_\infty$, which corresponds to the case $\alpha^2 = q^{-\nu} < q^{-1}$ in (7.3.1) and, therefore, condition (7.2.20) is satisfied when $0<\nu<1$.

In a similar fashion, consider the entire function

$$h_1(\omega) = \omega^{1-\nu}\ J_\nu^{(2)}(2\omega;q) \tag{7.5.4}$$

with the leading term $\omega\left(q^{\nu+1}\omega^2;q^2\right)_\infty$ as $|\omega|\to\infty$. This corresponds to $\alpha^2=q^{-\nu-1}<q^{-2}$ in (7.3.1) and condition (7.2.20) is satisfied when $0<\nu<1$. As a result we obtain the following extension of Theorem 7.1.

THEOREM 7.10. *Let $j_{\mu,k}(q)$ be the zeros of Jackson's q-Bessel function $J_\mu^{(2)}(x;q)$. The q-trigonometric system $\{\mathcal{E}_q(x;i\omega_k)\}_{k=-\infty}^{\infty}$ is complete in $\mathcal{L}_\rho^p(-1,1)$, where $1\le p<\infty$ and $\rho(x)$ is an integrable function, positive almost everywhere on $(-1,1)$, for $0<\nu<1$, if:*

(a) $\omega_k=\frac{1}{2}j_{\nu-1,k}(q)$, $k=\pm1,\pm2,\ldots$;

(b) $\omega_k=\frac{1}{2}j_{\nu,k}(q)$, $k=0,\pm1,\pm2,\ldots$.

REMARK 7.5. Theorem 7.9 is obviously the special case $\nu=1/2$ of Theorem 7.10 due to the relations (7.1.1)–(7.1.2) between the q-Bessel and q-trigonometric functions, but the systems in Theorem 7.10 are not orthogonal if $\nu\neq1/2$. An independent proof of the completeness in the orthogonal case $\nu=1/2$ is given in Section 7.1.

In a similar manner, one can prove the completeness of the basic trigonometric system on the zeros of the q-Bessel function on a q-quadratic grid considered in **[63]**; see also Exs. 5.6 and 5.7. We leave this and other examples to the reader.

7.6. Exercises for Chapter 7

(1) *Some properties of the zeros of Jackson's q-Bessel functions.* Let ν is real and $0<q<1$.
Prove that the basic Bessel functions $J_\nu^{(2)}(x;q)$ and $J_{\nu+1}^{(2)}(x;q)$ have no common zeros, except possibly $x=0$, when ν is real. **[56]**
Prove that all the zeros of $x^{-\nu}J_\nu^{(2)}(x;q)$ are real and simple for $\nu>-1$. Prove that there are infinitely many of them and that their only cluster point is the point at infinity. **[56]**
Prove that between any two consecutive positive zeros of the function $x^{-\nu}J_\nu^{(2)}(x;q)$, function $x^{-\nu-1}J_{\nu+1}^{(2)}(x;q)$ has precisely one zero when $\nu\ge-1$. **[56]**
Let $\nu>-1$. Prove that the positive zeros $j_{\nu,k}(q)$ of $J_\nu^{(2)}(x;q)$ increase with ν when $0<q<1$ and decrease with q if $e^{-4}<q<1$ but $q^{-1/4}j_{\nu,k}(q)$ increase with ν and decrease with q when $0<q<1$. **[64]**

(2) The basic sine $S_q(\eta;\omega)$ and cosine $C_q(\eta;\omega)$ functions are multiples of $J_{1/2}^{(2)}(2\omega;q)$ and $J_{-1/2}^{(2)}(2\omega;q)$ due to (7.1.1)–(7.1.2). Use the properties of zeros of the q-Bessel functions $J_\nu^{(2)}(r;q)$, see **[55]**, **[56]** and Ex. 7.1, in order to give independent proofs of Theorems 5.1–5.4.

(3) *Some properties of the q-Lommel polynomials.* Derive the following generating function [**56**]

$$\sum_{n=0}^{\infty} h_n^{\nu}(x;q)\ t^n = \sum_{k=0}^{\infty} (-2xtq^{\nu})^k\, q^{k(k-1)/2} \frac{(-t/2x;q)_k}{(2xt;q)_{k+1}}$$

and verify the explicit formula

$$h_n^{\nu}(x;q) = \sum_{k=0}^{[n/2]} (-1)^k\, q^{(\nu+k-1)k} \frac{(q;q)_{n-k}\,(q^{\nu};q)_{n-k}}{(q;q)_k\,(q;q)_{n-2k}\,(q^{\nu};q)_k}\ (2x)^{n-2k}$$

that implies the symmetry relation

$$h_n^{\nu}(-x;q) = (-1)^n\ h_n^{\nu}(x;q)\,.$$

Show that the q-Lommel polynomials are generated by the three-term recurrence relation

$$h_{n+1}^{\nu}(x;q) = \left(1-q^{\nu+n}\right)(2x)\, h_n^{\nu}(x;q) - q^{\nu+n-1} h_{n-1}^{\nu}(x;q)$$

$n > 0$ with

$$h_0^{\nu}(x;q) = 1, \qquad h_1^{\nu}(x;q) = (1-q^{\nu})\,(2x)$$

and verify that

$$h_2^{\nu}(x;q) = (1-q^{\nu})\left(1-q^{\nu+1}\right)(2x)^2 - q^{\nu},$$

$$h_3^{\nu}(x;q) = \left(1-q^{\nu+1}\right)(2x)\left[(1-q^{\nu})\left(1-q^{\nu+2}\right)(2x)^2 - (1+q)\,q^{\nu}\right].$$

See [**55**], [**56**] for more details.

(4) *Dual orthogonality relation.* Let $\{p_n(x)\}_{n=0}^{\infty}$ be a sequence of orthonormal polynomials, with measure of orthogonality μ, which satisfy a three term recurrence relation

$$xp_n(x) = a_n p_{n+1}(x) + b_n p_n(x) + a_{n-1} p_{n-1}(x)\,.$$

Assume that the sequence $\{a_n\}_{n=0}^{\infty}$ is bounded and the corresponding moment problem is determinate [**4**]. Prove that if the orthogonality measure μ has point masses at α and β (β may equal α) then

$$\sum_{n=0}^{\infty} p_n(\alpha)\, p_n(\beta) = \frac{\delta_{\alpha,\beta}}{\mu(\{\alpha\})}.$$

See [**60**]. [*Hint:* For a determinate moment problem $\sum_{n=0}^{\infty} p_n^2(x)$ converges if and only $\mu(\{x\}) > 0$, in which case the series converges to $1/\mu(\{x\})$; see [**4**], [**135**]. This settles the case $\alpha = \beta$. If $\alpha \neq \beta$ the Christoffel–Darboux formula [**151**], §3.2, implies

$$\sum_{n=0}^{\infty} p_n(\alpha)\, p_n(\beta) = \lim_{m\to\infty} a_m \frac{p_{m+1}(\alpha)\, p_m(\beta) - p_{m+1}(\beta)\, p_m(\alpha)}{\alpha - \beta},$$

which implies the result since $p_n(x) \to 0$ as $n \to \infty$ and a_n is bounded [**60**].]

(5) Derive the dual orthogonality relation for the q-Lommel polynomials **[60]**,

$$\sum_{n=0}^{\infty} \left(1-q^{\nu+n}\right) q^{-n\nu-n(n-1)/2}\ h_n^{\nu}\left(x_k;q\right)\ h_n^{\nu}\left(x_l;q\right) = \frac{j_{\nu-1,k}^2\left(q\right)}{A_k\left(\nu\right)}\ \delta_{kl},$$

where $x_p = \pm 1/j_{\nu-1,p}\left(q\right)$, the $j_{\mu,p}\left(q\right)$ are zeros of $z^{-\mu}J_{\mu}^{(2)}\left(z;q\right)$, and the step function $A_k\left(\nu\right)$ is defined by the partial fraction decomposition (7.1.8).

(6) *Special q-Lommel polynomials.* Derive the explicit orthogonality relation,

$$\begin{aligned}&\left(1+(-1)^{m+p}\right)\\&\times\sum_{n=1}^{\infty} h_m\left(\frac{1}{2\omega_n};\ q\right) h_p\left(\frac{1}{2\omega_n};\ q\right)\ \frac{1}{\omega_n^2\ \kappa\left(\omega_n\right)} = \frac{q^{(m+1)^2/2}}{1-q^{m+3/2}}\ \delta_{mp},\end{aligned}$$

and the dual orthogonality relation,

$$\sum_{m=0}^{\infty} h_m\left(\frac{1}{2\omega_n};q\right) h_m\left(\frac{1}{2\omega_l};q\right)\ \frac{1-q^{m+3/2}}{q^{(m+1)^2/2}} = \omega_n^2\ \kappa\left(\omega_n\right)\ \delta_{nl},$$

for the special q-Lommel polynomials $h_m\left(x;q\right) := h_m^{3/2}\left(x;q\right)$, where $0<\omega_1<\omega_2<\omega_3<\ldots$ are nonnegative zeros of the basic sine function $S_q\left(\eta;\omega\right)$ and $\kappa\left(\omega\right)$ is given by (7.1.31).

Derive the explicit orthogonality relation,

$$\begin{aligned}&\left(1+(-1)^{m+p}\right)\\&\times\sum_{n=1}^{\infty} \overline{h}_m\left(\frac{1}{2\varpi_n};\ q\right) \overline{h}_p\left(\frac{1}{2\varpi_n};\ q\right)\ \frac{1}{\varpi_n^2\ \kappa\left(\varpi_n\right)} = \frac{q^{m^2/2}}{1-q^{m+1/2}}\ \delta_{mp},\end{aligned}$$

and the dual orthogonality relation,

$$\sum_{m=0}^{\infty} \overline{h}_m\left(\frac{1}{2\varpi_n};q\right) \overline{h}_m\left(\frac{1}{2\varpi_l};q\right)\ \frac{1-q^{m}}{q^{m^2/2}} = \varpi_n^2\ \kappa\left(\varpi_n\right)\ \delta_{nl},$$

for another special q-Lommel polynomials $\overline{h}_m\left(x;q\right) := h_m^{1/2}\left(x;q\right)$, where $0<\varpi_1<\varpi_2<\varpi_3<\ldots$ are nonnegative zeros of the basic cosine function $C_q\left(\eta;\omega\right)$.

[*Hint*: Let $\nu = 3/2$ and $z_k = j_{1/2,k}\left(q\right) = 2\omega_k$. Obtain from (7.1.9) and (2.4.25)–(2.4.26) that

$$A_k\left(\frac{3}{2}\right) = 2q^{-1/2}\frac{J_{-1/2}^{(2)}\left(z_k;q\right)}{\dfrac{d}{dz}J_{1/2}^{(2)}\left(z_k;q\right)} = 4q^{-1/2}\frac{C_q\left(\eta;\omega_k\right)}{\dfrac{d}{d\omega}S_q\left(\eta;\omega_k\right)}$$

and then simplify the answer with the help of (5.5.14)–(5.5.15), (5.5.8)–(5.5.9), and (5.8.10); see also (9.4.7). The case $\nu = 1/2$ is similar.]

[*Note*: These explicit orthogonality relations for the special q-Lommel polynomials $h_m^{3/2}(x;q)$ and $h_m^{1/2}(x;q)$ were found in [**144**] and [**146**]; see also [**60**].]

(7) Show that

$$\begin{aligned} q^{n\nu+n(n-1)/2} J_{\nu+n}^{(2)}(x;q) &= \begin{vmatrix} h_n^{\nu}\left(\frac{1}{x};q\right) & h_{n-1}^{\nu+1}\left(\frac{1}{x};q\right) \\ J_{\nu-1}^{(2)}(x;q) & J_{\nu}^{(2)}(x;q) \end{vmatrix} \\ &= \begin{vmatrix} h_n^{\nu}\left(\frac{1}{x};q\right) + \alpha h_{n-1}^{\nu+1}\left(\frac{1}{x};q\right) & h_{n-1}^{\nu+1}\left(\frac{1}{x};q\right) \\ J_{\nu-1}^{(2)}(x;q) + \alpha J_{\nu}^{(2)}(x;q) & J_{\nu}^{(2)}(x;q) \end{vmatrix} \\ &= \left(h_n^{\nu}\left(\frac{1}{x};q\right) + \alpha h_{n-1}^{\nu+1}\left(\frac{1}{x};q\right)\right) J_{\nu}^{(2)}(x;q) \\ &\quad - h_{n-1}^{\nu+1}\left(\frac{1}{x};q\right)\left(J_{\nu-1}^{(2)}(x;q) + \alpha J_{\nu}^{(2)}(x;q)\right), \end{aligned}$$

where α is a constant.

(8) Prove Criterion 7.1. [*Note*: See [**3**] for the proof and bibliography.]

CHAPTER 8

Improved Asymptotics of Zeros

In Section 6.3 we derived the asymptotic formulas (6.3.14) and (6.3.15) for the zeros of the basic sine $S_q(\eta;\omega)$ and basic cosine $C_q(\eta;\omega)$ functions, respectively. In this chapter we shall find improved asymptotics for these zeros by a different method using the Lagrange inversion formula.

8.1. Interpretation of Zeros and Preliminary Results

The following mechanical interpretation of the functions $F(\omega)$ and $G(\omega)$ defined by (6.3.6)–(6.3.7) can be given. Introducing a new variable

$$v(\omega) = \int_0^\omega \kappa(s)\ ds, \tag{8.1.1}$$

one can write these functions as

$$F(\omega) = \sin v(\omega), \qquad G(\omega) = \cos v(\omega) \tag{8.1.2}$$

in view of (6.3.10); see also **[52]** and Ex. 6.5. These equations describe a circular motion with the angular velocity

$$\frac{dv}{d\omega} = \kappa(\omega), \tag{8.1.3}$$

where the $v(\omega)$ represents the total angle of rotation as a function of "time" ω. The zeros of the sine and cosine functions are known explicitly and due to (8.1.1) and (8.1.2) the following "quantization rules" hold

$$v(\omega_n) = \int_0^{\omega_n} \kappa(s)\ ds = \pi n, \tag{8.1.4}$$

$$v(\varpi_n) = \int_0^{\varpi_n} \kappa(s)\ ds = \frac{\pi}{2} + \pi n \tag{8.1.5}$$

with $n = 0, 1, 2, \ldots$ for the nonnegative zeros ω_n and ϖ_n of the $F(\omega)$ and $G(\omega)$, respectively. This gives also a geometric interpretation of these zeros in terms of "quantization" of the area under the graph of the positive function $\kappa(\omega)$. By (8.1.4)–(8.1.5) these zeros are completely defined in terms of the function $\kappa(\omega)$ given by (6.3.9).

On the other hand, one can easily verify that

$$\kappa(\omega) = \sum_{k=0}^{\infty} \frac{q^{k/2}}{1+\omega^2 q^k} = \frac{1}{2i}\frac{d}{d\omega}\log\frac{\left(-i\omega; q^{1/2}\right)_\infty}{\left(i\omega; q^{1/2}\right)_\infty}. \tag{8.1.6}$$

Therefore

$$v(\omega)=\int_0^{\omega}\kappa(s)\;ds=\frac{1}{2i}\log\frac{\left(-i\omega;q^{1/2}\right)_\infty}{\left(i\omega;q^{1/2}\right)_\infty} \tag{8.1.7}$$

and our functions $F(\omega)$ and $G(\omega)$ can be rewritten as

$$\begin{aligned} F(\omega) &= \sin\left(\frac{1}{2i}\log\frac{\left(-i\omega;q^{1/2}\right)_\infty}{\left(i\omega;q^{1/2}\right)_\infty}\right) \\ &= \frac{1}{(-\omega^2;q)_\infty^{1/2}}\sum_{k=0}^{\infty}(-1)^k\frac{q^{k(k+1/2)}}{\left(q^{1/2};q^{1/2}\right)_{2k+1}}\;\omega^{2k+1}, \end{aligned} \tag{8.1.8}$$

$$\begin{aligned} G(\omega) &= \cos\left(\frac{1}{2i}\log\frac{\left(-i\omega;q^{1/2}\right)_\infty}{\left(i\omega;q^{1/2}\right)_\infty}\right) \\ &= \frac{1}{(-\omega^2;q)_\infty^{1/2}}\sum_{k=0}^{\infty}(-1)^k\frac{q^{k(k-1/2)}}{\left(q^{1/2};q^{1/2}\right)_{2k}}\;\omega^{2k} \end{aligned} \tag{8.1.9}$$

by (5.4.1)–(5.4.2) and (6.3.6)–(6.3.7). In (8.1.7) we use the branch of the logarithm which takes the value $2\pi in$ at ω_n for each of the intervals $\omega_n \leq \omega < \omega_{n+1}$ with the cut along the positive real axis. One can view this substitution as a q-analog of the logarithmic scale.

It is important for the further consideration to find the large ω-asymptotics of the "phase function" $v(\omega)$. Introducing the basic exponential function

$$\mathcal{E}_q(\eta;i\omega)=C_q(\eta;\omega)+iS_q(\eta;\omega)=\frac{\left(-i\omega;q^{1/2}\right)_\infty}{\left(-q\omega^2;q^2\right)_\infty}, \tag{8.1.10}$$

see (2.3.10) and (2.4.19)–(2.4.20), one can rewrite the transformations (5.4.3)–(5.4.4) as

$$\begin{aligned} &\mathcal{E}_q(\eta;i\omega) \\ &\quad=\frac{\left(q^2;q^2\right)_\infty}{\left(q^{1/2};q^{1/2}\right)_\infty}\,\frac{\left(q^{1/2}\omega^2,q^{3/2}/\omega^2;q^2\right)_\infty+i\omega\left(q^{3/2}\omega^2,q^{1/2}/\omega^2;q^2\right)_\infty}{\left(-q\omega^2,-q/\omega^2;q^2\right)_\infty} \\ &\qquad\times\mathcal{E}_q(\eta;\frac{iq^{1/2}}{\omega}). \end{aligned} \tag{8.1.11}$$

The expression in the second line here can be viewed as an analog of the q-exponential function corresponding to the q-trigonometric functions discussed in [**47**]. The last equation can be rewritten as

$$\begin{aligned} \left(-i\omega;q^{1/2}\right)_\infty &= \frac{\left(q^2;q^2\right)_\infty}{\left(q^{1/2};q^{1/2}\right)_\infty} \\ &\quad\times\frac{\left(q^{1/2}\omega^2,q^{3/2}/\omega^2;q^2\right)_\infty+i\omega\left(q^{3/2}\omega^2,q^{1/2}/\omega^2;q^2\right)_\infty}{\left(iq^{1/2}/\omega;q^{1/2}\right)_\infty}, \end{aligned} \tag{8.1.12}$$

which gives an asymptotic behavior of the q-shifted factorials for the large values of the ω. Therefore

$$e^{2iv(\omega)} = \frac{\left(-i\omega; q^{1/2}\right)_\infty}{\left(i\omega; q^{1/2}\right)_\infty} = \frac{\left(q^{1/2}\omega^2, q^{3/2}/\omega^2; q^2\right)_\infty + i\omega\left(q^{3/2}\omega^2, q^{1/2}/\omega^2; q^2\right)_\infty}{\left(q^{1/2}\omega^2, q^{3/2}/\omega^2; q^2\right)_\infty - i\omega\left(q^{3/2}\omega^2, q^{1/2}/\omega^2; q^2\right)_\infty} \times \frac{\left(-iq^{1/2}/\omega; q^{1/2}\right)_\infty}{\left(iq^{1/2}/\omega; q^{1/2}\right)_\infty} \tag{8.1.13}$$

and the following "phase transformation" determines the large ω-asymptotic of the function $v(\omega)$:

$$v(\omega) = \frac{1}{2i}\log\frac{\left(q^{1/2}\omega^2, q^{3/2}/\omega^2; q^2\right)_\infty + i\omega\left(q^{3/2}\omega^2, q^{1/2}/\omega^2; q^2\right)_\infty}{\left(q^{1/2}\omega^2, q^{3/2}/\omega^2; q^2\right)_\infty - i\omega\left(q^{3/2}\omega^2, q^{1/2}/\omega^2; q^2\right)_\infty} + v\left(q^{1/2}/\omega\right). \tag{8.1.14}$$

In view of (8.1.3) and (8.1.12) this implies

$$\kappa(\omega) = \frac{1}{2i}\frac{d}{d\omega}\log\frac{\left(-i\omega, iq^{1/2}/\omega; q^{1/2}\right)_\infty}{\left(i\omega, -iq^{1/2}/\omega; q^{1/2}\right)_\infty} - \frac{q^{1/2}}{\omega^2}\kappa\left(\frac{q^{1/2}}{\omega}\right). \tag{8.1.15}$$

Comparing this transformation with (5.5.17), we arrive at the following differentiation formula

$$\frac{1}{2i}\frac{d}{d\omega}\log\frac{\left(-i\omega, iq^{1/2}/\omega; q^{1/2}\right)_\infty}{\left(i\omega, -iq^{1/2}/\omega; q^{1/2}\right)_\infty} = \frac{(q;q)_\infty^2}{\left(q^{1/2};q\right)_\infty^2}\frac{\left(-q^{1/2}\omega^2, -q^{1/2}/\omega^2; q\right)_\infty}{\left(-\omega^2, -q/\omega^2; q\right)_\infty} \tag{8.1.16}$$

and at the corresponding indefinite integral

$$\frac{(q;q)_\infty^2}{\left(q^{1/2};q\right)_\infty^2}\int\frac{\left(-q^{1/2}\omega^2, -q^{1/2}/\omega^2; q\right)_\infty}{\left(-\omega^2, -q/\omega^2; q\right)_\infty}\, d\omega = \frac{1}{2i}\log\frac{\left(-i\omega, iq^{1/2}/\omega; q^{1/2}\right)_\infty}{\left(i\omega, -iq^{1/2}/\omega; q^{1/2}\right)_\infty}. \tag{8.1.17}$$

Use of (8.1.6) gives an independent proof of these relations

$$\begin{aligned}\frac{d}{d\omega}\log\frac{\left(-i\omega, iq^{1/2}/\omega; q^{1/2}\right)_\infty}{\left(i\omega, -iq^{1/2}/\omega; q^{1/2}\right)_\infty} &= \frac{d}{d\omega}\log\frac{\left(-i\omega; q^{1/2}\right)_\infty}{\left(i\omega; q^{1/2}\right)_\infty} - \frac{d}{d\omega}\log\frac{\left(-iq^{1/2}/\omega; q^{1/2}\right)_\infty}{\left(iq^{1/2}/\omega; q^{1/2}\right)_\infty} \\ &= 2i\left(\kappa(\omega) + \frac{q^{1/2}}{\omega^2}\kappa\left(\frac{q^{1/2}}{\omega}\right)\right) = 2i\sum_{k=-\infty}^{\infty}\frac{q^{k/2}}{1+\omega^2q^k} \\ &= 2i\,\frac{\left(q, q, -q^{1/2}\omega^2, -q^{1/2}/\omega^2; q\right)_\infty}{\left(q^{1/2}, q^{1/2}, -\omega^2, -q/\omega^2; q\right)_\infty}\end{aligned} \tag{8.1.18}$$

by the Ramanujan ${}_1\psi_1$ summation formula (A.2.5).

8.2. Lagrange Inversion Formula

In the next sections we shall derive new asymptotic formulas which improve (6.3.14)–(6.3.15) and (6.3.44)–(6.3.45) with the help of (8.1.4)–(8.1.5) and (8.1.14). This will be done by means of the inverse function of the "phase function" $v(\omega)$ given by (8.1.7), which can be found by the Lagrange inversion formula. To make our presentation as self-contained as possible, we remind the reader this formula here and present its proof in Appendix B; see, for example, **[10]**, **[21]**, **[106]**, **[136]** for more details and applications.

Let $w = g(z)$ be regular at $z = z_0$:

$$w = g(z) = w_0 + \sum_{n=1}^{\infty} a_n (z - z_0)^n, \qquad a_1 = g'(z_0) \neq 0. \tag{8.2.1}$$

Then the inverse function $z = g^{-1}(w) = h(g(z))$ can be found by the *Lagrange inversion formula* as

$$z = h(w) = z_0 + \sum_{n=1}^{\infty} b_n (w - w_0)^n, \tag{8.2.2}$$

where

$$\begin{aligned} b_n &= \frac{1}{2\pi i} \int_{|z-z_0|=r} \frac{z g'(z)}{(g(z) - w_0)^{n+1}}\, dz \\ &= \frac{1}{n!} \lim_{z\to z_0} \frac{d^{n-1}}{dz^{n-1}} \left(\frac{z - z_0}{g(z) - w_0} \right)^n . \end{aligned} \tag{8.2.3}$$

Moreover, the first coefficients of expansion (8.2.2) are given explicitly by

$$b_1 = \frac{1}{a_1}, \quad b_2 = -\frac{a_2}{a_1^3}, \quad b_3 = \frac{1}{a_1^3} \left(2 \left(\frac{a_2}{a_1} \right)^2 - \frac{a_3}{a_1} \right). \tag{8.2.4}$$

We introduce the inverse of the $v(\omega)$ as the power series

$$\omega = f(v) = \omega_0 + \sum_{k=1}^{\infty} b_k\ (v - v_0)^k \tag{8.2.5}$$

with the coefficients

$$\begin{aligned} b_k &= \frac{f^{(k)}(v_0)}{k!} = \frac{1}{2\pi i} \int_{|\omega-\omega_0|=r} \frac{\omega \kappa(\omega)}{(v(\omega) - v_0)^{k+1}}\, d\omega \\ &= \frac{1}{k!} \lim_{\omega\to\omega_0} \frac{d^{k-1}}{d\omega^{k-1}} \left(\frac{\omega - \omega_0}{v(\omega) - v_0} \right)^k \end{aligned} \tag{8.2.6}$$

in view of (8.2.2)–(8.2.3). The first terms of this Lagrange expansion are

$$\omega = \omega_0 + \frac{1}{\kappa(\omega_0)} (v - v_0) - \frac{\kappa'(\omega_0)}{2\kappa^3(\omega_0)} (v - v_0)^2 \tag{8.2.7}$$

$$+\frac{1}{2\kappa^3(\omega_0)}\left(\left(\frac{\kappa'(\omega_0)}{\kappa(\omega_0)}\right)^2-\frac{\kappa''(\omega_0)}{3\kappa(\omega_0)}\right)(v-v_0)^3+...$$

due to (8.2.4).

8.3. Asymptotics of $\kappa'(\omega)$ and $\kappa''(\omega)$

In Section 5.5 we established asymptotic behavior of the function

$$\kappa(\omega)=\sum_{k=0}^{\infty}\frac{q^{k/2}}{1+\omega^2q^k}. \tag{8.3.1}$$

Formula (5.5.17) determines the asymptotic of this function for the large ω. According to (8.2.7) one needs to find asymptotic expressions for the first and the second derivatives of this function in order to derive improved asymptotics of zeros of the basic sine and cosine functions in the next section.

8.3.1. Asymptotic of $\kappa'(\omega)$. One can easily see from (8.3.1) that

$$\kappa'(\omega)=-2\omega\sum_{k=0}^{\infty}\frac{q^{3k/2}}{(1+\omega^2q^k)^2}, \tag{8.3.2}$$

which means that the $\kappa(\omega)$ is increasing for all negative values of the ω, decreasing for all positive ones, and attains its maximum at $\omega=0$.

The large ω-asymptotic of the $\kappa'(\omega)$ follows from

$$\begin{aligned}\kappa'(\omega)=&-\frac{(q;q)_\infty^2}{(q^{1/2};q)_\infty^2}\;\frac{(-q^{1/2}\omega^2,-q^{1/2}/\omega^2;q)_\infty}{\omega(-\omega^2,-q/\omega^2;q)_\infty}\\&-(q;q)_\infty^2\left(q^{1/2};q^{1/2}\right)_\infty^2\;\omega\frac{(q^{1/2}\omega^2,1/\omega^2;q^{1/2})_\infty}{(-\omega^2,-q/\omega^2;q)_\infty^2}\\&+\frac{2q^{1/2}}{\omega^3}\sum_{k=0}^{\infty}\frac{q^{k/2}}{(1+q^{1+k}/\omega^2)^2}.\end{aligned} \tag{8.3.3}$$

In order to verify this formula one can extend the sum in (8.3.2) to the corresponding ${}_2\psi_2$ series,

$$\sum_{k=-\infty}^{\infty}\frac{q^{3k/2}}{(1+\omega^2q^k)^2}=\frac{1}{(1+\omega^2)^2}\,{}_2\psi_2\left(\begin{matrix}-\omega^2,\ -\omega^2\\-q\omega^2,\ -q\omega^2\end{matrix};q,q^{3/2}\right), \tag{8.3.4}$$

which can be transformed by the Ex. 5.20(i) of [**46**] or (A.3.9),

$$\begin{aligned}&{}_2\psi_2\left(\begin{matrix}-\omega^2,\ -\omega^2\\-q\omega^2,-q\omega^2\end{matrix};q,q^{3/2}\right)\\&=\frac{(q,q,-q^{3/2}\omega^2,-q^{1/2}/\omega^2;q)_\infty}{(q^{1/2},q^{3/2},-q\omega^2,-q/\omega^2;q)_\infty}\,{}_2\psi_2\left(\begin{matrix}-\omega^2,\ -q^{1/2}\omega^2\\-q\omega^2,-q^{3/2}\omega^2\end{matrix};q,q\right).\end{aligned} \tag{8.3.5}$$

The last ${}_2\psi_2$ can be summed

$$
\begin{aligned}
{}_2\psi_2\left(\begin{matrix}-\omega^2, & -q^{1/2}\omega^2 \\ -q\omega^2, & -q^{3/2}\omega^2\end{matrix}; q, q\right) &= \sum_{k=-\infty}^{\infty} \frac{\left(-\omega^2,\ -q^{1/2}\omega^2; q\right)_k}{\left(-q\omega^2, -q^{3/2}\omega^2; q\right)_k}\, q^k \\
&= \sum_{k=-\infty}^{\infty} \frac{\left(-\omega^2; q^{1/2}\right)_{2k}}{\left(-q\omega^2; q^{1/2}\right)_{2k}} \left(q^{1/2}\right)^{2k} \\
&= \frac{1}{2}\left({}_1\psi_1\left(\begin{matrix}-\omega^2 \\ -q\omega^2\end{matrix}; q^{1/2}, q^{1/2}\right) + {}_1\psi_1\left(\begin{matrix}-\omega^2 \\ -q\omega^2\end{matrix}; q^{1/2}, -q^{1/2}\right)\right) \\
&= \frac{1}{2} \frac{\left(q^{1/2},\ q; q^{1/2}\right)_\infty}{\left(-q\omega^2, -q^{1/2}/\omega^2; q^{1/2}\right)_\infty} \\
&\quad \times \left(\frac{\left(-q^{1/2}\omega^2, -1/\omega^2; q^{1/2}\right)_\infty}{\left(q^{1/2},\ q^{1/2}; q^{1/2}\right)_\infty} + \frac{\left(q^{1/2}\omega^2, 1/\omega^2; q^{1/2}\right)_\infty}{\left(-q^{1/2},\ -q^{1/2}; q^{1/2}\right)_\infty}\right)
\end{aligned}
\tag{8.3.6}
$$

by a consequence of the Ramanujan ${}_1\psi_1$ summation formula (A.2.5). The original bilateral sum in (8.3.4) can be rewritten as

$$
\sum_{k=-\infty}^{\infty} \frac{q^{3k/2}}{\left(1+\omega^2 q^k\right)^2} = \sum_{k=0}^{\infty} \frac{q^{3k/2}}{\left(1+\omega^2 q^k\right)^2} + \frac{q^{1/2}}{\omega^4} \sum_{k=0}^{\infty} \frac{q^{k/2}}{\left(1+q^{1+k}/\omega^2\right)^2} \tag{8.3.7}
$$

and we arrive at (8.3.3) from (8.3.4)–(8.3.7).

8.3.2. Asymptotic of $\kappa''(\omega)$. The following formula holds

$$
\begin{aligned}
\kappa''(\omega) &= 3\,(q;q)_\infty^2 \left(q^{1/2}; q^{1/2}\right)_\infty^2 \frac{\left(q^{1/2}\omega^2,\ 1/\omega^2; q^{1/2}\right)_\infty}{\left(-\omega^2,\ -q/\omega^2; q\right)_\infty^2} \\
&\quad + \frac{(q;q)_\infty^2}{\left(q^{1/2}; q\right)_\infty^2} \frac{\left(-q^{1/2}\omega^2,\ -q^{1/2}/\omega^2; q\right)_\infty}{\omega^2\left(-\omega^2,\ -q/\omega^2; q\right)_\infty} \\
&\quad \times \left[3 - 2\frac{\left(q^{1/2}; q\right)_\infty^4 \left(q^2; q^2\right)_\infty^2}{(q;q)_\infty^4 \left(q; q^2\right)_\infty^2}\left(1 + 2^4 \sum_{k=1}^{\infty} \frac{q^{3k/2}}{\left(1+q^k\right)^3}\right)\right. \\
&\quad \left. + 2\frac{(q;q)_\infty^4}{(-q;q)_\infty^4} \frac{\left(\omega^2,\ q/\omega^2; q\right)_\infty^2}{\left(-\omega^2,\ -q/\omega^2; q\right)_\infty^2}\right] \\
&\quad - 6\frac{q^{1/2}}{\omega^4} \sum_{k=0}^{\infty} \frac{q^{k/2}}{\left(1+q^{1+k}/\omega^2\right)^2} + 8\frac{q^{3/2}}{\omega^6} \sum_{k=0}^{\infty} \frac{q^{3k/2}}{\left(1+q^{1+k}/\omega^2\right)^3},
\end{aligned}
\tag{8.3.8}
$$

which determines the asymptotic behavior of the $\kappa''(\omega)$ as $|\omega| \to \infty$.

From (8.3.2)

$$
\kappa''(\omega) = -2\sum_{k=0}^{\infty} \frac{q^{3k/2}}{\left(1+\omega^2 q^k\right)^2} + 8\omega^2 \sum_{k=0}^{\infty} \frac{q^{5k/2}}{\left(1+\omega^2 q^k\right)^3}\,. \tag{8.3.9}
$$

Mizan Rahman [**122**] has suggested the following transformation of this expression. The identity

$$\frac{\omega^2 q^{5k/2}}{\left(1+\omega^2 q^k\right)^3} = \frac{q^{3k/2}}{\left(1+\omega^2 q^k\right)^2} - \frac{q^{3k/2}}{\left(1+\omega^2 q^k\right)^3} \tag{8.3.10}$$

gives

$$\kappa''(\omega) = -3\frac{\kappa'(\omega)}{\omega} - 8\sum_{k=0}^{\infty} \frac{q^{3k/2}}{\left(1+\omega^2 q^k\right)^3} \tag{8.3.11}$$

and we can use equation (8.3.3) for the first term here. The second sum can be again extended to a bilateral series

$$\sum_{k=-\infty}^{\infty} \frac{q^{3k/2}}{\left(1+\omega^2 q^k\right)^3} = \frac{1}{\left(1+\omega^2\right)^3}\ {}_3\psi_3\left(\begin{matrix} -\omega^2, & -\omega^2, & -\omega^2 \\ -q\omega^2, & -q\omega^2, & -q\omega^2 \end{matrix} ; q, q^{3/2} \right). \tag{8.3.12}$$

Rahman's original idea [**122**] was to rewrite the last ${}_3\psi_3$ series as the following special case

$${}_8\psi_8\left(\begin{matrix} q\omega^2, -q\omega^2, -\omega^2, -\omega^2, -\omega^2, q^{1/2}\omega^2, \omega^2, -\omega^2 \\ \omega^2, -\omega^2, -q\omega^2, -q\omega^2, -q\omega^2, q^{1/2}\omega^2, q\omega^2, -q\omega^2 \end{matrix} ; q, q^{3/2} \right) \tag{8.3.13}$$

of very-well-poised ${}_8\psi_8$ series and then to apply the transformation formula (III.38) of [**46**] or (A.3.17). Here are the details. Let us consider the following transformation

$$\begin{aligned}
&{}_8\psi_8\left(\begin{matrix} q\omega^2, -q\omega^2, -\omega^2, -\omega^2, -\omega^2, q^{1/2}\omega^2, \varepsilon\omega^2, -\varepsilon\omega^2 \\ \omega^2, -\omega^2, -q\omega^2, -q\omega^2, -q\omega^2, q^{1/2}\omega^2, q\omega^2/\varepsilon, -q\omega^2/\varepsilon \end{matrix} ; q, \frac{q^{3/2}}{\varepsilon^2} \right) \\
&= {}_5\psi_5\left(\begin{matrix} q\omega^2, -\omega^2, -\omega^2, \varepsilon\omega^2, -\varepsilon\omega^2 \\ \omega^2, -q\omega^2, -q\omega^2, q\omega^2/\varepsilon, -q\omega^2/\varepsilon \end{matrix} ; q, \frac{q^{3/2}}{\varepsilon^2} \right) \\
&= \frac{\left(-\varepsilon\omega^2, -\varepsilon/\omega^2, q\omega^4, q/\omega^4, q; q\right)_\infty}{\left(-q\omega^2, -q\omega^2, -q\omega^2, q^{1/2}\omega^2, -1; q\right)_\infty} \\
&\times \frac{\left(-q/\varepsilon, -q/\varepsilon, -q/\varepsilon, -q\varepsilon, -q\varepsilon, -q\varepsilon, q^{1/2}\varepsilon, q^{1/2}/\varepsilon; q\right)_\infty}{\left(-q/\omega^2, -q/\omega^2, -q/\omega^2, q^{1/2}/\omega^2, q/\varepsilon\omega^2, q\omega^2/\varepsilon, -\varepsilon^2, q\varepsilon^2; q\right)_\infty} \\
&\times {}_5\varphi_4\left(\begin{matrix} \varepsilon^2, q\varepsilon, -\varepsilon, -\varepsilon, -\varepsilon^2 \\ \varepsilon, -q\varepsilon, -q\varepsilon, -q \end{matrix} ; q, \frac{q^{3/2}}{\varepsilon^2} \right) \\
&+ \frac{\left(\varepsilon\omega^2, \varepsilon/\omega^2, q\omega^4, q/\omega^4, q; q\right)_\infty}{\left(-q\omega^2, -q\omega^2, -q\omega^2, q^{1/2}\omega^2, -1; q\right)_\infty} \\
&\times \frac{\left(q/\varepsilon, q/\varepsilon, q/\varepsilon, q\varepsilon, q\varepsilon, q\varepsilon, -q^{1/2}\varepsilon, -q^{1/2}/\varepsilon; q\right)_\infty}{\left(-q/\omega^2, -q/\omega^2, -q/\omega^2, -q/\varepsilon\omega^2, -q\omega^2/\varepsilon, q^{1/2}/\omega^2, -\varepsilon^2, q\varepsilon^2; q\right)_\infty} \\
&\times {}_5\varphi_4\left(\begin{matrix} \varepsilon^2, -q\varepsilon, \varepsilon, \varepsilon, -\varepsilon^2 \\ -\varepsilon, q\varepsilon, q\varepsilon, -q \end{matrix} ; q, \frac{q^{3/2}}{\varepsilon^2} \right)
\end{aligned} \tag{8.3.14}$$

and take the limit $\varepsilon \to 1$. Then

$$\lim_{\varepsilon\to 1} {}_5\varphi_4\left(\begin{matrix}\varepsilon^2, -q\varepsilon, \varepsilon, \varepsilon, -\varepsilon^2 \\ -\varepsilon, q\varepsilon, q\varepsilon, -q\end{matrix}; q, \frac{q^{3/2}}{\varepsilon^2}\right) = 1 \tag{8.3.15}$$

and

$$\lim_{\varepsilon\to 1} {}_5\varphi_4\left(\begin{matrix}\varepsilon^2, q\varepsilon, -\varepsilon, -\varepsilon, -\varepsilon^2 \\ \varepsilon, -q\varepsilon, -q\varepsilon, -q\end{matrix}; q, \frac{q^{3/2}}{\varepsilon^2}\right) \tag{8.3.16}$$

$$= \lim_{\varepsilon\to 1}\left(1 + (1+\varepsilon)\sum_{k=1}^{\infty}\frac{(q\varepsilon^2;q)_{k-1}\,(-\varepsilon,-\varepsilon,-\varepsilon^2,q\varepsilon;q)_k}{(q\varepsilon;q)_{k-1}\,(-q\varepsilon,-q\varepsilon,-q,q;q)_k}\left(\frac{q^{3/2}}{\varepsilon^2}\right)^k\right)$$

$$= 1 + 2\sum_{k=1}^{\infty}\frac{(-1;q)_k^3}{(-q;q)_k^3}\, q^{3k/2}$$

$$= 1 + 2^4\sum_{k=1}^{\infty}\frac{q^{3k/2}}{(1+q^k)^3}.$$

Also

$$\sum_{k=-\infty}^{\infty}\frac{q^{3k/2}}{(1+\omega^2 q^k)^3} = \sum_{k=0}^{\infty}\frac{q^{3k/2}}{(1+\omega^2 q^k)^3} + \frac{q^{3/2}}{\omega^6}\sum_{k=0}^{\infty}\frac{q^{3k/2}}{(1+q^{1+k}/\omega^2)^3}. \tag{8.3.17}$$

All this together results in (8.3.8) after some substitutions. We leave the details to the reader.

8.3.3. Some Estimates. It is worth also noting the following useful estimates

$$\left|\kappa'(\omega)\right| < \frac{2\kappa(\omega)}{|\omega|} \tag{8.3.18}$$

and

$$\kappa''(\omega) < \frac{3\left|\kappa'(\omega)\right|}{|\omega|} < \frac{6\kappa(\omega)}{\omega^2}, \tag{8.3.19}$$

which follow directly from (8.3.1), (8.3.2) and (8.3.9). Indeed, by (8.3.1) and (8.3.2)

$$\begin{aligned}\frac{\left|\kappa'(\omega)\right|}{2|\omega|} &= \frac{1}{\omega^2}\sum_{k=0}^{\infty}\frac{\omega^2 q^k}{1+\omega^2 q^k}\frac{q^{k/2}}{1+\omega^2 q^k} \\ &< \frac{1}{\omega^2}\sum_{k=0}^{\infty}\frac{q^{k/2}}{1+\omega^2 q^k} = \frac{\kappa(\omega)}{\omega^2}.\end{aligned} \tag{8.3.20}$$

In a similar fashion in (8.3.9)

$$\sum_{k=0}^{\infty}\frac{q^{5k/2}}{(1+\omega^2 q^k)^3} = \frac{1}{\omega^2}\sum_{k=0}^{\infty}\frac{\omega^2 q^k}{1+\omega^2 q^k}\frac{q^{3k/2}}{(1+\omega^2 q^k)^2}$$

$$< \frac{1}{\omega^2} \sum_{k=0}^{\infty} \frac{q^{3k/2}}{\left(1+\omega^2 q^k\right)^2} \tag{8.3.21}$$

and

$$\kappa''(\omega) < (-2+8) \sum_{k=0}^{\infty} \frac{q^{3k/2}}{\left(1+\omega^2 q^k\right)^2} = \frac{-3\kappa'(\omega)}{\omega}, \tag{8.3.22}$$

which results in (8.3.19). Also,

$$\left|\kappa''(\omega)\right| < \frac{5\left|\kappa'(\omega)\right|}{|\omega|} < \frac{10\kappa(\omega)}{\omega^2}. \tag{8.3.23}$$

8.4. Improved Asymptotics

This section contains the main result of this chapter; see also [**147**]. Here we prove asymptotic formulas (6.3.44)–(6.3.45) and find one more term by a completely different method using the Lagrange inversion formula. Similar idea can be applied in order to establish next terms of these asymptotic expansions if needed.

In principle, Eqs. (8.1.4)–(8.1.5) give explicit formulas for the zeros ω_n and ϖ_n in terms of the inverse function (8.2.5), namely,

$$\omega_n = f(\pi n), \qquad \varpi_n = f\left(\frac{\pi}{2} + \pi n\right). \tag{8.4.1}$$

The problem is to find a convenient representation for the $f(v)$ valid for the large ω.

In the case of the basic sine function, consider $\omega_0 = \omega_n^{(0)} = q^{1/4-n}$. In view of (8.1.14) we get

$$v_n^{(0)} = v\left(\omega_n^{(0)}\right) = \pi n + v\left(\frac{q^{1/2}}{\omega_n^{(0)}}\right) \tag{8.4.2}$$

and by (8.2.7)

$$\begin{aligned}
\omega_n = f(\pi n) &= f\left(v_n^{(0)} - v\left(\frac{q^{1/2}}{\omega_n^{(0)}}\right)\right) \qquad (8.4.3)\\
&= \omega_n^{(0)} - \frac{1}{\kappa\left(\omega_n^{(0)}\right)}\, v\left(\frac{q^{1/2}}{\omega_n^{(0)}}\right) - \frac{\kappa'\left(\omega_n^{(0)}\right)}{2\kappa^3\left(\omega_n^{(0)}\right)}\, v^2\left(\frac{q^{1/2}}{\omega_n^{(0)}}\right)\\
&- \frac{1}{2\kappa^3\left(\omega_n^{(0)}\right)}\left(\left(\frac{\kappa'\left(\omega_n^{(0)}\right)}{\kappa\left(\omega_n^{(0)}\right)}\right)^2 - \frac{\kappa''\left(\omega_n^{(0)}\right)}{3\kappa\left(\omega_n^{(0)}\right)}\right) v^3\left(\frac{q^{1/2}}{\omega_n^{(0)}}\right)\\
&+ \dots,
\end{aligned}$$

where the series converges absolutely and uniformly in a certain closed disk centered at $v_n^{(0)}$.

In order to establish an asymptotic formula for the large positive zeros ω_n of the q-sine function one can consider the large n-asymptotic of the first terms in the convergent series (8.4.3). This can be done with the help of the Taylor expansions

$$\kappa(\omega) = \sum_{k=0}^{\infty} \frac{(-1)^k \omega^{2k}}{1-q^{k+1/2}} = \frac{1}{1-q^{1/2}} - \frac{\omega^2}{1-q^{3/2}} + \frac{\omega^4}{1-q^{5/2}} - \dots, \tag{8.4.4}$$

$$\begin{aligned} v(\omega) &= \sum_{k=0}^{\infty} \frac{(-1)^k \omega^{2k+1}}{(2k+1)\left(1-q^{k+1/2}\right)} \\ &= \frac{\omega}{1-q^{1/2}} - \frac{\omega^3}{3\left(1-q^{3/2}\right)} + \frac{\omega^5}{5\left(1-q^{5/2}\right)} - \dots \end{aligned} \tag{8.4.5}$$

valid for $|\omega| < 1$ as easy consequences of (8.1.1) and (8.1.6) and with the help of the following lemma.

LEMMA 8.1. *The following asymptotics hold*

$$\begin{aligned} &\frac{(q;q)_\infty^2}{\left(q^{1/2};q\right)_\infty^2} \frac{\left(-q^{1/2}\left(\omega_n^{(0)}\right)^2, -q^{1/2}/\left(\omega_n^{(0)}\right)^2; q\right)_\infty}{\left(-\left(\omega_n^{(0)}\right)^2, -q/\left(\omega_n^{(0)}\right)^2; q\right)_\infty} \\ &= \frac{q^{1/2}}{1-q^{1/2}} c_1^{-1}(q) \left(\omega_n^{(0)}\right)^{-1} \left(1 + \mathrm{O}\left(\left(\omega_n^{(0)}\right)^{-4}\right)\right), \end{aligned} \tag{8.4.6}$$

$$\begin{aligned} \kappa\left(\omega_n^{(0)}\right) &= \frac{q^{1/2}}{1-q^{1/2}} c_1^{-1}(q) \left(\omega_n^{(0)}\right)^{-1} \\ &\times \left(1 - \frac{c_1(q)}{\omega_n^{(0)}} + \frac{q}{1+q^{1/2}+q} \frac{c_1(q)}{\left(\omega_n^{(0)}\right)^3} + \mathrm{O}\left(\left(\omega_n^{(0)}\right)^{-4}\right)\right), \end{aligned} \tag{8.4.7}$$

$$\begin{aligned} \kappa'\left(\omega_n^{(0)}\right) &= -\frac{q^{1/2}}{1-q^{1/2}} c_1^{-1}(q) \left(\omega_n^{(0)}\right)^{-2} \\ &\times \left(1 - \frac{2c_1(q)}{\omega_n^{(0)}} + \frac{4q}{1+q^{1/2}+q} \frac{c_1(q)}{\left(\omega_n^{(0)}\right)^3} + \mathrm{O}\left(\left(\omega_n^{(0)}\right)^{-4}\right)\right), \end{aligned} \tag{8.4.8}$$

$$\begin{aligned} \kappa''\left(\omega_n^{(0)}\right) &= \frac{q^{1/2}}{1-q^{1/2}} \frac{\alpha(q)}{c_1(q)} \left(\omega_n^{(0)}\right)^{-3} \\ &\times \left(1 - \frac{6c_1(q)}{\alpha(q)} \frac{1}{\omega_n^{(0)}} + \mathrm{O}\left(\left(\omega_n^{(0)}\right)^{-3}\right)\right) \end{aligned} \tag{8.4.9}$$

as $n \to \infty$. *Here* $\omega_n^{(0)} = q^{1/4-n}$, $0 < q < 1$; $c_1(q)$ *is given by* (6.3.16), *and*

$$\alpha(q) = 3 - 2\frac{\left(q^{1/2};q\right)_\infty^4}{(q;q)_\infty^4}\frac{\left(q^2;q^2\right)_\infty^2}{\left(q;q^2\right)_\infty^2}\left(1 + 2^4\sum_{k=1}^{\infty}\frac{q^{3k/2}}{\left(1+q^k\right)^3}\right) + 2\frac{\left(q^{1/2};q^{1/2}\right)_\infty^4}{\left(-q^{1/2};q^{1/2}\right)_\infty^4}. \tag{8.4.10}$$

The same asymptotics are valid in the case $\varpi_n^{(0)} = q^{3/4-n}$.

PROOF. By the q-binomial formula

$$\frac{\left(-q^{1/2}/\omega^2;q\right)_\infty}{\left(-q/\omega^2;q\right)_\infty} = \sum_{k=0}^{\infty}\frac{\left(q^{-1/2};q\right)_k}{(q;q)_k}\left(-\frac{q}{\omega^2}\right)^k \tag{8.4.11}$$
$$= 1 + \frac{q^{1/2}}{1+q^{1/2}}\omega^{-2} + \mathrm{O}\left(\omega^{-4}\right), \quad |\omega| \to \infty.$$

Also

$$\frac{\left(-q^{1/2}\left(\omega_n^{(0)}\right)^2;q\right)_\infty}{\left(-\left(\omega_n^{(0)}\right)^2;q\right)_\infty} = \frac{\left(-q^{1-2n};q\right)_\infty}{\left(-q^{1/2-2n};q\right)_\infty} = \frac{\left(-q^{1-2n};q\right)_{2n}\left(-q;q\right)_\infty}{\left(-q^{1/2-2n};q\right)_{2n}\left(-q^{1/2};q\right)_\infty} \tag{8.4.12}$$

and by (I.9) of [**46**] or (1.4.4) and the q-binomial theorem

$$\frac{\left(-q^{1-2n};q\right)_{2n}}{\left(-q^{1/2-2n};q\right)_{2n}} = q^n\frac{(-1;q)_{2n}}{\left(-q^{1/2};q\right)_{2n}} = q^n\frac{\left(-q^{1/2+2n};q\right)_\infty}{\left(-q^{2n};q\right)_\infty}\frac{(-1;q)_\infty}{\left(-q^{1/2};q\right)_\infty}$$
$$= q^n\frac{(-1;q)_\infty}{\left(-q^{1/2};q\right)_\infty}\left(1 - \frac{q^{2n}}{1+q^{1/2}} + \mathrm{O}\left(q^{4n}\right)\right), \quad n \to \infty. \tag{8.4.13}$$

As a result

$$\frac{\left(-q^{1/2}\left(\omega_n^{(0)}\right)^2;q\right)_\infty}{\left(-\left(\omega_n^{(0)}\right)^2;q\right)_\infty} = 2q^{1/4}\frac{(-q;q)_\infty^2}{\left(-q^{1/2};q\right)_\infty^2} \tag{8.4.14}$$
$$\times\left(\omega_n^{(0)}\right)^{-1}\left(1 - \frac{q^{1/2}}{1+q^{1/2}}\left(\omega_n^{(0)}\right)^{-2} + \mathrm{O}\left(\left(\omega_n^{(0)}\right)^{-4}\right)\right), \quad n \to \infty.$$

Combining (8.4.11) and (8.4.14), one gets (8.4.6) with the help of (6.3.16).

Equation (8.4.7) follows now from (5.5.17), (8.4.4) and (8.4.6). In a similar manner, equation (8.4.8) follows from (8.3.3), where

$$\kappa_1(\omega) = \sum_{k=0}^{\infty}\frac{q^{k/2}}{\left(1+\omega^2 q^k\right)^2} = \frac{1}{1-q^{1/2}} - \frac{2\omega^2}{1-q^{3/2}} + \mathrm{O}\left(\omega^4\right) \tag{8.4.15}$$

as $|\omega| \to 0$.

The proof of (8.4.9) can be given on the basis of formula (8.3.8). By the q-binomial formula

$$\frac{\left(q/\omega^2;q\right)_\infty}{\left(-q/\omega^2;q\right)_\infty}=\sum_{k=0}^{\infty}\frac{(-1;q)_k}{(q;q)_k}\left(-\frac{q}{\omega^2}\right)^k \tag{8.4.16}$$

$$=1-\frac{2q}{1-q}\,\omega^{-2}+\frac{2q^2}{(1-q)^2}\,\omega^{-4}+\mathrm{O}\left(\omega^{-6}\right),\quad |\omega|\to\infty.$$

On the other hand

$$\frac{\left(\left(\omega_n^{(0)}\right)^2;q\right)_\infty}{\left(-\left(\omega_n^{(0)}\right)^2;q\right)_\infty}=\frac{\left(q^{1/2-2n};q\right)_\infty}{\left(-q^{1/2-2n};q\right)_\infty}=\frac{\left(q^{1/2-2n};q\right)_{2n}\left(q^{1/2};q\right)_\infty}{\left(-q^{1/2-2n};q\right)_{2n}\left(-q^{1/2};q\right)_\infty} \tag{8.4.17}$$

and by (I.9) of [**46**] or (1.4.4) and the q-binomial theorem

$$\frac{\left(q^{1/2-2n};q\right)_{2n}}{\left(-q^{1/2-2n};q\right)_{2n}}=\frac{\left(q^{1/2};q\right)_{2n}}{\left(-q^{1/2};q\right)_{2n}}=\frac{\left(-q^{1/2+2n};q\right)_\infty}{\left(q^{1/2+2n};q\right)_\infty}\frac{\left(q^{1/2};q\right)_\infty}{\left(-q^{1/2};q\right)_\infty} \tag{8.4.18}$$

$$=\frac{\left(q^{1/2};q\right)_\infty}{\left(-q^{1/2};q\right)_\infty}\left(1+\frac{2q^{1/2+2n}}{1-q}+\frac{2q^{1+4n}}{(1-q)^2}+\mathrm{O}\left(q^{6n}\right)\right)$$

as $n\to\infty$. Thus

$$\frac{\left(\left(\omega_n^{(0)}\right)^2;q\right)_\infty}{\left(-\left(\omega_n^{(0)}\right)^2;q\right)_\infty}=\frac{\left(q^{1/2};q\right)_\infty^2}{\left(-q^{1/2};q\right)_\infty^2} \tag{8.4.19}$$

$$\times\left(1+\frac{2q}{1-q}\left(\omega_n^{(0)}\right)^{-2}+\frac{2q^2}{(1-q)^2}\left(\omega_n^{(0)}\right)^{-4}+\mathrm{O}\left(\left(\omega_n^{(0)}\right)^{-6}\right)\right)$$

as $n\to\infty$ and combining (8.4.16) and (8.4.19), one gets

$$\frac{\left(\left(\omega_n^{(0)}\right)^2,q/\left(\omega_n^{(0)}\right)^2;q\right)_\infty}{\left(-\left(\omega_n^{(0)}\right)^2,-q/\left(\omega_n^{(0)}\right)^2;q\right)_\infty}=\frac{\left(q^{1/2};q\right)_\infty^2}{\left(-q^{1/2};q\right)_\infty^2}\left(1+\mathrm{O}\left(\left(\omega_n^{(0)}\right)^{-6}\right)\right) \tag{8.4.20}$$

as $n\to\infty$. Now asymptotic (8.4.9) follows from (8.3.8), (8.4.6) and (8.4.20). The case $\varpi_n^{(0)}=q^{3/4-n}$ can be considered in a similar fashion. □

The new improved asymptotic formula for the ω_n can be derived now from expansion (8.4.3) if we substitute

$$\frac{1}{\kappa\left(\omega_n^{(0)}\right)}\,v\left(\frac{q^{1/2}}{\omega_n^{(0)}}\right)=c_1(q)+c_1(q)^2\left(\omega_n^{(0)}\right)^{-1} \tag{8.4.21}$$

$$+ c_1(q)\left(c_1^2(q) - \frac{q}{3\left(1+q^{1/2}+q\right)}\right)\left(\omega_n^{(0)}\right)^{-2}$$
$$+ c_1^2(q)\left(c_1^2(q) - \frac{4q}{3\left(1+q^{1/2}+q\right)}\right)\left(\omega_n^{(0)}\right)^{-3} + \mathrm{O}\left(\left(\omega_n^{(0)}\right)^{-4}\right),$$

$$\frac{\kappa'\left(\omega_n^{(0)}\right)}{2\kappa^3\left(\omega_n^{(0)}\right)}\, v^2\left(\frac{q^{1/2}}{\omega_n^{(0)}}\right) \tag{8.4.22}$$
$$= -\frac{c_1^2(q)}{2\omega_n^{(0)}}\left(1 + c_1(q)\left(\omega_n^{(0)}\right)^{-1} - \frac{2q}{3\left(1+q^{1/2}+q\right)}\left(\omega_n^{(0)}\right)^{-2} + \mathrm{O}\left(\omega_n^{(0)}\right)^{-3}\right)$$

and

$$\frac{1}{2\kappa^3\left(\omega_n^{(0)}\right)}\left(\left(\frac{\kappa'\left(\omega_n^{(0)}\right)}{\kappa\left(\omega_n^{(0)}\right)}\right)^2 - \frac{\kappa''\left(\omega_n^{(0)}\right)}{3\kappa\left(\omega_n^{(0)}\right)}\right) v^3\left(\frac{q^{1/2}}{\omega_n^{(0)}}\right) \tag{8.4.23}$$
$$= -\frac{c_1^3(q)\left(\alpha(q)-3\right)}{6\left(\omega_n^{(0)}\right)^2}\left(1 + \frac{4\alpha(q)-9}{\alpha(q)-3}c_1(q)\left(\omega_n^{(0)}\right)^{-1} + \mathrm{O}\left(\omega_n^{(0)}\right)^{-2}\right)$$

as $n \to \infty$ in view of (8.4.4)–(8.4.9).

The case of the q-cosine function can be considered in a similar fashion. We summarize our results in the following main theorem of this chapter.

THEOREM 8.1. *Let $0 = \omega_0 < \omega_1 < \omega_2 < \omega_3 < \dots$ be positive zeros of $S_q(\eta;\omega)$ and let $\varpi_1 < \varpi_2 < \varpi_3 < \dots$ be positive zeros of $C_q(\eta;\omega)$ for $0 < q < 1$. Then*

$$\omega_n = \omega_n^{(0)} - c_1(q) - \frac{c_1^2(q)}{2\omega_n^{(0)}} \tag{8.4.24}$$
$$- \left(c_1^2(q)\left(2c_2(q)+3\right) - \frac{2q}{1+q^{1/2}+q}\right)\frac{c_1(q)}{6\left(\omega_n^{(0)}\right)^2} + \mathrm{O}\left(\left(\omega_n^{(0)}\right)^{-3}\right)$$

and

$$\varpi_n = \varpi_n^{(0)} - c_1(q) - \frac{c_1^2(q)}{2\varpi_n^{(0)}} \tag{8.4.25}$$
$$- \left(c_1^2(q)\left(2c_2(q)+3\right) - \frac{2q}{1+q^{1/2}+q}\right)\frac{c_1(q)}{6\left(\varpi_n^{(0)}\right)^2} + \mathrm{O}\left(\left(\varpi_n^{(0)}\right)^{-3}\right)$$

as $n \to \infty$. Here $\omega_n^{(0)} = q^{1/4-n}$, $\varpi_n^{(0)} = q^{3/4-n}$, and

$$c_1(q) = \frac{q^{1/4}}{2\left(1-q^{1/2}\right)}\,\frac{\left(q;q^2\right)_\infty^2}{\left(q^2;q^2\right)_\infty^2}, \tag{8.4.26}$$

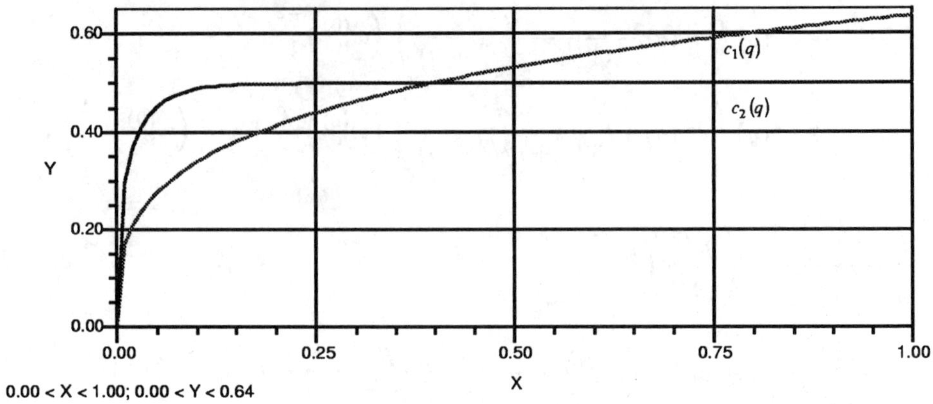

FIGURE 8.1. Functions $c_1(q)$ and $c_2(q)$, $0 \le q \le 1$.

$$c_2(q) = \frac{\left(q^{1/2};q\right)_\infty^4 \left(q^2;q^2\right)_\infty^2}{(q;q)_\infty^4 \left(q;q^2\right)_\infty^2} \left(1 + 2^4 \sum_{k=1}^{\infty} \frac{q^{3k/2}}{\left(1+q^k\right)^3}\right) - \frac{\left(q^{1/2};q^{1/2}\right)_\infty^4}{\left(-q^{1/2};q^{1/2}\right)_\infty^4}. \tag{8.4.27}$$

The functions $c_1(q)$ and $c_2(q)$ are nonnegative and increasing on $[0,1]$; see Figure 8.1.

Numerical examples demonstrating accuracy of these formulas are given in Appendix D at the end of the book. See also equations (8.5.3) and (8.5.5) below for the alternative forms of the constant $c_2(q)$. The proof of the monotonicity of the $c_1(q)$ will be given in Section 8.6.

8.5. Alternative Forms of $c_2(q)$

Expression (8.4.27) for the constant $c_2(q)$ in the asymptotic formulas (8.4.24)–(8.4.25) can be transformed to a single sum in the following manner. Due to (8.3.16) and (III.23) of [**46**] or (A.3.15)

$$
\begin{aligned}
&{}_5\varphi_4\left(\begin{matrix} \varepsilon^2, q\varepsilon, -\varepsilon, -\varepsilon, -\varepsilon^2 \\ \varepsilon, -q\varepsilon, -q\varepsilon, -q \end{matrix}; q, \frac{q^{3/2}}{\varepsilon^2}\right) \qquad (8.5.1)\\
&= {}_8\varphi_7\left(\begin{matrix} \varepsilon^2, q\varepsilon, -q\varepsilon, -\varepsilon, -\varepsilon, -\varepsilon, -\varepsilon^2, q^{1/2}\varepsilon \\ \varepsilon, -\varepsilon, -q\varepsilon, -q\varepsilon, -q\varepsilon, -q, q^{1/2}\varepsilon \end{matrix}; q, \frac{q^{3/2}}{\varepsilon^2}\right) \\
&= \frac{\left(q\varepsilon^2, q, -q^{3/2}/\varepsilon, -q^{3/2}/\varepsilon; q\right)_\infty}{\left(-q\varepsilon, -q\varepsilon, q^{3/2}, q^{3/2}/\varepsilon^2; q\right)_\infty} \\
&\quad \times {}_8\varphi_7\left(\begin{matrix} q^{1/2}, q^{5/4}, -q^{5/4}, q/\varepsilon, -q^{1/2}, -q^{1/2}/\varepsilon, -\varepsilon, -\varepsilon \\ q^{1/4}, -q^{1/4}, q^{1/2}\varepsilon, -q, -q\varepsilon, -q^{3/2}/\varepsilon, -q^{3/2}/\varepsilon \end{matrix}; q, q\right),
\end{aligned}
$$

and letting $\varepsilon \to 1^-$ one gets

$$1 + 2^4 \sum_{k=1}^{\infty} \frac{q^{3k/2}}{(1+q^k)^3} = \frac{\left(1-q^{1/2}\right)^2}{\left(1+q^{1/2}\right)^2} \frac{(q;q)_\infty^4 \left(q;q^2\right)_\infty^2}{\left(q^{1/2};q\right)_\infty^4 \left(q^2;q^2\right)_\infty^2} \tag{8.5.2}$$

$$\times\, {}_8\varphi_7 \left(\begin{matrix} q^{1/2}, q^{5/4}, -q^{5/4}, q, -q^{1/2}, -q^{1/2}, -1, -1 \\ q^{1/4}, -q^{1/4}, q^{1/2}, -q, -q, -q^{3/2}, -q^{3/2} \end{matrix} ; q, q \right)$$

$$= \frac{(q;q)_\infty^4 \left(q;q^2\right)_\infty^2}{\left(q^{1/2};q\right)_\infty^4 \left(q^2;q^2\right)_\infty^2} \left(\frac{\left(q^{1/2};q^{1/2}\right)_\infty^4}{\left(-q^{1/2};q^{1/2}\right)_\infty^4} + 4q^{1/2} \frac{\left(1+q^{1/2}+q\right)\left(1-q^{1/2}\right)^2}{\left(1+q^{1/2}\right)^2 (1+q)^2} \right.$$

$$\left. \times\, {}_8\varphi_7 \left(\begin{matrix} q^{3/2}, q^{7/4}, -q^{7/4}, q, -q, -q, -q^{1/2}, -q^{1/2} \\ q^{3/4}, -q^{3/4}, q^{3/2}, -q^{3/2}, -q^{3/2}, -q^2, -q^2 \end{matrix} ; q, q \right) \right)$$

by (II.25) of [**46**] or (A.2.11). From (8.4.27) and (8.5.2) we obtain a single series expansion for the constant $c_2(q)$ as follows

$$c_2(q) = 4q^{1/2} \left(1+q^{1/2}+q\right) \left(1-q^{1/2}\right)^2 \times \sum_{k=0}^{\infty} \frac{\left(1-q^{2k+3/2}\right) q^k}{\left(1-q^{3/2}\right)\left(1+q^{k+1/2}\right)^2 \left(1+q^{k+1}\right)^2}. \tag{8.5.3}$$

This formula is very convenient for numerical evaluation of this constant.

On the other hand, in the last line of (8.5.2)

$${}_8\varphi_7 \left(\begin{matrix} q^{3/2}, q^{7/4}, -q^{7/4}, q, -q, -q, -q^{1/2}, -q^{1/2} \\ q^{3/4}, -q^{3/4}, q^{3/2}, -q^{3/2}, -q^{3/2}, -q^2, -q^2 \end{matrix} ; q, q \right) \tag{8.5.4}$$

$$= \frac{(1+q)^2}{(1-q)\left(1-q^{3/2}\right)} \frac{\left(q;q^2\right)_\infty^2}{\left(q^2;q^2\right)_\infty^2} \, {}_4\varphi_3 \left(\begin{matrix} q, -q^{1/2}, -q^{1/2}, -q^{1/2} \\ -q^{3/4}, -q^{3/2}, -q^{3/2} \end{matrix} ; q, q^{3/2} \right)$$

by (III.23) of [**46**] or (A.3.15) and therefore

$$c_2(q) = 4q^{1/2} \frac{\left(q;q^2\right)_\infty^2}{\left(q^2;q^2\right)_\infty^2} \sum_{k=0}^{\infty} \frac{q^{3k/2}}{\left(1+q^{k+1/2}\right)^3}. \tag{8.5.5}$$

By (8.4.10) and (8.4.27)

$$\alpha(q) = 3 - 2c_2(q), \tag{8.5.6}$$

which allows to simplify the expression for this constant in view of (8.5.3) and (8.5.5).

Numerical analysis strongly indicates that

$$\lim_{q \to 1^-} c_2(q) = 1/2. \tag{8.5.7}$$

For example, Gosper's Macsyma program "**namesum**" gives $c_2(0.99999) \simeq 0.49999999999616$; see also Figure 8.1 for the graph of the $c_2(q)$.

The proof of this result can be given in the following matter. Consider the summand in (8.5.5) as function of a continuous variable, say s. Then

$$\frac{d}{ds}\frac{q^{3s/2}}{\left(1+q^{s+1/2}\right)^3}=\log q\; q^{3s/2}\frac{3\left(1-q^{s+1/2}\right)}{2\left(1+q^{s+1/2}\right)^4}<0 \tag{8.5.8}$$

for $0<q<1$ and this function is decreasing on $[0,\infty)$. Furthermore

$$\int_0^\infty \frac{q^{3s/2}\,ds}{\left(1+q^{s+1/2}\right)^3}=\sum_{k=0}^\infty\int_k^{k+1}\frac{q^{3s/2}\,ds}{\left(1+q^{s+1/2}\right)^3} \tag{8.5.9}$$

and

$$\frac{q^{3(k+1)/2}}{\left(1+q^{k+3/2}\right)^3}<\int_k^{k+1}\frac{q^{3s/2}\,ds}{\left(1+q^{s+1/2}\right)^3}<\frac{q^{3k/2}}{\left(1+q^{k+1/2}\right)^3}. \tag{8.5.10}$$

Thus

$$\begin{aligned}\sum_{k=0}^\infty\frac{q^{3k/2}}{\left(1+q^{k+1/2}\right)^3}&-\frac{1}{\left(1+q^{1/2}\right)^3}\\&<\int_0^\infty\frac{q^{3s/2}\,ds}{\left(1+q^{s+1/2}\right)^3}<\sum_{k=0}^\infty\frac{q^{3k/2}}{\left(1+q^{k+1/2}\right)^3}\end{aligned} \tag{8.5.11}$$

and

$$\int_0^\infty\frac{q^{3s/2}\,ds}{\left(1+q^{s+1/2}\right)^3}<\sum_{k=0}^\infty\frac{q^{3k/2}}{\left(1+q^{k+1/2}\right)^3}<\frac{1}{\left(1+q^{1/2}\right)^3}+\int_0^\infty\frac{q^{3s/2}\,ds}{\left(1+q^{s+1/2}\right)^3}.$$

Evaluating the integral

$$\begin{aligned}&\int_0^\infty\frac{q^{3s/2}\,ds}{\left(1+q^{s+1/2}\right)^3}\\&=\frac{1}{\log q^{-1}}\left(\frac14 q^{-3/4}\arctan q^{1/4}-q^{-1/2}\frac{1-q^{1/2}}{\left(1+q^{1/2}\right)^2}\right),\end{aligned} \tag{8.5.12}$$

we obtain the following lower and upper bounds for the sum under consideration

$$\begin{aligned}&\frac{1-q^{1/2}}{\log q^{-1}}\left(\frac14 q^{-3/4}\arctan q^{1/4}-q^{-1/2}\frac{1-q^{1/2}}{\left(1+q^{1/2}\right)^2}\right)\\&<\left(1-q^{1/2}\right)\sum_{k=0}^\infty\frac{q^{3k/2}}{\left(1+q^{k+1/2}\right)^3}\\&<\frac{1-q^{1/2}}{\left(1+q^{1/2}\right)^3}+\frac{1-q^{1/2}}{\log q^{-1}}\left(\frac14 q^{-3/4}\arctan q^{1/4}-q^{-1/2}\frac{1-q^{1/2}}{\left(1+q^{1/2}\right)^2}\right).\end{aligned} \tag{8.5.13}$$

By the Squeeze Theorem

$$\lim_{q\to 1^-}\left(1-q^{1/2}\right)\sum_{k=0}^{\infty}\frac{q^{3k/2}}{\left(1+q^{k+1/2}\right)^3} \tag{8.5.14}$$
$$=\lim_{q\to 1^-}\frac{1-q^{1/2}}{\log q^{-1}}\left(\frac{1}{4}q^{-3/4}\arctan q^{1/4}-q^{-1/2}\frac{1-q^{1/2}}{\left(1+q^{1/2}\right)^2}\right)=\frac{\pi}{32}.$$

From (8.5.5) and (6.3.3) or (8.4.26),

$$\frac{c_2(q)}{c_1(q)}=8q^{1/4}\left(1-q^{1/2}\right)\sum_{k=0}^{\infty}\frac{q^{3k/2}}{\left(1+q^{k+1/2}\right)^3}, \tag{8.5.15}$$

which implies (8.5.7) in view of (8.5.14) and (6.3.4).

The graph of the $c_2(q)$ on Figure 8.1 indicates that $\lim_{q\to 1^-} c_2'(q)=0$.

CONJECTURE 8.1. $\lim_{q\to 1^-} c_2'(q)=0$.

8.6. Monotonicity of $c_1(q)$

The graph of the $c_1(q)$ on Figure 8.1 reveals the monotonicity of this function on $[0,q]$. Here we give a rigorous proof of this result. Differentiating (8.4.26), one gets

$$\frac{dc_1(q)}{dq}=-2c_1(q)\left(\frac{d}{dq}\log\frac{(q^2;q^2)_\infty}{(q;q^2)_\infty}-\frac{\left(1+q^{1/2}\right)^2}{8q(1-q)}\right), \tag{8.6.1}$$

where

$$\frac{d}{dq}\log\frac{(q^2;q^2)_\infty}{(q;q^2)_\infty}=\frac{\frac{d}{dq}\left(\frac{(q^2;q^2)_\infty}{(q;q^2)_\infty}\right)}{\frac{(q^2;q^2)_\infty}{(q;q^2)_\infty}}=\frac{\sum_{k=1}^{\infty}\frac{k(k+1)}{2}q^{\frac{k(k+1)}{2}-1}}{\sum_{k=0}^{\infty}q^{\frac{k(k+1)}{2}}} \tag{8.6.2}$$

by the Gauss formula

$$\frac{(q^2;q^2)_\infty}{(q;q^2)_\infty}=\sum_{k=0}^{\infty}q^{\frac{k(k+1)}{2}}. \tag{8.6.3}$$

We first show that the expression in the right side of (8.6.2) can be reduced to the following single sum

$$\frac{d}{dq}\log\frac{(q^2;q^2)_\infty}{(q;q^2)_\infty}=\sum_{k=0}^{\infty}\frac{q^k}{\left(1+q^{k+1}\right)^2}. \tag{8.6.4}$$

Indeed, by the Jacobi triple product identity [**10**]

$$(q,-qx,-1/x;q)_\infty=\sum_{k=0}^{\infty}q^{\frac{k(k+1)}{2}}\left(x^k+x^{-k-1}\right) \tag{8.6.5}$$

and replacing x by $1/x$

$$(q,-q/x,-x;q)_\infty = \sum_{k=0}^{\infty} q^{\frac{k(k+1)}{2}} \left(x^k + x^{-k-1}\right). \tag{8.6.6}$$

The case $x = 1$ is

$$(q,-q,-q;q)_\infty = \sum_{k=0}^{\infty} q^{\frac{k(k+1)}{2}}. \tag{8.6.7}$$

From (8.6.5)–(8.6.7)

$$\begin{aligned}&(q,-x,-1/x;q)_\infty - (q,-1,-1;q)_\infty \\ &= \sum_{k=0}^{\infty} q^{\frac{k(k+1)}{2}} \left(x^{k+1} + x^{-k} + x^k + x^{-k-1} - 4\right).\end{aligned} \tag{8.6.8}$$

Now divide (8.6.8) by $(1-x)^2$ and let $x \to 1$ to get

$$\begin{aligned}&(q;q)_\infty \lim_{x\to1} \frac{(-x,-1/x;q)_\infty - (-1,-1;q)_\infty}{(1-x)^2} \\ &= \sum_{k=0}^{\infty} (2k\,(k+1)+1)\, q^{\frac{k(k+1)}{2}}.\end{aligned} \tag{8.6.9}$$

Introducing

$$\delta(x) = (-x;q)_\infty \tag{8.6.10}$$

one easily obtains

$$\lim_{x\to1} \frac{\delta(x)\,\delta(1/x) - \delta^2(1)}{(1-x)^2} = \delta(1)\,\delta''(1) - \left(\delta'(1)\right)^2 + \delta(1)\,\delta'(1), \tag{8.6.11}$$

where

$$\frac{\delta'(x)}{\delta(x)} = \sum_{k=0}^{\infty} \frac{q^k}{1+xq^k} \tag{8.6.12}$$

and

$$\left(\frac{\delta'(x)}{\delta(x)}\right)' = \frac{\delta(x)\,\delta''(x) - \left(\delta'(x)\right)^2}{\delta^2(x)} = -\sum_{k=0}^{\infty} \frac{q^{2k}}{(1+xq^k)^2}. \tag{8.6.13}$$

Thus

$$\delta(1)\,\delta''(1) - \left(\delta'(1)\right)^2 + \delta(1)\,\delta'(1) = (\delta(1))^2 \sum_{k=0}^{\infty} \frac{q^k}{(1+q^k)^2} \tag{8.6.14}$$

and

$$\sum_{k=1}^{\infty} \frac{k\,(k+1)}{2} q^{\frac{k(k+1)}{2}-1} = (q;q)_\infty (-q;q)_\infty^2 \sum_{k=0}^{\infty} \frac{q^k}{(1+q^{k+1})^2}, \tag{8.6.15}$$

which completes the proof of (8.6.4) by (8.6.2) and (8.6.3). See also Ex. 8.7 for another proof of (8.6.4).

Substitution of (8.6.4) into (8.6.1) results in

$$\frac{dc_1(q)}{dq} = -2c_1(q)\left(\sum_{k=0}^{\infty}\frac{q^k}{(1+q^{k+1})^2} - \frac{\left(1+q^{1/2}\right)^2}{8q(1-q)}\right), \tag{8.6.16}$$

and in order to prove the monotonicity of the $c_1(q)$ we need to show that

$$\sum_{k=0}^{\infty}\frac{q^k}{(1+q^{k+1})^2} < \frac{\left(1+q^{1/2}\right)^2}{8q(1-q)} \tag{8.6.17}$$

for $0 < q < 1$.

The last inequality can be proven in the following manner. One gets

$$\frac{q^k}{(1+q^{k+1})^2} < \frac{1}{q}\frac{q^{k+1}}{(1+q^{k+1})(1+q^{k+2})} = \frac{1}{q(1-q)}\Delta\left(\frac{1}{1+q^{k+1}}\right), \tag{8.6.18}$$

where $\Delta l(k) = l(k+1) - l(k)$. Therefore

$$\sum_{k=0}^{\infty}\frac{q^k}{(1+q^{k+1})^2} < \frac{1}{q(1-q)}\sum_{k=0}^{\infty}\Delta\left(\frac{1}{1+q^{k+1}}\right) = \frac{1}{1-q^2}. \tag{8.6.19}$$

The final step is to show that, in fact,

$$\sum_{k=0}^{\infty}\frac{q^k}{(1+q^{k+1})^2} < \frac{1}{1-q^2} \le \frac{\left(1+q^{1/2}\right)^2}{8q(1-q)} \tag{8.6.20}$$

on $[0,1]$. The last inequality is equivalent to

$$\left(1-q^{1/2}\right)^2\left(1+4q^{1/2}+q\right) \ge 0, \tag{8.6.21}$$

which holds for all $0 < q \le 1$. As a result

$$\frac{dc_1(q)}{dq} > 0, \qquad 0 < q \le 1, \tag{8.6.22}$$

which completes our proof of the monotonicity of the $c_1(q)$.

In a similar fashion

$$\frac{q^k}{(1+q^{k+1})^2} > \frac{q^k}{(1+q^k)(1+q^{k+1})} = \frac{1}{(1-q)}\Delta\left(\frac{1}{1+q^k}\right), \tag{8.6.23}$$

$$\sum_{k=0}^{\infty}\frac{q^k}{(1+q^{k+1})^2} > \frac{1}{2(1-q)}, \tag{8.6.24}$$

and in view of (8.6.19) and (8.6.24)

$$\frac{1}{2(1-q)} < \sum_{k=0}^{\infty}\frac{q^k}{(1+q^{k+1})^2} < \frac{1}{1-q^2}. \tag{8.6.25}$$

By the Squeeze Theorem then

$$\lim_{q\to 1^-}(1-q)\sum_{k=0}^{\infty}\frac{q^k}{(1+q^{k+1})^2} = \frac{1}{2}. \tag{8.6.26}$$

Numerical analysis strongly indicates that $\lim_{q\to 1^-} c_1'(q) = 1/2\pi$. For example, Gosper's Macsyma program "**namesum**" gives

$$c_1'(0.99999) \simeq 0.50000641104998/\pi.$$

This motivates the following conjecture.

CONJECTURE 8.2. $\lim_{q\to 1^-} c_1'(q) = 1/2\pi$.

8.7. Exercises for Chapter 8

(1) Rewrite relation (8.1.12) as

$$\frac{\left(q^{1/2};q^{1/2}\right)_\infty}{\left(q^2;q^2\right)_\infty}\left(-i\omega, iq^{1/2}/\omega; q^{1/2}\right)_\infty$$
$$= \left(q^{1/2}\omega^2, q^{3/2}/\omega^2; q^2\right)_\infty + i\omega\left(q^{3/2}\omega^2, q^{1/2}/\omega^2; q^2\right)_\infty$$

and derive this result as a special case of the Ex. 5.21 of [**46**]. Obtain another identity

$$\frac{\left(q^{1/2};q^{1/2}\right)^2_\infty}{\left(q^2;q^2\right)^2_\infty}\left(-\omega^2, -q/\omega^2; q\right)_\infty$$
$$= \left(q^{1/2}\omega^2, q^{3/2}/\omega^2; q^2\right)^2_\infty + \omega^2\left(q^{3/2}\omega^2, q^{1/2}/\omega^2; q^2\right)^2_\infty.$$

(2) Obtain Eqs. (8.2.4) by a direct series manipulation and as a consequence of the Lagrange inversion formula (8.2.2)–(8.2.3).

(3) Show how the limiting case $q \to 1^-$ of Eqs. (8.1.8)–(8.1.9) gives the classical sine and cosine functions in the following manner. Use (8.1.7),

$$v(\omega) = \frac{1}{2i}\log\frac{\left(-i\omega; q^{1/2}\right)_\infty}{\left(i\omega; q^{1/2}\right)_\infty} = \frac{1}{2i}\log\left(e_{q^{1/2}}(i\omega)\, E_{q^{1/2}}(i\omega)\right),$$

where the Jackson q-exponential functions,

$$e_q(z) = \sum_{n=0}^{\infty}\frac{z^n}{(q;q)_n} = \frac{1}{(z;q)_\infty}, \qquad |z|<1,$$
$$E_q(z) = \sum_{n=0}^{\infty} q^{n(n-1)/2}\frac{z^n}{(q;q)_n} = (-z;q)_\infty,$$

have the limits

$$\lim_{q\to 1^-} e_q((1-q)z) = \lim_{q\to 1^-} E_q((1-q)z) = e^z,$$

see, for example, [**46**] or Ex. 1.5, to get

$$\lim_{q\to 1^-} v\left(\left(1-q^{1/2}\right)\omega\right)$$

$$= \frac{1}{2i} \log \left(\lim_{q \to 1^-} e_{q^{1/2}} \left(i \left(1 - q^{1/2}\right) \omega \right) \, E_{q^{1/2}} \left(i \left(1 - q^{1/2}\right) \omega \right) \right)$$
$$= \frac{1}{2i} \log \left(\exp\left(2i\omega\right)\right) = \omega$$

and finally

$$\lim_{q \to 1^-} F\left(\left(1 - q^{1/2}\right)\omega\right) = \sin \omega,$$
$$\lim_{q \to 1^-} G\left(\left(1 - q^{1/2}\right)\omega\right) = \cos \omega.$$

(4) Show that

$$\frac{(q;q)_n}{(1-q)^n} = 1\,(1+q)\,\left(1+q+q^2\right) \dots \left(1+q+\dots+q^{n-1}\right)$$
$$= n! \left(1 - \frac{n\,(n-1)}{4}\,(1-q) + \mathrm{O}\left((1-q)^2\right)\right)$$

as $q \to 1^-$ and obtain, formally,

$$e_q\left((1-q)\,z\right) = e^z \left(1 + \frac{1}{4}\,(1-q)\,z^2\right) + \mathrm{O}\left((1-q)^2\right),$$
$$E_q\left((1-q)\,z\right) = e^z \left(1 - \frac{1}{4}\,(1-q)\,z^2\right) + \mathrm{O}\left((1-q)^2\right)$$

and

$$e_q\left((1-q)\,z\right) E_q\left((1-q)\,z\right) = e^{2z} + \mathrm{O}\left((1-q)^2\right), \quad q \to 1^-.$$

As a result obtain

$$v\left(\left(1-q^{1/2}\right)\omega\right)$$
$$= \frac{1}{2i} \log \left(e_{q^{1/2}} \left(i \left(1 - q^{1/2}\right) \omega \right) \, E_{q^{1/2}} \left(i \left(1 - q^{1/2}\right) \omega \right) \right)$$
$$= \frac{1}{2i} \log \left(\exp\left(2i\omega\right) \left(1 + \mathrm{O}\left(\left(1 - q^{1/2}\right)^2\right)\right)\right)$$
$$= \omega + \mathrm{O}\left(\left(1 - q^{1/2}\right)^2\right),$$

formally, as $q \to 1^-$. Show that in view of (8.1.4)–(8.1.5), this automatically leads us to the following asymptotic formulas

$$\omega_n = \pi n \left(1 - q^{1/2}\right) + \mathrm{O}\left(\left(1 - q^{1/2}\right)^3\right),$$
$$\varpi_n = \left(\frac{\pi}{2} + \pi n\right)\left(1 - q^{1/2}\right) + \mathrm{O}\left(\left(1 - q^{1/2}\right)^3\right)$$

valid for small zeros of the basic sine and cosine functions, respectively, in the limit $q \to 1^-$. Derive these asymptotics by applying Sturm–Liouville theory to equation (6.3.13).

(5) Prove Conjectures 8.1 and 8.2.

(6) Show that

$$\frac{(a;q)_\infty - (b;q)_\infty}{a-b} = -(qa;q)_\infty \; {}_2\varphi_1\left(\begin{matrix} q, \; b \\ qa \end{matrix}\; ; q, q\right)$$

and

$$\frac{d}{da}(a;q)_\infty = -(a;q)_\infty \sum_{k=0}^\infty \frac{q^k}{1-aq^k}.$$

(7) Show that

$$\sum_{n=0}^\infty \frac{(-1)^n (n+1) q^n}{1-q^{n+1}}$$
$$= \sum_{k=0}^\infty \frac{(2k+1) q^{2k}}{1-q^{2k+1}} - \sum_{k=0}^\infty \frac{(2k+2) q^{2k+1}}{1-q^{2k+2}}$$
$$= \sum_{n=0}^\infty \frac{q^n}{(1+q^{n+1})^2}, \qquad |q|<1$$

and using this identity prove that

$$\frac{d}{dq} \log \frac{(q^2;q^2)_\infty}{(q;q^2)_\infty} = \sum_{k=0}^\infty \frac{q^k}{(1+q^{k+1})^2}.$$

(8) Show that

$$\sum_{k=0}^\infty \frac{(2k+1) q^{2k+1}}{1-q^{2k+1}} = \sum_{n=0}^\infty \frac{nq^n}{1+q^n}$$
$$= \frac{1}{24}\left(16q \left(q^2;q^2\right)_\infty^4 \left(-q^2;q^2\right)_\infty^8 + \left(q^2;q^2\right)_\infty^4 \left(-q;q^2\right)_\infty^8 - 1\right).$$

[*Remark*: More Lambert series identities can be found in [**34**], [**102**], [**103**], [**104**] and [**109**].]

(9) Prove that

$$\sum_{n=-\infty}^\infty q^{n^2} = \left(-q, -q, q^2; q^2\right)_\infty, \qquad \text{(Gauss)} \tag{8.7.1}$$

$$\sum_{n=-\infty}^\infty (-1)^n q^{n^2} = \left(q, q, q^2; q^2\right)_\infty, \qquad \text{(Gauss)} \tag{8.7.2}$$

$$\sum_{n=-\infty}^\infty (-1)^n q^{n(3n+1)/2} = (q;q)_\infty, \qquad \text{(Euler)} \tag{8.7.3}$$

$$\sum_{n=0}^{\infty} q^{n(n+1)/2} = \frac{\left(q^2;q^2\right)_\infty}{\left(q;q^2\right)_\infty}, \qquad \text{(Gauss)} \tag{8.7.4}$$

$$\sum_{n=0}^{\infty} (-1)^n (2n+1) q^{n(n+1)/2} = (q;q)_\infty^3 . \qquad \text{(Jacobi)} \tag{8.7.5}$$

[*Hint*: Identities (8.7.1)–(8.7.2) are special cases $z = \pm 1$ of the Jacobi triple product identity (A.2.4); note also that (8.7.2) follows from (8.7.1) by changing q to $-q$. To get (8.7.3) replace $q \to q^{3/2}$ and set $z = -q^{1/2}$. For (8.7.4) and (8.7.5), write (A.2.4) with $z \to qz$ and then $q \to q^{1/2}$ as

$$\sum_{n=0}^{\infty} q^{n(n+1)/2} \left(z^n + z^{-n-1}\right) = (1 + 1/z)(q, -q, -q/z; q)_\infty .$$

Set $z = 1$ to obtain (8.7.4); divide by $z+1$ and let $z \to -1$ to get (8.7.5). [**10**]]

CHAPTER 9

Some Expansions in Basic Fourier Series

Our main objective in this chapter is to present explicit expansions of many elementary and q-functions in basic Fourier series. This leads, in a natural way, to a new class of formulas never investigated before from the analytical and/or numerical point of view **[26]**, **[146]**. In our presentation most of the material can be read independently from Chapters 7 and 8, but we assume that the reader is familiar with the investigation of the basic Fourier series in Chapters 5 and 6 and with the evaluation of certain integrals in Chapter 4.

9.1. Expansions of Some Polynomials

We continue to study expansion (6.7.8) of the basic analogs of the Legendre polynomials in q-Fourier series.

9.1.1. Preliminary Information. The continuous q-ultraspherical polynomials $C_m(x;\beta|q)$ are defined as

$$C_m(\cos\theta;\beta|q)=\sum_{k=0}^{m}\frac{(\beta;q)_k\,(\beta;q)_{m-k}}{(q;q)_k\,(q;q)_{m-k}}\,e^{i(m-2k)\theta}, \tag{9.1.1}$$

see, for example, **[46]**. In Section 6.7 we have established the following q-Fourier expansion,

$$\begin{aligned} C_m\left(x;q^{1/2}\middle|\,q\right) &= \pi\,\frac{\left(q^{1/2};q\right)_\infty}{(q;q)_\infty} \\ &\times\sum_{n=-\infty}^{\infty}(-i)^m\,q^{m^2/4}\,\frac{\omega_n^{-1/2}}{k(\omega_n)\left(-q\omega_n^2;q^2\right)_\infty}\,J^{(2)}_{m+1/2}(2\omega_n;q)\;\mathcal{E}_q(x;i\omega_n) \end{aligned} \tag{9.1.2}$$

with $m>0$, for a special case of these polynomials, namely, $C_m\left(x;q^{1/2}\middle|\,q\right)$, which are basic analogs of the Legendre polynomials. Here $J^{(2)}_\nu(x;q)$ is Jackson's q-Bessel function given by (4.5.3). Expansion (9.1.2) is an "inversion" of the following special case,

$$\begin{aligned} \mathcal{E}_q(x;i\omega_n) &= \frac{(q;q)_\infty}{\left(q^{1/2};q\right)_\infty}\frac{\omega_n^{-1/2}}{\left(-q\omega_n^2;q^2\right)_\infty} \\ &\times\sum_{m=0}^{\infty}i^m\left(1-q^{m+1/2}\right)\,q^{m^2/4}\,J^{(2)}_{m+1/2}(2\omega_n;q)\;C_m\left(x;q^{1/2}\middle|\,q\right), \end{aligned} \tag{9.1.3}$$

of Ismail and Zhang's formula (4.5.2).

As two special cases of the formula (9.1.2), the basic Fourier expansions for the linear and quadratic functions,

$$x=\left(q^{1/4}+q^{-1/4}\right)\sum_{n=1}^{\infty}\frac{(-1)^{n-1}}{\kappa\left(\omega_n\right)\,\omega_n}\sqrt{\frac{\left(-q\omega_n^2;q^2\right)_\infty}{\left(-\omega_n^2;q^2\right)_\infty}}\,S_q\left(x;\omega_n\right) \tag{9.1.4}$$

and

$$\begin{aligned} x^2=&\frac{\left(1+q^{1/2}\right)^2}{4\left(1+q^{1/2}+q\right)}-\frac{\left(1+q^{1/2}\right)\left(1-q^2\right)}{2q} \\ &\times\sum_{n=1}^{\infty}\frac{(-1)^{n-1}}{\kappa\left(\omega_n\right)\,\omega_n^2}\sqrt{\frac{\left(-q\omega_n^2;q^2\right)_\infty}{\left(-\omega_n^2;q^2\right)_\infty}}\,C_q\left(x;\omega_n\right), \end{aligned} \tag{9.1.5}$$

have been discussed in Section 6.8; see also [**48**] and Chapter 11 for numerical investigation of the convergence of these series using the Mascyma computer algebra system.

9.1.2. Relations with q-Lommel Polynomials. We shall continue the study of formula (9.1.2) here, emphasizing now the relation with the q-Lommel polynomials $h_{n,\nu}(x;q)=h_n^\nu(x;q)$ introduced by Ismail in [**56**]. These polynomials arise from the expansion

$$\begin{aligned} &q^{n\nu+n(n-1)/2}J_{\nu+n}^{(2)}(x;q) \\ &\quad=h_n^\nu\left(\frac{1}{x};q\right)J_\nu^{(2)}(x;q)-h_{n-1}^{\nu+1}\left(\frac{1}{x};q\right)J_{\nu-1}^{(2)}(x;q). \end{aligned} \tag{9.1.6}$$

They are generated by the three-term recurrence relation

$$h_{n+1}^\nu(x;q)=\left(1-q^{\nu+n}\right)(2x)\,h_n^\nu(x;q)-q^{\nu+n-1}h_{n-1}^\nu(x;q), \tag{9.1.7}$$

$n>0$ with

$$h_0^\nu(x;q)=1,\qquad h_1^\nu(x;q)=\left(1-q^\nu\right)(2x). \tag{9.1.8}$$

The q-Lommel polynomials $h_n^\nu(x;q)$ are orthogonal with respect to a purely discrete measure whose masses are located at the reciprocals of the zeros of $J_{\nu-1}^{(2)}(x;q)$ [**56**]; see (7.1.7)–(7.1.8). Ismail gave also an explicit formula for them,

$$h_n^\nu(x;q)=\sum_{k=0}^{[n/2]}(-1)^k q^{(\nu+k-1)k}\frac{(q;q)_{n-k}\left(q^\nu;q\right)_{n-k}}{(q;q)_k\,(q;q)_{n-2k}\left(q^\nu;q\right)_k}(2x)^{n-2k}, \tag{9.1.9}$$

which implies the symmetry relation,

$$h_n^\nu(-x;q)=(-1)^n h_n^\nu(x;q). \tag{9.1.10}$$

The q-spherical Bessel function $J^{(2)}_{m+1/2}(2\omega_n;q)$ appearing in (9.1.2) can be reduced to the q-Lommel polynomials and $J^{(2)}_{\pm 1/2}(2\omega_n;q)$ by (9.1.6),

$$q^{m^2/2} J^{(2)}_{m+1/2}(2\omega_n;q) = h^{1/2}_m\left(\frac{1}{2\omega_n};\, q\right) J^{(2)}_{1/2}(2\omega_n;q) - h^{3/2}_{m-1}\left(\frac{1}{2\omega_n};\, q\right) J^{(2)}_{-1/2}(2\omega_n;q). \tag{9.1.11}$$

But due to (2.4.25)–(2.4.26), namely,

$$S_q(\eta;\omega) = \frac{(q;q)_\infty}{(q^{1/2};q)_\infty} \frac{\omega^{1/2}}{(-q\omega^2;q^2)_\infty} J^{(2)}_{1/2}(2\omega;q), \tag{9.1.12}$$

$$C_q(\eta;\omega) = \frac{(q;q)_\infty}{(q^{1/2};q)_\infty} \frac{\omega^{1/2}}{(-q\omega^2;q^2)_\infty} J^{(2)}_{-1/2}(2\omega;q), \tag{9.1.13}$$

the basic sine $S_q(\eta;\omega)$ and basic cosine $C_q(\eta;\omega)$ functions are just multiples of $J^{(2)}_{1/2}(2\omega;q)$ and $J^{(2)}_{-1/2}(2\omega;q)$. (This is a direct analog of the well-known fact that the classical Bessel functions of the form $J_{m+1/2}(z)$ can be reduced in the same manner to elementary functions [**163**].) Also, by (5.2.30) and (5.8.10),

$$J^{(2)}_{1/2}(2\omega_n;q) = 0, \tag{9.1.14}$$

$$J^{(2)}_{-1/2}(2\omega_n;q) = (-1)^n \frac{(q^{1/2};q)_\infty}{(q;q)_\infty} (-q\omega_n^2;q^2)_\infty\, \omega_n^{-1/2} \sqrt{\frac{(-\omega_n^2;q^2)_\infty}{(-q\omega_n^2;q^2)_\infty}}. \tag{9.1.15}$$

Therefore

$$J^{(2)}_{m+1/2}(2\omega_n;q) = (-1)^{n-1} q^{-m^2/2} \frac{(q^{1/2};q)_\infty}{(q;q)_\infty} (-q\omega_n^2;q^2)_\infty \times \omega_n^{-1/2}\, h^{3/2}_{m-1}\left(\frac{1}{2\omega_n};\, q\right) \sqrt{\frac{(-\omega_n^2;q^2)_\infty}{(-q\omega_n^2;q^2)_\infty}}, \tag{9.1.16}$$

$m > 0$, which reduces the infinite series representation (4.5.3) to a finite sum in the case of the q-spherical Bessel function at the points $x = 2\omega_n$.

It is worth mentioning also that the orthogonality properties (6.7.26)–(6.7.27) for the q-Bessel functions under consideration imply the explicit orthogonality relation,

$$\left(1+(-1)^{m+p}\right) \times \sum_{n=1}^{\infty} h_m\left(\frac{1}{2\omega_n};\, q\right) h_p\left(\frac{1}{2\omega_n};\, q\right) \frac{1}{\omega_n^2\, \kappa(\omega_n)} = \frac{q^{(m+1)^2/2}}{1-q^{m+3/2}}\, \delta_{mp}, \tag{9.1.17}$$

and the dual orthogonality relation,

$$\sum_{m=0}^{\infty} h_m\left(\frac{1}{2\omega_n};q\right) h_m\left(\frac{1}{2\omega_l};q\right) \frac{1-q^{m+3/2}}{q^{(m+1)^2/2}} = \omega_n^2\, \kappa\left(\omega_n\right)\, \delta_{nl}, \qquad (9.1.18)$$

found in [**146**] for the special q-Lommel polynomials

$$h_m\left(x;q\right) := h_m^{3/2}\left(x;q\right). \qquad (9.1.19)$$

This shorter notation will be used throughout the book. The measure can also be found explicitly for the special case $\nu = 1/2$ of the q-Lommel polynomials; see Eqs. (9.1.30) and (9.1.34)–(9.1.35) below. The proof is given in Ex. 7.6.

In view of (9.1.16), the Ismail and Zhang formula (9.1.3) takes the form

$$\begin{aligned}\mathcal{E}_q\left(x;i\omega_n\right) &= \frac{(-1)^{n-1}}{\omega_n} \sum_{m=1}^{\infty} i^m \left(1-q^{m+1/2}\right) q^{-m^2/4} \\ &\times h_{m-1}\left(\frac{1}{2\omega_n};\, q\right) \sqrt{\frac{\left(-\omega_n^2;q^2\right)_\infty}{\left(-q\omega_n^2;q^2\right)_\infty}}\, C_m\left(x;q^{1/2}\middle|\, q\right), \qquad (9.1.20)\end{aligned}$$

and, finally, the q-Fourier expansion (9.1.2) can be written in terms of the q-Lommel polynomials as

$$\begin{aligned}C_m\left(x;q^{1/2}\middle|\, q\right) &= \sum_{n=-\infty}^{\infty}{}' \left(-i\right)^m \frac{q^{-m^2/4}}{\kappa\left(\omega_n\right)\omega_n} h_{m-1}\left(\frac{1}{2\omega_n};\, q\right) \\ &\times (-1)^{n-1} \sqrt{\frac{\left(-q\omega_n^2;q^2\right)_\infty}{\left(-\omega_n^2;q^2\right)_\infty}}\, \mathcal{E}_q\left(x;i\omega_n\right), \qquad m>0. \qquad (9.1.21)\end{aligned}$$

(The prime in the sum here means that the term with $n = 0$ should be omitted; we could also formally assume that $h_{-1} = 0$.)

The "odd" and "even" parts of this expansion are

$$\begin{aligned}C_{2m+1}\left(x;q^{1/2}\middle|\, q\right) &= 2\, q^{-m(m+1)-1/4} \sum_{n=1}^{\infty} \frac{(-1)^{m+n-1}}{\kappa\left(\omega_n\right)\omega_n} h_{2m}\left(\frac{1}{2\omega_n};\, q\right) \\ &\times \sqrt{\frac{\left(-q\omega_n^2;q^2\right)_\infty}{\left(-\omega_n^2;q^2\right)_\infty}}\, S_q\left(x;\omega_n\right) \qquad (9.1.22)\end{aligned}$$

and

$$\begin{aligned}C_{2m}\left(x;q^{1/2}\middle|\, q\right) &= 2q^{-m^2} \sum_{n=1}^{\infty} \frac{(-1)^{m+n-1}}{\kappa\left(\omega_n\right)\omega_n} h_{2m-1}\left(\frac{1}{2\omega_n};\, q\right) \\ &\times \sqrt{\frac{\left(-q\omega_n^2;q^2\right)_\infty}{\left(-\omega_n^2;q^2\right)_\infty}}\, C_q\left(x;\omega_n\right). \qquad (9.1.23)\end{aligned}$$

The series in (9.1.22) and (9.1.23) converge absolutely and uniformly in x by the Weierstrass M-Test and the Limit Comparison Test on every closed subinterval of $(-\eta, \eta)$ and $(-\eta_1, \eta_1)$, respectively, where $\eta = x(1/4)$, $\eta_1 = x(1/2)$ and $x(\varepsilon) = (q^{\varepsilon} + q^{-\varepsilon})/2$. This can be easily verified with the help of Lemma 6.7 and asymptotic formula (6.3.14); by Theorem 6.12 these expansions hold in the interiors of the ellipses of convergence $E_{1/4}$ and $E_{3/4}$ in the complex x-plane, respectively.

9.1.3. Further Examples. The above formulas are convenient for expansion of the elementary powers $x, x^2, x^3, x^4, \dots$ in the basic Fourier series. Eqs. (9.1.4) and (9.1.5) are obviously the special cases $m = 0$ and $m = 1$ of (9.1.22) and (9.1.23), respectively. Moreover, the case $m = 1$ of (9.1.22) gives rise to

$$
\begin{aligned}
x^3 = {} & \frac{\left(1+q^{1/2}\right)^3}{4q^{3/4}} \sum_{n=1}^{\infty} \frac{(-1)^{n-1}}{\kappa(\omega_n)\ \omega_n} \sqrt{\frac{\left(-q\omega_n^2; q^2\right)_\infty}{\left(-\omega_n^2; q^2\right)_\infty}}\ S_q(x;\omega_n) \\
& - \frac{\left(1+q^{1/2}\right)\left(1-q^2\right)\left(1-q^3\right)}{4q^{9/4}} \\
& \times \sum_{n=1}^{\infty} \frac{(-1)^{n-1}}{\kappa(\omega_n)\ \omega_n^3} \sqrt{\frac{\left(-q\omega_n^2; q^2\right)_\infty}{\left(-\omega_n^2; q^2\right)_\infty}}\ S_q(x;\omega_n). \qquad (9.1.24)
\end{aligned}
$$

The case $m = 2$ of (9.1.23) results in

$$
\begin{aligned}
x^4 = {} & \frac{\left(1+q^{1/2}\right)^2 (1-q)^2 \left(2-q^{1/2}+2q\right)}{16\left(1-q^{3/2}\right)\left(1-q^{5/2}\right)} \\
& - \frac{\left(1+q^{1/2}\right)^3 \left(1-q^2\right)\left(1-2q^{1/2}+4q-2q^{3/2}+q^2\right)}{8q^{5/2}} \\
& \times \sum_{n=1}^{\infty} \frac{(-1)^{n-1}}{\kappa(\omega_n)\ \omega_n^2} \sqrt{\frac{\left(-q\omega_n^2; q^2\right)_\infty}{\left(-\omega_n^2; q^2\right)_\infty}}\ C_q(x;\omega_n) \\
& + \frac{\left(1+q^{1/2}\right)\left(1-q^2\right)\left(1-q^3\right)\left(1-q^4\right)}{8q^4} \\
& \times \sum_{n=1}^{\infty} \frac{(-1)^{n-1}}{\kappa(\omega_n)\ \omega_n^4} \sqrt{\frac{\left(-q\omega_n^2; q^2\right)_\infty}{\left(-\omega_n^2; q^2\right)_\infty}}\ C_q(x;\omega_n). \qquad (9.1.25)
\end{aligned}
$$

In a similar fashion, one can obtain expansions of the higher order powers in the basic Fourier series using a computer algebra system. Convenient relation between (9.1.22)–(9.1.23) and q-analogs of the Bernoulli polynomials will be discussed in the next chapter; see (10.2.32)–(10.2.33). Expansions of some "generalized power functions" in the q-Fourier series are found in Section 9.3; see (9.3.19) and (9.3.22).

Eqs. (9.1.22)–(9.1.23) give us the explicit expansion of the analog of the Legendre polynomials, namely, $C_m\left(x; q^{1/2} \middle| q\right)$, in the q-Fourier series. In order to find expansion for the continuous q-ultraspherical polynomials

$C_m(x;\gamma|\, q)$ of the general form, one can just use Rogers' connection coefficient formula,

$$C_m(x;\gamma|\, q) \tag{9.1.26}$$
$$= \sum_{k=0}^{[m/2]} \beta^k \frac{(\gamma/\beta;q)_k\,(\gamma;q)_{m-k}\,\left(1-\beta q^{m-2k}\right)}{(q;q)_k\,(q\beta;q)_{m-k}\,(1-\beta)}\, C_{m-2k}(x;\beta|\, q)$$

(see, for example, (7.6.14) of [**46**]), for $\beta = q^{1/2}$ and expansions (9.1.22)–(9.1.23) together with the explicit representation for the q-Lommel polynomials (9.1.9); see Ex. 9.2 where, in particular, the q-Fourier expansions of the continuous q-Hermite polynomials are found.

In a similar manner, the connection coefficient formulas for the Askey–Wilson polynomials found in [**15**] and [**68**], see also (7.6.2)–(7.6.3) of [**46**], give rise to a q-Fourier expansion for these general classical orthogonal polynomials. We leave the details to the reader.

9.1.4. Expansions in "Modified" Basic Fourier Series. Let us discuss also somewhat similar expansions in the "modified" q-Fourier series (5.7.9)–(5.7.13). One can use the same arguments as above replacing ω_n by ϖ_n and eliminating the constant term. Analog of (9.1.2) has the form

$$C_m\left(x;q^{1/2}\middle|\, q\right) = \pi\frac{\left(q^{1/2};q\right)_\infty}{(q;q)_\infty}\sum_{n=-\infty}^{\infty}(-i)^m\, q^{m^2/4}\,\frac{\varpi_n^{-1/2}}{k(\varpi_n)\,(-q\varpi_n^2;q^2)_\infty}$$
$$\times\ J_{m+1/2}^{(2)}(2\varpi_n;q)\ \mathcal{E}_q(x;i\varpi_n),\quad m\geq 0, \tag{9.1.27}$$

which is an "inversion" of

$$\mathcal{E}_q(x;i\varpi_n) = \frac{(q;q)_\infty}{\left(q^{1/2};q\right)_\infty}\frac{\varpi_n^{-1/2}}{(-q\varpi_n^2;q^2)_\infty} \tag{9.1.28}$$
$$\times\sum_{m=0}^{\infty} i^m\left(1-q^{m+1/2}\right)q^{m^2/4}\,J_{m+1/2}^{(2)}(2\varpi_n;q)\ C_m\left(x;q^{1/2}\middle|\, q\right).$$

The relation with the q-Lommel polynomials is

$$J_{m+1/2}^{(2)}(2\varpi_n;q) = (-1)^{n-1}\,q^{-m^2/2}\frac{\left(q^{1/2};q\right)_\infty}{(q;q)_\infty}\left(-q\varpi_n^2;q^2\right)_\infty \tag{9.1.29}$$
$$\times\,\varpi_n^{-1/2}\,h_m^{1/2}\left(\frac{1}{2\varpi_n};q\right)\sqrt{\frac{(-\varpi_n^2;q^2)_\infty}{(-q\varpi_n^2;q^2)_\infty}},$$

$m\geq 0$; cf. (9.1.16). We use a shorter notation,

$$\overline{h}_m(x;q) := h_m^{1/2}(x;q), \tag{9.1.30}$$

for this special q-Lommel polynomials throughout the book.

As a result, expansion (9.1.27) can be written as

$$C_m\left(x;q^{1/2}\big|\,q\right)=\sum_{n=-\infty}^{\infty}(-i)^m\ \frac{q^{-m^2/4}}{\kappa\left(\varpi_n\right)\varpi_n}\ \overline{h}_m\left(\frac{1}{2\varpi_n};\ q\right)$$

$$\times\left(-1\right)^{n-1}\sqrt{\frac{\left(-q\varpi_n^2;q^2\right)_\infty}{\left(-\varpi_n^2;q^2\right)_\infty}}\ \mathcal{E}_q\left(x;i\varpi_n\right), \qquad (9.1.31)$$

$m\geq 0$. The "odd" and "even" parts of this expansion are

$$C_{2m+1}\left(x;q^{1/2}\big|\,q\right)=2\ q^{-m(m+1)-1/4}\sum_{n=1}^{\infty}\frac{\left(-1\right)^{m+n-1}}{\kappa\left(\varpi_n\right)\varpi_n}\ \overline{h}_{2m+1}\left(\frac{1}{2\varpi_n};\ q\right)$$

$$\times\sqrt{\frac{\left(-q\varpi_n^2;q^2\right)_\infty}{\left(-\varpi_n^2;q^2\right)_\infty}}\ S_q\left(x;\varpi_n\right) \qquad (9.1.32)$$

and

$$C_{2m}\left(x;q^{1/2}\big|\,q\right)=2q^{-m^2}\sum_{n=1}^{\infty}\frac{\left(-1\right)^{m+n-1}}{\kappa\left(\varpi_n\right)\varpi_n}\ \overline{h}_{2m}\left(\frac{1}{2\varpi_n};\ q\right)$$

$$\times\sqrt{\frac{\left(-q\varpi_n^2;q^2\right)_\infty}{\left(-\varpi_n^2;q^2\right)_\infty}}\ C_q\left(x;\varpi_n\right). \qquad (9.1.33)$$

The series in (9.1.32) and (9.1.33) converge absolutely and uniformly in x on every closed subinterval of $(-\eta_1,\eta_1)$ and $(-\eta,\eta)$, respectively, where $\eta=x\,(1/4)$, $\eta_1=x\,(1/2)$ and $x\,(\varepsilon)=\left(q^{\varepsilon}+q^{-\varepsilon}\right)/2$. This can be shown with the help of Lemma 6.7 and (6.3.15). It is not hard to see that Theorem 6.12 is also true for the "modified q-Fourier series". By the analytic continuation then expansions (9.1.32) and (9.1.33) hold inside the ellipses $E_{3/4}$ and $E_{1/4}$ in the complex x-plane, respectively.

It is worth noting the explicit orthogonality relation,

$$\left(1+(-1)^{m+p}\right) \qquad (9.1.34)$$

$$\times\sum_{n=1}^{\infty}\overline{h}_m\left(\frac{1}{2\varpi_n};\ q\right)\overline{h}_p\left(\frac{1}{2\varpi_n};\ q\right)\ \frac{1}{\varpi_n^2\ \kappa\left(\varpi_n\right)}=\frac{q^{m^2/2}}{1-q^{m+1/2}}\ \delta_{mp},$$

and the dual orthogonality relation,

$$\sum_{m=0}^{\infty}\overline{h}_m\left(\frac{1}{2\varpi_n};q\right)\overline{h}_m\left(\frac{1}{2\varpi_l};q\right)\ \frac{1-q^m}{q^{m^2/2}}=\varpi_n^2\ \kappa\left(\varpi_n\right)\ \delta_{nl}, \qquad (9.1.35)$$

for the special q-Lommel polynomials (9.1.30) found in [**146**]; see also Ex. 7.6.

Formulas (9.1.31)–(9.1.33) give us the possibility to find the expansions of the elementary powers, similar to (9.1.4), (9.1.5), (9.1.24) and (9.1.25). For example,

$$1=2\sum_{n=1}^{\infty}\frac{\left(-1\right)^{n-1}}{\kappa\left(\varpi_n\right)\ \varpi_n}\sqrt{\frac{\left(-q\varpi_n^2;q^2\right)_\infty}{\left(-\varpi_n^2;q^2\right)_\infty}}\ C_q\left(x;\varpi_n\right), \qquad (9.1.36)$$

$$x = \frac{1-q}{q^{1/4}} \sum_{n=1}^{\infty} \frac{(-1)^{n-1}}{\kappa(\varpi_n)\ \varpi_n^2} \sqrt{\frac{(-q\varpi_n^2;q^2)_\infty}{(-\varpi_n^2;q^2)_\infty}}\ S_q(x;\varpi_n), \tag{9.1.37}$$

$$\begin{aligned} x^2 = & \frac{\left(1+q^{1/2}\right)^2}{2q^{1/2}} \sum_{n=1}^{\infty} \frac{(-1)^{n-1}}{\kappa(\varpi_n)\ \varpi_n} \sqrt{\frac{(-q\varpi_n^2;q^2)_\infty}{(-\varpi_n^2;q^2)_\infty}}\ C_q(x;\varpi_n) \\ & - \frac{(1-q)\left(1-q^2\right)}{2q} \sum_{n=1}^{\infty} \frac{(-1)^{n-1}}{\kappa(\varpi_n)\ \varpi_n^3} \sqrt{\frac{(-q\varpi_n^2;q^2)_\infty}{(-\varpi_n^2;q^2)_\infty}}\ C_q(x;\varpi_n), \end{aligned} \tag{9.1.38}$$

and

$$\begin{aligned} x^3 = & \frac{\left(1-q^{1/2}\right)\left(1+q^{1/2}\right)^3\left(1-q^{1/2}+3q-q^{3/2}+q^2\right)}{4q^{7/2}} \\ & \times \sum_{n=1}^{\infty} \frac{(-1)^{n-1}}{\kappa(\varpi_n)\ \varpi_n^2} \sqrt{\frac{(-q\varpi_n^2;q^2)_\infty}{(-\varpi_n^2;q^2)_\infty}}\ S_q(x;\varpi_n) \\ & - \frac{(1-q)\left(1-q^2\right)\left(1-q^3\right)}{4q^{9/4}} \\ & \times \sum_{n=1}^{\infty} \frac{(-1)^{n-1}}{\kappa(\varpi_n)\ \varpi_n^4} \sqrt{\frac{(-q\varpi_n^2;q^2)_\infty}{(-\varpi_n^2;q^2)_\infty}}\ S_q(x;\varpi_n). \end{aligned} \tag{9.1.39}$$

We leave the details to the reader.

9.2. Basic Cosine and Sine Functions

The Fourier series of the cosine function $\cos\omega x$ on the interval $(-\pi,\pi)$ has the form

$$\cos\omega x = \frac{2}{\pi}\sin\omega\pi\left[\frac{1}{2\omega}+\sum_{n=1}^{\infty}(-1)^n\frac{\omega\cos nx}{\omega^2-n^2}\right]. \tag{9.2.1}$$

Here we shall establish a q-analog of this formula, namely, the basic Fourier expansion for $C_q(x;\omega)$, as

$$\begin{aligned} C_q(x;\omega) & = 2\ S_q(\eta;\omega) \\ & \times \left[\frac{1}{2\kappa(0)\,\omega}+\sum_{n=1}^{\infty}\frac{(-1)^n\,\omega}{\kappa(\omega_n)\left(\omega^2-\omega_n^2\right)}\sqrt{\frac{(-q\omega_n^2;q^2)_\infty}{(-\omega_n^2;q^2)_\infty}}\ C_q(x;\omega_n)\right]. \end{aligned} \tag{9.2.2}$$

Let us first derive this q-Fourier series formally and then prove its uniform convergence. The integral (5.2.32) gives the values of the q-Fourier coefficients in (9.2.2) when $\omega' = \omega_n$ are nonnegative zeros of (5.2.30). Using the uniform bounds and asymptotics for the basis functions from Lemma 6.7 and the leading term, $\lim_{n\to\infty} q^n\omega_n = q^{1/4}$, in the asymptotic formula (6.3.14) for the zeros ω_n, one can see that when $\omega \neq \omega_n$ this series converges absolutely and uniformly on every closed subinterval of $(-\eta_1,\eta_1)$

with $\eta_1 = \left(q^{1/2} + q^{-1/2}\right)/2$ by the Weierstrass M-Test and the Limit Comparison Test. So, it represents $C_q(x;\omega)$ on $[-1,1]$ due to Theorem 6.3. By Theorem 6.12 expansion (9.2.2) holds then inside the ellipse of convergence $E_{3/4}$ in the complex x-plane.

The Fourier series for the sine function $\sin\omega x$ on the interval $(-\pi,\pi)$ has the form

$$\sin\omega x = \frac{2}{\pi}\sin\omega\pi\sum_{n=1}^{\infty}(-1)^n\frac{n\sin nx}{\omega^2-n^2}. \tag{9.2.3}$$

An analog of this formula for the basic sine function $S_q(x;\omega)$ is

$$S_q(x;\omega) = 2\ S_q(\eta;\omega) \tag{9.2.4}$$
$$\times\sum_{n=1}^{\infty}\frac{(-1)^n\,\omega_n}{\kappa(\omega_n)\left(\omega^2-\omega_n^2\right)}\sqrt{\frac{(-q\omega_n^2;q^2)_\infty}{(-\omega_n^2;q^2)_\infty}}\ S_q(x;\omega_n)\,.$$

Indeed, the integral (5.2.33) results in the q-Fourier coefficients in this series when $\omega' = \omega_n$ are nonnegative zeros of (5.2.30). When $\omega \neq \omega_n$ the series converges absolutely and uniformly on every closed subinterval of $(-\eta,\eta)$ with $\eta = \left(q^{1/4}+q^{-1/4}\right)/2$, so it represents $S_q(x;\omega)$ on $[-1,1]$ and, by analytic continuation, inside the ellipse of convergence $E_{1/4}$.

In a similar manner one can obtain expansions of the q-trigonometric functions in the series (5.7.9)–(5.7.13). Expansion for $C_q(x;\omega)$ has the form

$$C_q(x;\omega) = 2\ C_q(\eta;\omega) \tag{9.2.5}$$
$$\times\left[\sum_{n=1}^{\infty}\frac{(-1)^n\,\varpi_n}{\kappa(\varpi_n)\left(\omega^2-\varpi_n^2\right)}\sqrt{\frac{(-q\varpi_n^2;q^2)_\infty}{(-\varpi_n^2;q^2)_\infty}}\ C_q(x;\varpi_n)\right].$$

The q-Fourier coefficients can be evaluated with the help of (5.2.32) when $\omega' = \varpi_n$ are positive zeros of $C_q(\eta;\omega)$. If $\omega \neq \varpi_n$ the series converges absolutely and uniformly on every closed subinterval of $(-\eta,\eta)$ by the Weierstrass M-Test and the Limit Comparison Test. Thus, it represents $C_q(x;\omega)$ on $[-1,1]$ due to Theorem 6.3. By the analytic continuation expansion (9.2.5) holds inside the ellipse of convergence $E_{1/4}$.

Expansion of the basic sine function $S_q(x;\omega)$ is

$$S_q(x;\omega) = 2\ C_q(\eta;\omega) \tag{9.2.6}$$
$$\times\sum_{n=1}^{\infty}\frac{(-1)^n\,\omega}{\kappa(\varpi_n)\left(\omega^2-\varpi_n^2\right)}\sqrt{\frac{(-q\varpi_n^2;q^2)_\infty}{(-\varpi_n^2;q^2)_\infty}}\ S_q(x;\varpi_n)$$

in view of the integral (5.2.33). When $\omega \neq \varpi_n$ the series converges absolutely and uniformly on every closed subinterval of $(-\eta_1,\eta_1)$ with $\eta_1 = \left(q^{1/2}+q^{-1/2}\right)/2$, so it represents $S_q(x;\omega)$ on $[-1,1]$ and, by the analytic continuation, inside the ellipse $E_{3/4}$.

9.3. Basic Exponential Function

In this section we present an explicit q-Fourier expansion for the basic exponential function $\mathcal{E}_q(x;\alpha)$ and some related formulas. Certain integrals evaluated by Ismail and Stanton [**71**] play an important role in this analysis.

9.3.1. Main Result. Expansion of the exponential function $\exp(\alpha x)$ in Fourier series on $(-\pi,\pi)$ is

$$e^{\alpha x} = \frac{e^{\alpha\pi}-e^{-\alpha\pi}}{\pi}\left[\frac{1}{2\alpha}+\sum_{n=1}^{\infty}\frac{(-1)^n}{\alpha^2+n^2}\left(\alpha\cos nx - n\sin nx\right)\right]. \tag{9.3.1}$$

Expansion for $\mathcal{E}_q(x;\alpha)$ in the basic Fourier series has a similar form

$$\begin{aligned}\mathcal{E}_q(x;\alpha) &= \frac{(-\alpha;q^{1/2})_\infty-(\alpha;q^{1/2})_\infty}{(q\alpha^2;q^2)_\infty} \\ &\times\left[\frac{1}{2\kappa(0)\,\alpha}+\sum_{n=1}^{\infty}\frac{(-1)^n}{\kappa(\omega_n)\left(\alpha^2+\omega_n^2\right)}\right. \\ &\times\left.\sqrt{\frac{(-q\omega_n^2;q^2)_\infty}{(-\omega_n^2;q^2)_\infty}}\left(\alpha C_q(x;\omega_n)-\omega_n S_q(x;\omega_n)\right)\right].\end{aligned} \tag{9.3.2}$$

It follows directly from the basic Fourier series (9.2.2) and (9.2.4) for the $C_q(x;\omega)$ and $S_q(x;\omega)$ and the analog of Euler's formula (2.4.1); or one can use integral (4.4.6).

Formula (9.3.2) admits further generalization, namely,

$$\begin{aligned}&\mathcal{E}_q(\cos\theta;\alpha)\,\frac{\left(q^{1/2}e^{2i\theta},\ q^{1/2}e^{-2i\theta};\ q\right)_\infty}{\left(\gamma e^{2i\theta},\ \gamma e^{-2i\theta};\ q\right)_\infty}\,C_m(\cos\theta;\gamma|q) \\ &= \frac{\pi\left(\gamma,\ \gamma q^{m+1};q\right)_\infty}{(q;q)_m\left(q,\ \gamma^2q^m;q\right)_\infty\left(q\alpha^2;q^2\right)_\infty} \\ &\times\sum_{n=-\infty}^{\infty}\frac{\left(i\alpha\omega_n q^{(m+1)/2};q\right)_\infty}{k(\omega_n)\left(-q\omega_n^2;q^2\right)_\infty}\alpha^m q^{m^2/4}\left(iq^{(1-m)/2}\omega_n/\alpha;q\right)_m\mathcal{E}_q(\cos\theta;i\omega_n) \\ &\times {}_2\varphi_2\left(\begin{matrix}iq^{(m+1)/2}\omega_n/\alpha, & -iq^{(m+1)/2}\alpha/\omega_n \\ \gamma q^{m+1}, & iq^{(m+1)/2}\alpha\omega_n\end{matrix}\ ;\ q,\ iq^{(m+1)/2}\alpha\gamma\omega_n\right),\end{aligned} \tag{9.3.3}$$

where $C_m(\cos\theta;\gamma|q)$ are the continuous q-ultraspherical polynomials. We first find the corresponding q-Fourier coefficients and then prove the uniform convergence of this series.

Ismail and Stanton [**71**] evaluated the following integral

$$\begin{aligned}&\int_0^\pi \mathcal{E}_q(\cos\theta;\alpha)\,\mathcal{E}_q(\cos\theta;\beta)\,C_m(\cos\theta;\gamma|q)\,\frac{\left(e^{2i\theta},\ e^{-2i\theta};\ q\right)_\infty}{\left(\gamma e^{2i\theta},\ \gamma e^{-2i\theta};\ q\right)_\infty}\,d\theta \\ &= \frac{2\pi\left(\gamma,\gamma q^{m+1},-\alpha\beta q^{(m+1)/2};q\right)_\infty\left(-q^{(1-m)/2}\beta/\alpha;q\right)_m}{(q;q)_m\left(q,\ \gamma^2q^m;q\right)_\infty\left(q\alpha^2,\ q\beta^2;q^2\right)_\infty}\,\alpha^m q^{m^2/4}\end{aligned}$$

$$\times {}_2\varphi_2\left(\begin{matrix} -q^{(m+1)/2}\alpha/\beta, & -q^{(m+1)/2}\beta/\alpha \\ \gamma q^{m+1}, & -q^{(m+1)/2}\alpha\beta \end{matrix}; q, -q^{(m+1)/2}\alpha\beta\gamma\right). \quad (9.3.4)$$

The proof is given in Section 4.3. The special case $\beta = 0$ takes the form

$$\int_0^\pi \mathcal{E}_q(\cos\theta;\alpha)\,\frac{\left(e^{2i\theta},\ e^{-2i\theta};\ q\right)_\infty}{\left(\gamma e^{2i\theta},\ \gamma e^{-2i\theta};\ q\right)_\infty}\,C_m(\cos\theta;\gamma|q)\ d\theta \quad (9.3.5)$$

$$= \frac{2\pi\left(\gamma,\gamma q^{m+1};q\right)_\infty}{(q;q)_m\left(q,\gamma^2q^m;q\right)_\infty\left(q\alpha^2;q^2\right)_\infty}\alpha^m q^{m^2/4}\,{}_0\varphi_1\left(\begin{matrix} - \\ \gamma q^{m+1}\end{matrix}; q, \alpha^2\gamma q^{m+1}\right).$$

Expansion (9.3.3) follows formally from the definition of the complex form of q-Fourier series (5.7.5)–(5.7.6) and the integral (9.3.4). When $0 < |\alpha| < 1$ and $0 \le \gamma < 1$ the series in the right side of (9.3.3) converges absolutely and uniformly on $[-1, 1]$ which completes the proof by Theorem 6.3. Here are the details.

9.3.2. Proof of Uniform Convergence. Let $0 < \gamma < 1$ and $0 < |\alpha| < 1$. By (III.4) of [**46**] or (A.3.4) one can rewrite (9.3.3) as

$$\mathcal{E}_q(\cos\theta;\alpha)\,\frac{\left(q^{1/2}e^{2i\theta},\ q^{1/2}e^{-2i\theta};\ q\right)_\infty}{\left(\gamma e^{2i\theta},\ \gamma e^{-2i\theta};\ q\right)_\infty}\,C_m(\cos\theta;\gamma|q) \quad (9.3.6)$$

$$= \frac{\pi\left(\gamma,\ \gamma q^{m+1},\alpha^2;q\right)_\infty}{(q;q)_m\left(q,\ \gamma^2q^m;q\right)_\infty\left(q\alpha^2;q^2\right)_\infty}\ \alpha^m q^{m^2/4}$$

$$\times\sum_{n=-\infty}^{\infty}\frac{\left(iq^{(1-m)/2}\omega_n/\alpha;q\right)_m}{k(\omega_n)\ \left(-q\omega_n^2;q^2\right)_\infty}\ \mathcal{E}_q(\cos\theta;i\omega_n)$$

$$\times\ {}_2\varphi_1\left(\begin{matrix} iq^{(m+1)/2}\omega_n/\alpha,\ iq^{(m+1)/2}\gamma\omega_n/\alpha \\ \gamma q^{m+1}\end{matrix};\ q,\ \alpha^2\right).$$

We have already discussed the uniform bounds for $\mathcal{E}_q(\cos\theta;i\omega_n)$ and the large ω-asymptotic of $k(\omega_n)$ in (6.7.20) and (6.7.22), respectively.

Introduce

$$C_n = \left(iq^{(1-m)/2}\omega_n/\alpha;q\right)_m \quad (9.3.7)$$

$$\times\ {}_2\varphi_1\left(\begin{matrix} iq^{(m+1)/2}\omega_n/\alpha, iq^{(m+1)/2}\gamma\omega_n/\alpha \\ \gamma q^{m+1}\end{matrix};q,\alpha^2\right).$$

Then

$$|C_n| \le \left(-q^{(1-m)/2}|\omega_n/\alpha|;q\right)_m$$

$$\times\sum_{k=0}^{\infty}\frac{\left(-q^{(m+1)/2}|\omega_n/\alpha|,\ -q^{(m+1)/2}|\gamma\omega_n/\alpha|;q\right)_k}{\left(q,\gamma q^{m+1};q\right)_k}\,|\alpha|^{2k}$$

$$< \left(-q^{(1-m)/2}|\omega_n/\alpha|;q\right)_m\ \frac{\left(-q^{(m+1)/2}|\gamma\omega_n/\alpha|;q\right)_\infty}{\left(\gamma q^{m+1};q\right)_\infty}$$

$$\times \sum_{k=0}^{\infty} \frac{\left(-q^{(m+1)/2}|\omega_n/\alpha|;q\right)_k}{(q;q)_k} |\alpha|^{2k}$$

and by the q-binomial theorem

$$|C_n| < \left(-q^{(1-m)/2}|\omega_n/\alpha|;q\right)_m \times \frac{\left(-q^{(m+1)/2}|\alpha\omega_n|,\ -q^{(m+1)/2}|\gamma\omega_n/\alpha|;q\right)_\infty}{\left(|\alpha|^2, \gamma q^{m+1};q\right)_\infty}. \tag{9.3.8}$$

Therefore the n-th term in the series (9.3.6) can be estimated as

$$\left|\frac{C_n\ \mathcal{E}_q(\cos\theta; i\omega_n)}{k(\omega_n)\ \left(-q\omega_n^2;q^2\right)_\infty}\right| \tag{9.3.9}$$
$$< A\ \frac{\left(-q^{1/4-\varepsilon}\ |\omega_n|\,;q^{1/2}\right)_\infty}{\kappa(\omega_n)\ \left(-\omega_n^2;q\right)_\infty}\ \left(-q^{(1-m)/2}|\omega_n/\alpha|;q\right)_m$$
$$\times \left(-q^{(m+1)/2}|\alpha\omega_n|,\ -q^{(m+1)/2}|\gamma\omega_n/\alpha|;q\right)_\infty = A\ a_n$$

for $-\left(q^{\varepsilon}+q^{-\varepsilon}\right)/2 < x < \left(q^{\varepsilon}+q^{-\varepsilon}\right)/2$ and $0 < \varepsilon < 1/2$. The value of the positive constant A here can be easily found explicitly but it is not essential for our consideration. Suppose $n \geq 0$, the case of negative n is similar. Consider

$$b_n = B\ \left(\gamma q^{-2\varepsilon}\right)^n, \quad B = \frac{\left(q^{1/2};q\right)_\infty^2}{(q;q)_\infty^2}$$

with $\gamma q^{-2\varepsilon} < 1$ and $\sum_{n=0}^{\infty} b_n < \infty$. In view of (5.5.17), (6.3.14), and (I.7) of **[46]** or (1.4.1),

$$\lim_{n\to\infty} \frac{a_n}{b_n} = \lim_{n\to\infty} \left[\frac{\left(-q^{1/4-\varepsilon}\omega_n;q^{1/2}\right)_\infty \left(-q^{(1-m)/2}|\omega_n/\alpha|;q\right)_\infty}{\left(-q^{1/2}\omega_n^2;q\right)_\infty \left(-q^{(1+m)/2}|\omega_n/\alpha|;q\right)_\infty} \right.$$
$$\left. \times\ \frac{\left(-q^{(m+1)/2}|\alpha\omega_n|, -q^{(m+1)/2}|\gamma\omega_n/\alpha|;q\right)_\infty}{\left(\gamma q^{-2\varepsilon}\right)^n}\right]$$
$$= \left(-q^{1/2-\varepsilon};q^{1/2}\right)_\infty \left(-q^{(1-m)/2+1/4}/|\alpha|;q\right)_m$$
$$\times\ \frac{\left(-q^{(m+1)/2+1/4}|\alpha|, -q^{(m+1)/2+1/4}|\gamma/\alpha|;q\right)_\infty}{(-q;q)_\infty}$$
$$\times \lim_{n\to\infty} \left[\frac{\left(-q^{1/4-\varepsilon}\omega_n;q^{1/2}\right)_{2n} \left(-q^{(1-m)/2}|\omega_n/\alpha|;q\right)_n}{\left(-q^{1/2}\omega_n^2;q\right)_{2n} \left(-q^{(1+m)/2}|\omega_n/\alpha|;q\right)_n} \right.$$
$$\left. \times\ \frac{\left(-q^{(m+1)/2}|\alpha\omega_n|, -q^{(m+1)/2}|\gamma\omega_n/\alpha|;q\right)_n}{\left(\gamma q^{-2\varepsilon}\right)^n}\right]$$
$$= \left(-q^{1/2-\varepsilon};q^{1/2}\right)_\infty \left(-q^{(1-m)/2+1/4}/|\alpha|;q\right)_m$$

$$\times \frac{\left(-q^{(m+1)/2+1/4}|\alpha|, -q^{(m+1)/2+1/4}|\gamma/\alpha|; q\right)_\infty}{(-q;q)_\infty}$$

$$\times \lim_{n\to\infty} \frac{\left(-q^{1/4+\varepsilon-n}/\omega_n; q^{1/2}\right)_{2n} \left(-q^{(1+m)/2-n}|\alpha/\omega_n|; q\right)_n}{\left(-q^{1/2-2n}/\omega_n^2; q\right)_{2n} \left(-q^{(1-m)/2-n}|\alpha/\omega_n|; q\right)_n}$$

$$\times \lim_{n\to\infty} \left(-q^{(1-m)/2-n}/|\alpha\omega_n|, -q^{(1-m)/2-n}|\alpha/\gamma\omega_n|; q\right)_n$$

$$= \frac{\left(-q^{\varepsilon}, -q^{1/2-\varepsilon}, -q^{1/4+m/2}|\alpha|, -q^{1/4-m/2}/|\alpha|; q^{1/2}\right)_\infty}{(-q,-1;q)_\infty}$$

$$\times \frac{\left(-q^{3/4+m/2}|\gamma/\alpha|, -q^{1/4-m/2}|\alpha/\gamma|; q\right)_\infty}{\left(-q^{1/4-m/2}|\alpha|, -q^{3/4+m/2}/|\alpha|; q\right)_\infty}.$$

As a result

$$0 < \lim_{n\to\infty} \frac{a_n}{b_n} < \infty$$

and the series (9.3.6) converges absolutely and uniformly on $[-1, 1]$ when $0 < \gamma < 1$ due to the Weierstrass M-Test and the Limit Comparison Test. This completes the proof of the pointwise convergence of the q-Fourier series (9.3.3) by Theorem 6.3.

9.3.3. Some Special Cases. It is worth noting a few special cases of a general expansion (9.3.3). When $m = 0$ and $\gamma = q^{1/2}$ we obtain the q-Fourier expansion (9.3.2) again. Case $\alpha = 0$ and $\gamma = q^{1/2}$ corresponds to the expansion of the q-Legendre polynomials $C_m\left(x; q^{1/2}|q\right)$ in basic Fourier series (9.1.21). When $\gamma = 0$ expansion (9.3.3) takes the form

$$\mathcal{E}_q(\cos\theta;\alpha)\left(q^{1/2}e^{2i\theta},\ q^{1/2}e^{-2i\theta};\ q\right)_\infty H_m(\cos\theta|q) \tag{9.3.10}$$
$$= \frac{\pi}{(q;q)_\infty\,(q\alpha^2;q^2)_\infty}\,\alpha^m q^{m^2/4}$$
$$\times \sum_{n=-\infty}^{\infty} \frac{\left(i\alpha\omega_n q^{(m+1)/2};q\right)_\infty}{k(\omega_n)\left(-q\omega_n^2;q^2\right)_\infty} \left(iq^{(1-m)/2}\omega_n/\alpha;q\right)_m \mathcal{E}_q(\cos\theta;i\omega_n),$$

where

$$H_m(\cos\theta|q) = (q;q)_m\ C_m(\cos\theta;0|q) \tag{9.3.11}$$

are the continuous q-Hermite polynomials. This expansion follows also directly from the integral (4.1.2).

Two special cases of (9.3.10), namely, $m = 0$,

$$\mathcal{E}_q(\cos\theta;\alpha)\left(q^{1/2}e^{2i\theta},\ q^{1/2}e^{-2i\theta};\ q\right)_\infty \tag{9.3.12}$$
$$= \frac{\pi}{(q;q)_\infty\,(q\alpha^2;q^2)_\infty} \sum_{n=-\infty}^{\infty} \frac{\left(i\alpha\omega_n q^{1/2};q\right)_\infty}{k(\omega_n)\left(-q\omega_n^2;q^2\right)_\infty}\ \mathcal{E}_q(\cos\theta;i\omega_n),$$

and $\alpha = 0$,

$$\left(q^{1/2}e^{2i\theta},\ q^{1/2}e^{-2i\theta};\ q\right)_\infty H_m(\cos\theta|q) \tag{9.3.13}$$

$$= \frac{\pi}{(q;q)_\infty} \sum_{n=-\infty}^{\infty} \frac{q^{m^2/4}(-i\omega_n)^m}{k(\omega_n)(-q\omega_n^2;q^2)_\infty} \mathcal{E}_q(\cos\theta; i\omega_n),$$

are of interest to note. Use of the generating function (3.4.15) in (9.3.12) leads to (9.3.13) again. Roger's generating relation for the continuous q-Hermite polynomials (6.1.2) and (9.3.13) results in the following expansion,

$$\frac{(q, q^{1/2}e^{2i\theta}, q^{1/2}e^{-2i\theta}; q)_\infty}{\pi(\alpha e^{i\theta},\ \alpha e^{-i\theta};\ q)_\infty} = \sum_{n=-\infty}^{\infty} \frac{\varepsilon_q(-i\alpha\omega_n q^{1/4})\,\mathcal{E}_q(\cos\theta; i\omega_n)}{k(\omega_n)(-q\omega_n^2;q^2)_\infty}, \quad (9.3.14)$$

where $|\alpha| < 1$ and

$$\varepsilon_q(x) = \sum_{m=0}^{\infty} \frac{q^{m(m-1)/4}}{(q;q)_m} x^m \quad (9.3.15)$$

is an analog of the exponential function on a q-linear grid; see Section 2.5.

Multiplying both sides of (9.3.13) by $r^m\, H_m(\cos\varphi|q)\,/\,(q;q)_m$ and then summing over m from zero to infinity with the help of the Poisson kernel for the continuous q-Hermite polynomials (3.5.4) the generating function (3.4.15), we obtain an independent proof of the bilinear generating relation for the basic exponential functions (6.5.2):

$$\sum_{n=-\infty}^{\infty} \frac{(-qr^2\omega_n^2;q^2)_\infty}{(-q\omega_n^2;q^2)_\infty\, k(\omega_n)} \mathcal{E}_q(\cos\theta; i\omega_n)\,\mathcal{E}_q(\cos\varphi; -ir\omega_n)$$

$$= \frac{(q, r^2, q^{1/2}e^{2i\theta},\ q^{1/2}e^{-2i\theta};\ q)_\infty}{\pi(re^{i\theta+i\varphi}, re^{i\theta-i\varphi}, re^{i\varphi-i\theta}, re^{-i\theta-i\varphi}; q)_\infty}, \quad |r| < 1. \quad (9.3.16)$$

A method of summation of the basic Fourier series on the basis of this explicit "Poisson-type" kernel was discussed in Section 6.5. Some integrals related to the kernels (9.3.14) and (9.3.16) will be considered in Section 9.8.

Eq. (9.3.3) gives us also the possibility to find q-Fourier expansions of "generalized power functions". The case $m = 0$ and $\alpha = 0$ results in

$$\frac{(q,\ \gamma^2,\ q^{1/2}e^{2i\theta},\ q^{1/2}e^{-2i\theta};\ q)_\infty}{\pi(\gamma,\ \gamma q,\ \gamma e^{2i\theta},\ \gamma e^{-2i\theta};\ q)_\infty} \quad (9.3.17)$$

$$= \sum_{n=-\infty}^{\infty} \frac{\mathcal{E}_q(\cos\theta; i\omega_n)}{k(\omega_n)(-q\omega_n^2;q^2)_\infty}\, {}_0\varphi_1\left(\begin{matrix} - \\ \gamma q \end{matrix}; q, -\gamma q\omega_n^2\right)$$

$$= \frac{1}{k(0)} + 2\sum_{n=1}^{\infty} \frac{C_q(\cos\theta; \omega_n)}{k(\omega_n)(-q\omega_n^2;q^2)_\infty}\, {}_0\varphi_1\left(\begin{matrix} - \\ \gamma q \end{matrix};\ q,\ -\gamma q\omega_n^2\right),$$

where the left side is, essentially, the quotient of the weight functions for the two systems of the continuous q-ultraspherical polynomials, $C_m(\cos\theta; \gamma|q)$ and $C_m(\cos\theta; q^{1/2}|q)$. When $\gamma = 0$ we arrive at the generating relation (6.5.9).

Substituting $\gamma = q^{m+1/2}$ in (9.3.17), with the help of (1.1.3)–(1.1.6), (4.5.3), and (9.1.16), one gets the q-Fourier cosine expansion for the even "generalized power function" of the form,

$$\left(q^{1/2}e^{2i\theta},\ q^{1/2}e^{-2i\theta};\ q\right)_m = \prod_{k=0}^{m-1}\left(\left(1+q^{k+1/2}\right)^2 - 4q^{k+1/2}x^2\right), \tag{9.3.18}$$

$x = \cos\theta$, namely,

$$\begin{aligned}\prod_{k=0}^{m-1}&\left(\left(1+q^{k+1/2}\right)^2 - 4q^{k+1/2}x^2\right) \qquad (9.3.19)\\ &= \frac{(q;q)_{2m}}{\left(q^{1/2},\ q^{3/2};q\right)_m} + 2\frac{(q;q)_{2m}}{\left(q^{1/2};q\right)_m}\, q^{-m^2/2}\\ &\times\sum_{n=1}^{\infty}\frac{(-1)^{n-1}}{\kappa(\omega_n)\,\omega_n^{m+1}}\, h_{m-1}\left(\frac{1}{2\omega_n};\ q\right)\sqrt{\frac{(-q\omega_n^2;q^2)_\infty}{(-\omega_n^2;q^2)_\infty}}\, C_q(x;\omega_n)\end{aligned}$$

with $m > 1$.

The case $m = 1$ and $\alpha = 0$ of (9.3.3) gives rise to the following q-Fourier expansion,

$$\begin{aligned}&\frac{\left(q,\ \gamma^2 q,\ q^{1/2}e^{2i\theta},\ q^{1/2}e^{-2i\theta};\ q\right)_\infty}{\pi\left(\gamma q,\ \gamma q^2,\ \gamma e^{2i\theta},\ \gamma e^{-2i\theta};\ q\right)_\infty}\cos\theta \qquad (9.3.20)\\ &= \sum_{n=-\infty}^{\infty}\frac{q^{1/4}\left(-i\omega_n\right)\ \mathcal{E}_q\left(\cos\theta; i\omega_n\right)}{2k(\omega_n)\left(-q\omega_n^2;q^2\right)_\infty}\ {}_0\varphi_1\left(\begin{matrix}-\\ \gamma q^2\end{matrix}\ q, -\gamma q^2\omega_n^2\right)\\ &= \sum_{n=1}^{\infty}\frac{q^{1/4}\omega_n\ S_q\left(\cos\theta;\omega_n\right)}{k(\omega_n)\left(-q\omega_n^2;q^2\right)_\infty}\ {}_0\varphi_1\left(\begin{matrix}-\\ \gamma q^2\end{matrix}\ ; q, -\gamma q^2\omega_n^2\right).\end{aligned}$$

Substituting $\gamma = q^{m-1/2}$ we obtain, in a similar manner, a "counterpart" of (9.3.19), the q-Fourier sine expansion for the odd "generalized power function" of the form,

$$\left(q^{1/2}e^{2i\theta},\ q^{1/2}e^{-2i\theta};\ q\right)_{m-1}\cos\theta = x\prod_{k=0}^{m-2}\left(\left(1+q^{k+1/2}\right)^2 - 4q^{k+1/2}x^2\right), \tag{9.3.21}$$

$x = \cos\theta$, namely,

$$\begin{aligned}x\prod_{k=0}^{m-2}&\left(\left(1+q^{k+1/2}\right)^2 - 4q^{k+1/2}x^2\right) \qquad (9.3.22)\\ &= \frac{(q;q)_{2m-1}}{\left(q^{1/2};q\right)_m}\, q^{-m^2/2+1/4}\\ &\times\sum_{n=1}^{\infty}\frac{(-1)^{n-1}}{\kappa(\omega_n)\,\omega_n^{m}}\, h_{m-1}\left(\frac{1}{2\omega_n};\ q\right)\sqrt{\frac{(-q\omega_n^2;q^2)_\infty}{(-\omega_n^2;q^2)_\infty}}\, S_q(x;\omega_n)\end{aligned}$$

with $m > 0$. One can obtain expansions (9.1.4)–(9.1.5) and (9.1.24)–(9.1.25) as the special cases $m = 1$ and $m = 2$ of (9.3.19) and (9.3.22). These formulas are very convenient for expansion of the elementary powers of the higher degrees in basic Fourier series.

On the other hand, the connection relation [68],

$$\left(\alpha e^{i\chi}, \alpha e^{-i\chi}; q\right)_m = \sum_{k=0}^{m} q^k \frac{\left(q^{-m}; q\right)_k \left(\alpha a, \alpha/a; q\right)_m}{\left(q, \alpha a, q^{1-m} a/\alpha; q\right)_k} \left(a e^{i\chi}, a e^{-i\chi}; q\right)_k, \tag{9.3.23}$$

see also (II.12) of [46] or (A.2.8), with $a = q^{1/2}$ and $\chi = 2\theta$ gives rise to one parameter extensions of (9.3.19) and (9.3.22).

9.3.4. Expansions in "Modified" Basic Fourier Series. Expansion of the $\mathcal{E}_q(x;\alpha)$ in the "modified" basic Fourier series (5.7.9), which will be needed in the next chapter, has the form

$$\mathcal{E}_q(x;\alpha) = \frac{(-\alpha; q^{1/2})_\infty + (\alpha; q^{1/2})_\infty}{(q\alpha^2; q^2)_\infty} \tag{9.3.24}$$

$$\times \sum_{n=1}^{\infty} \frac{(-1)^{n-1}}{\kappa(\varpi_n)(\alpha^2 + \varpi_n^2)} \sqrt{\frac{(-q\varpi_n^2; q^2)_\infty}{(-\varpi_n^2; q^2)_\infty}} \left(\varpi_n C_q(x;\varpi_n) + \alpha S_q(x;\varpi_n)\right).$$

It follows directly from the series (9.2.4), (9.2.5) and the q-analog of Euler's formula (2.4.1). An extension of (9.3.24) is

$$\mathcal{E}_q(\cos\theta;\alpha) \frac{\left(q^{1/2}e^{2i\theta},\ q^{1/2}e^{-2i\theta};\ q\right)_\infty}{\left(\gamma e^{2i\theta},\ \gamma e^{-2i\theta};\ q\right)_\infty} C_m(\cos\theta;\gamma|q) \tag{9.3.25}$$

$$= \frac{\pi\left(\gamma,\ \gamma q^{m+1}; q\right)_\infty}{(q;q)_m\,(q,\ \gamma^2 q^m; q)_\infty\,(q\alpha^2;q^2)_\infty}$$

$$\times \sum_{n=-\infty}^{\infty} \frac{\left(i\alpha\varpi_n q^{(m+1)/2}; q\right)_\infty \alpha^m q^{m^2/4}}{k(\omega_n)\left(-q\varpi_n^2; q^2\right)_\infty} \left(iq^{(1-m)/2}\varpi_n/\alpha; q\right)_m \mathcal{E}_q(\cos\theta; i\varpi_n)$$

$$\times\, {}_2\varphi_2\left(\begin{matrix} iq^{(m+1)/2}\varpi_n/\alpha, & -iq^{(m+1)/2}\alpha/\varpi_n \\ \gamma q^{m+1}, & iq^{(m+1)/2}\alpha\varpi_n \end{matrix}\ ;\ q,\ iq^{(m+1)/2}\alpha\gamma\varpi_n\right).$$

This follows formally from the definition of the "modified" q-Fourier series (5.7.12) and the integral (9.3.4). When $0 < |\alpha| < 1$ and $0 \le \gamma < 1$ this series converges absolutely and uniformly on $[-1,1]$ what can be shown by the same method as in the proof of convergence of (9.3.3). This completes the proof of (9.3.25) by Theorem 6.3.

The q-Fourier expansion (9.3.25) has many interesting special and limiting cases. We leave the details to the reader and discuss the q-Fourier expansions of "generalized power functions" only. The special case $m = 0$ and $\alpha = 0$ is

$$\frac{\left(q,\ \gamma^2,\ q^{1/2}e^{2i\theta},\ q^{1/2}e^{-2i\theta};\ q\right)_\infty}{\pi\left(\gamma,\ \gamma q,\ \gamma e^{2i\theta},\ \gamma e^{-2i\theta};\ q\right)_\infty} \tag{9.3.26}$$

$$= \sum_{n=-\infty}^{\infty} \frac{\mathcal{E}_q(\cos\theta; i\varpi_n)}{k(\varpi_n)(-q\varpi_n^2;q^2)_\infty} \, {}_0\varphi_1\left(\begin{matrix} - \\ \gamma q \end{matrix}; q, -\gamma q \varpi_n^2\right)$$

$$= 2\sum_{n=1}^{\infty} \frac{C_q(\cos\theta; \varpi_n)}{k(\varpi_n)(-q\varpi_n^2;q^2)_\infty} \, {}_0\varphi_1\left(\begin{matrix} - \\ \gamma q \end{matrix}; q, -\gamma q \varpi_n^2\right).$$

Substituting $\gamma = q^{m+1/2}$, one gets the q-Fourier cosine expansion for the even "generalized powers"

$$\prod_{k=0}^{m-1}\left(\left(1+q^{k+1/2}\right)^2 - 4q^{k+1/2}x^2\right) = 2\frac{(q;q)_{2m}}{\left(q^{1/2};q\right)_m} q^{-m^2/2} \tag{9.3.27}$$

$$\times \sum_{n=1}^{\infty} \frac{(-1)^{n-1}}{\kappa(\varpi_n)\varpi_n^{m+1}} \bar{h}_m\left(\frac{1}{2\varpi_n}; q\right) \sqrt{\frac{(-q\varpi_n^2;q^2)_\infty}{(-\varpi_n^2;q^2)_\infty}} C_q(x;\varpi_n).$$

In a similar fashion, letting $m=1$, $\alpha=0$ in (9.3.25) and then substituting $\gamma = q^{m-1/2}$, we obtain the q-Fourier sine expansion for the odd "generalized powers"

$$x\prod_{k=0}^{m-2}\left(\left(1+q^{k+1/2}\right)^2 - 4q^{k+1/2}x^2\right) = \frac{(q;q)_{2m-1}}{\left(q^{1/2};q\right)_m} q^{-m^2/2+1/4} \tag{9.3.28}$$

$$\times \sum_{n=1}^{\infty} \frac{(-1)^{n-1}}{\kappa(\varpi_n)\varpi_n^{m}} \bar{h}_m\left(\frac{1}{2\varpi_n}; q\right) \sqrt{\frac{(-q\varpi_n^2;q^2)_\infty}{(-\varpi_n^2;q^2)_\infty}} S_q(x;\varpi_n),$$

with $m > 0$. These formulas are very convenient for deriving the expansions (9.1.36)–(9.1.39).

9.4. Basic Cosecant and Cotangent Functions

The partial fraction decompositions for the cosecant and cotangent functions are

$$\frac{1}{\sin\omega} = \frac{1}{\omega} + \sum_{n=1}^{\infty}(-1)^n\left(\frac{1}{\omega-\pi n} + \frac{1}{\omega+\pi n}\right) \tag{9.4.1}$$

and

$$\cot\omega = \frac{1}{\omega} + \sum_{n=1}^{\infty}\left(\frac{1}{\omega-\pi n} + \frac{1}{\omega+\pi n}\right). \tag{9.4.2}$$

The special cases $x=0$ and $x=\eta=\left(q^{1/4}+q^{-1/4}\right)/2$ of the expansion of $C_q(x;\omega)$ in the basic Fourier series (9.2.2), result in natural q-extensions of these classical results

$$\frac{1}{S_q(\eta;\omega)} = \frac{1}{2\kappa(0)\,\omega} \tag{9.4.3}$$

$$+\sum_{n=1}^{\infty}\frac{(-1)^n}{\kappa(\omega_n)}\left(\frac{1}{\omega-\omega_n}+\frac{1}{\omega+\omega_n}\right)\sqrt{\frac{(-q\omega_n^2;q^2)_\infty}{(-\omega_n^2;q^2)_\infty}}$$

and

$$\begin{aligned}\operatorname{Cot}_q(\omega) &:= \frac{C_q(\eta;\omega)}{S_q(\eta;\omega)} \\ &= \frac{1}{\kappa(0)\,\omega} + \sum_{n=1}^{\infty} \frac{1}{\kappa(\omega_n)} \left(\frac{1}{\omega-\omega_n} + \frac{1}{\omega+\omega_n} \right),\end{aligned} \tag{9.4.4}$$

respectively.

These partial fraction decompositions can also be established with the help of Mittag–Leffler's theorem; see, for example, [**53**] and [**84**]. Indeed, functions $f(\omega) = 1/S_q(\eta;\omega)$ and $g(\omega) = C_q(\eta;\omega)/S_q(\eta;\omega)$ are meromorphic functions in ω with simple poles at the zeros of $S_q(\eta;\omega)$, $\omega = 0, \pm\omega_1, \pm\omega_2, \pm\omega_3, \ldots$. The residues at these poles $\omega = \omega_n$ are

$$\begin{aligned}\operatorname*{Res}_{\omega=\omega_n} f(\omega) &= \lim_{\omega\to\omega_n} \frac{\omega-\omega_n}{S_q(\eta;\omega)} = \lim_{\omega\to\omega_n} \frac{1}{\dfrac{d}{d\omega} S_q(\eta;\omega)} \\ &= \pi \frac{\left(q^{1/2};q\right)_\infty^2}{(q;q)_\infty^2} \frac{(-1)^n}{k(\omega_n)} \sqrt{\frac{(-\omega_n^2;q^2)_\infty}{(-q\omega_n^2;q^2)_\infty}}\end{aligned} \tag{9.4.5}$$

and

$$\begin{aligned}\operatorname*{Res}_{\omega=\omega_n} g(\omega) &= \lim_{\omega\to\omega_n} (\omega-\omega_n) \frac{C_q(\eta;\omega)}{S_q(\eta;\omega)} \\ &= \lim_{\omega\to\omega_n} \frac{C_q(\eta;\omega)}{\dfrac{d}{d\omega} S_q(\eta;\omega)} \\ &= \pi \frac{\left(q^{1/2};q\right)_\infty^2}{(q;q)_\infty^2} \frac{1}{k(\omega_n)} \frac{\left(-\omega_n^2;q^2\right)_\infty}{\left(-q\omega_n^2;q^2\right)_\infty},\end{aligned} \tag{9.4.6}$$

respectively. We have used here (5.8.10) and the following consequence of (5.5.14)–(5.5.15),

$$k(\omega_n) = \pi \frac{\left(q^{1/2};q\right)_\infty^2}{(q;q)_\infty^2} C_q(\eta;\omega_n) \frac{d}{d\omega} S_q(\eta;\omega_n). \tag{9.4.7}$$

To complete the proof of (9.4.3) and (9.4.4) on the basis of the Mittag–Leffler theorem, one has to show that the still undetermined additive entire functions in the partial fraction decompositions for $f(\omega)$ and $g(\omega)$ are identically zero.

Special cases $x = 0$ of (9.2.5) and $x = \eta$ of (9.2.6) result in the following partial fraction decompositions

$$\frac{1}{C_q(\eta;\omega)} = \sum_{n=1}^{\infty} \frac{(-1)^n}{\kappa(\varpi_n)} \left(\frac{1}{\omega-\varpi_n} - \frac{1}{\omega+\varpi_n} \right) \sqrt{\frac{(-q\varpi_n^2;q^2)_\infty}{(-\varpi_n^2;q^2)_\infty}} \tag{9.4.8}$$

and

$$\operatorname{Tan}_q(\omega) := \frac{S_q(\eta;\omega)}{C_q(\eta;\omega)} = -\sum_{n=1}^{\infty} \frac{1}{\kappa(\varpi_n)} \left(\frac{1}{\omega - \varpi_n} + \frac{1}{\omega + \varpi_n} \right), \tag{9.4.9}$$

respectively.

9.5. Some Consequences of Parseval's Identity

Parseval's formula,

$$\begin{aligned} 2k(0)\, a_0^2 &+ \sum_{n=1}^{\infty} k(\omega_n)\, \left(a_n^2 + b_n^2\right) \\ &= \sum_{n=-\infty}^{\infty} 2k(\omega_n)\, |c_n|^2 = \int_{-1}^{1} |f(x)|^2\, \rho(x)\, dx, \end{aligned} \tag{9.5.1}$$

holds for basic Fourier series due to the completeness of the q-trigonometric system $\{\mathcal{E}_q(x; i\omega_n)\}_{n=-\infty}^{\infty}$; see Section 6.2. Here c_n are the q-Fourier coefficients of $f(x)$ defined by (5.7.6), $2k(\omega_n)$ and $\rho(x)$ are the $\mathcal{L}^2$-norm and the weight function in the orthogonality relations (5.2.26)–(5.2.28), respectively. Eq. (9.5.1) gives rise to some new series. We consider only a few examples here.

Expansion of the basic cosine function $C_q(x;\omega)$ in basic Fourier series (9.2.2) and (9.5.1) result in the following partial fraction decomposition

$$\begin{aligned} \frac{1}{2k(0)\,\omega^2} &+ \sum_{n=1}^{\infty} \frac{\omega^2}{k(\omega_n)\,(\omega^2-\omega_n^2)^2}\, \frac{(-\omega_n^2;q^2)_\infty}{(-q\omega_n^2;q^2)_\infty} \\ &= \frac{(q;q)_\infty^4}{4\pi^2\,(q^{1/2};q)_\infty^4} \left[\frac{k(\omega)}{S_q^2(\eta;\omega)} + \frac{\pi\,(q^{1/2};\,q)_\infty^2}{\omega\,(q;\,q)_\infty^2}\, \frac{C_q(\eta;\omega)}{S_q(\eta;\omega)} \right] \end{aligned} \tag{9.5.2}$$

in view of the integral (5.5.10), namely,

$$\begin{aligned} \int_0^{\pi} C_q^2(\cos\theta;\omega)\, &\left(e^{2i\theta}, e^{-2i\theta};\, q\right)_{1/2}\, d\theta \\ &= k(\omega) + \frac{\pi\,(q^{1/2};q)_\infty^2}{\omega\,(q;q)_\infty^2}\, C_q(\eta;\omega)\, S_q(\eta;\omega). \end{aligned} \tag{9.5.3}$$

Expansion of the basic sine function $S_q(x;\omega)$ in basic Fourier series (9.2.4) together with (9.5.1) gives another partial fraction decomposition

$$\begin{aligned} \sum_{n=1}^{\infty} &\frac{\omega_n^2}{k(\omega_n)\,(\omega^2-\omega_n^2)^2}\, \frac{(-\omega_n^2;q^2)_\infty}{(-q\omega_n^2;q^2)_\infty} \\ &= \frac{(q;q)_\infty^4}{4\pi^2\,(q^{1/2};q)_\infty^4} \left[\frac{k(\omega)}{S_q^2(\eta;\omega)} - \frac{\pi\,(q^{1/2};\,q)_\infty^2}{\omega\,(q;\,q)_\infty^2}\, \frac{C_q(\eta;\omega)}{S_q(\eta;\omega)} \right] \end{aligned} \tag{9.5.4}$$

due to the integral (5.5.11),

$$\int_0^{\pi} S_q^2(\cos\theta;\omega)\ \left(e^{2i\theta},\ e^{-2i\theta};\ q\right)_{1/2}\ d\theta \tag{9.5.5}$$
$$= k(\omega) - \frac{\pi\left(q^{1/2};q\right)_\infty^2}{\omega\,(q;q)_\infty^2}\ C_q(\eta;\omega)\,S_q(\eta;\omega).$$

Subtracting (9.5.4) from (9.5.2) we obtain expansion (9.4.4) once again. This gives an independent proof of this formula. The addition of two equations (9.5.2) and (9.5.4) leads to a new result,

$$\frac{1}{2k(0)\,\omega^2} + \sum_{n=1}^{\infty} \frac{\left(\omega^2+\omega_n^2\right)}{k(\omega_n)\left(\omega^2-\omega_n^2\right)^2}\ \frac{\left(-\omega_n^2;q^2\right)_\infty}{\left(-q\omega_n^2;q^2\right)_\infty} \tag{9.5.6}$$
$$= \frac{(q;q)_\infty^4}{2\pi^2\left(q^{1/2};q\right)_\infty^4}\ \frac{k(\omega)}{S_q^2(\eta;\omega)}.$$

Writing

$$\frac{\left(\omega^2+\omega_n^2\right)}{\left(\omega^2-\omega_n^2\right)^2} = \frac{1}{2}\left[\frac{1}{\left(\omega-\omega_n\right)^2} + \frac{1}{\left(\omega+\omega_n\right)^2}\right],$$

one gets, finally,

$$\sum_{n=-\infty}^{\infty} \frac{1}{\kappa(\omega_n)\left(\omega-\omega_n\right)^2} = \frac{\left(-\omega^2;q^2\right)_\infty}{\left(-q\omega^2;q^2\right)_\infty}\ \frac{\kappa(\omega)}{S_q^2(\eta;\omega)}, \tag{9.5.7}$$

as a q-analog of

$$\sum_{n=-\infty}^{\infty} \frac{1}{\left(\omega-\pi n\right)^2} = \frac{1}{\sin^2\omega}. \tag{9.5.8}$$

It is worth noting that, as in the classical case, expansion (9.5.7) can be obtain by a formal differentiation of the both sides of (9.4.4) using (6.3.6)–(6.3.7) and (6.3.10).

In a similar fashion, expansion of the basic exponential function $\mathcal{E}_q(x;\alpha)$ in basic Fourier series (9.3.2) results in the following partial fraction decomposition

$$\frac{1}{2\kappa(0)\,\alpha} + \sum_{n=1}^{\infty} \frac{\alpha}{\kappa(\omega_n)\left(\alpha^2+\omega_n^2\right)} = \frac{1}{2}\,\frac{\left(-\alpha;q^{1/2}\right)_\infty + \left(\alpha;q^{1/2}\right)_\infty}{\left(-\alpha;q^{1/2}\right)_\infty - \left(\alpha;q^{1/2}\right)_\infty}. \tag{9.5.9}$$

Here we have used integral (4.4.6),

$$\int_0^{\pi} \mathcal{E}_q(\cos\theta;\alpha)\ \mathcal{E}_q(\cos\theta;\beta)\ \left(e^{2i\theta},\ e^{-2i\theta};\ q\right)_{1/2}\ d\theta \tag{9.5.10}$$
$$= \pi\frac{\left(q^{1/2};q\right)_\infty^2}{(q;q)_\infty^2}\ \frac{\left(-\alpha,\ -\beta;q^{1/2}\right)_\infty - \left(\alpha,\ \beta;q^{1/2}\right)_\infty}{(\alpha+\beta)\left(q\alpha^2,\ q\beta^2;q^2\right)_\infty}.$$

Expansion (9.4.4) for the basic cotangent function appears once again if one changes α by $i\omega$. This gives another proof of this partial fraction decomposition.

Parseval's identity (9.5.1) for expansions (9.1.4) and (9.1.5) results in

$$\sum_{n=1}^{\infty} \frac{1}{\kappa\left(\omega_n\right)\ \omega_n^2} = \frac{q^{1/2}}{2\left(1-q^{3/2}\right)} \tag{9.5.11}$$

and

$$\sum_{n=1}^{\infty} \frac{1}{\kappa\left(\omega_n\right)\ \omega_n^4} = \frac{q^2}{2\left(1-q^{3/2}\right)^2\left(1-q^{5/2}\right)}, \tag{9.5.12}$$

respectively. We have used (9.1.5) and (9.1.25) in order to evaluate the integrals; see [**48**] and Chapter 11 for numerical investigation of the convergence of these series. Relations (9.5.11) and (9.5.12) are q-analogs of Euler's sums

$$\sum_{n=1}^{\infty} \frac{1}{n^2} = \frac{\pi^2}{6}, \qquad \sum_{n=1}^{\infty} \frac{1}{n^4} = \frac{\pi^4}{90}, \tag{9.5.13}$$

respectively. See further discussion of these and related series in Section 10.6.

We can also derive (9.5.11) if we rewrite (9.5.7) in the form

$$\sum_{n=1}^{\infty} \frac{2}{\kappa\left(\omega_n\right)\left(\omega-\omega_n\right)^2} = \frac{\left(-\omega^2;q^2\right)_\infty}{\left(-q\omega^2;q^2\right)_\infty} \frac{\kappa\left(\omega\right)}{S_q^2(\eta;\omega)} - \frac{1}{\kappa\left(0\right)\omega^2} \tag{9.5.14}$$

and then take the limit of the both sides as $\omega \to 0$. The termwise limit of the left side gives the series in (9.5.11) up to a constant multiple. In order to take the limit in the right side of (9.5.14), we use the Laurent expansion

$$\frac{\left(-\omega^2;q^2\right)_\infty}{\left(-q\omega^2;q^2\right)_\infty} \frac{\kappa\left(\omega\right)}{S_q^2(\eta;\omega)} = \frac{1-q^{1/2}}{\omega^2} + \frac{q^{1/2}}{1-q^{3/2}} + O\left(\omega^2\right) \tag{9.5.15}$$

as $\omega \to 0$, which can be obtain from (5.5.4) and the second equation in Ex. 5.1. The summation formula (9.5.11) follows also from (9.5.9) in a similar fashion, using the constant term in the Laurent expansion of the right side at $\alpha = 0$.

It is worth mentioning, finally, that the Parseval identity (9.5.1) for the expansions (9.1.22)–(9.1.23) gives the $\mathcal{L}^2$-norm of the q-Lommel polynomials (9.1.19) in the orthogonality relation (9.1.17). Similar considerations hold in the case of the basic Fourier series (5.7.12); we leave the details to the reader.

9.6. More Expansions

Here we derive a few more q-Fourier series related to the Al-Salam and Chihara polynomials and to the continuous q-ultraspherical polynomials.

9.6.1. Al-Salam and Chihara Polynomials. The Al-Salam and Chihara polynomials

$$p_m(x;a,b) = p_m(x;a,b|q) = p_m(x;a,b,0,0|q) \tag{9.6.1}$$

are given by (4.7.1); see **[7]**, **[15]**, and **[86]**. They have the generating function (4.7.2). Ismail and Stanton **[71]** evaluated integral (4.7.6), which implies the following q-Fourier expansion,

$$\begin{aligned}
&\mathcal{E}_q(\cos\theta;\alpha)\,\frac{\left(q^{1/2}e^{2i\theta},\ q^{1/2}e^{-2i\theta};\ q\right)_\infty}{\left(a^2e^{2i\theta},\ a^2e^{-2i\theta};\ q^2\right)_\infty}\,p_m(\cos\theta;a,-a|q) \qquad (9.6.2)\\
&= \frac{\pi}{(q,-a^2q^m;q)_\infty\,(q\alpha^2;q^2)_\infty}\\
&\times \sum_{n=-\infty}^{\infty}\frac{\left(i\alpha\omega_n q^{(m+1)/2};q\right)_\infty}{k(\omega_n)\left(-q\omega_n^2;q^2\right)_\infty}\alpha^m q^{m^2/4}\left(iq^{(1-m)/2}\omega_n/\alpha;q\right)_m\mathcal{E}_q(\cos\theta;i\omega_n)\\
&\times {}_2\varphi_2\left(\begin{matrix} iq^{(m+1)/2}\omega_n/\alpha, & -iq^{(m+1)/2}\alpha/\omega_n\\ -q, & iq^{(m+1)/2}\alpha\omega_n\end{matrix};q,\ iq^{(m+1)/2}\alpha a^2\omega_n\right),
\end{aligned}$$

somewhat similar to (9.3.3). Using the generating relation (4.7.2) in (9.6.2), we arrive at

$$\begin{aligned}
&\mathcal{E}_q(\cos\theta;\alpha)\,\frac{\left(q\alpha^2,\ a^2t^2;\ q^2\right)_\infty\left(q,q^{1/2}e^{2i\theta},\ q^{1/2}e^{-2i\theta};\ q\right)_\infty}{\pi\left(a^2e^{2i\theta},\ a^2e^{-2i\theta};\ q^2\right)_\infty\left(te^{i\theta},\ te^{-i\theta};\ q\right)_\infty} \qquad (9.6.3)\\
&= \sum_{n=-\infty}^{\infty}\frac{1}{k(\omega_n)\ \left(-q\omega_n^2;q^2\right)_\infty}\,\mathcal{E}_q(\cos\theta;i\omega_n)\\
&\times \sum_{m=0}^{\infty}\frac{\left(iq^{(1-m)/2}\omega_n/\alpha;q\right)_m\left(i\alpha\omega_n q^{(m+1)/2};q\right)_\infty}{(q;q)_m\left(-a^2q^m;q\right)_\infty}\ (\alpha t)^m\,q^{m^2/4}\\
&\times {}_2\varphi_2\left(\begin{matrix} iq^{(m+1)/2}\omega_n/\alpha, & -iq^{(m+1)/2}\alpha/\omega_n\\ -q, & iq^{(m+1)/2}\alpha\omega_n\end{matrix};q,iq^{(m+1)/2}\alpha a^2\omega_n\right).
\end{aligned}$$

9.6.2. Continuous q-Ultraspherical Polynomials. Using Rogers' generating function for the continuous q-ultraspherical polynomials,

$$\sum_{m=0}^{\infty}C_m(\cos\theta;\gamma|q)\ t^m = \frac{\left(\gamma te^{i\theta},\ \gamma te^{-i\theta};q\right)_\infty}{\left(te^{i\theta},\ te^{-i\theta};q\right)_\infty}, \tag{9.6.4}$$

$|t|<1$; see, for example, **[46]**; and (9.3.3), one gets the following expansion

$$\begin{aligned}
&\mathcal{E}_q(\cos\theta;\alpha)\,\frac{\left(q,\ q^{1/2}e^{2i\theta},\ q^{1/2}e^{-2i\theta},\ \gamma te^{i\theta},\ \gamma te^{-i\theta};\ q\right)_\infty}{\pi\left(\gamma,\ \gamma e^{2i\theta},\ \gamma e^{-2i\theta},\ te^{i\theta},\ te^{-i\theta};\ q\right)_\infty} \qquad (9.6.5)\\
&= \sum_{n=-\infty}^{\infty}\frac{\left(q\alpha^2;q^2\right)_\infty^{-1}}{k(\omega_n)\ \left(-q\omega_n^2;q^2\right)_\infty}\,\mathcal{E}_q(\cos\theta;i\omega_n)
\end{aligned}$$

$$\times \sum_{m=0}^{\infty} \frac{\left(iq^{(1-m)/2}\omega_n/\alpha;q\right)_m \left(\gamma q^{m+1},\ i\alpha\omega_n q^{(m+1)/2};q\right)_\infty}{(q;q)_m\ (\gamma^2 q^m;q)_\infty} q^{m^2/4}(\alpha t)^m$$

$$\times {}_2\varphi_2\left(\begin{matrix} iq^{(m+1)/2}\omega_n/\alpha, & -iq^{(m+1)/2}\alpha/\omega_n \\ \gamma q^{m+1}, & iq^{(m+1)/2}\alpha\omega_n \end{matrix} ;q, iq^{(m+1)/2}\alpha\gamma\omega_n\right).$$

The continuous q-ultraspherical polynomials $C_m(\cos\theta;\gamma|q)$ have also the following bilinear generating relation,

$$\sum_{m=0}^{\infty} \frac{(q;q)_m}{(\gamma;q)_m} C_m(\cos\theta;\gamma|q)\ C_m(\cos\varphi;\gamma|q)\ t^m \tag{9.6.6}$$

$$\times {}_2\varphi_1\left(\begin{matrix} \gamma,\ \gamma^2 q^m \\ \gamma q^{m+1} \end{matrix} ;\ q,\ t^2\right)$$

$$= \frac{\left(\gamma te^{i\theta+i\varphi}, \gamma te^{i\theta-i\varphi}, \gamma te^{i\varphi-i\theta}, \gamma te^{-i\theta-i\varphi};q\right)_\infty}{\left(te^{i\theta+i\varphi}, te^{i\theta-i\varphi}, te^{i\varphi-i\theta}, te^{-i\theta-i\varphi};q\right)_\infty},$$

$|t| < 1$; see **[24]**, **[70]**, and **[115]**. Expansions (9.3.3) and (9.6.6) give

$$\frac{\left(q, \gamma te^{i\theta+i\varphi}, \gamma te^{i\theta-i\varphi}, \gamma te^{i\varphi-i\theta}, \gamma te^{-i\theta-i\varphi};\ q\right)_\infty}{\pi\left(\gamma, te^{i\theta+i\varphi}, te^{i\theta-i\varphi}, te^{i\varphi-i\theta}, te^{-i\theta-i\varphi};\ q\right)_\infty} \tag{9.6.7}$$

$$\times \frac{\left(q^{1/2}e^{2i\theta}, q^{1/2}e^{-2i\theta};q\right)_\infty}{\left(\gamma e^{2i\theta}, \gamma e^{-2i\theta};q\right)_\infty} \mathcal{E}_q(\cos\theta;\alpha)$$

$$= \sum_{n=-\infty}^{\infty} \frac{\left(q\alpha^2;q^2\right)_\infty^{-1}}{k(\omega_n)\ \left(-q\omega_n^2;q^2\right)_\infty} \mathcal{E}_q(\cos\theta;i\omega_n)$$

$$\times \sum_{m=0}^{\infty} \frac{\left(iq^{(1-m)/2}\omega_n/\alpha;q\right)_m \left(\gamma q^{m+1}, i\alpha\omega_n q^{(m+1)/2};q\right)_\infty}{(\gamma^2 q^m;q)_\infty (\gamma;q)_m} q^{m^2/4}(\alpha t)^m$$

$$\times\ C_m(\cos\varphi;\gamma|q)\ {}_2\varphi_1\left(\begin{matrix} \gamma,\ \gamma^2 q^m \\ \gamma q^{m+1} \end{matrix} ;\ q,\ t^2\right)$$

$$\times {}_2\varphi_2\left(\begin{matrix} iq^{(m+1)/2}\omega_n/\alpha, & -iq^{(m+1)/2}\alpha/\omega_n \\ \gamma q^{m+1}, & iq^{(m+1)/2}\alpha\omega_n \end{matrix} ;q, iq^{(m+1)/2}\alpha\gamma\omega_n\right).$$

9.7. Even More Expansions

In this section we shall establish several more explicit q-Fourier expansions with the help of the Askey–Wilson integral (A.4.1).

Consider the basic Fourier series (5.7.5),

$$f(x) = \sum_{n=-\infty}^{\infty} c_n\ \mathcal{E}_q(x;i\omega_n) \tag{9.7.1}$$

for some function $f(x)$ defined at $(-1,1)$ with the coefficients given by

$$c_n = \frac{1}{2k(\omega_n)} \int_0^\pi f(\cos\theta)\ \mathcal{E}_q(\cos\theta;-i\omega_n)\left(e^{2i\theta}, e^{-2i\theta};q\right)_{1/2} d\theta, \tag{9.7.2}$$

where $\omega_n = 0, \pm\omega_1, \pm\omega_2, \pm\omega_3, ...$ and $\omega_0 = 0 < \omega_1 < \omega_2 < \omega_3 < ...$ are nonnegative zeros of the basic sine function $S_q(\eta;\omega)$; the normalization constants $k(\omega_n)$ are defined by (5.5.7).

In view of (2.3.27)–(2.3.28), one gets

$$2k(\omega_n)\, c_n = \frac{\left(i\omega_n; q^{1/2}\right)_\infty}{\left(-q\omega_n^2; q^2\right)_\infty} \sum_{k=0}^{\infty} \frac{(i\omega_n)^k}{(q;q)_k} \tag{9.7.3}$$
$$\times \int_0^\pi f(\cos\theta) \left(q^{1/4}e^{i\theta}, q^{1/4}e^{-i\theta}; q^{1/2}\right)_k \left(e^{2i\theta}, e^{-2i\theta}; q\right)_{1/2} d\theta.$$

Denote this integral as I_k and transform all infinite products to the base $q^{1/2}$ as follows

$$I_k = \int_0^\pi \frac{f(\cos\theta)}{\left(q^{1/4}e^{i\theta}, q^{1/4}e^{-i\theta}, -q^{1/4}e^{i\theta}, -q^{1/4}e^{-i\theta}; q^{1/2}\right)_\infty} \tag{9.7.4}$$
$$\times \frac{\left(e^{2i\theta}, e^{-2i\theta}; q^{1/2}\right)_\infty}{\left(-q^{1/4}e^{i\theta}, -q^{1/4}e^{-i\theta}, q^{1/4+k/2}e^{i\theta}, q^{1/4+k/2}e^{-i\theta}; q^{1/2}\right)_\infty} d\theta$$

with the help of (I.25) and (I.28) of [**46**] or (1.4.12) and (1.4.13), respectively.

If we now choose

$$f(\cos\theta) \tag{9.7.5}$$
$$= \frac{\left(q^{1/4}e^{i\theta}, q^{1/4}e^{-i\theta}, -q^{1/4}e^{i\theta}, -q^{1/4}e^{-i\theta}, -q^{1/4}e^{i\theta}, -q^{1/4}e^{-i\theta}; q^{1/2}\right)_\infty}{\left(ae^{i\theta}, ae^{-i\theta}, be^{i\theta}, be^{-i\theta}, ce^{i\theta}, ce^{-i\theta}; q^{1/2}\right)_\infty}$$

where $\max(|a|, |b|, |c|) < 1$, then the resulting integral in (9.7.4) can be evaluated by means of the Askey–Wilson integral (A.4.1) with respect to the base $q^{1/2}$ as follows

$$I_k = \int_0^\pi \frac{\left(e^{2i\theta}, e^{-2i\theta}; q^{1/2}\right)_\infty}{\left(ae^{i\theta}, ae^{-i\theta}, be^{i\theta}, be^{-i\theta}, ce^{i\theta}, ce^{-i\theta}; q^{1/2}\right)_\infty} \tag{9.7.6}$$
$$\times \frac{d\theta}{\left(q^{1/4+k/2}e^{i\theta}, q^{1/4+k/2}e^{-i\theta};\right)_\infty}$$
$$= \frac{2\pi \left(abcq^{1/4}; q^{1/2}\right)_\infty \left(aq^{1/4}, bq^{1/4}, cq^{1/4}; q^{1/2}\right)_k}{\left(q^{1/2}, aq^{1/4}, bq^{1/4}, cq^{1/4}, ab, ac, bc; q^{1/2}\right)_\infty \left(abcq^{1/4}; q^{1/2}\right)_k}$$

and we arrive at the following formal q-Fourier expansion

$$\frac{\left(q^{1/4}e^{i\theta}, q^{1/4}e^{-i\theta}, -q^{1/4}e^{i\theta}, -q^{1/4}e^{-i\theta}, -q^{1/4}e^{i\theta}, -q^{1/4}e^{-i\theta}; q^{1/2}\right)_\infty}{\left(ae^{i\theta}, ae^{-i\theta}, be^{i\theta}, be^{-i\theta}, ce^{i\theta}, ce^{-i\theta}; q^{1/2}\right)_\infty}$$
$$= \frac{\pi \left(abcq^{1/4}; q^{1/2}\right)_\infty}{\left(q^{1/2}, aq^{1/4}, bq^{1/4}, cq^{1/4}, ab, ac, bc; q^{1/2}\right)_\infty}$$
$$\times \sum_{n=-\infty}^{\infty} \frac{\left(i\omega_n; q^{1/2}\right)_\infty}{\left(-q\omega_n^2; q^2\right)_\infty \ k(\omega_n)}\, \mathcal{E}_q(\cos\theta; i\omega_n)$$

$$\times {}_3\varphi_2\left(\begin{matrix} aq^{1/4},\ bq^{1/4},\ cq^{1/4} \\ -q^{1/2},\ abcq^{1/4} \end{matrix};q^{1/2},\ i\omega_n\right). \tag{9.7.7}$$

Analytic continuation of the ${}_3\varphi_2$ series can be obtain by means of Eq. (4.5.2) of [**46**].

Now consider a few special cases of (9.7.7). When $c = 0$ this expansion takes a simpler form

$$\frac{\left(q^{1/4}e^{i\theta}, q^{1/4}e^{-i\theta}, -q^{1/4}e^{i\theta}, -q^{1/4}e^{-i\theta}, -q^{1/4}e^{i\theta}, -q^{1/4}e^{-i\theta}; q^{1/2}\right)_\infty}{\left(ae^{i\theta}, ae^{-i\theta}, be^{i\theta}, be^{-i\theta}; q^{1/2}\right)_\infty}$$

$$= \frac{\pi}{\left(q^{1/2}, aq^{1/4}, bq^{1/4}, ab; q^{1/2}\right)_\infty}$$

$$\times \sum_{n=-\infty}^{\infty} \frac{\left(i\omega_n; q^{1/2}\right)_\infty}{\left(-q\omega_n^2; q^2\right)_\infty\ k(\omega_n)}\ \mathcal{E}_q(\cos\theta; i\omega_n)$$

$$\times {}_2\varphi_1\left(\begin{matrix} aq^{1/4},\ bq^{1/4} \\ -q^{1/2} \end{matrix};q^{1/2},\ i\omega_n\right). \tag{9.7.8}$$

Analytic continuation of the ${}_2\varphi_1$ series can be given with the help of transformation (III.1) of [**46**] or (A.3.1), which implies

$$\frac{\left(q^{1/4}e^{i\theta}, q^{1/4}e^{-i\theta}, -q^{1/4}e^{i\theta}, -q^{1/4}e^{-i\theta}, -q^{1/4}e^{i\theta}, -q^{1/4}e^{-i\theta}; q^{1/2}\right)_\infty}{\left(ae^{i\theta}, ae^{-i\theta}, be^{i\theta}, be^{-i\theta}; q^{1/2}\right)_\infty}$$

$$= \frac{\pi}{\left(q^{1/2}, -q^{1/2}, aq^{1/4}, ab; q^{1/2}\right)_\infty}$$

$$\times \sum_{n=-\infty}^{\infty} \frac{\left(iaq^{1/4}\omega_n; q^{1/2}\right)_\infty}{\left(-q\omega_n^2; q^2\right)_\infty\ k(\omega_n)}\ \mathcal{E}_q(\cos\theta; i\omega_n)$$

$$\times {}_2\varphi_1\left(\begin{matrix} -q^{1/4}/b,\ i\omega_n \\ iaq^{1/4}\omega_n \end{matrix};q^{1/2},\ bq^{1/4}\right). \tag{9.7.9}$$

The case $b = -q^{1/4}$ is

$$\frac{\left(q^{1/2}e^{2i\theta}, q^{1/2}e^{-2i\theta}; q\right)_\infty}{\left(ae^{i\theta}, ae^{-i\theta}; q^{1/2}\right)_\infty} \tag{9.7.10}$$

$$= \frac{\pi}{(q, qa^2; q)_\infty} \sum_{n=-\infty}^{\infty} \frac{\left(iaq^{1/4}\omega_n; q^{1/2}\right)_\infty}{\left(-q\omega_n^2; q^2\right)_\infty\ k(\omega_n)}\ \mathcal{E}_q(\cos\theta; i\omega_n).$$

When $a = 0$ one gets the special case $m = 0$ of (9.3.13).

Another interesting special case occurs from (9.7.8) with $b = -a$ when

$${}_2\varphi_1\left(\begin{matrix} aq^{1/4},\ -aq^{1/4} \\ -q^{1/2} \end{matrix};q^{1/2},\ i\omega_n\right) = \sum_{k=0}^{\infty} \frac{(qa^2; q)_k}{(q;q)_k}\ (i\omega_n)^k = \frac{(iqa^2\omega_n; q)_\infty}{(i\omega_n; q)_\infty} \tag{9.7.11}$$

by the q-binomial theorem. As a result

$$\frac{\left(q^{1/2}e^{2i\theta}, q^{1/2}e^{-2i\theta}; q\right)_\infty}{\left(a^2e^{2i\theta}, a^2e^{-2i\theta}; q\right)_\infty} \left(-q^{1/4}e^{i\theta}, -q^{1/4}e^{-i\theta}; q^{1/2}\right)_\infty \tag{9.7.12}$$
$$= \frac{\pi}{\left(q^{1/2}, -a^2; q^{1/2}\right)_\infty \left(q^{1/2}a^2; q\right)_\infty}$$
$$\times \sum_{n=-\infty}^{\infty} \frac{\left(iq^{1/2}\omega_n, iqa^2\omega_n; q\right)_\infty}{\left(-q\omega_n^2; q^2\right)_\infty \, k(\omega_n)} \, \mathcal{E}_q(\cos\theta; i\omega_n).$$

The case $a = q^{1/4}$ is

$$\left(-q^{1/4}e^{i\theta}, -q^{1/4}e^{-i\theta}; q^{1/2}\right)_\infty \tag{9.7.13}$$
$$= \frac{\pi}{(q;q)_\infty^2} \sum_{n=-\infty}^{\infty} \frac{\left(iq^{3/2}\omega_n; q\right)_\infty \, \mathcal{E}_q(\cos\theta; i\omega_n)}{\left(-iq^{1/2}\omega_n; q\right)_\infty \, k(\omega_n)}.$$

It is of interest to compare this expansion with the case $m = 0$ of (9.3.13).

In a similar fashion, one can use the Nassrallah–Rahman integral (A.4.2)–(A.4.4) in order to produce more explicit q-Fourier expansions. We leave the details to the reader; see, for example, Ex. 9.7.

9.8. Miscellaneous Results

Expansion (9.3.14) and the orthogonality property of the basic trigonometric system (5.6.5) result in the following connection relation,

$$\varepsilon_q\left(i\alpha\omega_n q^{1/4}\right) = \frac{(q;q)_\infty \left(-q\omega_n^2; q^2\right)_\infty}{2\pi} \tag{9.8.1}$$
$$\times \int_0^\pi \mathcal{E}_q(\cos\theta; i\omega_n) \, \frac{\left(e^{2i\theta},\ e^{-2i\theta};\ q\right)_\infty}{\left(\alpha e^{i\theta},\ \alpha e^{-i\theta};\ q\right)_\infty} \, d\theta,$$

between the analogs of the exponential function on q-linear and q-quadratic grids. Using (9.3.14) twice, with the help of the orthogonality property of the basic trigonometric system, one gets

$$\int_0^\pi \frac{\left(e^{2i\theta},\ e^{-2i\theta};\ q^{1/2}\right)_\infty}{\left(\alpha e^{i\theta},\ \alpha e^{-i\theta}\ \beta e^{i\theta},\ \beta e^{-i\theta};\ q\right)_\infty} \, d\theta \tag{9.8.2}$$
$$= \frac{2\pi^2}{(q;q)_\infty^2} \sum_{n=-\infty}^{\infty} \frac{\varepsilon_q\left(i\alpha\omega_n q^{1/4}\right) \, \varepsilon_q\left(-i\beta\omega_n q^{1/4}\right)}{k(\omega_n)\left(-q\omega_n^2; q^2\right)_\infty^2}.$$

The same consideration for (9.3.16) gives us a possibility to evaluate another interesting integral,

$$\int_0^\pi \frac{\left(e^{2i\theta},\ e^{-2i\theta};\ q^{1/2}\right)_\infty}{\left(\alpha e^{i\theta+i\varphi}, \alpha e^{i\theta-i\varphi}, \alpha e^{i\varphi-i\theta}, \alpha e^{-i\theta-i\varphi};\ q\right)_\infty} \tag{9.8.3}$$
$$\times \frac{d\theta}{\left(\beta e^{i\theta+i\phi}, \beta e^{i\theta-i\phi}, \beta e^{i\phi-i\theta}, \beta e^{-i\theta-i\phi}; q\right)_\infty}$$

$$= \frac{2\pi^2}{\left(q, q, \alpha^2, \beta^2; q\right)_\infty} \sum_{n=-\infty}^{\infty} \frac{\left(-q\alpha^2\omega_n^2,\ -q\beta^2\omega_n^2; q^2\right)_\infty}{k\left(\omega_n\right)\left(-q\omega_n^2; q^2\right)_\infty^2} \times \mathcal{E}_q\left(\cos\varphi; i\alpha\omega_n\right)\, \mathcal{E}_q\left(\cos\phi; -i\beta\omega_n\right).$$

It is of interest to compare these expansions with the classical Askey–Wilson integral (A.4.1).

The limiting case $\beta \to 1^-$ of the last formula leads to the bilinear generating relation (9.3.16) by (6.6.12).

Substituting $\phi = \varphi$ and integrating both sides of (9.8.3) over φ with the help of (9.5.10), we arrive at the following double integral

$$\int_0^\pi \int_0^\pi \frac{\left(e^{2i\theta},\ e^{-2i\theta};\ q\right)_\infty}{\left(\alpha e^{i\theta+i\varphi}, \alpha e^{i\theta-i\varphi}, \alpha e^{i\varphi-i\theta}, \alpha e^{-i\theta-i\varphi};\ q\right)_\infty} \tag{9.8.4}$$

$$\times \frac{\left(e^{2i\varphi},\ e^{-2i\varphi};\ q\right)_\infty}{\left(\alpha e^{i\theta+i\varphi}, \alpha e^{i\theta-i\varphi}, \alpha e^{i\varphi-i\theta}, \alpha e^{-i\theta-i\varphi};\ q\right)_\infty}$$

$$\times \frac{\left(q^{1/2}e^{2i\theta},\ q^{1/2}e^{-2i\theta};\ q\right)_\infty}{\left(q^{1/2}e^{2i\varphi},\ q^{1/2}e^{-2i\varphi};\ q\right)_\infty}\, d\theta\, d\varphi$$

$$= \frac{2\pi^3}{\left(q, q, \alpha^2, \beta^2; q\right)_\infty (\alpha - \beta)} \frac{\left(q^{1/2}; q\right)_\infty^2}{(q;q)_\infty^2}$$

$$\times \sum_{n=-\infty}^{\infty} \frac{\left(-i\alpha\omega_n,\ i\beta\omega_n; q^{1/2}\right)_\infty - \left(i\alpha\omega_n,\ -i\beta\omega_n; q^{1/2}\right)_\infty}{i\, k\left(\omega_n\right)\omega_n\ \left(-q\omega_n^2; q^2\right)_\infty^2}.$$

9.9. Exercises for Chapter 9

(1) Show that the series in (9.1.22) and (9.1.23) converge absolutely and uniformly in x on every closed subinterval of $(-\eta, \eta)$ and of $(-\eta_1, \eta_1)$, respectively, where $\eta = x(1/4)$, $\eta_1 = x(1/2)$, $x(\varepsilon) = \left(q^\varepsilon + q^{-\varepsilon}\right)/2$.
Show that the series in (9.1.32) and (9.1.33) converge absolutely and uniformly in x on every closed subinterval of $(-\eta_1, \eta_1)$ and of $(-\eta, \eta)$, respectively.
Show that the series in (9.2.2) and (9.2.4) converge absolutely and uniformly on every closed subinterval of $(-\eta_1, \eta_1)$ and of $(-\eta, \eta)$, respectively.
Show that the series in (9.2.5) and (9.2.6) converge absolutely and uniformly on every closed subinterval of $(-\eta, \eta)$ and of $(-\eta_1, \eta_1)$, respectively. [*Hint*: Use Lemma 6.7 and the asymptotic formula (6.3.14).]

(2) Prove Rogers' connection coefficient formula (9.1.26) and obtain as special cases that

$$\frac{H_{2m+1}(x|q)}{(q;q)_{2m+1}} = \sum_{k=0}^{m} q^{k/2} \frac{1 - q^{2m-2k+3/2}}{1 - q^{1/2}} \frac{C_{2m-2k+1}\left(x; q^{1/2}|q\right)}{(q;q)_k \left(q^{3/2}; q\right)_{2m-k+1}}$$

and

$$\frac{H_{2m}(x|q)}{(q;q)_{2m}} = \sum_{k=0}^{m} q^{k/2} \frac{1-q^{2m-2k+1/2}}{1-q^{1/2}} \frac{C_{2m-2k}\left(x;q^{1/2}|q\right)}{(q;q)_k \left(q^{3/2};q\right)_{2m-k}}.$$

Use these formulas together with (9.1.22) and (9.1.23) in order to derive the following q-Fourier expansions

$$\frac{H_{2m+1}(x|q)}{(q;q)_{2m+1}} = 2\sum_{n=1}^{\infty} \frac{(-1)^{n-1}}{\kappa(\omega_n)\,\omega_n} \sqrt{\frac{\left(-q\omega_n^2;q^2\right)_\infty}{\left(-\omega_n^2;q^2\right)_\infty}}\; S_q(x;\omega_n)$$
$$\times \sum_{k=0}^{m} (-1)^{m-k} \frac{1-q^{2m-2k+3/2}}{1-q^{1/2}} \frac{q^{k/2-(m-k)(m-k+1)-1/4}}{(q;q)_k \left(q^{3/2};q\right)_{2m-k+1}}$$
$$\times h_{2m-2k}\left(\frac{1}{2\omega_n};q\right)$$

and

$$\frac{H_{2m}(x|q)}{(q;q)_{2m}} = 2\sum_{n=1}^{\infty} \frac{(-1)^{n-1}}{\kappa(\omega_n)\,\omega_n} \sqrt{\frac{\left(-q\omega_n^2;q^2\right)_\infty}{\left(-\omega_n^2;q^2\right)_\infty}}\; C_q(x;\omega_n)$$
$$\times \sum_{k=0}^{m-1} (-1)^{m-k} \frac{1-q^{2m-2k+1/2}}{1-q^{1/2}} \frac{q^{k/2-(m-k)^2}}{(q;q)_k \left(q^{3/2};q\right)_{2m-k}}$$
$$\times h_{2m-2k-1}\left(\frac{1}{2\omega_n};q\right)$$

for the continuous q-Hermite polynomials.
Extend this result to the continuous q-ultraspherical polynomials

$$C_m(x;\beta|q)$$
$$= \sum_{n=-\infty}^{\infty}{}' \frac{(-i)^m (-1)^{n-1}}{\kappa(\omega_n)\,\omega_n} \sqrt{\frac{\left(-q\omega_n^2;q^2\right)_\infty}{\left(-\omega_n^2;q^2\right)_\infty}}\; \mathcal{E}_q(\cos\theta; i\omega_n)$$
$$\times \sum_{k=0}^{[m/2]} (-1)^k q^{k/2-(m-2k)^2/4} \frac{\left(\beta q^{-1/2};q\right)_k (\beta;q)_{m-k}}{(q;q)_k \left(q^{3/2};q\right)_{m-k}}$$
$$\times \frac{1-q^{m-2k+1/2}}{1-q^{1/2}}\, h_{m-2k-1}\left(\frac{1}{2\omega_n};q\right).$$

(The prime in the sum here means that the term with $n=0$ should be omitted; we assume also that $h_{-1}=0$.)

(3) Derive the orthogonality property for the q-quadratic exponentials (5.6.5) by a direct integral evaluation with the help of expansion (9.1.21), the orthogonality relation for special q-ultraspherical polynomials (6.7.11), and the dual orthogonality relation for special q-Lommel polynomials (9.1.18).

(4) Derive (9.3.2) and (9.3.24) as special cases of (9.3.3) and (9.3.25), respectively. [*Hint*: Sum corresponding ${}_2\varphi_2$ series as in Section 4.4 and then simplify.]

(5) Show that

$$\mathcal{E}_q(x,y;\alpha) = \frac{(-\alpha;q^{1/2})_\infty - (\alpha;q^{1/2})_\infty}{(q\alpha^2;q^2)_\infty}\,\mathcal{E}_q(y;\alpha)$$
$$\times\left[\frac{1}{2\kappa(0)\,\alpha} + \sum_{n=1}^{\infty}\frac{(-1)^n}{\kappa(\omega_n)(\alpha^2+\omega_n^2)}\frac{\left(-q\omega_n^2;q^2\right)_\infty^{1/2}}{\left(-\omega_n^2;q^2\right)_\infty^{1/2}}\right.$$
$$\times\left(\frac{\alpha C_q(y;\omega_n)+\omega_n S_q(y;\omega_n)}{C_q^2(y;\omega_n)+S_q^2(y;\omega_n)}\,C_q(x,y;\omega_n)\right.$$
$$\left.\left.+\frac{\alpha S_q(y;\omega_n)-\omega_n C_q(y;\omega_n)}{C_q^2(y;\omega_n)+S_q^2(y;\omega_n)}\,S_q(x,y;\omega_n)\right)\right].$$

(6) Suppose that the q-Fourier series

$$f(x) = \sum_{n=-\infty}^{\infty} c_n\,\mathcal{E}_q(x;i\omega_n), \qquad x=\cos\theta$$

converges uniformly on $[-1,1]$. Show that

$$2k(\omega_n)\,c_n = \int_0^\pi f(\cos\theta)\,\mathcal{E}_q(\cos\theta;-i\omega_n)\left(e^{2i\theta},e^{-2i\theta};q\right)_{1/2}d\theta$$
$$= \frac{1}{(-q\omega_n^2;q^2)_\infty}\sum_{k=0}^{\infty} q^{k^2/4}\frac{(-i\omega_n)^k}{(q;q)_k}$$
$$\times\int_0^\pi f(\cos\theta)\;H_k(\cos\theta|q)\;\left(e^{2i\theta},e^{-2i\theta};q\right)_{1/2}\;d\theta.$$

Derive (9.3.13) as a special case.

(7) Show that if the q-Fourier series

$$f(x) = \sum_{n=-\infty}^{\infty} c_n\,\mathcal{E}_q(x;i\omega_n), \qquad x=\cos\theta$$

converges uniformly on $[-1,1]$, then

$$2k(\omega_n)\,c_n = \frac{1}{\left(-q^{1/2};q^{1/2}\right)_\infty(-q\omega_n^2;q^2)_\infty}$$
$$\times\sum_{k=0}^{\infty} q^{k(k+1)/4}\frac{\left(-\omega_n^2;q\right)_k}{\left(q^{1/2};q^{1/2}\right)_k}$$
$$\times\int_0^\pi f(\cos\theta)\left(i\omega_n q^{1/4+k/2}e^{i\theta},i\omega_n q^{1/4+k/2}e^{-i\theta};q^{1/2}\right)_\infty$$
$$\times\left(e^{2i\theta},e^{-2i\theta};q\right)_{1/2}\;d\theta.$$

Derive formally the q-Fourier series for the function

$$f(x) = \left(\gamma e^{i\theta}, \gamma e^{-i\theta}; q^{1/2}\right)_\infty^{-1}, \qquad |\gamma| < 1$$

and investigate its convergence. Consider the special case $\gamma = q^{1/4}$.

(8) Show that

$$\begin{aligned}
&\mathcal{E}_q(\cos\theta;\alpha)\,\mathcal{E}_q(\cos\theta;\beta)\,\left(q^{1/2}e^{2i\theta}, q^{1/2}e^{-2i\theta}; q\right)_\infty \\
&= \frac{\pi}{(q;q)_\infty} \sum_{n=-\infty}^{\infty} \frac{\mathcal{E}_q(\cos\theta; i\omega_n)}{k(\omega_n)\left(-q\omega_n^2; q^2\right)_\infty} \\
&\times \sum_{m=0}^{\infty} q^{m^2/2} (\alpha\beta)^m \frac{\left(i\alpha\omega_n q^{(m+1)/2}; q\right)_\infty \left(iq^{(1-m)/2}\omega_n/\alpha; q\right)_m}{\left(q\alpha^2, q\beta^2; q^2\right)_\infty (q;q)_m}
\end{aligned}$$

and

$$\begin{aligned}
&\mathcal{E}_q(\cos\theta;\alpha)\,\frac{\left(q^{1/2}e^{2i\theta}, q^{1/2}e^{-2i\theta}; q\right)_\infty}{\left(ae^{i\theta}, ae^{-i\theta}; q\right)_\infty} \\
&= \frac{\pi}{(q;q)_\infty} \sum_{n=-\infty}^{\infty} \frac{\mathcal{E}_q(\cos\theta; i\omega_n)}{k(\omega_n)\left(-q\omega_n^2; q^2\right)_\infty} \\
&\times \sum_{m=0}^{\infty} q^{m^2/4} (\alpha a)^m \frac{\left(i\alpha\omega_n q^{(m+1)/2}; q\right)_\infty \left(iq^{(1-m)/2}\omega_n/\alpha; q\right)_m}{\left(q\alpha^2; q^2\right)_\infty (q;q)_m}.
\end{aligned}$$

(9) Show that

$$\begin{aligned}
&\mathcal{E}_q(\cos\theta;\alpha)\,\frac{\left(q^{1/2}e^{2i\theta}, q^{1/2}e^{-2i\theta}; q\right)_\infty}{\left(ae^{2i\theta}, ae^{-2i\theta}, be^{2i\theta}, be^{-2i\theta}; q^2\right)_\infty} \\
&= \frac{\pi}{(q;q)_\infty} \sum_{n=-\infty}^{\infty} \frac{\mathcal{E}_q(\cos\theta; i\omega_n)}{k(\omega_n)\left(-q\omega_n^2; q^2\right)_\infty} \frac{\left(iq^{1/2}\alpha\omega_n; q\right)_\infty}{\left(q\alpha^2, ab; q^2\right)_\infty} \\
&\times \sum_{m,n=0}^{\infty} (-1)^n q^{m^2/4} \frac{(-i\alpha\omega_n)^m (ab)^{(m+n)/2}}{(q;q)_n (q^2;q^2)_m} H_{m+n}\left(\left.\frac{a+b}{2\sqrt{ab}}\right| q^2\right) \\
&\times \frac{\left(iq^{1/2}\omega_n/\alpha, -iq^{1/2}\alpha/\omega_n; q\right)_m}{\left(iq^{1/2}\alpha\omega_n; q\right)_m}.
\end{aligned}$$

[*Hint*: Use (4.9.5).]

(10) Investigate convergence of the series (9.7.7)–(9.7.13).

(11) Obtain expansions for the basic cotangent and basic tangent functions (9.4.4) and (9.4.9) as special cases of the partial fraction decomposition (7.1.8).

(12) The Poisson kernel for the continuous q-ultraspherical polynomials is given by [**46**]

$$\sum_{m=0}^{\infty} \frac{(q;q)_m (1-\beta q^m)}{(\beta^2;q)_m (1-\beta)} C_m(x;\beta|q)\, C_m(y;\beta|q)\ t^m$$
$$= \frac{\left(\beta, t^2, \beta t e^{i\theta+i\varphi}, \beta t e^{-i\theta-i\varphi}, q\beta t e^{i\theta-i\varphi}, q\beta t e^{i\varphi-i\theta}; q\right)_\infty}{\left(\beta^2, q\beta t^2, t e^{i\theta+i\varphi}, t e^{i\theta-i\varphi}, t e^{i\varphi-i\theta}, t e^{-i\theta-i\varphi}; q\right)_\infty}$$
$$\times {}_8W_7\left(\beta t^2; \beta, q t e^{i\theta+i\varphi}, q t e^{-i\theta-i\varphi}, t e^{i\theta-i\varphi}, t e^{i\varphi-i\theta}; q, \beta\right),$$

where $x = \cos\theta$, $y = \cos\varphi$, and $\max(|q|, |\beta|, |t|) < 1$. Verify that this kernel is positive when $-1 < q, t < 1$ and $-1 \le x, y \le 1$ if $0 \le \beta < 1$; see [**46**] for more details and the bibliography. Consider the special case $\beta = q^{1/2}$ of this kernel

$$T_r(x,y) := \sum_{m=0}^{\infty} \frac{C_m\left(x;q^{1/2}|q\right) C_m\left(y;q^{1/2}|q\right)}{d_m^2} r^m, \qquad |r| < 1,$$

where

$$d_m^2 = 2\pi \frac{\left(q^{1/2};q\right)_\infty^2}{(q;q)_\infty^2} \left(1 - q^{m+1/2}\right)$$

is the $\mathcal{L}^2$-norm of the q-Legendre polynomials $C_m\left(x;q^{1/2}|q\right)$ in the orthogonality relation (6.7.11), and introduce the following extension of the basic exponential function $\mathcal{E}_q(x;i\omega)$,

$$\mathcal{F}_q(x;r;i\omega) := \int_{-1}^{1} T_r(x,y)\ \mathcal{E}_q(y;i\omega)\ \rho(y)\, dy$$
$$= \frac{(q;q)_\infty\, \omega^{-1/2}}{(-q\omega^2;q^2)_\infty \left(q^{1/2};q\right)_\infty}$$
$$\times \sum_{m=0}^{\infty} (ir)^m \left(1 - q^{m+1/2}\right) q^{m^2/4}\ J_{m+1/2}^{(2)}(2\omega;q)\, C_m\left(x;q^{1/2}\middle| q\right),$$

where $\rho(x)$ is the weight function in the orthogonality relation (5.6.5) and we have used the Ismail and Zhang formula (6.7.7). Show that

$$\lim_{r\to 1^-} \mathcal{F}_q(x;r;i\omega) = \mathcal{E}_q(x;i\omega).$$

Consider $\omega = \omega_n$, where $\omega_{-n} = -\omega_n$ and $\omega_0 = 0 < \omega_1 < \omega_2 < \omega_3 < \ldots$ are nonnegative zeros of the basic sine function $S_q(\eta;\omega)$, and show that

$$\mathcal{F}_q(x;r;i\omega_n) = \frac{(-1)^{n-1}}{\omega_n} \sum_{m=1}^{\infty} (ir)^m \left(1 - q^{m+1/2}\right)\ q^{-m^2/4}$$

$$\times h_{m-1}\left(\frac{1}{2\omega_n}; q\right)\sqrt{\frac{(-\omega_n^2;q^2)_\infty}{(-q\omega_n^2;q^2)_\infty}}\, C_m\left(x;q^{1/2}\middle| q\right), \quad n\neq 0,$$

$$\mathcal{F}_q(x;r;i0)=1,$$

where $h_m(s;q)$ are the special q-Lommel polynomials (9.1.19). Introduce the following kernel

$$K_r(x,y) := \sum_{n=0}^{\infty} \frac{\mathcal{E}_q(x;i\omega)\,\mathcal{F}_q(x;r;-i\omega_n)}{2k(\omega_n)},$$

where $k(\omega_n)$ is the $\mathcal{L}^2$-norm in the orthogonality relation (5.6.5), and show formally that

$$K_r(x,y) = T_r(x,y).$$

Investigate convergence in the definition of the K-kernel. Introduce a method of summation of the basic Fourier series related to the K-kernel and investigate its convergence.

CHAPTER 10

Basic Bernoulli and Euler Polynomials and Numbers and q-Zeta Function

Certain q-Fourier expansions found in the previous chapter give us a possibility to introduce analogs of the Bernoulli polynomials and numbers, the Euler polynomials and numbers, and the Riemann zeta function [146]. We shall study some of their properties that are close to the classical ones.

10.1. Bernoulli Polynomials, Numbers and Their q-Extensions

The classical *Bernoulli polynomials* can be introduced by means of the generating function

$$\frac{\alpha e^{\alpha x}}{e^{\alpha}-1}=1+\sum_{m=1}^{\infty} B_m(x)\,\frac{\alpha^m}{m!}, \qquad |\alpha|<2\pi. \tag{10.1.1}$$

They have the following Fourier expansions

$$B_{2m-1}(x)=(-1)^m\,\frac{2\,(2m-1)!}{(2\pi)^{2m-1}}\sum_{n=1}^{\infty}\frac{\sin 2\pi nx}{n^{2m-1}} \tag{10.1.2}$$

and

$$B_{2m}(x)=(-1)^{m-1}\,\frac{2\,(2m)!}{(2\pi)^{2m}}\sum_{n=1}^{\infty}\frac{\cos 2\pi nx}{n^{2m}} \tag{10.1.3}$$

with $m=1,2,3,\ldots$; see, for example, [1], [3], [10], [37], [152], and [164].

The *Bernoulli numbers* are defined as

$$B_m=B_m(0)\,. \tag{10.1.4}$$

One can see that $B_0=1$, $B_1=-1/2$, $B_2=1/6$, $B_4=-1/30$, $B_6=1/42$; $B_{2m+1}=0$ when $m=1,2,\ldots$. These numbers appear as coefficients in Taylor's expansion

$$\begin{aligned}\frac{\omega}{2}\cot\frac{\omega}{2}&=1+\sum_{m=1}^{\infty}(-1)^m\,\frac{B_{2m}}{(2m)!}\,\omega^{2m} \qquad (10.1.5)\\ &=1-\frac{1}{12}\,\omega^2-\frac{1}{720}\,\omega^4-\frac{1}{30240}\,\omega^6-\ldots\,.\end{aligned}$$

Also, the special case $x=0$ of (10.1.1) is

$$\frac{\alpha}{e^{\alpha}-1}=1+\sum_{m=1}^{\infty}B_m\,\frac{\alpha^m}{m!} \tag{10.1.6}$$

and, in view of (10.1.1) and (10.1.6), we arrive at the explicit expression for the Bernoulli polynomials

$$B_m(x)=\sum_{k=0}^{m} B_{m-k}\,C_m^k\,x^k,\qquad C_m^k=\frac{m!}{k!\,(m-k)!} \tag{10.1.7}$$

in terms of Bernoulli's numbers and binomial coefficients. See **[1]**, **[10]**, **[37]**, **[152]**, and **[164]** for a detailed account of other properties of the Bernoulli polynomials and numbers.

The basic Fourier series (9.3.2) give us the possibility to introduce q-analogs of the Bernoulli polynomials and numbers in the following manner. Let us start from the classical case $q=1$, rewrite expansion (9.3.1) as

$$\begin{aligned}\frac{\alpha\pi e^{\alpha x}}{e^{\alpha\pi}-e^{-\alpha\pi}}&\\ =\frac{1}{2}&+\sum_{n=1}^{\infty}(-1)^n\frac{\alpha^2}{\alpha^2+n^2}\cos nx+\sum_{n=1}^{\infty}(-1)^{n-1}\frac{\alpha n}{\alpha^2+n^2}\sin nx\end{aligned} \tag{10.1.8}$$

and substitute

$$\frac{\alpha^2}{\alpha^2+n^2}=\sum_{m=1}^{\infty}(-1)^{m-1}\frac{\alpha^{2m}}{n^{2m}},\qquad \frac{\alpha n}{\alpha^2+n^2}=\sum_{m=1}^{\infty}(-1)^{m-1}\frac{\alpha^{2m-1}}{n^{2m-1}}.$$

Thus,

$$\frac{\alpha\pi e^{\alpha x}}{e^{\alpha\pi}-e^{-\alpha\pi}}=\frac{1}{2}+\sum_{m=1}^{\infty}\alpha^{2m-1}\,b_{2m-1}(x)+\sum_{m=1}^{\infty}\alpha^{2m}\,b_{2m}(x), \tag{10.1.9}$$

where

$$b_{2m-1}(x)=(-1)^{m-1}\sum_{n=1}^{\infty}(-1)^{n-1}\frac{\sin nx}{n^{2m-1}}, \tag{10.1.10}$$

$$b_{2m}(x)=(-1)^{m}\sum_{n=1}^{\infty}(-1)^{n-1}\frac{\cos nx}{n^{2m}} \tag{10.1.11}$$

are, essentially, Bernoulli's polynomials

$$b_m(x)=\frac{(2\pi)^m}{2\,(m!)}\,B_m\left(\frac{x+\pi}{2\pi}\right), \tag{10.1.12}$$

see **[3]**, or just compare (10.1.2)–(10.1.3) with (10.1.10)–(10.1.11). The multiples of the Bernoulli numbers appear here when one substitutes $x=-\pi$ due to (10.1.4).

We can now repeat this consideration in the q-case, starting from the expansion (9.3.2). It gives the following *generating relation*

$$\frac{\alpha(q\alpha^2;q^2)_\infty\,\mathcal{E}_q(x;\alpha)}{(-\alpha;q^{1/2})_\infty-(\alpha;q^{1/2})_\infty} \tag{10.1.13}$$

$$= \frac{1}{2}\left(1-q^{1/2}\right) + \sum_{m=1}^{\infty} \alpha^{2m-1}\, B_{2m-1}(x;q) + \sum_{m=1}^{\infty} \alpha^{2m}\, B_{2m}(x;q),$$

where

$$B_{2m-1}(x;q) \tag{10.1.14}$$

$$:= (-1)^{m-1} \sum_{n=1}^{\infty} \frac{(-1)^{n-1}}{\kappa(\omega_n)\, {\omega_n}^{2m-1}} \sqrt{\frac{(-q\omega_n^2;q^2)_\infty}{(-\omega_n^2;q^2)_\infty}}\, S_q(x;\omega_n),$$

$$B_{2m}(x;q) \tag{10.1.15}$$

$$:= (-1)^{m} \sum_{n=1}^{\infty} \frac{(-1)^{n-1}}{\kappa(\omega_n)\, {\omega_n}^{2m}} \sqrt{\frac{(-q\omega_n^2;q^2)_\infty}{(-\omega_n^2;q^2)_\infty}}\, C_q(x;\omega_n)$$

are q-analogs of the polynomials (10.1.10)–(10.1.11). Both equations can be written in a unified form

$$B_m(x;q) = -\pi \frac{\left(q^{1/2};q\right)_\infty^2}{(q;q)_\infty^2} \sum_{n=-\infty}^{\infty}{}' \frac{\mathcal{E}_q(x;i\omega_n)\,\mathcal{E}_q(\eta;-i\omega_n)}{2k(\omega_n)\,(i\omega_n)^m}, \tag{10.1.16}$$

where the prime in the sum means that the term with $n = 0$ should be omitted. This follows from the q-Euler formula (5.6.4) and (5.8.12) with the help of (5.5.8). By Theorem 6.12 the series (10.1.16) converges inside the ellipse $E_{m/2-1/4}$ in the complex x-plane. Eq. (10.1.16) is obviously an analog of the Fourier expansion

$$B_m(x) = -m! \sum_{n=-\infty}^{\infty}{}' \frac{\exp(2\pi i n x)}{(2\pi i n)^m}, \tag{10.1.17}$$

compare with (10.1.2)–(10.1.3), valid for the classical Bernoulli polynomials.

In order to prove that the functions $B_m(x;q)$ defined in (10.1.14)–(10.1.15) are, indeed, polynomials of exact degrees m in x, let us use the explicit representation (9.1.9) for the q-Lommel polynomials in expansions (9.3.19) and (9.3.22),

$$x \prod_{k=0}^{m-2}\left(\left(1+q^{k+1/2}\right)^2 - 4q^{k+1/2}x^2\right) \tag{10.1.18}$$

$$= (-1)^{m-1} \frac{(q;q)_{2m-1}}{\left(q^{1/2};q\right)_m} q^{-m^2/2+1/4}$$

$$\times \sum_{k=0}^{[(m-1)/2]} q^{k(k+1/2)} \frac{(q;q)_{m-k-1}\left(q^{3/2};q\right)_{m-k-1}}{(q;q)_k\,(q;q)_{m-2k-1}\left(q^{3/2};q\right)_k} B_{2m-2k-1}(x;q)$$

and

$$\prod_{k=0}^{m-1}\left(\left(1+q^{k+1/2}\right)^2 - 4q^{k+1/2}x^2\right) - \frac{(q;q)_{2m}}{\left(q^{1/2},\,q^{3/2};q\right)_m} \tag{10.1.19}$$

$$= 2\,(-1)^m \frac{(q;q)_{2m}}{\left(q^{1/2};q\right)_m}\, q^{-m^2/2}$$
$$\times \sum_{k=0}^{[(m-1)/2]} q^{k(k+1/2)} \frac{(q;q)_{m-k-1}\left(q^{3/2};q\right)_{m-k-1}}{(q;q)_k\,(q;q)_{m-2k-1}\left(q^{3/2};q\right)_k} B_{2m-2k}(x;q)$$

for the odd and even degrees, respectively. Eqs. (10.1.18) and (10.1.19) can be thought as systems of linear equations for $B_m(x;q)$ with triangular matrices which can be solved by the method of elimination. The left sides of (10.1.18) and (10.1.19) are polynomials of exact degrees $2m-3$ and $2m-2$, respectively. This completes the proof.

We shall call $B_m(x;q)$ as the *q-Bernoulli polynomials*[1][**146**]. One can also introduce the *q-Bernoulli numbers* as

$$B_m(q) := B_m(-\eta;q), \quad \eta = \left(q^{1/4} + q^{-1/4}\right)/2 \tag{10.1.20}$$

in accordance with (10.1.4) and (10.1.12). By using the generating function (10.1.13) and the limiting relation

$$\lim_{q\to 1^-} \mathcal{E}_q(x;(1-q)\alpha/2) = e^{\alpha x} \tag{10.1.21}$$

it can be shown that the classical Bernoulli polynomials and numbers are the $q \to 1^-$ limits of (10.1.14)–(10.1.15) and (10.1.20), respectively. We leave the details to the reader; see Ex. 10.1.

10.2. Some Properties of q-Bernoulli Polynomials

Let us discuss several properties of the q-Bernoulli polynomials and numbers that are close to the classical ones. The explicit form of the first five q-Bernoulli's polynomials follows directly from the generating relation (10.1.13) and expansions (9.1.4)–(9.1.5) and (9.1.24)–(9.1.25):

$$B_0(x;q) = \frac{1}{2}\left(1-q^{1/2}\right), \quad B_1(x;q) = q^{1/4}\,\frac{1-q^{1/2}}{(q;q)_1}\,x, \tag{10.2.1}$$

$$B_2(x;q) = 2q\,\frac{1-q^{1/2}}{(q;q)_2}\left[x^2 - \frac{\left(1+q^{1/2}\right)^2}{4\left(1+q^{1/2}+q\right)}\right], \tag{10.2.2}$$

$$B_3(x;q) = 4q^{9/4}\,\frac{1-q^{1/2}}{(q;q)_3}\,x\left[x^2 - \frac{\left(1+q^{1/2}\right)^2}{4q^{1/2}}\right], \tag{10.2.3}$$

$$B_4(x;q) = 8q^4\,\frac{1-q^{1/2}}{(q;q)_4} \tag{10.2.4}$$
$$\times \left[x^4 - \frac{\left(1+q^{1/2}\right)^2\left(1-2q^{1/2}+4q-2q^{3/2}+q^2\right)}{4q^{3/2}}\,x^2\right.$$

[1] At the moment it is not clear what is the best normalization for the q-Bernoulli polynomials, so the author has chosen the simplest one.

$$+ \frac{\left(1+q^{1/2}\right)\left(1+q^{3/2}\right)(1-q)^2\left(1+2q+q^{3/2}+2q^2+q^3\right)}{16q^{3/2}\left(1-q^{3/2}\right)\left(1-q^{5/2}\right)}\Bigg].$$

The corresponding q-Bernoulli numbers defined by (10.1.20) are

$$B_0(q) = \frac{1}{2}\left(1-q^{1/2}\right), \qquad B_1(q) = -\frac{1}{2}, \tag{10.2.5}$$

$$B_2(q) = \frac{q^{1/2}}{2\left(1-q^{3/2}\right)}, \qquad B_3(q) = 0, \tag{10.2.6}$$

$$B_4(q) = -\frac{q^2}{2\left(1-q^{3/2}\right)^2\left(1-q^{5/2}\right)}. \tag{10.2.7}$$

It can be shown that as in the classical case,

$$B_{2m+1}(q) = 0, \quad m = 1, 2, 3\ldots . \tag{10.2.8}$$

Indeed, substituting $x = -\eta$ in the generating relation (10.1.13) and taking into account that

$$\mathcal{E}_q(-\eta;\alpha) = \frac{(\alpha;q^{1/2})_\infty}{(q\alpha^2;q^2)_\infty} \tag{10.2.9}$$

by (2.3.10), one gets

$$\alpha\,\frac{\left(-\alpha;q^{1/2}\right)_\infty + \left(\alpha;q^{1/2}\right)_\infty}{\left(-\alpha;q^{1/2}\right)_\infty - \left(\alpha;q^{1/2}\right)_\infty} \tag{10.2.10}$$

$$= 1 - q^{1/2} + 2\sum_{m=2}^{\infty} \alpha^{2m-1}\, B_{2m-1}(q) + 2\sum_{m=1}^{\infty} \alpha^{2m}\, B_{2m}(q).$$

The left side here is an even function of α, so (10.2.8) takes place.

Equating coefficients of α^{2n+1} in both sides of

$$\alpha\,\frac{(-\alpha;q^{1/2})_\infty + (\alpha;q^{1/2})_\infty}{(-\alpha;q^{1/2})_\infty - (\alpha;q^{1/2})_\infty}\left((-\alpha;q^{1/2})_\infty - (\alpha;q^{1/2})_\infty\right)$$

$$= \alpha\left((-\alpha;q^{1/2})_\infty + (\alpha;q^{1/2})_\infty\right) \tag{10.2.11}$$

by (10.2.10) and (II.2) of **[46]** or (A.2.2), we obtain the *recurrence relation* for the q-Bernoulli numbers,

$$2\sum_{k=0}^{n} B_{2k}(q)\,\frac{q^{(n-k)(n-k+1/2)}}{\left(q^{1/2};q^{1/2}\right)_{2n-2k+1}} = \frac{q^{n(n-1/2)}}{\left(q^{1/2};q^{1/2}\right)_{2n}}, \tag{10.2.12}$$

which is convenient for their evaluation; see Ex. 10.2. This is an analog of the formula

$$\sum_{k=0}^{n-1} C_n^k\, B_k = 0, \qquad C_n^k = \frac{n!}{k!\,(n-k)!}. \tag{10.2.13}$$

Substituting $\alpha = i\omega$ in (10.2.10), we obtain a q-analog of Taylor's expansion (10.1.5) as

$$\omega \operatorname{Cot}_q(\omega) = 1 - q^{1/2} + 2\sum_{m=1}^{\infty} (-1)^m B_{2m}(q)\ \omega^{2m}. \tag{10.2.14}$$

This follows also from (9.5.9) in a similar manner.

Our generating function (10.1.13) can be rewritten in the form

$$\frac{\alpha(q\alpha^2;q^2)_\infty\ \mathcal{E}_q(-\eta;\alpha)}{(-\alpha;q^{1/2})_\infty - (\alpha;q^{1/2})_\infty}\ (q\alpha^2;q^2)_\infty\ \mathcal{E}_q(x;\alpha)\ \frac{1}{(\alpha;q^{1/2})_\infty}$$

in view of (10.2.9). Expanding this function in the powers of α with the help of (10.1.13) with $x = -\eta$, (10.1.20), (3.4.15), and (II.1) of [**46**] or (A.2.1); we get a representation of the q-Bernoulli polynomials

$$B_m(x;q) = \sum_{n=0}^{m} B_{m-n}(q) \sum_{k=0}^{n} \frac{q^{k^2/4}}{(q^{1/2};q^{1/2})_{n-k}\,(q;q)_k}\ H_k(x|q) \tag{10.2.15}$$

in terms of the q-Bernoulli numbers and the continuous q-Hermite polynomials.

On the other hand, an *explicit expression*

$$B_n(x;q) = \sum_{k=0}^{n} \frac{B_{n-k}(q)}{(q;q)_k}\ \varphi_k(-x;q), \tag{10.2.16}$$

where the polynomials $\varphi_k(x;q)$ are given by

$$\varphi_n(\cos\theta;q) = \left(q^{1/4}e^{i\theta}, q^{1/4}e^{-i\theta}; q^{1/2}\right)_n, \qquad x = \cos\theta, \tag{10.2.17}$$

follows directly from (10.1.13), (10.1.20) and (2.3.27)–(2.3.28) as a q-analog of (10.1.7). One can also see that the summation formula (3.7.15),

$$\varphi_k(-x;q) = \sum_{k=0}^{n} q^{k^2/4} \frac{(q;q)_n}{(q;q)_k\left(q^{1/2};q^{1/2}\right)_{n-k}}\ H_k(x|q), \tag{10.2.18}$$

allows to simplify our expression (10.2.15) for the q-Bernoulli polynomials.

The generating relation (10.1.13) implies the following *contour integral representation*

$$B_m(x;q) = \frac{1}{2\pi i}\int_{|\alpha|=r<\omega_1} \frac{(q\alpha^2;q^2)_\infty\ \mathcal{E}_q(x;\alpha)}{(-\alpha;q^{1/2})_\infty - (\alpha;q^{1/2})_\infty}\ \alpha^{-m}\,d\alpha \tag{10.2.19}$$

for the q-Bernoulli polynomials. The special case $x = -\eta$ is

$$B_m(q) = \frac{1}{2\pi i}\int_{|\alpha|=r<\omega_1} \frac{(\alpha;q^{1/2})_\infty}{(-\alpha;q^{1/2})_\infty - (\alpha;q^{1/2})_\infty}\ \alpha^{-m}\,d\alpha \tag{10.2.20}$$

by (10.1.20) and (10.2.9). On the other hand, the generating relation (10.2.10) and (10.2.8) imply

$$B_{2m}(q) = \frac{1}{4\pi i} \int_{|\alpha|=r<\omega_1} \frac{(-\alpha; q^{1/2})_\infty + (\alpha; q^{1/2})_\infty}{(-\alpha; q^{1/2})_\infty - (\alpha; q^{1/2})_\infty} \alpha^{-2m} \, d\alpha \qquad (10.2.21)$$

in a similar fashion.

The q-trigonometric functions satisfy the difference-differentiation formulas

$$\frac{\delta}{\delta x} C_q(x;\omega) = -\frac{2q^{1/4}}{1-q} \omega \, S_q(x;\omega), \qquad (10.2.22)$$

$$\frac{\delta}{\delta x} S_q(x;\omega) = \frac{2q^{1/4}}{1-q} \omega \, C_q(x;\omega) \qquad (10.2.23)$$

by (2.4.17) and (2.4.18), where the operator $\delta/\delta x$ is the standard Askey–Wilson divided difference operator of the first order defined by (2.3.15). Applying the operator $\delta/\delta x$ to the both sides of (10.1.14) and (10.1.15) we obtain the following *difference-differentiation formula*

$$\frac{\delta}{\delta x} B_m(x;q) = \frac{2q^{1/4}}{1-q} B_{m-1}(x;q), \qquad (10.2.24)$$

as an analog of

$$B'_m(x) = m \, B_{m-1}(x). \qquad (10.2.25)$$

Formula (10.2.24) follows also from (10.2.15), (10.2.16) or (10.2.19) if one uses, respectively, the difference-differentiation formulas

$$\frac{\delta}{\delta x} H_k(x|q) = 2q^{(1-k)/2} \frac{1-q^k}{1-q} H_{k-1}(x|q) \qquad (10.2.26)$$

$$\frac{\delta}{\delta x} \varphi_n(x;q) = -2q^{1/4} \frac{1-q^n}{1-q} \varphi_{n-1}(x;q), \qquad (10.2.27)$$

or

$$\frac{\delta}{\delta x} \mathcal{E}_q(x;\alpha) = \frac{2q^{1/4}\alpha}{1-q} \mathcal{E}_q(x;\alpha). \qquad (10.2.28)$$

See (3.4.18), (3.4.34) and (2.3.13), respectively.

Expansions (10.1.14)–(10.1.15) and the orthogonality relations for the basic trigonometric system (5.2.26)–(5.2.28) together with (10.1.20) and (5.8.10) result in an integral evaluation

$$\int_0^\pi B_m(\cos\theta;q) \, B_n(\cos\theta;q) \left(e^{2i\theta}, \, e^{-2i\theta}; \, q\right)_{1/2} d\theta \qquad (10.2.29)$$

$$= (-1)^{m-1} \, \pi \frac{\left(q^{1/2};q\right)_\infty^2}{(q;q)_\infty^2} B_{m+n}(q).$$

This implies that the q-Bernoulli polynomials are orthogonal for $m+n$ odd due to (10.2.8), which is also true due to the *symmetry property*

$$B_m(-x;q) = (-1)^m \, B_m(x;q). \qquad (10.2.30)$$

Eq. (10.2.29) is an analog of the integral evaluation

$$\int_0^1 B_m(x)\, B_n(x)\; dx = (-1)^{m-1} \frac{m!\, n!}{(m+n)!}\, B_{m+n}, \tag{10.2.31}$$

$m, n = 1, 2, 3, \ldots$ for the classical Bernoulli polynomials [**1**].

Expansions (9.1.22)–(9.1.23) and (9.1.9) lead to the following relations between the q-Legendre polynomials $C_m\left(x; q^{1/2} \middle| q\right)$ and the q-Bernoulli polynomials

$$\begin{aligned} C_{2m+1}\left(x; q^{1/2} \middle| q\right) &= 2q^{-m(m+1)-1/4} \\ &\times \sum_{k=0}^{m} q^{k(k+1/2)} \frac{(q;q)_{2m-k}\left(q^{3/2};q\right)_{2m-k}}{(q;q)_k\,(q;q)_{2m-2k}\left(q^{3/2};q\right)_k}\, B_{2m-2k+1}(x;q) \end{aligned} \tag{10.2.32}$$

and

$$\begin{aligned} C_{2m}\left(x; q^{1/2} \middle| q\right) &= 2q^{-m^2} \\ &\times \sum_{k=0}^{m-1} q^{k(k+1/2)} \frac{(q;q)_{2m-k-1}\left(q^{3/2};q\right)_{2m-k-1}}{(q;q)_k\,(q;q)_{2m-2k-1}\left(q^{3/2};q\right)_k}\, B_{2m-2k}(x;q) \end{aligned} \tag{10.2.33}$$

for the odd and even degrees, respectively.

On the other hand, expansions of the q-Bernoulli polynomials in terms of the q-Legendre polynomials are

$$\begin{aligned} B_{2m+1}(x;q) &= \sum_{n=0}^{m} C_{2n+1}\left(x; q^{1/2} \middle| q\right) q^{-n(n+1)-1/4}\left(1 - q^{2n+3/2}\right) \\ &\times \sum_{k=0}^{n} q^{k(k+1/2)} \frac{(q;q)_{2n-k}\left(q^{3/2};q\right)_{2n-k}}{(q;q)_k\,(q;q)_{2n-2k}\left(q^{3/2};q\right)_k}\, B_{2m+2n-2k+2}(q) \end{aligned} \tag{10.2.34}$$

and

$$\begin{aligned} B_{2m}(x;q) &= -\sum_{n=0}^{m-1} C_{2n}\left(x; q^{1/2} \middle| q\right) q^{-n^2}\left(1 - q^{2n+1/2}\right) \\ &\times \sum_{k=0}^{n-1} q^{k(k+1/2)} \frac{(q;q)_{2n-k-1}\left(q^{3/2};q\right)_{2n-k-1}}{(q;q)_k\,(q;q)_{2n-2k-1}\left(q^{3/2};q\right)_k}\, B_{2m+2n-2k}(q) \end{aligned} \tag{10.2.35}$$

due to (10.2.32)–(10.2.33), (10.2.29), and (6.7.11).

10.3. Extension of q-Bernoulli Polynomials

An important property

$$B_m(x+1) - B_m(x) = m\, x^{m-1} \tag{10.3.1}$$

follows directly from the generating function (10.1.1) for the classical Bernoulli polynomials. In order to establish a q-analog of this formula we need to introduce an extension of the q-Bernoulli polynomials by means of the generating relation [**146**]

$$\frac{\alpha(q\alpha^2;q^2)_\infty\,\mathcal{E}_q(x,y;\alpha)}{(-\alpha;q^{1/2})_\infty-(\alpha;q^{1/2})_\infty} \tag{10.3.2}$$
$$=\frac{1}{2}\left(1-q^{1/2}\right)+\sum_{m=1}^{\infty}\alpha^{2m-1}\,B_{2m-1}(x,y;q)+\sum_{m=1}^{\infty}\alpha^{2m}\,B_{2m}(x,y;q),$$

where $\mathcal{E}_q(x,y;\alpha)$ is the q-quadratic exponential in two independent variables x and y; see Section 2.3. This function appears in the generating relation

$$\sum_{m=0}^{\infty}\frac{q^{m^2/4}}{(q;q)_m}\,\alpha^m H_m(x,y|q)=\left(q\alpha^2;q^2\right)_\infty\mathcal{E}_q(x,y;\alpha), \tag{10.3.3}$$

where $H_m(x,y|q)$ are the q-analogs of the classical Hermite polynomials $H_m(x+y)$ introduced in [**140**]; see Section 3.7 for more details including the explicit representation of these polynomials in terms of terminating ${}_6\varphi_5$ series. Using the same arguments as in (10.2.15) with the help of (10.3.3) one gets

$$B_m(x,y;q)=\sum_{n=0}^{m}B_{m-n}(q)\sum_{k=0}^{n}\frac{q^{k^2/4}}{(q^{1/2};q^{1/2})_{n-k}\,(q;q)_k}\,H_k(x,y|q). \tag{10.3.4}$$

It is natural to view the generalized q-Bernoulli polynomials $B_m(x,y;q)$ here as an extension of $B_m(x+y)$. The following properties hold

$$B_m(x,y;q)=B_m(y,x;q),\qquad B_m(x,0;q)=B_m(x;q). \tag{10.3.5}$$

Substituting $y=\pm\eta$ in the generating relation (10.3.2) and then subtracting the results we obtain

$$\alpha\,\mathcal{E}_q(x;\alpha)=\sum_{n=1}^{\infty}\alpha^n\left(B_n(x,\eta;q)-B_n(x,-\eta;q)\right) \tag{10.3.6}$$

with the help of the addition theorem,

$$\mathcal{E}_q(x,y;\alpha)=\mathcal{E}_q(x;\alpha)\;\mathcal{E}_q(y;\alpha), \tag{10.3.7}$$

the symmetry relation

$$\mathcal{E}_q(-x;\alpha)=\mathcal{E}_q(x;-\alpha) \tag{10.3.8}$$

and (10.2.9). By (2.3.23)

$$\mathcal{E}_q(x;\alpha)=\sum_{n=0}^{\infty}\frac{q^{n/2}\,\alpha^n}{(q;q)_n}\,\phi_n(x;q), \tag{10.3.9}$$

where polynomials $\phi_n(x;q)$ are given by (2.3.24)–(2.3.25). Substituting (10.3.9) to the left hand side of (10.3.6) and equating the coefficients of α^n we finally obtain

$$B_n(x,\eta;q)-B_n(x,-\eta;q)=\frac{q^{(n-1)/2}}{(q;q)_{n-1}}\,\phi_{n-1}(x;q) \tag{10.3.10}$$

as an interesting analog of (10.3.1).

In a similar manner, rewriting the left side of the generating relation (10.3.2) in the form

$$\frac{\alpha(q\alpha^2;q^2)_\infty\,\mathcal{E}_q\left(x;\alpha\right)}{(-\alpha;q^{1/2})_\infty-(\alpha;q^{1/2})_\infty}\,\mathcal{E}_q\left(y;\alpha\right)$$

and then using (10.1.13) and (10.3.9), one gets

$$B_n\left(x,y;q\right)=\sum_{k=0}^{n}\frac{q^{(n-k)/2}}{(q;q)_{n-k}}\,B_k\left(x;q\right)\;\phi_{n-k}\left(y;q\right) \tag{10.3.11}$$

as a q-analog of

$$B_n\left(x+y\right)=\sum_{k=0}^{n}C_n^k\;B_k\left(x\right)\;y^{n-k}. \tag{10.3.12}$$

This shows that $B_n\left(x,y;q\right)$ are polynomials in x and y. The first q-Bernoulli polynomials $B_n\left(x,y;q\right)$ are

$$B_0\left(x,y;q\right)=\frac{1}{2}\left(1-q^{1/2}\right), \tag{10.3.13}$$

$$B_1\left(x,y;q\right)=q^{1/4}\frac{1-q^{1/2}}{(q;q)_1}\left(x+y\right), \tag{10.3.14}$$

$$B_2\left(x,y;q\right)=2q\frac{1-q^{1/2}}{(q;q)_2} \tag{10.3.15}$$

$$\times\left[x^2+y^2+\frac{1+q}{q^{1/2}}\;xy-\frac{\left(1+q^{1/2}\right)^2}{4\left(1+q^{1/2}+q\right)}\right],$$

$$B_3\left(x,y;q\right)=4q^{9/4}\frac{1-q^{1/2}}{(q;q)_3} \tag{10.3.16}$$

$$\times\left[x^3+y^3+\frac{1+q+q^2}{q}\;xy\left(x+y\right)-\frac{\left(1+q^{1/2}\right)^2}{4q^{1/2}}\;\left(x+y\right)\right],$$

$$B_4\left(x,y;q\right)=8q^4\frac{1-q^{1/2}}{(q;q)_4} \tag{10.3.17}$$

$$\times\left[x^4+y^4+\frac{1+q+q^2+q^3}{q^{3/2}}\;xy\left(x+y\right)\right.$$

$$+\frac{\left(1+q^2\right)\left(1+q+q^2\right)}{q^2}\;x^2y^2$$

$$-\frac{\left(1+q^{1/2}\right)^2\left(1-2q^{1/2}+4q-2q^{3/2}+q^2\right)}{4q^{3/2}}\left(x^2+y^2\right)$$

$$+\frac{\left(1+q+q^2\right)\left(1+q^{1/2}\right)^2\left(1-3q^{1/2}+q\right)}{4q^{5/2}}\;xy$$

$$+\frac{\left(1+q^{1/2}\right)\left(1+q^{3/2}\right)(1-q)^2\left(1+2q+q^{3/2}+2q^2+q^3\right)}{16q^{3/2}\left(1-q^{3/2}\right)\left(1-q^{5/2}\right)}\Bigg].$$

This follows directly from (10.3.11); we leave the details to the reader.

We have discussed here some properties of the q-Bernoulli polynomials and numbers. They deserve a more detailed consideration. Other q-analogs of the Bernoulli polynomials and numbers were studied in **[6]**, **[27]**, **[28]**, **[29]**, **[67]**, **[85]**, **[134]**, **[158]**, and **[162]**.

10.4. Basic Euler Polynomials and Numbers

The classical *Euler polynomials* can be introduced by means of the generating function

$$\frac{2e^{\alpha x}}{e^{\alpha}+1}=1+\sum_{m=1}^{\infty}E_m(x)\,\frac{\alpha^m}{m!},\qquad |\alpha|<\pi. \tag{10.4.1}$$

They have the following Fourier expansions

$$E_{2m-1}(x)=(-1)^m\,\frac{4\,(2m-1)!}{\pi^{2m}}\sum_{n=1}^{\infty}\frac{\cos\pi\,(2n+1)\,x}{(2n+1)^{2m}} \tag{10.4.2}$$

and

$$E_{2m}(x)=(-1)^m\,\frac{4\,(2m)!}{\pi^{2m+1}}\sum_{n=1}^{\infty}\frac{\sin\pi\,(2n+1)\,x}{(2n+1)^{2m+1}} \tag{10.4.3}$$

with $m=1,2,3,\ldots$; see, for example, **[1]** and **[37]**. The *Euler numbers* are defined by

$$E_m=2^mE_m(1/2)\,. \tag{10.4.4}$$

One can see that $E_0=1$, $E_2=-1$, $E_4=5$; $E_{2m+1}=0$ when $m=0,1,2,\ldots$. These numbers appear as coefficients in Taylor's expansion

$$\frac{1}{\cos\omega}=1+\sum_{m=1}^{\infty}(-1)^m\,\frac{E_{2m}}{(2m)!}\,\omega^{2m} \tag{10.4.5}$$

$$=1+\frac{1}{2}\,\omega^2+\frac{5}{24}\,\omega^4+\frac{61}{720}\,\omega^6+\ldots.$$

Also, the special case $x=1/2$ of (10.4.1) is

$$\frac{2e^{\alpha/2}}{e^{\alpha}+1}=1+\sum_{m=1}^{\infty}E_m\,\frac{(\alpha/2)^m}{m!} \tag{10.4.6}$$

and, in view of (10.4.1) and (10.4.6), we arrive at the explicit expression for the Euler polynomials

$$E_m(x)=\sum_{k=0}^{m}2^{-k}E_k\,C_m^k\,(x-1/2)^{m-k},\quad C_m^k=\frac{m!}{k!\,(m-k)!} \tag{10.4.7}$$

in terms of Euler's numbers and the binomial coefficients. See, for example, [**1**], [**37**], and [**152**] for an account of further properties of the Euler polynomials and numbers.

The basic Fourier series (9.3.24) gives us the possibility to introduce q-analogs of the Euler polynomials and numbers. Let us rewrite (9.3.24) as

$$\frac{(q\alpha^2;q^2)_\infty\ \mathcal{E}_q(x;\alpha)}{(-\alpha;q^{1/2})_\infty+(\alpha;q^{1/2})_\infty} \tag{10.4.8}$$
$$=\sum_{n=1}^{\infty}\frac{(-1)^{n-1}\,\varpi_n}{\kappa(\varpi_n)(\alpha^2+\varpi_n^2)}\sqrt{\frac{(-q\varpi_n^2;q^2)_\infty}{(-\varpi_n^2;q^2)_\infty}}\ C_q(x;\varpi_n)$$
$$+\sum_{n=1}^{\infty}\frac{(-1)^{n-1}\,\alpha}{\kappa(\varpi_n)(\alpha^2+\varpi_n^2)}\sqrt{\frac{(-q\varpi_n^2;q^2)_\infty}{(-\varpi_n^2;q^2)_\infty}}\ S_q(x;\varpi_n)$$

and substitute

$$\frac{\varpi_n}{\alpha^2+\varpi_n^2}=\sum_{m=0}^{\infty}(-1)^m\,\frac{\alpha^{2m}}{\varpi_n^{2m+1}},\qquad \frac{\alpha}{\alpha^2+\varpi_n^2}=\sum_{m=1}^{\infty}(-1)^{m-1}\,\frac{\alpha^{2m-1}}{\varpi_n{}^{2m}},$$

when $|\alpha|<\varpi_1$. As a result, we arrive at the following generating relation

$$\frac{(q\alpha^2;q^2)_\infty\ \mathcal{E}_q(x;\alpha)}{(-\alpha;q^{1/2})_\infty+(\alpha;q^{1/2})_\infty} \tag{10.4.9}$$
$$=\frac{1}{2}+\sum_{m=1}^{\infty}\alpha^{2m-1}\ E_{2m-1}(x;q)+\sum_{m=1}^{\infty}\alpha^{2m}\ E_{2m}(x;q),$$

where

$$E_{2m-1}(x;q) \tag{10.4.10}$$
$$:=(-1)^{m-1}\sum_{n=1}^{\infty}\frac{(-1)^{n-1}}{\kappa(\varpi_n)\,\varpi_n^{2m}}\sqrt{\frac{(-q\varpi_n^2;q^2)_\infty}{(-\varpi_n^2;q^2)_\infty}}\ S_q(x;\varpi_n),$$

$$E_{2m}(x;q) \tag{10.4.11}$$
$$:=(-1)^{m}\sum_{n=1}^{\infty}\frac{(-1)^{n-1}}{\kappa(\varpi_n)\,\varpi_n^{2m+1}}\sqrt{\frac{(-q\varpi_n^2;q^2)_\infty}{(-\varpi_n^2;q^2)_\infty}}\ C_q(x;\varpi_n)$$

are q-analogs of the Euler polynomials (10.4.2)–(10.4.3). Both equations can be written in a unified form

$$E_m(x;q)=\pi\frac{(q^{1/2};q)_\infty^2}{(q;q)_\infty^2}\sum_{n=-\infty}^{\infty}\frac{\mathcal{E}_q(x;i\varpi_n)\,\mathcal{E}_q(\eta;-i\varpi_n)}{2k(\varpi_n)\,(i\varpi_n)^{m+1}} \tag{10.4.12}$$

in view of (5.6.4), (5.8.13), and (5.5.8). By Theorem 6.12 the series converges inside the ellipse $E_{m/2+1/4}$ in the complex x-plane.

If we introduce the *q-Euler numbers* as

$$E_m(q)=E_m(0;q), \tag{10.4.13}$$

the generating relation (10.4.9) takes the form

$$\frac{(q\alpha^2;q^2)_\infty}{(-\alpha;q^{1/2})_\infty+(\alpha;q^{1/2})_\infty}=\frac{1}{2}+\sum_{m=1}^{\infty}\alpha^m\ E_m(q), \tag{10.4.14}$$

due to $\mathcal{E}_q(0;\alpha)=1$, and $E_{2m+1}(q)=0$, $m=0,1,2,\dots$. By using the generating relations (10.4.9) and (10.4.14) one can show that the classical Euler polynomials (10.4.2)–(10.4.3) and numbers (10.4.4) are the $q\to1^-$ limits of (10.4.10)–(10.4.11) and (10.4.13), respectively; see Ex. 10.1.

Equating the coefficients of α^n in the both sides of (10.4.9) with the help of (10.4.14) and (10.3.9) we obtain

$$E_n(x;q)=\sum_{k=0}^{n}E_k(q)\ \frac{q^{(n-k)/2}}{(q;q)_{n-k}}\ \phi_{n-k}(x;q) \tag{10.4.15}$$

as an analog of (10.4.7). The last equation shows that the functions $E_n(x;q)$ defined by (10.4.10)–(10.4.11) are, indeed, polynomials of degree n in x. We shall call them the *q-Euler polynomials*[2][**146**].

10.5. Some Properties of q-Euler Polynomials

Let us discuss some properties of the q-Euler polynomials and numbers that are similar to the classical ones. In view of (10.4.9)–(10.4.11) or (10.4.15), the following *symmetry relation* holds

$$E_n(-x;q)=(-1)^n\ E_n(x;q). \tag{10.5.1}$$

The first five q-Euler's numbers can be found directly from the generating relation (10.4.14):

$$E_0(q)=\frac{1}{2},\qquad E_1(q)=0, \tag{10.5.2}$$

$$E_2(q)=-\frac{q^{1/2}}{2\left(1-q^{1/2}\right)^2(1+q)},\qquad E_3(q)=0, \tag{10.5.3}$$

$$E_4(q)=\frac{q\left(1+2q-q^{3/2}+2q^2+q^3\right)}{2\left(1-q^{1/2}\right)^3\left(1-q^{3/2}\right)(1+q)^2\left(1+q^2\right)}. \tag{10.5.4}$$

The corresponding q-Euler polynomials are

$$E_0(x;q)=\frac{1}{2},\qquad E_1(x;q)=\frac{q^{1/4}}{(q;q)_1}\ x, \tag{10.5.5}$$

$$E_2(x;q)=\frac{2q}{(q;q)_2}\left[x^2-\frac{\left(1+q^{1/2}\right)^2}{4q^{1/2}}\right], \tag{10.5.6}$$

$$E_3(x;q)=\frac{q^{5/4}}{(q;q)_3} \tag{10.5.7}$$

[2]At the moment it is not clear what is the best normalization for the q-Euler polynomials, so the author has chosen the simplest one.

$$\times\, x\left[4q\, x^2-\frac{\left(1+q^{1/2}\right)^2}{q^{1/2}}\left(1-q^{1/2}+3q-q^{3/2}+q^2\right)\right],$$

$$E_4\left(x;q\right)=\frac{2q^2}{(q;q)_4}\left[4q^2x^4-\frac{\left(1+q^{1/2}\right)^2}{q^{1/2}}\right. \tag{10.5.8}$$

$$\times\left(1-q^{1/2}+3q-3q^{3/2}+6q^2-3q^{5/2}+3q^3-q^{7/2}+q^4\right)x^2$$

$$\left.+\frac{\left(1+q^{1/2}\right)^3\left(1+q^{3/2}\right)\left(1+2q-q^{3/2}+2q^2+q^3\right)}{4q}\right].$$

This follows directly from (10.4.15) and (10.5.2)–(10.5.4).

Equating the coefficients of α^{2n} in the both sides of

$$\frac{(q\alpha^2;q^2)_\infty}{(-\alpha;q^{1/2})_\infty+(\alpha;q^{1/2})_\infty}\left((-\alpha;q^{1/2})_\infty+(\alpha;q^{1/2})_\infty\right)=(q\alpha^2;q^2)_\infty$$

by (10.4.14) and (II.2) of [**46**] or (A.2.2), we obtain the *recurrence relation* for the q-Euler numbers,

$$2\sum_{k=0}^{n}E_{2k}\left(q\right)\frac{q^{(n-k)(n-k-1/2)}}{\left(q^{1/2};q^{1/2}\right)_{2n-2k}}=(-1)^n\frac{q^{n^2}}{\left(q^2;q^2\right)_n}, \tag{10.5.9}$$

which is a q-analog of

$$\sum_{k=0}^{n}E_{2k}\; C_{2n}^{2k}=0. \tag{10.5.10}$$

Substituting $\alpha=i\omega$ in (10.4.14) we arrive at the following analog of (10.4.5):

$$\frac{1}{C_q(\eta;\omega)}=1+2\sum_{m=1}^{\infty}(-1)^m\;E_{2m}\left(q\right)\;\omega^{2m}. \tag{10.5.11}$$

The q-exponential function satisfies (10.2.28) and applying the Askey–Wilson operator (2.3.15) to the both sides of (10.3.9) one gets

$$\frac{\delta}{\delta x}\;\phi_n\left(x;q\right)=2q^{-1/4}\frac{1-q^n}{1-q}\;\phi_{n-1}\left(x;q\right) \tag{10.5.12}$$

With the help of the last equation we obtain from (10.4.15) the following *difference-differentiation formula*

$$\frac{\delta}{\delta x}\;E_m\left(x;q\right)=\frac{2q^{1/4}}{1-q}\;E_{m-1}\left(x;q\right), \tag{10.5.13}$$

which is an analog of

$$E_m'\left(x\right)=m\;E_{m-1}\left(x\right). \tag{10.5.14}$$

The property

$$E_m\left(x+1\right)+E_m\left(x\right)=2x^m \tag{10.5.15}$$

follows directly from the generating function (10.4.1) for the classical Euler polynomials. In order to establish a q-analog of this formula we need to introduce an extension of the q-Euler polynomials by means of the generating relation

$$\frac{(q\alpha^2;q^2)_\infty\,\mathcal{E}_q(x,y;\alpha)}{(-\alpha;q^{1/2})_\infty+(\alpha;q^{1/2})_\infty} \tag{10.5.16}$$
$$=\frac{1}{2}+\sum_{m=1}^{\infty}\alpha^{2m-1}\,E_{2m-1}(x,y;q)+\sum_{m=1}^{\infty}\alpha^{2m}\,E_{2m}(x,y;q),$$

where $\mathcal{E}_q(x,y;\alpha)$ is the q-quadratic exponential in two independent variables x and y, which is an analog of $\exp[\alpha(x+y)]$. Thus, it is natural to view the generalized q-Euler polynomials $E_m(x,y;q)$ here as an extension of $E_m(x+y)$. The following properties hold

$$E_m(x,y;q)=E_m(y,x;q),\qquad E_m(x,0;q)=E_m(x;q). \tag{10.5.17}$$

Substituting $y=\pm\eta$ in (10.5.16) and then adding the results we obtain

$$\mathcal{E}_q(x;\alpha)=1+\sum_{n=1}^{\infty}\alpha^n\left(E_n(x,\eta;q)+E_n(x,-\eta;q)\right) \tag{10.5.18}$$

with the help of the addition theorem (10.3.7), the symmetry relation (10.3.8) and (10.2.9). Using the expansion (10.3.9) in the left hand side of (10.5.18) and equating the coefficients of α^n we obtain an analog of (10.5.15):

$$E_n(x,\eta;q)+E_n(x,-\eta;q)=\frac{q^{n/2}}{(q;q)_n}\,\phi_n(x;q). \tag{10.5.19}$$

In a similar manner, using (10.3.7) in (10.5.16) one can derive the following extension of (10.4.15),

$$E_n(x,y;q)=\sum_{k=0}^{n}\frac{q^{(n-k)/2}}{(q;q)_{n-k}}\,E_k(x;q)\;\phi_{n-k}(y;q), \tag{10.5.20}$$

which is the q-analog of

$$E_n(x+y)=\sum_{k=0}^{n}C_n^k\,E_k(x)\;y^{n-k}. \tag{10.5.21}$$

This consideration shows that $E_n(x,y;q)$ are polynomials in x and y. They satisfy the difference-differentiation formula (10.5.13). The first q-Euler polynomials $E_n(x,y;q)$ are

$$E_0(x,y;q)=\frac{1}{2},\qquad E_1(x,y;q)=\frac{q^{1/4}}{(q;q)_1}\,(x+y), \tag{10.5.22}$$

$$E_2(x,y;q)=\frac{2q}{(q;q)_2}\left[x^2+y^2+\frac{1+q}{q^{1/2}}\,xy-\frac{\left(1+q^{1/2}\right)^2}{4q^{1/2}}\right], \tag{10.5.23}$$

$$E_3(x,y;q)=\frac{4q^{9/4}}{(q;q)_3}\left[x^3+y^3+\frac{1+q+q^2}{q}\,xy\,(x+y)\right. \tag{10.5.24}$$

$$- \frac{\left(1+q^{1/2}\right)^2}{4q^{3/2}} \left(1 - q^{1/2} + 3q - q^{3/2} + q^2\right) \ (x+y) \Bigg],$$

$$E_4(x,y;q) = \frac{8q^4}{(q;q)_4} \tag{10.5.25}$$

$$\times \Bigg[x^4 + y^4 + \frac{1+q+q^2+q^3}{q^{3/2}} \, xy\,(x+y)$$

$$+ \frac{\left(1+q^2\right)\left(1+q+q^2\right)}{q^2} \, x^2y^2 - \frac{\left(1+q^{1/2}\right)^2}{4q^{5/2}} \left(x^2+y^2\right)$$

$$\times \left(1 - q^{1/2} + 3q - 3q^{3/2} + 6q^2 - 3q^{5/2} + 3q^3 - q^{7/2} + q^4\right)$$

$$- \frac{1+q+q^2}{4q^{5/2}} \left(1+q^{1/2}\right)^2 \left(1 - 2q^{1/2} + 5q - 2q^{3/2} + q^2\right) xy$$

$$+ \frac{\left(1+q^{1/2}\right)^3 \left(1+q^{3/2}\right)\left(1+2q-q^{3/2}+2q^2+q^3\right)}{16q^3} \Bigg].$$

The left side of the generating relation (10.5.16) can be rewritten as

$$\frac{(q\alpha^2;q^2)_\infty}{(-\alpha;q^{1/2})_\infty + (\alpha;q^{1/2})_\infty} \, (q\alpha^2;q^2)_\infty \, \mathcal{E}_q(x,y;\alpha) \, \frac{1}{(q\alpha^2;q^2)_\infty}$$

and expanding this function in the powers of α, we obtain the following representation of the q-Euler polynomials

$$E_m(x,y;q) = \sum_{n=0}^{m} E_{m-n}(q) \sum_{k=0}^{[n/2]} \frac{q^{k(k-n+1)+n^2/4}}{(q^2;q^2)_k \, (q;q)_{n-2k}} H_{n-2k}(x,y|q) \tag{10.5.26}$$

in terms of the q-Euler numbers and the continuous q-Hermite polynomials. When $y=0$ one gets (10.4.15) again with the help of (3.7.13).

Expansions (10.4.10)–(10.4.11) and the orthogonality relations for the basic trigonometric system (5.2.26)–(5.2.28) together with (5.8.11) result in the following integral evaluation

$$\int_0^\pi E_m(\cos\theta;q) \ E_n(\cos\theta;q) \ \left(e^{2i\theta}, \ e^{-2i\theta}; \ q\right)_{1/2} \ d\theta \tag{10.5.27}$$

$$= (-1)^{[(m+n)/2]} \ \pi \frac{\left(q^{1/2};q\right)_\infty^2}{(q;q)_\infty^2} \ E_{m+n+1}(\eta;q) \, .$$

This implies that the q-Euler polynomials are orthogonal when $m+n$ is odd due to (10.4.14); that is also true due to the symmetry property (10.5.1). Equation (10.5.27) is an analog of the integral evaluation

$$\int_0^1 E_m(x) \, E_n(x) \ dx = (-1)^m \, 2 \frac{m! \ n!}{(m+n+1)!} \ E_{m+n+1}(1) \, , \tag{10.5.28}$$

$m,n = 0,1,2,\dots$, for the classical Euler polynomials [**1**].

Expansions (9.3.27)–(9.3.28) together with the explicit representation (9.1.9) for the q-Lommel polynomials give

$$\prod_{k=0}^{m-1}\left(\left(1+q^{k+1/2}\right)^2-4q^{k+1/2}x^2\right)=2\,(-1)^m\,\frac{(q;q)_{2m}}{\left(q^{1/2};q\right)_m}\,q^{-m^2/2} \tag{10.5.29}$$
$$\times\sum_{k=0}^{[m/2]}q^{k(k-1/2)}\frac{(q;q)_{m-k}\left(q^{1/2};q\right)_{m-k}}{(q;q)_k\,(q;q)_{m-2k}\left(q^{1/2};q\right)_k}\,E_{2m-2k}\,(x;q)\,,$$

and

$$x\prod_{k=0}^{m-2}\left(\left(1+q^{k+1/2}\right)^2-4q^{k+1/2}x^2\right)=\frac{(q;q)_{2m-1}}{\left(q^{1/2};q\right)_m}\,q^{-m^2/2+1/4} \tag{10.5.30}$$
$$\times(-1)^{m-1}\sum_{k=0}^{[m/2]}q^{k(k-1/2)}\frac{(q;q)_{m-k}\left(q^{1/2};q\right)_{m-k}}{(q;q)_k\,(q;q)_{m-2k}\left(q^{1/2};q\right)_k}\,E_{2m-2k-1}\,(x;q)$$

for even and odd degrees, respectively.

In a similar manner, expansions (9.1.32)–(9.1.33) and (9.1.9) lead to relations between the q-Legendre polynomials $C_m\left(x;q^{1/2}\middle|\,q\right)$ and the q-Euler polynomials

$$C_{2m}\left(x;q^{1/2}\middle|\,q\right)=2q^{-m^2} \tag{10.5.31}$$
$$\times\sum_{k=0}^{m}q^{k(k-1/2)}\frac{(q;q)_{2m-k}\left(q^{1/2};q\right)_{2m-k}}{(q;q)_k\,(q;q)_{2m-2k}\left(q^{1/2};q\right)_k}\,E_{2m-2k}\,(x;q)$$

and

$$C_{2m+1}\left(x;q^{1/2}\middle|\,q\right)=2q^{-m(m+1)-1/4} \tag{10.5.32}$$
$$\times\sum_{k=0}^{m}q^{k(k-1/2)}\frac{(q;q)_{2m-k+1}\left(q^{1/2};q\right)_{2m-k+1}}{(q;q)_k\,(q;q)_{2m-2k+1}\left(q^{1/2};q\right)_k}\,E_{2m-2k+1}\,(x;q)\,.$$

On the other hand, using these formulas and the orthogonality relation (6.7.11) one can find expansions of the q-Euler polynomials in terms of the q-Legendre polynomials; cf. (10.2.34)–(10.2.35).

We have discussed here some properties of the q-Euler polynomials and numbers. They deserve more detailed consideration. Other analogs of Euler's numbers were discussed in **[28]**, **[133]** and **[134]**.

10.6. Extensions of Riemann Zeta and Related Functions

The *Riemann zeta function* is, usually, introduced as the Dirichlet series of the form

$$\zeta\,(z)=\sum_{n=1}^{\infty}n^{-z}. \tag{10.6.1}$$

This series converges absolutely and uniformly for Re $z > 1$ and, therefore, it defines a holomorphic function in the half-plane Re $z > 1$. For the analytic continuation of this function in an entire complex plane, other properties and applications, see, for example, **[1]**, **[10]**, **[36]**, **[37]**, **[84]**, **[98]**, **[155]**, and **[164]**. In view of the relation,

$$\zeta(2m) = (-1)^{m-1} \frac{(2\pi)^{2m}}{2(2m)!} B_{2m}, \qquad m > 0 \tag{10.6.2}$$

(cf. (10.6.1) and (9.1.3) for $x = 0$), the zeta function can be thought as an extension of Bernoulli's numbers to an arbitrary complex index.

Certain q-Fourier expansions found in the previous chapter give us a possibility to introduce analogs of the zeta and related functions **[144]**, **[146]**. Eq. (10.1.15), considered at $x = -\eta$, and (5.8.10) give rise to a natural extension of Riemann's zeta function as

$$\zeta_q(z) = \sum_{n=1}^{\infty} \frac{1}{\kappa(\omega_n)\ {\omega_n}^z}. \tag{10.6.3}$$

The right side here is an absolutely and uniformly convergent series of analytic functions in any domain Re $z > 1$ and consequently the series is an analytic function in such a domain. Indeed, for $z = x + iy$ consider

$$\left| \frac{{\omega_n}^{-z}}{\kappa(\omega_n)} \right| = \frac{{\omega_n}^{-x}}{\kappa(\omega_n)} = M_n, \tag{10.6.4}$$

and

$$\frac{M_{n+1}}{M_n} = \left(\frac{\omega_n}{\omega_{n+1}} \right)^x \frac{\kappa(\omega_n)}{\kappa(\omega_{n+1})}. \tag{10.6.5}$$

In view of (6.7.23),

$$\lim_{n\to\infty} \frac{\kappa(\omega_n)}{\kappa(\omega_{n+1})} = q^{-1} \frac{\lim_{n\to\infty} \kappa(\omega_n)\, q^{-n}}{\lim_{n\to\infty} \kappa(\omega_{n+1})\, q^{-n-1}} = q^{-1}, \tag{10.6.6}$$

and with the help of the asymptotic formula

$$\lim_{m\to\infty} q^m \omega_m = q^{1/4} \tag{10.6.7}$$

one gets

$$\lim_{n\to\infty} \frac{M_{n+1}}{M_n} = q^{\mathrm{Re}\ z-1} < 1, \qquad \mathrm{Re}\ z > 1. \tag{10.6.8}$$

This completes the proof by the Ration Test. Another proof using convergence of a general Dirichlet series will be given in the next section.

We shall call the function $\zeta_q(z)$ given by (10.6.3) the *q-zeta function* **[144]**, **[146]**.

Let us investigate some properties of this function. The series (10.6.3) can also be written as a contour integral of the form

$$\frac{1}{2\pi i} \int_C \mathrm{Cot}_q(\omega)\ \omega^{-z}\, d\omega = \zeta_q(z) \tag{10.6.9}$$

by Cauchy's residue theorem and (9.4.7). Here C is a contour which starts at the infinity in the upper half-plane, encircles the line $\omega > \omega_1$ in the positive direction excluding the origin, and returns to the infinity in the lower half-plane.

Moreover, in view of (10.1.15) or (10.1.16), (10.1.20) and (5.8.12) one gets

$$\zeta_q(2m) = (-1)^{m-1} B_{2m}(q), \tag{10.6.10}$$

which is an analog of the classical relation (10.6.2) between the zeta function and the Bernoulli numbers. This relation can be thought of as a consequence of Parseval's identity (9.5.1) for the expansions (10.1.14) and (10.1.15); the corresponding integral can be evaluated by (10.2.29). On the other hand, due to the orthogonality relation (9.1.17) and (10.6.3), formula (10.6.10) gives explicit representation for the moments of the q-Lommel polynomials (9.1.19) in terms of the q-Bernoulli numbers. One can write Gram's determinant expression for these polynomials in terms of the q-zeta function under consideration; see Ex. 10.7.

As special cases, due to (9.5.11) and (9.5.12), we have

$$\zeta_q(2) = \frac{q^{1/2}}{2\left(1-q^{3/2}\right)} \tag{10.6.11}$$

and

$$\zeta_q(4) = \frac{q^2}{2\left(1-q^{3/2}\right)^2\left(1-q^{5/2}\right)}, \tag{10.6.12}$$

which are analogs of the classical Euler's results (9.5.13), namely,

$$\zeta(2) = \frac{\pi^2}{6}, \qquad \zeta(4) = \frac{\pi^4}{90}. \tag{10.6.13}$$

Euler considered also the following series

$$\phi(z) = \sum_{n=1}^{\infty} \frac{(-1)^{n-1}}{n^z}, \tag{10.6.14}$$

see [19] for a nice review on Euler's discovery of the main properties of the zeta and related functions. It is connected to the zeta function as

$$\phi(z) = \left(1-2^{1-z}\right)\zeta(z). \tag{10.6.15}$$

The special values are

$$\phi(2) = \frac{\pi^2}{12}, \qquad \phi(4) = \frac{7\pi^4}{720} \tag{10.6.16}$$

in view of (10.6.13) and (10.6.15).

Eq. (10.1.15) at $x=0$ gives us the possibility to introduce a q-extension of $\phi(z)$ as

$$\phi_q(z) = \sum_{n=1}^{\infty} \frac{(-1)^{n-1}}{\kappa(\omega_n)\,\omega_n{}^z} \sqrt{\frac{\left(-q\omega_n^2;q^2\right)_\infty}{\left(-\omega_n^2;q^2\right)_\infty}} \tag{10.6.17}$$

and to evaluate the following sums

$$\phi_q(2m) = (-1)^m \, B_{2m}(0;q) \tag{10.6.18}$$

in terms of the values of the q-Bernoulli polynomials $B_{2m}(x;q)$ at $x = 0$. These values can be found from (10.2.16)–(10.2.17) as follows

$$B_n(0;q) = \sum_{k=0}^{n} \frac{\left(-q^{1/2};q\right)_k}{(q;q)_k} \, B_{n-k}(q). \tag{10.6.19}$$

Extensions of (10.6.16) are

$$\phi_q(2) = \frac{q}{2(1+q)\left(1-q^{3/2}\right)} \tag{10.6.20}$$

and

$$\phi_q(4) = \frac{q^{5/2}\left(1+2q+q^{3/2}+2q^2+q^3\right)}{2(1+q)^2\left(1+q^2\right)\left(1-q^{3/2}\right)^2\left(1-q^{5/2}\right)} \tag{10.6.21}$$

in view of (10.2.2), (10.2.4), and (10.6.18). From (10.6.11)–(10.6.12) and (10.6.20)–(10.6.21) one gets

$$\phi_q(2)/\zeta_q(2) = \frac{q^{1/2}}{1+q} \to \frac{1}{2}, \tag{10.6.22}$$

$$\phi_q(4)/\zeta_q(4) = q^{1/2}\,\frac{1+2q+q^{3/2}+2q^2+q^3}{(1+q)^2\left(1+q^2\right)} \to \frac{7}{8} \tag{10.6.23}$$

as $q \to 1^-$; cf. (10.6.15) for $z = 2$ and $z = 4$, respectively.

Let us introduce also the following functions

$$\xi_q(z) = \sum_{n=1}^{\infty} \frac{1}{\kappa(\varpi_n)\ \varpi_n{}^z}, \tag{10.6.24}$$

and

$$\chi_q(z) = \sum_{n=1}^{\infty} \frac{(-1)^{n-1}}{\kappa(\varpi_n)\ \varpi_n{}^z} \sqrt{\frac{\left(-q\varpi_n^2;q^2\right)_\infty}{\left(-\varpi_n^2;q^2\right)_\infty}} \tag{10.6.25}$$

as q-analogs of

$$\xi(z) = \sum_{n=0}^{\infty} \frac{1}{(2n+1)^z}, \qquad \chi(z) = \sum_{n=0}^{\infty} \frac{(-1)^n}{(2n+1)^z}, \tag{10.6.26}$$

respectively. Letting $x = \eta$ and $x = 0$ in (10.4.10) and (10.4.11), respectively, we arrive at the following relations with the q-Euler polynomials and numbers

$$\xi_q(2m) = (-1)^{m-1} E_{2m-1}(\eta;q), \tag{10.6.27}$$

$$\chi_q(2m+1) = (-1)^m E_{2m}(q). \tag{10.6.28}$$

In particular,

$$\xi_q(2) = \frac{1}{2\left(1-q^{1/2}\right)}, \tag{10.6.29}$$

$$\xi_q(4) = \frac{q^{1/2}}{2\left(1-q^{1/2}\right)^2\left(1-q^{3/2}\right)}, \tag{10.6.30}$$

$$\chi_q(1) = \frac{1}{2}, \quad \chi_q(3) = \frac{q^{1/2}}{2\left(1-q^{1/2}\right)^2(1+q)}, \tag{10.6.31}$$

which are q-analogs of the classical sums [**1**]

$$\xi(2) = \frac{\pi^2}{8}, \qquad \xi(4) = \frac{\pi^4}{96}, \tag{10.6.32}$$

$$\chi(1) = \frac{\pi}{4}, \qquad \chi(3) = \frac{\pi^3}{32}. \tag{10.6.33}$$

Functions $\zeta_q(z)$, $\phi_q(z)$, $\xi_q(z)$, and $\chi_q(z)$ deserve a more detailed consideration.

10.7. Analytic Continuation of q-Zeta Function

Analytic continuation of the q-zeta function is an interesting open problem. In this section we shall make a modest step in this direction.

10.7.1. Basic Zeta Function as Dirichlet series. An extension of the zeta function introduced above is

$$\zeta_q(z) = \sum_{n=1}^{\infty} \frac{1}{\kappa(\omega_n)\ \omega_n{}^z}, \tag{10.7.1}$$

where $\{\omega_n\}_{n=1}^{\infty}$ are the positive zeros of the q-sine function $S_q(\eta;\omega)$ and the $\kappa(\omega)$ is defined by (5.5.9). The right side here is an absolutely and uniformly convergent series of analytic functions in any domain Re $z > 1$ and consequently the series is an analytic function in such a domain. Here we give, first of all, an independent proof of this result using elementary facts about convergence of the Dirichlet series of a general form [**21**], [**106**]

$$\sum_{n=1}^{\infty} a_n e^{-\lambda_n z}, \tag{10.7.2}$$

where the a_n are complex numbers and the exponents λ_n are nonnegative real numbers satisfying the conditions

$$\lambda_n < \lambda_{n+1}, \quad n = 1, 2, 3, \dots, \qquad \lim_{n\to\infty} \lambda_n = \infty. \tag{10.7.3}$$

The q-zeta function (10.7.1) is the Dirichlet series of the form (10.7.2) with

$$a_n = \frac{1}{\kappa(\omega_n)}, \qquad \lambda_n = \log \omega_n. \tag{10.7.4}$$

In view of the asymptotic formula

$$\omega_n = q^{1/4-n} + \mathrm{O}(1), \qquad n \to \infty \tag{10.7.5}$$

see (6.3.1); one gets

$$\lambda_n = \log \omega_n = (n - 1/4)\log q^{-1} + \mathrm{O}(q^n), \qquad n \to \infty \tag{10.7.6}$$

and both conditions (10.7.3) are obviously satisfied when $0 < q < 1$.

By the definition, a Dirichlet series of a general form (10.7.2) is said to have *abscissa of convergence* C, or *half-plane of convergence* $\operatorname{Re} z > C$, if the series converges at every point of the half-plane $\operatorname{Re} z > C$, but diverges at any point of the half-plane $\operatorname{Re} z < C$. Similarly, a Dirichlet series (10.7.2) is said to have *abscissa of absolute convergence* A, or *half-plane of absolute convergence* $\operatorname{Re} z > A$, if the series converges absolutely at every point of the half-plane $\operatorname{Re} z > A$, but not at any point of the half-plane $\operatorname{Re} z < A$; see [**21**], [**106**] and Appendix B.3 for more details. In particular, Theorem B.10 provides a generalization of the Cauchy–Hadamard Theorem for the power series — it states that for a Dirichlet series (10.7.2) satisfying conditions (10.7.3) and the limiting condition

$$\limsup_{n\to\infty} \frac{\log n}{\lambda_n} = 0, \tag{10.7.7}$$

the numbers A and C are given by the formula

$$A = C = \limsup_{n\to\infty} \frac{\log |a_n|}{\lambda_n}. \tag{10.7.8}$$

One can look at [**21**], [**106**] for the proof of this result.

In the case of the q-zeta function (10.7.1) we get

$$\limsup_{n\to\infty} \frac{\log n}{\lambda_n} = \lim_{n\to\infty} \frac{\log n}{n \log q^{-1} + \mathrm{O}(1)} = \frac{1}{\log q^{-1}} \lim_{n\to\infty} \frac{\log n}{n} = 0 \tag{10.7.9}$$

due to (10.7.6). Thus, the condition (10.7.7) is satisfied. We shall show in this section that

$$\kappa(\omega_n) = 2\frac{\left(q^2; q^2\right)_\infty^2}{\left(q; q^2\right)_\infty^2} q^n \left(1 + \mathrm{o}(1)\right), \qquad n \to \infty, \tag{10.7.10}$$

see a more general result in (10.7.13) below. Therefore

$$\log |a_n| = -\log \kappa(\omega_n) = n \log q^{-1} + \mathrm{O}(1), \qquad n \to \infty \tag{10.7.11}$$

and by (10.7.6) and (10.7.8)

$$A = C = \lim_{n\to\infty} \frac{\log |a_n|}{\lambda_n} = \lim_{n\to\infty} \frac{n \log q^{-1} + \mathrm{O}(1)}{n \log q^{-1} + \mathrm{O}(1)} = 1. \tag{10.7.12}$$

As a result, by Theorem B.10 the series (10.7.1) for the q-zeta function $\zeta_q(z)$ converges absolutely and uniformly in the half-plane $\operatorname{Re} z > 1$ and it defines an analytic function in such a half-plane. This series diverges in the half-plane $\operatorname{Re} z < 1$.

10.7.2. Analytic Continuation in Larger Domain. Analytic continuation of the $\zeta_q(z)$ to the entire complex plane is an interesting open problem. We shall show here that, as in the classical case, the $\zeta_q(z)$ has a simple pole at $z = 1$ and it has no other singularities in the half-plane $\operatorname{Re} z > 0$. With the help of the improved asymptotic for the zeros of the basic sine function found in Section 8.4, we will be able to show that, in addition, the q-zeta function under consideration has simple poles at $z = -1$ and $z = -2$ and it has no other singularities in the half-plane $\operatorname{Re} z > -3$.

We first establish the following asymptotic formula

$$\begin{aligned}\kappa(\omega_n) = {} & \frac{q^{1/2}}{\left(1-q^{1/2}\right)c_1(q)}\left(\omega_n^{(0)}\right)^{-1} + \frac{q^{1/2}c_1(q)\,c_2(q)}{1-q^{1/2}}\left(\omega_n^{(0)}\right)^{-3} \\ & + \frac{q^{1/2}}{2\left(1-q^{1/2}\right)}\left(c_1^2(q)\left(2c_2(q)-3\right)+\frac{2q}{1+q^{1/2}+q}\right)\left(\omega_n^{(0)}\right)^{-4} \\ & + \mathrm{O}\left(\left(\omega_n^{(0)}\right)^{-5}\right), \qquad n\to\infty,\end{aligned} \tag{10.7.13}$$

where $\omega_n^{(0)} = q^{1/4-n}$, functions $c_1(q)$ and $c_2(q)$ are given by (8.4.26) and (8.5.5), respectively. This asymptotic formula is of independent interest.

Indeed, Taylor's formula

$$\begin{aligned}\kappa(\omega_n) = {} & \kappa\left(\omega_n^{(0)}\right) + \kappa'\left(\omega_n^{(0)}\right)\left(\omega_n-\omega_n^{(0)}\right) \\ & + \frac{1}{2}\kappa''\left(\omega_n^{(0)}\right)\left(\omega_n-\omega_n^{(0)}\right)^2 + \ldots\end{aligned} \tag{10.7.14}$$

and asymptotics (8.4.7)–(8.4.9) and (8.4.24) result in (10.7.13) after some simplification.

Eq. (10.7.13) implies

$$\begin{aligned}\frac{1}{\kappa(\omega_n)} = {} & \frac{\left(1-q^{1/2}\right)c_1(q)}{q^{1/2}}\,\omega_n^{(0)}\left(1-\frac{c_1^2(q)\,c_2(q)}{\left(\omega_n^{(0)}\right)^2}\right. \\ & \left. -\left(c_1^2(q)\left(2c_2(q)-3\right)+\frac{2q}{1+q^{1/2}+q}\right)\frac{c_1(q)}{2\left(\omega_n^{(0)}\right)^3} + \mathrm{O}\left(\left(\omega_n^{(0)}\right)^{-4}\right)\right)\end{aligned} \tag{10.7.15}$$

as $n\to\infty$.

In a similar fashion, with the help of the binomial formula and (8.4.24)

$$\omega_n^{-z} = \left(\omega_n^{(0)}\right)^{-z}\left(1+\frac{zc_1(q)}{\omega_n^{(0)}}+\frac{z(z+2)c_1(q)}{2\left(\omega_n^{(0)}\right)^2}\right. \tag{10.7.16}$$

$$+\frac{c_1(q)\, z\left(c_1^2(q)(z+1)(z+5)+c_3(q)\right)}{6\left(\omega_n^{(0)}\right)^3}+\mathrm{O}\left(\left(\omega_n^{(0)}\right)^{-4}\right)\Bigg)$$

as $n \to \infty$, where by the definition

$$c_3(q)=c_1^2(q)\left(2c_2(q)+3\right)-\frac{2q}{1+q^{1/2}+q} \tag{10.7.17}$$

and functions $c_1(q)$ and $c_2(q)$ are given by (8.4.26) and (8.5.5), respectively.

As a result

$$\frac{1}{\kappa(\omega_n)\;\omega_n^z}=\frac{\left(1-q^{1/2}\right)c_1(q)}{q^{1/2}}\left(\omega_n^{(0)}\right)^{1-z} \tag{10.7.18}$$

$$\times\left(1+\frac{c_1(q)\, z}{\omega_n^{(0)}}+\frac{c_1^2(q)\, a_1(z)}{2\left(\omega_n^{(0)}\right)^2}+\frac{c_1(q)\, a_2(z)}{6\left(\omega_n^{(0)}\right)^3}+\mathrm{O}\left(\left(\omega_n^{(0)}\right)^{-4}\right)\right)$$

as $n \to \infty$, where

$$a_1(z)=z^2+2z-2c_2(q) \tag{10.7.19}$$

and

$$a_2(z)=c_1^2(q)(z+1)\left(z^2+5z-6c_2(q)\right) \tag{10.7.20}$$

$$+z\left(c_1^2(q)\left(2c_2(q)+3\right)-\frac{2q}{1+q^{1/2}+q}\right)+9c_1^2(q)-\frac{6q}{1+q^{1/2}+q}.$$

Introducing

$$b_n=\frac{1}{\kappa(\omega_n)\;\omega_n^z}-\frac{1}{2}\frac{\left(q;q^2\right)_\infty^2}{\left(q^2;q^2\right)_\infty^2}\,q^{n(z-1)-z/4} \tag{10.7.21}$$

$$-\frac{1}{4\left(1-q^{1/2}\right)}\frac{\left(q;q^2\right)_\infty^4}{\left(q^2;q^2\right)_\infty^4}\,z\,q^{z(n-1/4)}$$

$$-\frac{q^{1/4}}{16\left(1-q^{1/2}\right)^2}\frac{\left(q;q^2\right)_\infty^6}{\left(q^2;q^2\right)_\infty^6}\,a_1(z)\,q^{(z+1)(n-1/4)}$$

$$-\frac{1}{24\left(1-q^{1/2}\right)}\frac{\left(q;q^2\right)_\infty^4}{\left(q^2;q^2\right)_\infty^4}\,a_2(z)\,q^{(z+2)(n-1/4)}$$

$$=\mathrm{O}\left(\left(\left(\omega_n^{(0)}\right)^{-z-3}\right)\right),\qquad n\to\infty,$$

one can see that the series

$$\sum_{n=1}^{\infty}|b_n|<\infty \tag{10.7.22}$$

converges absolutely and uniformly when Re $z > -3$. Also

$$\sum_{n=1}^{\infty} \frac{1}{\left(\omega_n^{(0)}\right)^{z-1}} = q^{(1-z)/4} \sum_{n=1}^{\infty} q^{n(z-1)} = \frac{q^{3(z-1)/4}}{1-q^{z-1}} \tag{10.7.23}$$

when Re $z > 1$ and as a result we arrive at the following series representation for the q-zeta function,

$$\begin{aligned} \zeta_q(z) &= \frac{1}{2} \frac{\left(q;q^2\right)_\infty^2}{\left(q^2;q^2\right)_\infty^2} \frac{q^{3z/4-1}}{1-q^{z-1}} + \frac{1}{4\left(1-q^{1/2}\right)} \frac{\left(q;q^2\right)_\infty^4}{\left(q^2;q^2\right)_\infty^4} \frac{zq^{3z/4}}{1-q^z} \\ &+ \frac{1}{16\left(1-q^{1/2}\right)^2} \frac{\left(q;q^2\right)_\infty^6}{\left(q^2;q^2\right)_\infty^6} \frac{a_1(z)\, q^{3z/4+1}}{1-q^{z+1}} \\ &+ \frac{1}{24\left(1-q^{1/2}\right)} \frac{\left(q;q^2\right)_\infty^4}{\left(q^2;q^2\right)_\infty^4} \frac{a_2(z)\, q^{3(z+2)/4}}{1-q^{z+2}} \\ &+ \sum_{n=1}^{\infty} \left(\frac{1}{\kappa(\omega_n)\ \omega_n^z} - \frac{1}{2} \frac{\left(q;q^2\right)_\infty^2}{\left(q^2;q^2\right)_\infty^2}\ q^{n(z-1)-z/4} \right. \\ &- \frac{1}{4\left(1-q^{1/2}\right)} \frac{\left(q;q^2\right)_\infty^4}{\left(q^2;q^2\right)_\infty^4}\, z\, q^{z(n-1/4)} \\ &- \frac{q^{1/4}}{16\left(1-q^{1/2}\right)^2} \frac{\left(q;q^2\right)_\infty^6}{\left(q^2;q^2\right)_\infty^6}\, a_1(z)\, q^{(z+1)(n-1/4)} \\ &\left. - \frac{1}{24\left(1-q^{1/2}\right)} \frac{\left(q;q^2\right)_\infty^4}{\left(q^2;q^2\right)_\infty^4}\, a_2(z)\, q^{(z+2)(n-1/4)} \right), \end{aligned} \tag{10.7.24}$$

where the series defines a holomorphic function in the half-plane Re $z > -3$.

We summarize our findings in the following theorem.

THEOREM 10.1. *The q-zeta function under consideration is a meromorphic function in the half-plane* Re $z > -3$. *The $\zeta_q(z)$ has simple poles at $z = 1, -1, -2$ with the residues*

$$\operatorname*{Res}_{z=1} \zeta_q(z) = \frac{q^{-1/4}}{2\log q^{-1}} \frac{\left(q;q^2\right)_\infty^2}{\left(q^2;q^2\right)_\infty^2}, \tag{10.7.25}$$

$$\operatorname*{Res}_{z=-1} \zeta_q(z) = -\frac{q^{1/4}}{16\log q^{-1}} \frac{\left(q;q^2\right)_\infty^6}{\left(q^2;q^2\right)_\infty^6} \frac{2c_2(q)+1}{\left(1-q^{1/2}\right)^2} \tag{10.7.26}$$

and

$$\begin{aligned} \operatorname*{Res}_{z=-2} \zeta_q(z) &= \frac{1}{24\log q^{-1}\left(1-q^{1/2}\right)} \frac{\left(q;q^2\right)_\infty^4}{\left(q^2;q^2\right)_\infty^4} \\ &\times \left(c_1^2(q)\left(2c_2(q)+9\right) - \frac{2q}{1+q^{1/2}+q} \right), \end{aligned} \tag{10.7.27}$$

where the $c_1(q)$ and $c_2(q)$ are given by are given by (8.4.26) *and* (8.5.5), *respectively. It has no other singularities in the half-plane* Re $z > -3$.

The corresponding q-Euler constant can be defined as

$$\lim_{z\to 1}\left(\zeta_q(z)-\frac{1}{2}\frac{\left(q;q^2\right)_\infty^2}{\left(q^2;q^2\right)_\infty^2}\frac{q^{3z/4-1}}{1-q^{z-1}}\right) \tag{10.7.28}$$

$$=\lim_{m\to\infty}\left(\sum_{n=1}^{m}\frac{1}{\kappa(\omega_n)\,\omega_n}-\frac{1}{2}q^{-1/4}\frac{\left(q;q^2\right)_\infty^2}{\left(q^2;q^2\right)_\infty^2}m\right)=\gamma_q,$$

which can be viewed as an analog of the classical result

$$\lim_{z\to 1}\left(\zeta(z)-\frac{1}{1-z}\right)=\lim_{m\to\infty}\left(\sum_{n=1}^{m}\frac{1}{n}-\log m\right)=\gamma. \tag{10.7.29}$$

From (10.7.24)

$$\zeta_q(0)=\lim_{z\to 0}\zeta_q(z) \tag{10.7.30}$$

$$=-\frac{1}{2(1-q)}\frac{\left(q;q^2\right)_\infty^2}{\left(q^2;q^2\right)_\infty^2}+\sum_{n=1}^{\infty}\left(\frac{1}{\kappa(\omega_n)}-\frac{1}{2}\frac{\left(q;q^2\right)_\infty^2}{\left(q^2;q^2\right)_\infty^2}q^{-n}\right).$$

This method can also be applied in order to investigate an analytic continuation of the similar series introduced in the previous section.

It is hoped that our observations in the last two sections will lead, eventually, to a q-extension of the theory of the zeta and related functions. Other q-analogs of the zeta function were discussed by Al-Salam [**6**], Satoh [**133**], Tsumura [**157**], [**158**], [**159**], [**160**] and in the papers by Cherednik [**30**], [**31**], and [**32**].

10.8. Exercises for Chapter 10

(1) Use the generating function (10.1.13) and the limiting relation (10.1.21) in order to show that the classical Bernoulli polynomials and numbers are the $q\to 1^-$ limits of (10.1.14)–(10.1.15) and (10.1.20), respectively.
Show that the classical Euler polynomials (10.4.2)–(10.4.3) and numbers (10.4.4) are the $q\to 1^-$ limits of (10.4.10)–(10.4.11) and (10.4.13), respectively.

(2) Verify (10.2.1)–(10.2.7) by using (10.2.12), (10.2.16) and Ex. 3.5.

(3) In the classical case, Eq. (10.1.7) leads directly to a trivial identity, $B_m = B_m$, when $x=0$. Substitute $x=-\eta$ in (10.2.15),

$$B_m(q)=\sum_{n=0}^{m}B_{m-n}(q)\sum_{k=0}^{n}\frac{q^{k^2/4}}{\left(q^{1/2};q^{1/2}\right)_{n-k}(q;q)_k}H_k(-\eta|q),$$

and use

$$H_k(-\eta|q) = (-1)^k \frac{(q;q)_k}{\left(q^{1/2};q^{1/2}\right)_k} q^{-k/4}$$

in order to get

$$B_m(q) = \sum_{n=0}^{m} B_{m-n}(q) \sum_{k=0}^{n} \frac{(-1)^k q^{k(k-1)/4}}{\left(q^{1/2};q^{1/2}\right)_k (q^{1/2};q^{1/2})_{n-k}},$$

which leads again to the trivial identity $B_m(q) = B_m(q)$ due to

$$\sum_{k=0}^{n} \frac{(-1)^k q^{k(k-1)/4}}{\left(q^{1/2};q^{1/2}\right)_k (q^{1/2};q^{1/2})_{n-k}} = \delta_{n,0}.$$

Derive the last relation expanding both sides of the trivial identity $1 = \left(\alpha;q^{1/2}\right) / \left(\alpha;q^{1/2}\right)$ in the powers of α with the help of (A.2.1)–(A.2.2).

(4) Show that

$$\begin{aligned} B_{2m}(0;q) &= \sum_{n=0}^{2m} B_{2m-n}(q) \sum_{k=0}^{[n/2]} \frac{(-1)^k q^{k^2}}{\left(q^{1/2};q^{1/2}\right)_{n-2k} (q^2;q^2)_k} \\ &= \sum_{n=0}^{m} B_{2m-2n}(q) \sum_{k=0}^{n} \frac{(-1)^k q^{k^2}}{\left(q^{1/2};q^{1/2}\right)_{2n-2k} (q^2;q^2)_k} \\ &+ B_1(q) \sum_{k=0}^{m-1} \frac{(-1)^k q^{k^2}}{\left(q^{1/2};q^{1/2}\right)_{2m-2k-1} (q^2;q^2)_k}. \end{aligned}$$

(5) Derive (10.3.11) from (10.3.12) and derive (10.5.19) from (10.5.20).

(6) Using the definition of the q-shift operator given in Ex. 3.14 and relations (10.3.4) and (10.5.26) show that

$$T_q(y) B_n(x;q) = B_n(x \dot{+} y;q) = B_n(x,y;q),$$
$$T_q(y) E_n(x;q) = E_n(x \dot{+} y;q) = E_n(x,y;q).$$

(7) *Gram's determinants for special q-Lommel polynomials.* Show that the moments

$$C_m = \int_E x^m \, d\mu, \qquad E = Supp\ \mu$$

for the special q-Lommel polynomials $h_m(x;q) = h_m^\nu(x;q)$ are given by

$$C_{2k} = 2^{1-2k} \sum_{n=1}^{\infty} \frac{1}{\kappa(\omega_n)\ \omega_n{}^{2k+2}} = 2^{1-2k} \zeta_q(2k+2),$$

$$C_{2k+1} = 0,$$

where $\zeta_q(z)$ is the q-zeta function.

Derive the following Gram determinant representations for these polynomials

$$h_{2k}(x;q) = A_{2k}$$
$$\times \begin{vmatrix} 2\zeta_q(2) & 0 & \dots & 0 & 2^{1-2k}\zeta_q(2k+2) \\ 0 & 2^{-1}\zeta_q(4) & \dots & 2^{1-2k}\zeta_q(2k+2) & 0 \\ \dots & & & & \\ 0 & 2^{1-2k}\zeta_q(2k+2) & \dots & 2^{3-4k}\zeta_q(4k) & 0 \\ 1 & x & \dots & x^{2k-1} & x^{2k} \end{vmatrix}$$

and

$$h_{2k+1}(x;q) = A_{2k+1}$$
$$\times \begin{vmatrix} 2\zeta_q(2) & 0 & \dots & 2^{1-2k}\zeta_q(2k+2) & 0 \\ 0 & 2^{-1}\zeta_q(4) & \dots & 0 & 2^{-1-2k}\zeta_q(2k+4) \\ \dots & & & & \\ 2^{1-2k}\zeta_q(2k+2) & 0 & \dots & 2^{1-4k}\zeta_q(4k+2) & 0 \\ 1 & x & \dots & x^{2k} & x^{2k+1} \end{vmatrix},$$

where A_m are normalizing constants. From here find $h_0(x;q)$, $h_1(x;q)$, and $h_2(x;q)$ using special values $\zeta_q(2)$ and $\zeta_q(4)$ of the q-zeta function.
Find the moments and Gram's determinant representations for the special q-Lommel polynomials $\overline{h}_m(x;q) = h_m^{1/2}(x;q)$.

(8) *Fractional integrals and Bernoulli's polynomials.* Following Weyl's approach [**166**] define a fractional integral of order α of a function f which is integrable and of period 2π to be $I_\alpha f$ given by

$$(I_\alpha f)(x) \sim {\sum_{n=-\infty}^{\infty}}' c_n \frac{e^{inx}}{(in)^\alpha}, \qquad \text{if} \quad f(x) \sim \sum_{n=-\infty}^{\infty} c_n\, e^{inx},$$

where $i^\alpha = e^{i\alpha\pi/2}$ and the prime in the sum means that the term with $n=0$ should be omitted. Show that

$$(I_\alpha f)(x) = \frac{1}{2\pi}\int_0^{2\pi} f(y)\ \mathcal{K}_\alpha(x-y)\ dy,$$

where

$$\mathcal{K}_\alpha(x) = {\sum_{n=-\infty}^{\infty}}' \frac{e^{inx}}{(in)^\alpha}.$$

Prove the semigroup property $I_\alpha I_\beta = I_\beta I_\alpha = I_{\alpha+\beta}$ for $\alpha,\ \beta > 0$.
Verify that

$$(I_1 f)(x) = \int_0^x f(y)\ dy,$$

where $f(x) \sim \sum_{n=-\infty}^{\infty} c_n\, e^{inx}$ with $c_0 = 0$. Show that if $\alpha > m$ then

$$\frac{d^m}{dx^m} I_\alpha = I_{\alpha-m}.$$

Verify that when m is a positive integer $\mathcal{K}_m(x)$ is a constant multiple of the Bernoulli polynomials $B_m(x)$.

(9) *Some q-fractional integrals.* Introduce the following kernel [**67**]

$$\mathcal{K}_\alpha(x,y|q) := \sum_{n=-\infty}^{\infty}{}' \frac{\mathcal{E}_q(x;i\omega_n)\,\mathcal{E}_q(y;-i\omega_n)}{2k(\omega_n)(i\omega_n)^\alpha} = \sum_{n=-\infty}^{\infty}{}' \frac{\mathcal{E}_q(x,-y;i\omega_n)}{2k(\omega_n)(i\omega_n)^\alpha}, \qquad \alpha>0$$

with $i^\alpha = e^{i\alpha\pi/2}$, where the prime in the sum means that the term with $n=0$ should be omitted, and investigate convergence of this series.

Prove the following symmetry relations

$$\mathcal{K}_\alpha(x,y|q) = \mathcal{K}_\alpha(-y,-x|q) = e^{i\pi\alpha}\mathcal{K}_\alpha(y,x|q).$$

Show that if $\alpha,\ \beta>0$ then

$$\int_{-1}^{1} \mathcal{K}_\alpha(x,y|q)\ \mathcal{E}_q(y;i\omega_n)\ \rho(y)\,dy = \frac{1}{(i\omega_n)^\alpha}\ \mathcal{E}_q(x;i\omega_n),$$

$$\int_{-1}^{1} \mathcal{K}_\alpha(x,y|q)\ \mathcal{K}_\beta(y,z|q)\ \rho(y)\,dy = \mathcal{K}_{\alpha+\beta}(x,z|q)$$

and

$$\frac{\delta}{\delta x}\mathcal{K}_\alpha(x,y|q) = \frac{2q^{1/4}}{1-q}\ \mathcal{K}_{\alpha-1}(x,y|q), \qquad \alpha>1,$$

where $\rho(x)$ is the weight function in the orthogonality relation (5.6.5) and the operator $\mathcal{D}_q=\delta/\delta x$ is the standard Askey–Wilson divided difference operator of the first order defined by (2.3.15).

Let

$$f(x) = \sum_{n=-\infty}^{\infty} c_n\ \mathcal{E}_q(x;i\omega_n),$$

where the series converges absolutely and uniformly on $[-1,1]$. Show that

$$\lim_{\alpha\to 0^+}\int_{-1}^{1} \mathcal{K}_\alpha(x,y|q)\ f(y)\ \rho(y)\,dy = f(x),$$

when $-1\le x\le 1$.

Introduce the following "q-fractional" integral operator [**67**]

$$(I_\alpha f)(x) := \left(\frac{1-q}{2q^{1/4}}\right)^\alpha \int_{-1}^{1} \mathcal{K}_\alpha(x,y|q)\ f(y)\ \rho(y)\,dy,$$

prove the semigroup property

$$I_\alpha I_\beta = I_\beta I_\alpha = I_{\alpha+\beta}, \qquad \alpha,\ \beta>0,$$

and show that

$$\mathcal{D}_q I_\alpha = I_{\alpha-1}, \qquad \mathcal{D}_q^m I_\alpha = I_{\alpha-m}, \qquad \alpha > m.$$

Suppose that function f has the q-Fourier series expansion that converges absolutely and uniformly in the interior and on the ellipse $E_{1/2}$ given by (6.9.1)–(6.9.2) with $\varepsilon = 1/2$. Show that

$$\mathcal{D}_q I_1 = E,$$

where E is the identity operator, for all such functions f.
Define a version of the q-Bernoulli polynomials as [**67**]

$$\mathcal{B}_m(x, y; q) := \mathcal{K}_m(x, y|q), \qquad m = 1, 2, 3, ...$$

and investigate their properties. Show that

$$B_m(x; q) = -\pi \frac{\left(q^{1/2}; q\right)_\infty^2}{(q; q)_\infty^2} \mathcal{B}_m(x, \eta; q).$$

Show that

$$\sum_{m=0}^{\infty} \alpha^m \mathcal{B}_m(x, y; q) = \alpha \sum_{n=-\infty}^{\infty} \frac{\mathcal{E}_q(x; i\omega_n)\, \mathcal{E}_q(y; -i\omega_n)}{2k(\omega_n)(i\omega_n - \alpha)},$$

when $|\alpha| < \omega_1$ and derive from here as a special case the generating relation (10.1.13) for $y = \eta$ with the help of the q-Fourier expansion (9.3.2).
[*Note*: Explicit forms of the kernels $\mathcal{K}_\alpha(x, y|q)$ when $\alpha = 1, 2, 3$ have been recently evaluated in [**67**] by formal series manipulations. A rigorous proof needs to be found.]

CHAPTER 11

Numerical Investigation of Basic Fourier Series

Our main objective in this chapter is to discuss several computational aspects of the theory of basic Fourier series. This includes numerical evaluation of the zeros of basic trigonometric functions, study of their bounds and asymptotics, and numerical examples demonstrating convergence of the q-Fourier series. Most of this material appeared in our joint paper with Bill Gosper [**48**], who wrote the special Macsyma program "namesum" for numerical evaluation of infinite sums and infinite products.

11.1. Eigenvalues

11.1.1. Preliminaries. In this section we shall find numerical solutions of the transcendental equations

$$S_q(\eta;\omega)=\frac{(-i\omega;q^{1/2})_\infty-(i\omega;q^{1/2})_\infty}{2i(-q\omega^2;q^2)_\infty}=0, \tag{11.1.1}$$

$$C_q(\eta;\omega)=\frac{(-i\omega;q^{1/2})_\infty+(i\omega;q^{1/2})_\infty}{2(-q\omega^2;q^2)_\infty}=0 \tag{11.1.2}$$

for different values of the parameter q and discuss some properties of these zeros. The following series representations,

$$S_q(\eta;\omega)=\frac{1}{(-q\omega^2;q^2)_\infty}\sum_{n=0}^{\infty}(-1)^n\frac{q^{n(n+1/2)}}{\left(q^{1/2};q^{1/2}\right)_{2n+1}}\,\omega^{2n+1}, \tag{11.1.3}$$

$$C_q(\eta;\omega)=\frac{1}{(-q\omega^2;q^2)_\infty}\sum_{n=0}^{\infty}(-1)^n\frac{q^{n(n-1/2)}}{\left(q^{1/2};q^{1/2}\right)_{2n}}\,\omega^{2n}, \tag{11.1.4}$$

which arise from (11.1.1) and (11.1.2) with the help of Euler's summation formula (II.2) of [**46**] or (A.2.2), converge very fast and thus they are convenient for numerical investigation of these functions.

There are also important relations with the Jackson q-Bessel functions $J_\nu^{(2)}(r;q)$ given by (2.4.28), namely,

$$S_q(\eta;\omega)=\frac{(q;q)_\infty}{\left(q^{1/2};q\right)_\infty}\,\frac{\omega^{1/2}}{(-q\omega^2;q^2)_\infty}\,J_{1/2}^{(2)}(2\omega;q), \tag{11.1.5}$$

$$C_q(\eta;\omega)=\frac{(q;q)_\infty}{\left(q^{1/2};q\right)_\infty}\,\frac{\omega^{1/2}}{(-q\omega^2;q^2)_\infty}\,J_{-1/2}^{(2)}(2\omega;q), \tag{11.1.6}$$

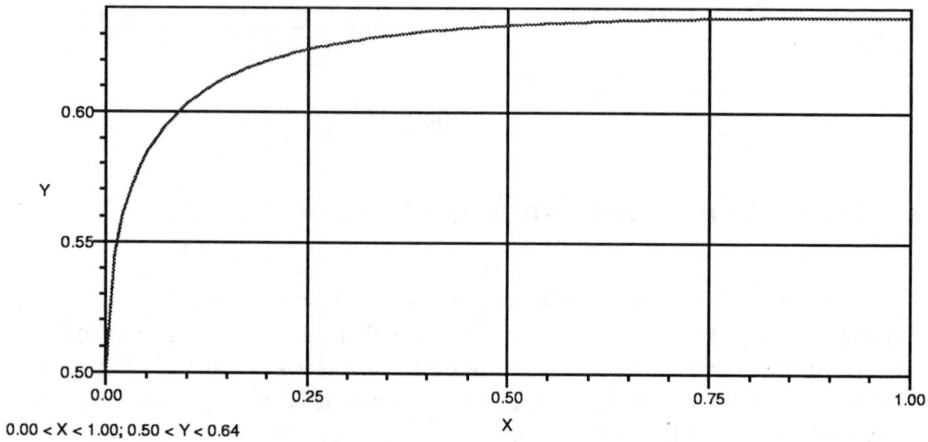

FIGURE 11.1. Function $q^{-1/4}c_1(q)$, $0 \leq q \leq 1$.

see (2.4.25)–(2.4.26), which give another possibility for numerical evaluation of these functions.

The main properties of the zeros of the q-trigonometric functions under consideration were discussed in Sections 5.4, 6.3 and 8.4; see also [**55**], [**56**], [**26**], and [**147**]. We remind the reader that when $0 < q < 1$ all the zeros of $S_q(\eta;\omega)$ and $C_q(\eta;\omega)$ are real. Also these zeros are simple, the positive zeros of the basic sine function $S_q(\eta;\omega)$ are interlaced with those of the basic cosine function $C_q(\eta;\omega)$; see Theorems 5.1–5.4.

11.1.2. Tables of Zeros. We used Newton's method [**48**] in order to evaluate first 10 to 15 zeros ω_n and ϖ_n of the $S_q(\eta;\omega)$ and $C_q(\eta;\omega)$, respectively, for different values of parameter q with the double precision. Numerical results are presented Appendix C for different values of the basic parameter q.

11.1.3. Asymptotics of Zeros. Asymptotics of the large zeros ω_n and ϖ_n are given by Eqs. (8.4.24) and (8.4.25), respectively. Numerical examples demonstrating accuracy of these asymptotic formulas are presented in Appendix D.

11.1.4. Monotonicity of Zeros. Monotonicity of the zeros of Jackson's q-Bessel function $J_\nu^{(2)}(x;q)$ was investigated by Ismail and Muldoon [**64**].

THEOREM 11.1. *Let $\nu > -1$. The positive zeros $j_{\nu,k}(q)$ of $J_\nu^{(2)}(x;q)$ increase with ν when $0 < q < 1$ and decrease with q if $e^{-4} < q < 1$ but $q^{-1/4}j_{\nu,k}(q)$ increase with ν and decrease with q when $0 < q < 1$.*

Due to (11.1.5)–(11.1.6) this theorem describes monotonicity of the zeros of basic trigonometric functions, ω_n and ϖ_n. Our numerical results confirm

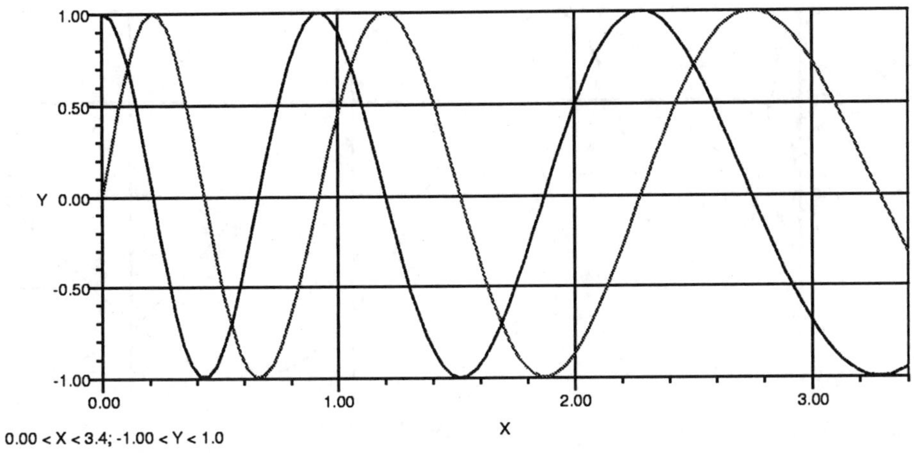

FIGURE 11.2. Functions $F(\omega)$ and $G(\omega)$ for $q = 0.75$.

Ismail–Muldoon's theorem for $\nu = \pm 1/2$. One can observe also that the asymptotic formulas (6.3.14)–(6.3.16) reveal the same monotonic behavior. The graph of $c_1(q)$ is presented on Figures 6.1 and 8.1. Function $q^{-1/4}c_1(q)$ is also nonnegative and increasing on $[0,1]$; see Figure 11.1.

11.1.5. Graphing. It was shown in [**48**] that the graphs of the $S_q(\eta;\omega)$ and $C_q(\eta;\omega)$ look much more elegant if we choose the following normalization for these functions,

$$F(\omega) = \sqrt{\frac{(-q\omega^2;q^2)_\infty}{(-\omega^2;q^2)_\infty}}\, S_q(\eta;\omega), \tag{11.1.7}$$

$$G(\omega) = \sqrt{\frac{(-q\omega^2;q^2)_\infty}{(-\omega^2;q^2)_\infty}}\, C_q(\eta;\omega), \tag{11.1.8}$$

see (6.3.6)–(6.3.7).

Functions $F(\omega)$ and $G(\omega)$ have the same zeros as our original functions, $S_q(\eta;\omega)$ and $C_q(\eta;\omega)$, but they obey nice properties similar to those for the usual sine and cosine functions. Indeed, both functions, $F(\omega)$ and $G(\omega)$, are bounded and vary from -1 to 1. Moreover, all extrema of function $F(\omega)$ ($G(\omega)$) are located at the zeros of $G(\omega)$ ($F(\omega)$), function $F(\omega)$ ($G(\omega)$) is monotone between any two successive zeros of $G(\omega)$ ($F(\omega)$); see Section 6.3.2 for more details.

Figures 11.2 to 11.4 present the graphs of $F(\omega)$ and $G(\omega)$ for $q = 0.75$, $q = 0.5$ and $q = 0.25$, respectively.

11.1.6. Points of Inflection. The graphs of the $F(\omega)$ and $G(\omega)$ show that the concavity of these q-sine and cosine functions changes somewhere before the zeros ω_n and ϖ_n, respectively; see Figures 11.2–11.4. Here we

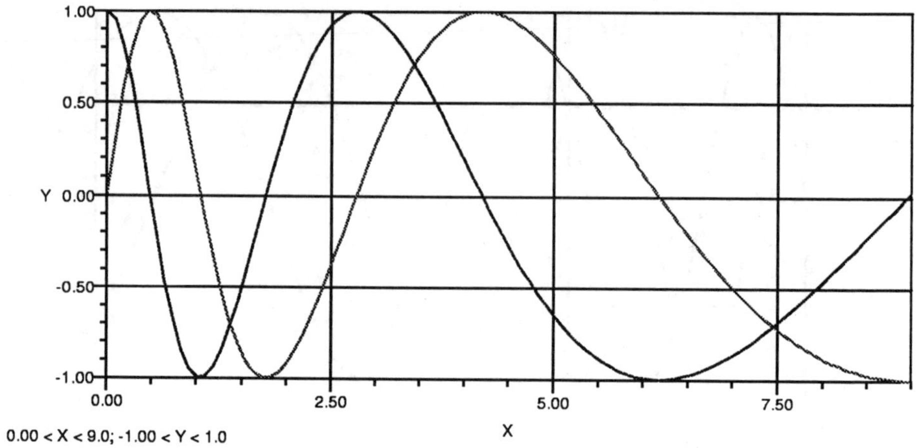

FIGURE 11.3. Functions $F(\omega)$ and $G(\omega)$ for $q = 0.5$.

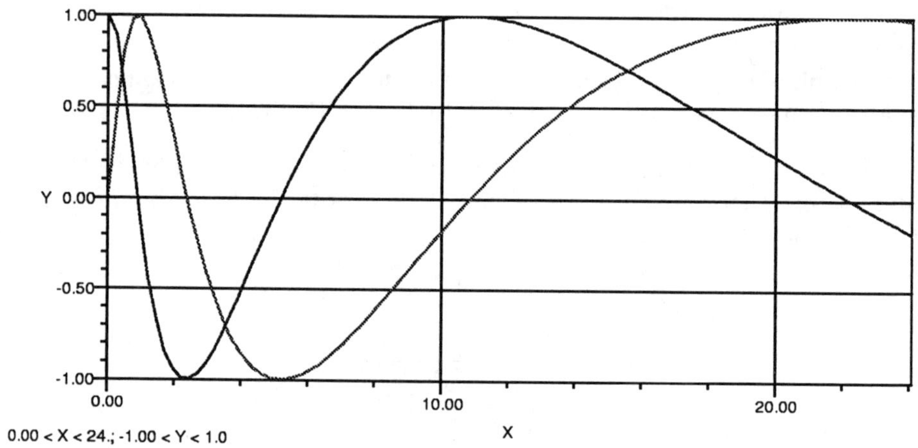

FIGURE 11.4. Functions $F(\omega)$ and $G(\omega)$ for $q = 0.25$.

shall discuss briefly the corresponding points of inflection. Detailed analysis and numerical study is a subject of the future investigation. Consider the case of the basic sine function $F(\omega)$. In view of (6.3.10),

$$F''(\omega) = \kappa'(\omega) G(\omega) - \kappa^2(\omega) F(\omega), \tag{11.1.9}$$

and the location of the points of inflection can be found as the roots of the equation

$$\frac{F(\omega)}{G(\omega)} = \frac{\kappa'(\omega)}{\kappa^2(\omega)} = -\left(\frac{1}{\kappa(\omega)}\right)'. \tag{11.1.10}$$

Also

$$\frac{d}{d\omega}\left(\frac{F(\omega)}{G(\omega)}\right)=\frac{\kappa(\omega)}{G^2(\omega)}>0 \tag{11.1.11}$$

and the q-tangent function $F(\omega)/G(\omega)$ is increasing from $-\infty$ to $+\infty$ on each of the intervals (ϖ_n, ϖ_{n+1}). On the other hand, function $\kappa'(\omega)/\kappa^2(\omega)$ is negative and bounded for all $\omega > 0$. Thus, equation (11.1.10) has at least one solution on each of the subintervals (ϖ_n, ω_n).

A similar analysis shows that the points of inflection of the basic cosine function $G(\omega)$ are solutions of

$$\frac{G(\omega)}{F(\omega)}=-\frac{\kappa'(\omega)}{\kappa^2(\omega)}=\left(\frac{1}{\kappa(\omega)}\right)' \tag{11.1.12}$$

and that this equation has at least one root on each of the intervals (ω_{n-1}, ϖ_n).

Let us consider "asymptotic solutions" of Eq. (11.1.10) in the form

$$\omega=\beta q^{-n}, \qquad q^{3/4}<\beta<q^{1/4}. \tag{11.1.13}$$

Substitute (11.1.13) into (11.1.10) and take the limit as $n \to \infty$. Using the limits

$$\lim_{n\to\infty}\frac{F(\beta q^{-n})}{G(\beta q^{-n})}=\beta\frac{\left(q^{3/2}\beta^2, q^{1/2}/\beta^2; q^2\right)_\infty}{\left(q^{1/2}\beta^2, q^{3/2}/\beta^2; q^2\right)_\infty} \tag{11.1.14}$$

and

$$\begin{aligned}
&-\frac{(q;q)_\infty^2}{\left(q^{1/2};q\right)_\infty^2}\beta\lim_{n\to\infty}\frac{\kappa'(\beta q^{-n})}{\kappa^2(\beta q^{-n})} \\
&=\frac{\left(-\beta^2,-q/\beta^2;q\right)_\infty}{\left(-q^{1/2}\beta^2,-q^{1/2}/\beta^2;q\right)_\infty} \\
&+\left(q^{1/2};q\right)_\infty^2\,\beta^2\frac{\left(q^{1/2},q^{1/2},q^{1/2}\beta^2,1/\beta^2;q^{1/2}\right)_\infty}{\left(-q^{1/2}\beta^2,-q^{1/2}/\beta^2;q\right)_\infty^2},
\end{aligned} \tag{11.1.15}$$

which follow from (5.4.3)–(5.4.4) and (5.5.17), (8.3.3) in a similar fashion as in Section 6.3.3, we obtain the following transcendental equation

$$\begin{aligned}
&-\frac{(q;q)_\infty^2}{\left(q^{1/2};q\right)_\infty^2}\beta^2\frac{\left(q^{3/2}\beta^2, q^{1/2}/\beta^2; q^2\right)_\infty}{\left(q^{1/2}\beta^2, q^{3/2}/\beta^2; q^2\right)_\infty} \\
&=\frac{\left(-\beta^2,-q/\beta^2;q\right)_\infty}{\left(-q^{1/2}\beta^2,-q^{1/2}/\beta^2;q\right)_\infty} \\
&+\left(q^{1/2};q\right)_\infty^2\,\beta^2\frac{\left(q^{1/2},q^{1/2},q^{1/2}\beta^2,1/\beta^2;q^{1/2}\right)_\infty}{\left(-q^{1/2}\beta^2,-q^{1/2}/\beta^2;q\right)_\infty^2}
\end{aligned} \tag{11.1.16}$$

for the parameter β. One can easily see that this equation has at least one solution for $q^{3/4} < \beta < q^{1/4}$.

Similar arguments hold in the case of the q-cosine function. The corresponding "asymptotic equation" has the form

$$\frac{(q;q)^2_\infty}{\left(q^{1/2};q\right)^2_\infty}\frac{\left(q^{1/2}\beta^2,q^{3/2}/\beta^2;q^2\right)_\infty}{\left(q^{3/2}\beta^2,q^{1/2}/\beta^2;q^2\right)_\infty} \tag{11.1.17}$$
$$= \frac{\left(-\beta^2,-q/\beta^2;q\right)_\infty}{\left(-q^{1/2}\beta^2,-q^{1/2}/\beta^2;q\right)_\infty}$$
$$+\left(q^{1/2};q\right)^2_\infty \beta^2 \frac{\left(q^{1/2},q^{1/2},q^{1/2}\beta^2,1/\beta^2;q^{1/2}\right)_\infty}{\left(-q^{1/2}\beta^2,-q^{1/2}/\beta^2;q\right)^2_\infty}.$$

This consideration motivates the following conjecture.

CONJECTURE 11.1. *The leading terms of the large asymptotics of the points of inflection χ_n and $\overline{\chi}_n$ of the basic sine $F(\omega)$ and cosine $G(\omega)$ functions, respectively, are defined by*

$$\lim_{n\to\infty} q^n\chi_n = \beta_0, \qquad q^{3/4} < \beta_0 < q^{1/4} \tag{11.1.18}$$

and

$$\lim_{n\to\infty} q^n\overline{\chi}_n = \overline{\beta}_0, \qquad q^{1/4} < \overline{\beta}_0 < q^{-1/4}. \tag{11.1.19}$$

Here β_0 and $\overline{\beta}_0$ are solutions of the transcendental equations (11.1.16) *and* (11.1.17), *respectively.*

11.2. Euler–Rayleigh Method

In this section we apply the Euler–Rayleigh method in order to find convergence sequences of successively improving lower and upper bounds for the first positive zeros of the basic sine and cosine functions. See **[65]** for an excellent account and the bibliography on the Euler–Rayleigh method.

11.2.1. Description of the Method. We follow **[65]**. Suppose that an entire function $f(z)$ has the power series expansion

$$f(z) = 1 + \sum_{k=1}^{\infty} a_k z^k \tag{11.2.1}$$

and an infinite product representation

$$f(z) = \prod_{k=1}^{\infty}\left(1 - \frac{z}{z_k}\right) \tag{11.2.2}$$

where $\sum_{k=1}^{\infty}|z_k|^{-1} < \infty$. Then the following procedure can be used to express the power sums

$$S_m = \sum_{k=1}^{\infty} z_k^{-m} \tag{11.2.3}$$

in terms of the coefficients a_k.

Logarithmic differentiation of (11.2.2) leads to

$$\frac{f'(z)}{f(z)} = \sum_{k=1}^{\infty} \frac{1}{z - z_k}, \tag{11.2.4}$$

whence

$$S_1 = -\frac{f'(0)}{f(0)} = -a_1. \tag{11.2.5}$$

Further differentiation of (11.2.4) results in

$$\frac{f''(z)}{f(z)} = \left(\frac{f'(z)}{f(z)}\right)^2 - \sum_{k=1}^{\infty} \frac{1}{(z - z_k)^2}, \tag{11.2.6}$$

so we obtain

$$S_2 = -2a_2 - a_1 S_1 = -2a_2 - a_1^2. \tag{11.2.7}$$

This process can be repeated and the general result is

$$S_n = -n a_n - \sum_{k=1}^{n-1} a_k S_{n-k}. \tag{11.2.8}$$

This result can also be obtained by comparing the coefficients of powers of z on both sides of

$$f'(z) = -f(z) \sum_{m=0}^{\infty} S_{m+1}\, z^m, \qquad |z| < z_1. \tag{11.2.9}$$

The last equation arises from (11.2.4) in view of

$$\sum_{k=1}^{\infty} \frac{1}{z - z_k} = -\sum_{k=1}^{\infty} \frac{1}{z_k\,(1 - z/z_k)} = -\sum_{m=0}^{\infty} z^m \sum_{k=1}^{\infty} \frac{1}{z_k^{m+1}}.$$

Formula (11.2.8) is actually a generalization to entire functions of Newton's formula for sums of powers of roots of a polynomial in terms of symmetric functions of the roots (or coefficients of the polynomial).

If the zeros satisfy

$$0 < |z_1| < |z_2| < |z_3| < \ldots, \tag{11.2.10}$$

we have

$$z_1 = \lim_{m\to\infty} \frac{S_m}{S_{m+1}}. \tag{11.2.11}$$

This is equivalent to the statement

$$\lim_{n\to\infty} \frac{1 + (z_1/z_2)^n + (z_1/z_3)^n + \ldots}{1 + (z_1/z_2)^{n+1} + (z_1/z_3)^{n+1} + \ldots} = 1, \tag{11.2.12}$$

which follows from

$$\lim_{n\to\infty} \left(\left|\frac{z_1}{z_2}\right|^n + \left|\frac{z_1}{z_3}\right|^n + \ldots \right) = 0.$$

This, in turn, follows from the lemma [**65**].

LEMMA 11.1. *Let* $1 > y_1 \geq y_2 \geq y_3 \geq \ldots \geq 0$ *and suppose that*

$$\sum_{k=1}^{\infty} y_k < \infty. \tag{11.2.13}$$

Then

$$\lim_{r \to \infty} \sum_{k=1}^{\infty} y_k^r = 0. \tag{11.2.14}$$

PROOF. Let $\varepsilon > 0$ be given. There exists an m, independent of r, such that

$$y_{m+1} + y_{m+2} + \ldots < \varepsilon/2.$$

Hence for all $r \geq 1$,

$$y_{m+1}^r + y_{m+2}^r + \ldots < \varepsilon/2.$$

Also, for $r > (\log \varepsilon - \log(2m)) / \log(y_1)$, we have

$$y_1^r + \ldots + y_m^r < \varepsilon/2,$$

so for such r, we have $\sum_{k=1}^{\infty} y_k^r < \varepsilon$ and the lemma is proved. □

In the case where the zeros are real and positive the relation between x_1 and the sums S_m can be made much more specific.

LEMMA 11.2. *With the notations of this section, suppose that*

$$0 < x_1 < x_2 < x_3 < \ldots . \tag{11.2.15}$$

Then

$$S_m^{-1/m} < x_1 < S_m / S_{m+1}, \qquad m = 1, 2, \ldots , \tag{11.2.16}$$

where the lower bounds increase and the upper bounds decrease to x_1 *as* $m \to \infty$.

PROOF. The inequalities (11.2.16) follow easily from the definition of the S_n. Indeed,

$$S_m - \frac{1}{x_1^m} = \sum_{k=2}^{\infty} \frac{1}{x_k^m} > 0,$$

which implies $x_1 > S_m^{-1/m}$; and

$$\sum_{k=1}^{\infty} \left(\frac{x_1}{x_k}\right) \frac{1}{x_k^m} < \sum_{k=1}^{\infty} \frac{1}{x_k^m}, \quad \text{or} \quad x_1 S_{m+1} < S_m,$$

which leads to $x_1 < S_m / S_{m+1}$.

That the lower bounds increase can be seen by using $S_m^{-1/m} < S_m / S_{m+1}$, namely,

$$(S_{m+1})^{-1/(m+1)} > (S_m)^{-1/m}$$

implies

$$\frac{S_{m+1}}{S_m S_m^{1/m}} < 1.$$

That the upper bounds decrease is a consequence of the Cauchy–Schwarz inequality,

$$\left(\sum_{k=1}^{\infty} u_k\, v_k\right)^2 \leq \sum_{k=1}^{\infty} u_k^2 \sum_{k=1}^{\infty} v_k^2$$

when

$$u_k = \frac{1}{x_k^{m/2}}, \qquad u_k = \frac{1}{x_k^{m/2+1}},$$

and the lemma is proved. □

The situation where all except one of the zeros are positive, the remaining one being negative, is discussed in [**65**]; see Ex. 11.1.

We shall use Lemma 11.2 in order to find the lower and upper bounds for the first zeros of the basic trigonometric functions $S_q(\eta;\omega)$ and $C_q(\eta;\omega)$ given by the power series (11.1.3) and (11.1.4), respectively.

11.2.2. Bounds for First Zeros. Let ω_1 be the first zero of $S_q(\eta;\omega)$ and let ϖ_1 be the first zero of $C_q(\eta;\omega)$. Applying the Euler–Rayleigh method to the following functions

$$\begin{aligned} f(x) &= \left(1-q^{1/2}\right)\left(-q\omega^2;q\right)_\infty \frac{S_q(\eta;\omega)}{\omega} \\ &= \sum_{n=0}^{\infty} (-1)^n \frac{q^{n(n+1/2)}}{\left(q^{1/2};q^{1/2}\right)_{2n+1}}\, \omega^{2n} \end{aligned} \tag{11.2.17}$$

and

$$\begin{aligned} g(x) &= \left(-q\omega^2;q\right)_\infty C_q(\eta;\omega) \\ &= \sum_{n=0}^{\infty} (-1)^n \frac{q^{n(n-1/2)}}{\left(q^{1/2};q^{1/2}\right)_{2n}}\, \omega^{2n}, \end{aligned} \tag{11.2.18}$$

where $x=\omega^2$, one gets the following lower and upper bounds for ω_1 and ϖ_1 :

m	Lower Bounds for ω_1 :
1	$q^{-3/4}\left(\left(1-q\right)\left(1-q^{3/2}\right)\right)^{1/2}$
2	$q^{-3/4}\left(1-q^{3/2}\right)^{1/2}\left(\dfrac{\left(1-q^2\right)\left(1-q^{5/2}\right)}{1+2q-q^{5/2}}\right)^{1/4}$
3	$q^{-3/4}\left(\dfrac{\left(1-q^{1/2}\right)\left(1-q^{3/2}\right)^2\left(1-q^{5/2}\right)\left(1-q^3\right)\left(1-q^{7/2}\right)}{P(q)}\right)^{1/6}$

m	Upper Bounds for ω_1 :
1	$q^{-3/4}\left(\frac{\left(1-q^{3/2}\right)\left(1-q^{5/2}\right)(1+q)}{1+2q-q^{5/2}}\right)^{1/2}$
2	$q^{-3/4}\left(\frac{\left(1-q^{1/2}\right)\left(1-q^{3}\right)\left(1-q^{7/2}\right)\left(1+2q-q^{5/2}\right)}{\left(1-q^{2}\right)P(q)}\right)^{1/2}$
3	$q^{-3/4}\left(\frac{\left(1-q^{4}\right)\left(1-q^{5/2}\right)\left(1-q^{9/2}\right)P(q)}{\left(1-q^{1/2}\right)\left(1+q^{3/2}\right)Q(q)}\right)^{1/2}$

Here

$$P(q)=1-q^{1/2}+3q-2q^{3/2}+2q^{2}-q^{5/2}+2q^{3}-2q^{7/2}-q^{9/2}+q^{5}$$

and

$$\begin{aligned}Q(q) &= 1+4q+q^{3/2}+6q^{2}+2q^{5/2}+9q^{3}+q^{7/2}+10q^{4}+6q^{5}-4q^{11/2}\\ &+6q^{6}-4q^{13/2}+2q^{7}-6q^{15/2}+q^{8}-3q^{17/2}+2q^{9}+3q^{10}-q^{23/2}.\end{aligned}$$

m	Lower Bounds for ϖ_1 :
1	$q^{-1/4}\left(\left(1-q^{1/2}\right)(1-q)\right)^{1/2}$
2	$q^{-1/4}\left(1-q^{1/2}\right)^{1/2}\left(\frac{\left(1-q^{3/2}\right)\left(1-q^{2}\right)}{1+2q-q^{3/2}}\right)^{1/4}$
3	$q^{-1/4}\left(1-q^{1/2}\right)^{1/2}\left(\frac{\left(1-q^{3/2}\right)\left(1-q^{3}\right)\left(1-q^{5/2}\right)}{\overline{P}(q)}\right)^{1/6}$

m	Upper Bounds for ϖ_1 :
1	$q^{-1/4}\left(\frac{\left(1-q^{1/2}\right)\left(1-q^{3/2}\right)(1+q)}{1+2q-q^{3/2}}\right)^{1/2}$
2	$q^{-1/4}\left(\frac{\left(1-q^{1/2}\right)\left(1-q^{3}\right)\left(1-q^{5/2}\right)\left(1+2q-q^{3/2}\right)}{\left(1-q^{2}\right)\overline{P}(q)}\right)^{1/2}$
3	$q^{-1/4}\left(\frac{\left(1-q^{4}\right)\left(1-q^{7/2}\right)\overline{P}(q)}{\left(1+q^{3/2}\right)\overline{Q}(q)}\right)^{1/2}$

Here

$$\overline{P}(q)=1+3q-q^{3/2}+3q^{2}-q^{5/2}+3q^{3}-3q^{7/2}+q^{4}$$

and

$$\begin{aligned}\overline{Q}(q) &= 1+q^{1/2}+5q+3q^{3/2}+9q^{2}+4q^{5/2}+13q^{3}+4q^{7/2}+14q^{4}\\ &+10q^{5}-q^{11/2}+8q^{6}-5q^{13/2}+3q^{7}-3q^{15/2}+3q^{8}-q^{17/2}.\end{aligned}$$

According to Theorem 5.4 one gets $\varpi_1 < \omega_1$. It is also of interest noting that comparing the first upper bound for ϖ_1 with the first lower bound for ω_1 we obtain

$$\begin{aligned}\varpi_1 &< q^{-1/4}\left(\frac{\left(1-q^{1/2}\right)\left(1-q^{3/2}\right)(1+q)}{1+2q-q^{3/2}}\right)^{1/2}\\ &< q^{-3/4}\left((1-q)\left(1-q^{3/2}\right)\right)^{1/2} < \omega_1,\end{aligned}$$

because the inequality $2q^2 - q^{3/2} - q - 1 < 0$ holds for all $0 < q < 1$.

11.2.3. Numerical Examples. Numerical examples demonstrating accuracy of the Euler–Rayleigh method are presented in Appendix E, where we consider bounds for the first zero of the $S_q(\eta;\omega)$ and $C_q(\eta;\omega)$ when $q = 0.1$ and $q = 0.5$. Our numerical analysis has shown that the convergence become very slow when parameter q approaches to 1.

11.3. Lower and Upper Bounds of Zeros

In this section we construct monotone convergent sequences of the lower and upper bounds for all positive zeros of the basic sine and cosine functions. The following theorem holds [**147**].

THEOREM 11.2. *Let* $\{\omega_n\}_{n=1}^{\infty}$ *be positive zeros of* $S_q(\eta;\omega)$ *and* $\{\varpi_n\}_{n=1}^{\infty}$ *be positive zeros of* $C_q(\eta;\omega)$ *arranged in ascending order of magnitude. Choose* $\omega_n < U_n^{(0)} < \varpi_{n+1}$,

$$L_n^{(k)} = U_n^{(k-1)} - \frac{F\left(U_n^{(k-1)}\right)}{F'\left(U_n^{(k-1)}\right)}, \tag{11.3.1}$$

$$U_n^{(k)} = U_n^{(k-1)} - \frac{\left|F\left(U_n^{(k-1)}\right)\right|}{\kappa\left(L_n^{(k)}\right)}, \tag{11.3.2}$$

and $\varpi_n < \overline{U}_n^{(0)} < \omega_n$,

$$\overline{L}_n^{(k)} = \overline{U}_n^{(k-1)} - \frac{G\left(\overline{U}_n^{(k-1)}\right)}{G'\left(\overline{U}_n^{(k-1)}\right)}, \tag{11.3.3}$$

$$\overline{U}_n^{(k)} = \overline{U}_n^{(k-1)} - \frac{\left|G\left(\overline{U}_n^{(k-1)}\right)\right|}{\kappa\left(\overline{L}_n^{(k)}\right)} \tag{11.3.4}$$

for all positive integer $k = 1, 2, 3, \ldots$ *. Then*

$$L_n^{(1)} < \ldots < L_n^{(k-1)} < L_n^{(k)} < \omega_n < U_n^{(k)} < U_n^{(k-1)} < \ldots < U_n^{(0)}, \tag{11.3.5}$$

$$\lim_{k\to\infty} L_n^{(k)} = \lim_{k\to\infty} U_n^{(k)} = \omega_n, \tag{11.3.6}$$

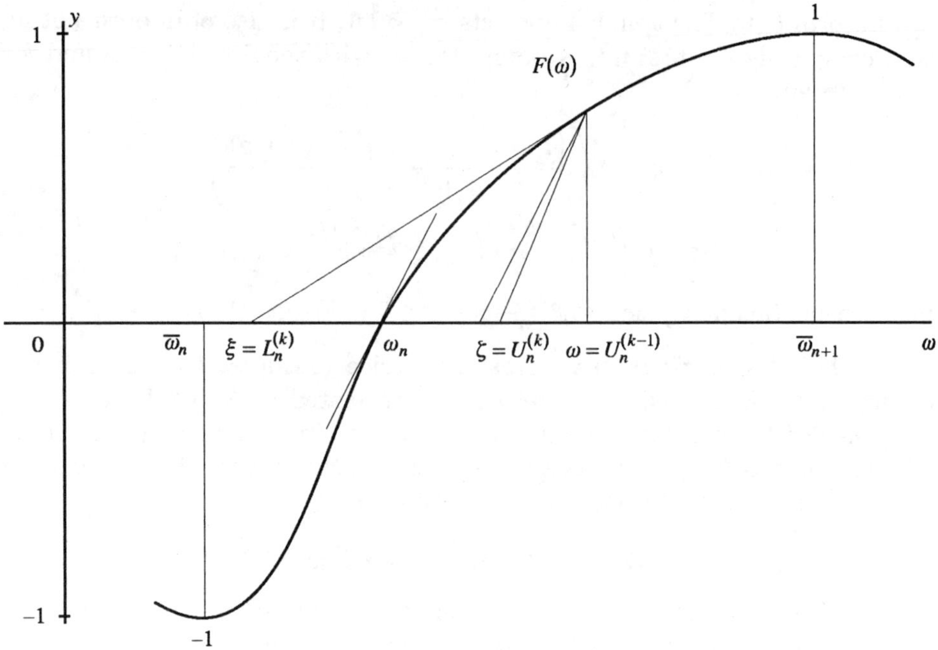

FIGURE 11.5. Geometric interpretation of the lower $\xi = \omega - F(\omega)/F'(\omega) = L_n^{(k)}$ and upper $\zeta = \omega - |F(\omega)|/\kappa(\xi) = U_n^{(k)}$ bounds of the zero ω_n in Theorem 11.2, $n = 2m$.

and

$$\overline{L}_n^{(1)} < \ldots < \overline{L}_n^{(k-1)} < \overline{L}_n^{(k)} < \varpi_n < \overline{U}_n^{(k)} < \overline{U}_n^{(k-1)} < \ldots < \overline{U}_n^{(0)}, \tag{11.3.7}$$

$$\lim_{k\to\infty} \overline{L}_n^{(k)} = \lim_{k\to\infty} \overline{U}_n^{(k)} = \varpi_n. \tag{11.3.8}$$

PROOF. Consider the case of the basic sine function. Let $\omega \in (\omega_n, \varpi_{n+1})$ and $\xi \in (\varpi_n, \omega_n)$. The same arguments as in Section 6.3.3; see (6.3.21)–(6.3.23) with $m = n$; result in

$$\frac{|F(\omega)|}{\kappa(\xi)} < \frac{|F(\omega)|}{\kappa(\omega_n)} < \omega - \omega_n < \frac{F(\omega)}{F'(\omega)}, \tag{11.3.9}$$

or solving for ω_n,

$$\omega - \frac{F(\omega)}{F'(\omega)} < \omega_n < \omega - \frac{|F(\omega)|}{\kappa(\xi)}. \tag{11.3.10}$$

Substituting $\omega = U_n^{(k-1)}$ and $\xi = L_n^{(k)}$ (see Figure 11.5) one gets

$$L_n^{(k)} < \omega_n < U_n^{(k)}, \qquad k = 1, 2, 3, \ldots. \tag{11.3.11}$$

The monotonicity of the sequence of the upper bounds $\left\{U_n^{(k)}\right\}_{k=0}^{\infty}$ follows directly from (11.3.2),

$$U_n^{(k-1)} - U_n^{(k)} = \frac{\left|F\left(U_n^{(k-1)}\right)\right|}{\kappa\left(L_n^{(k)}\right)} > 0. \tag{11.3.12}$$

On the other hand, the function

$$L(\omega) = \omega - \frac{F(\omega)}{F'(\omega)}, \qquad L\left(U_n^{(k-1)}\right) = L_n^{(k)} \tag{11.3.13}$$

defined by (11.3.1) is monotone on (ω_n, ϖ_{n+1}) because its derivative

$$\frac{dL(\omega)}{d\omega} = \frac{F(\omega)F''(\omega)}{(F'(\omega))^2} = \frac{F(\omega)\left(\kappa'(\omega)G(\omega) - \kappa^2(\omega)F(\omega)\right)}{(F'(\omega))^2} \tag{11.3.14}$$

does not change the sign on this interval. Hence, the sequence $\left\{U_n^{(k)}\right\}_{k=0}^{\infty}$ is decreasing and bounded below, while the $\left\{L_n^{(k)}\right\}_{k=1}^{\infty}$ is increasing and bounded above. By the Monotone Convergence Theorem the following limits exist

$$\lim_{k\to\infty} L_n^{(k)} = L_n \le \omega_n, \qquad \lim_{k\to\infty} U_n^{(k)} = U_n \ge \omega_n. \tag{11.3.15}$$

Finally, taking the limit $k \to \infty$ in (11.3.1)–(11.3.2) one gets

$$L_n = U_n = \omega_n. \tag{11.3.16}$$

This proves (11.3.5)–(11.3.6). Similar arguments hold for the case of the basic cosine function. We leave the details to the reader. □

Construction the lower and upper bounds for the zeros of the basic sine and cosine functions in Theorem 11.2 is based on a simple geometric principle similar to geometric interpretation of Newton's method (see Figure 11.6). Notice if, e.g., $L_n^{(k)}$ and $U_n^{(k-1)}$ satisfy (11.3.1), then $L_n^{(k)}$ is the ω-intercept of the tangent line to $y = F(\omega)$ at the point $\left(U_n^{(k-1)}, F\left(U_n^{(k-1)}\right)\right)$. Also, notice that $U_n^{(k)}$ defined by (11.3.2) is the ω-intercept of the line passing through the same point $\left(U_n^{(k-1)}, F\left(U_n^{(k-1)}\right)\right)$ with the slope $(-1)^n \kappa\left(L_n^{(k)}\right)$, such that $\kappa\left(L_n^{(k)}\right) > \kappa(\omega_n) = |F'(\omega_n)|$. Eqs. (11.3.3)–(11.3.4) admit a similar geometric interpretation.

If the convexity of $F(\omega)$ does not change on the interval $\left(L_n^{(1)}, U_n^{(0)}\right)$ we can choose the ω-intercepts of the chords passing through the points $\left(L_n^{(k)}, F\left(L_n^{(k)}\right)\right)$ and $\left(U_n^{(k-1)}, F\left(U_n^{(k-1)}\right)\right)$ as another sequence of the upper bounds (see Figure 11.7). This consideration leads to the following theorem.

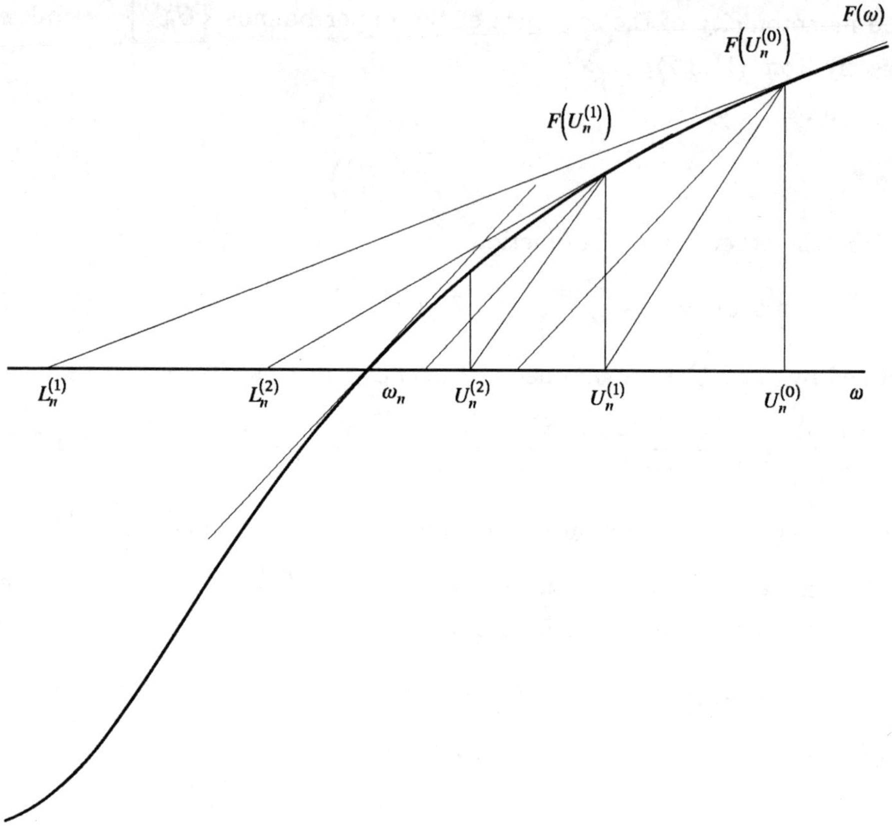

FIGURE 11.6. First upper $U_n^{(k)}$ and lower $L_n^{(k)}$ bounds of the zero ω_n in Theorem 11.2, $n = 2m$.

THEOREM 11.3. *Let $\{\omega_n\}_{n=1}^{\infty}$ be positive zeros of $S_q(\eta;\omega)$ and $\{\varpi_n\}_{n=1}^{\infty}$ be positive zeros of $C_q(\eta;\omega)$ arranged in ascending order of magnitude. Choose $\omega_n < U_n^{(0)} < \varpi_{n+1}$,*

$$L_n^{(k)} = U_n^{(k-1)} - \frac{F\left(U_n^{(k-1)}\right)}{F'\left(U_n^{(k-1)}\right)}, \tag{11.3.17}$$

$$U_n^{(k)} = U_n^{(k-1)} - F\left(U_n^{(k-1)}\right) \frac{U_n^{(k-1)} - L_n^{(k)}}{F\left(U_n^{(k-1)}\right) - F\left(L_n^{(k)}\right)}, \tag{11.3.18}$$

and $\varpi_n < \overline{U}_n^{(0)} < \omega_n$,

$$\overline{L}_n^{(k)} = \overline{U}_n^{(k-1)} - \frac{G\left(\overline{U}_n^{(k-1)}\right)}{G'\left(\overline{U}_n^{(k-1)}\right)}, \tag{11.3.19}$$

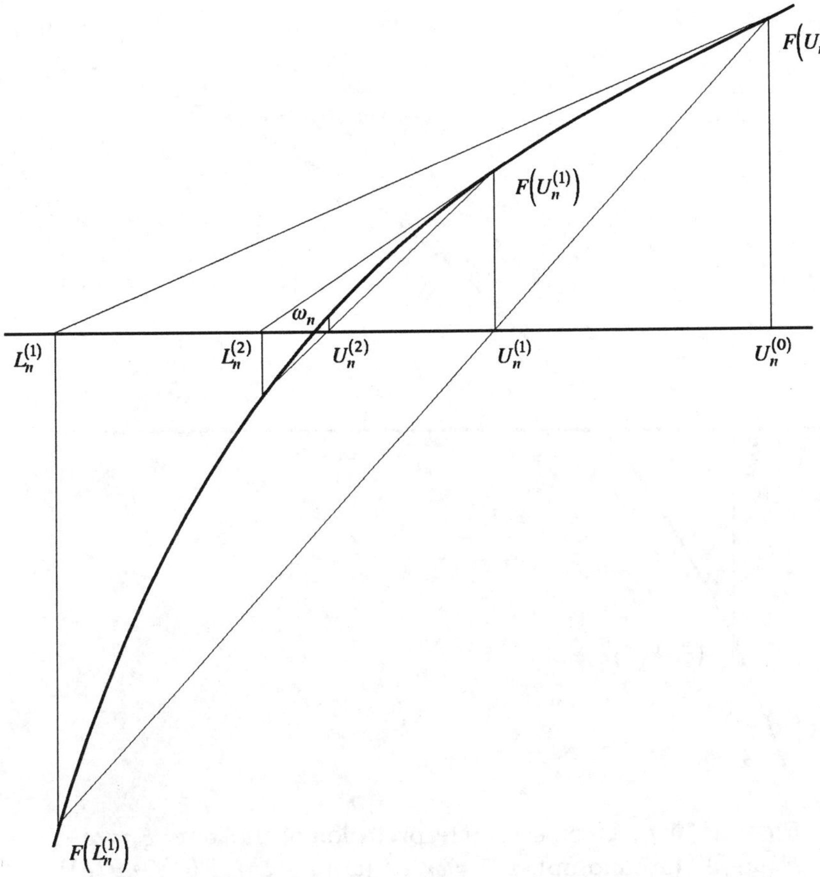

FIGURE 11.7. First upper $U_n^{(k)}$ and lower $L_n^{(k)}$ bounds of the zero ω_n in Theorem 11.3, $n = 2m$.

$$\overline{U}_n^{(k)} = \overline{U}_n^{(k-1)} - G\left(\overline{U}_n^{(k-1)}\right) \frac{\overline{U}_n^{(k-1)} - \overline{L}_n^{(k)}}{G\left(\overline{U}_n^{(k-1)}\right) - G\left(\overline{L}_n^{(k)}\right)} \tag{11.3.20}$$

for $k = 1, 2, 3, \ldots$. *If* $F''\left(L_n^{(1)}\right) F''\left(U_n^{(0)}\right) > 0$, *then*

$$L_n^{(1)} < \ldots < L_n^{(k-1)} < L_n^{(k)} < \omega_n < U_n^{(k)} < U_n^{(k-1)} < \ldots < U_n^{(0)}, \tag{11.3.21}$$

$$\lim_{k\to\infty} L_n^{(k)} = \lim_{k\to\infty} U_n^{(k)} = \omega_n, \tag{11.3.22}$$

and if $G''\left(L_n^{(1)}\right) G''\left(U_n^{(0)}\right) > 0$, *then*

$$\overline{L}_n^{(1)} < \ldots < \overline{L}_n^{(k-1)} < \overline{L}_n^{(k)} < \varpi_n < \overline{U}_n^{(k)} < \overline{U}_n^{(k-1)} < \ldots < \overline{U}_n^{(0)}, \tag{11.3.23}$$

$$\lim_{k\to\infty} \overline{L}_n^{(k)} = \lim_{k\to\infty} \overline{U}_n^{(k)} = \varpi_n. \tag{11.3.24}$$

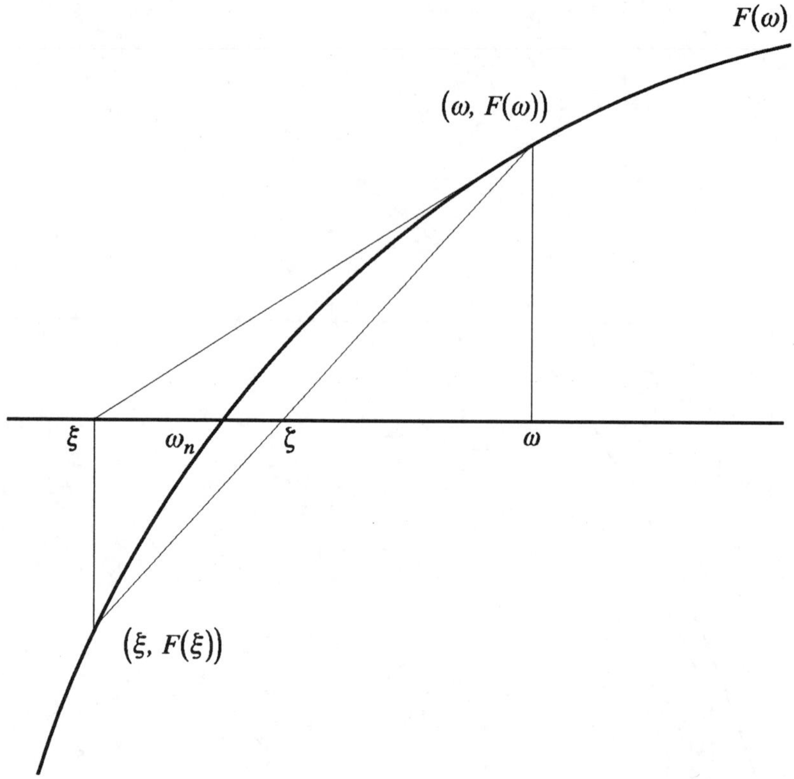

FIGURE 11.8. Geometric interpretation of the lower $\xi = \omega - F(\omega)/F'(\omega)$ and upper $\zeta = \omega - F(\omega)(\omega - \xi)/(F(\omega) - F(\xi))$ bounds of the zero ω_n in Theorem 11.3, $n = 2m$.

PROOF. We supply the details of the proof only for the case of the basic sine function. The proof for the q-cosine function is similar. One can replace (11.3.10) by

$$\omega - \frac{F(\omega)}{F'(\omega)} < \omega_n < \omega - F(\omega)\frac{\omega - \xi}{F(\omega) - F(\xi)}, \tag{11.3.25}$$

when $\omega \in (\omega_n, \varpi_{n+1})$, $\xi \in (\varpi_n, \omega_n)$ and $F''(\xi)\, F''(\omega) > 0$. The second inequality holds due to the convexity of the basic sine function. Consider, for example, the case of even zeros $\omega_n = \omega_{2m}$ when F is concave on (ξ, ω) and $F'' < 0$ (see Figure 11.8). The case of odd zeros can be discussed in a similar fashion. By the definition of concave function

$$F(c) > \frac{F(\omega) - F(\xi)}{\omega - \xi}(c - \omega) + F(\omega), \qquad c \in (\xi, \omega), \tag{11.3.26}$$

which means that the chord through the points $(\xi, F(\xi))$ and $(\omega, F(\omega))$ lies below the graph of F on (ξ, ω). The ω- intercept of this chord is

$$\zeta = \omega - F(\omega)\,\frac{\omega - \xi}{F(\omega) - F(\xi)} \tag{11.3.27}$$

and

$$F(\zeta) > 0, \qquad \zeta > \omega_n, \tag{11.3.28}$$

which justifies the second inequality in (11.3.25).

Substituting $\omega = U_n^{(k-1)}$ and $\xi = L_n^{(k)}$ in (11.3.25) we get

$$L_n^{(k)} < \omega_n < U_n^{(k)}, \qquad k = 1, 2, 3, \dots . \tag{11.3.29}$$

The monotonicity property (11.3.21) can be justified in the same manner as in the proof of Theorem 11.2. Taking the limit $k \to \infty$ in (11.3.17)–(11.3.18) one gets (11.3.22). We leave the details to the reader. □

Numerical examples of the monotone sequences of the lower and upper bounds for the zeros of basic sine and cosine functions, constructed in Theorems 11.2 and 11.3, are presented in Appendix F.

11.4. Eigenfunctions

Our main objective in this section is to present a few graphs of the eigenfunctions, or basic spherical harmonics, corresponding to the eigenvalues discussed in Section 11.1.

11.4.1. Graphing q-Spherical Harmonics. Analytic continuation of the basic cosine $C_q(x;\omega)$ and sine $S_q(x;\omega)$ functions, considered in Section 2.4 in detail, is given by Eqs. (2.4.8)–(2.4.13).

We have used all the representations (2.4.8) to (2.4.13) in order to graph the orthogonal functions $S_q(x;\omega_n)$ and $C_q(x;\omega_n)$ for various eigenvalues ω_n given in Appendix C. We have found that although Eqs. (2.4.8)–(2.4.9) involve only calculations with the real numbers, they are very inconvenient for the large eigenvalues. Indeed, the termwise limits of both ${}_2\varphi_2$'s in these equations are zero and the infinite products diverge as ω approaches infinity, which creates a computational problem for the large eigenvalues. But these representations are convenient for sufficiently small ω and they work then at the endpoints $x = \pm 1$ and even outside the orthogonality interval $[-1, 1]$.

Representations (2.4.12)–(2.4.13) look as the most convenient ones for the large values eigenvalues ω, but the use of both (2.4.10)–(2.4.11) and (2.4.12)–(2.4.13) involve computations with the complex numbers which are usually much slower than computations with the real numbers. Another problem is that there are removable singularities at the endpoints $x = \pm 1$ where one has to take the corresponding limits $\theta \to 0$ and $\theta \to \pi$ in (2.4.10)–(2.4.13). This leads to some nested sums which creates an additional computational problem. At the present stage we did not yet implement evaluation of the values of eigenfunctions at the endpoints $x = \pm 1$ and outside the

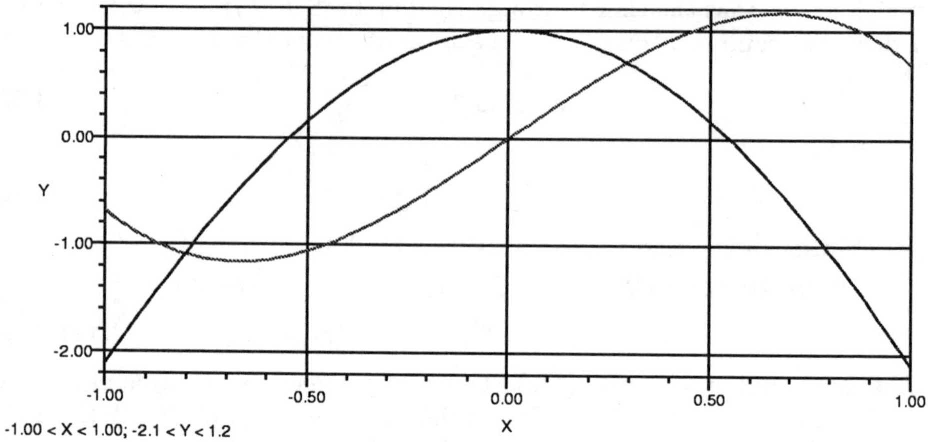

FIGURE 11.9. Functions $S_q(x;\omega_1)$ and $C_q(x;\omega_1)$ for $q = 0.1$.

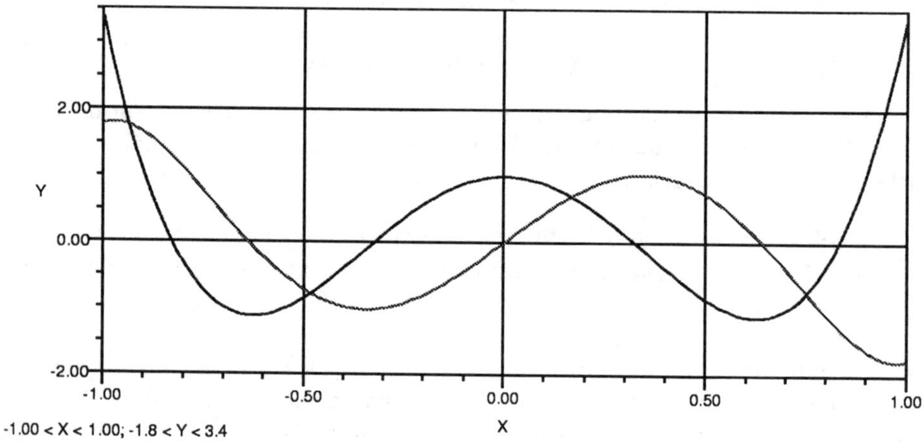

FIGURE 11.10. Functions $S_q(x;\omega_2)$ and $C_q(x;\omega_2)$ for $q = 0.1$.

orthogonality interval $[-1, 1]$ in our computational algorithm based on both (2.4.10)–(2.4.11) and (2.4.12)–(2.4.13).

The graphs of the first functions $S_q(x;\omega_n)$ and $C_q(x;\omega_n)$ with $n = 1, 2, 3, 4, 5$ are presented in Figures 11.9 to 11.13 for $q = 0.1$. One can see that these functions oscillate thus resemble essentially the behavior of the usual sine and cosine functions in the interval of their orthogonality. There is no periodicity property anymore and the absolute values of the extrema increase in the directions to the endpoints from the origin.

As another example, we present the graphs of the functions $S_q(x;\omega_n)$ and $C_q(x;\omega_n)$ with $n = 1, 2, 3, 4, 5$ in Figures 11.14 to 11.18 when $q = 0.5$.

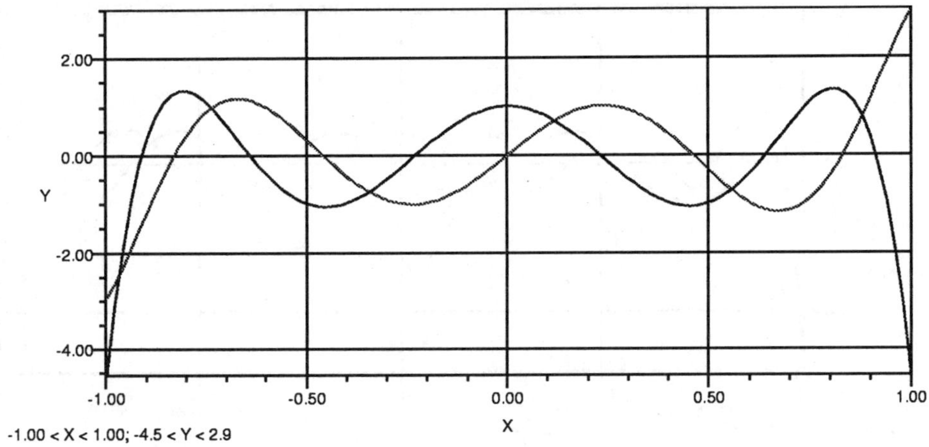

FIGURE 11.11. Functions $S_q(x;\omega_3)$ and $C_q(x;\omega_3)$ for $q = 0.1$.

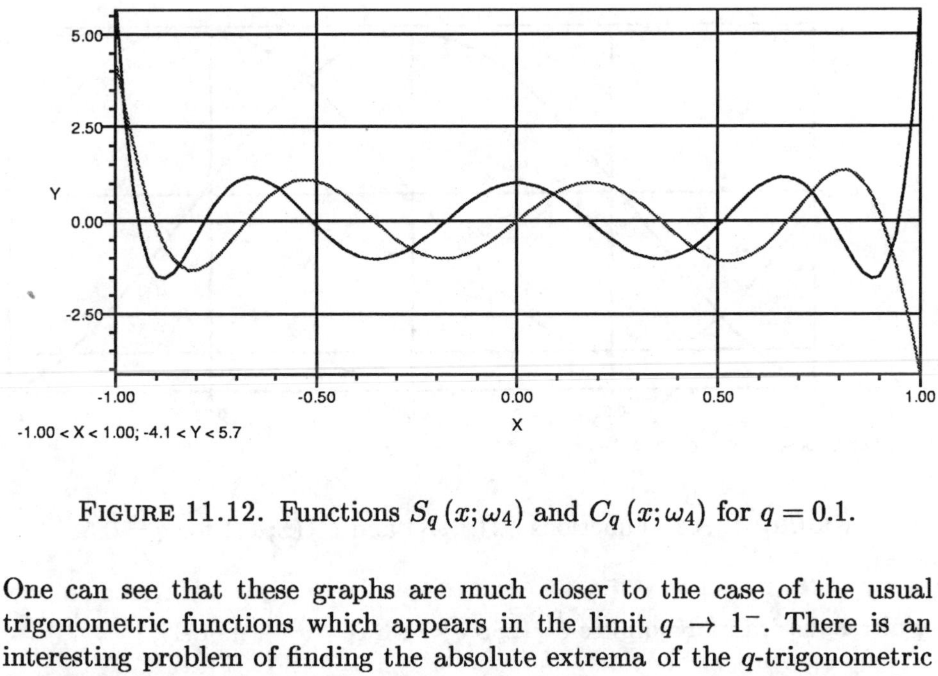

FIGURE 11.12. Functions $S_q(x;\omega_4)$ and $C_q(x;\omega_4)$ for $q = 0.1$.

One can see that these graphs are much closer to the case of the usual trigonometric functions which appears in the limit $q \to 1^-$. There is an interesting problem of finding the absolute extrema of the q-trigonometric functions in their interval of orthogonality. These examples also show that the basic sine $S_q(x;\omega_n)$ and cosine $C_q(x;\omega_n)$ functions have the same numbers of x-zeros inside the orthogonality interval as the classical sine and cosine functions.

11.4.2. Weight Functions and Orthogonality Relations. The weight function in the orthogonality relations (5.2.26)–(5.2.29) for the basic trigonometric functions coincide with the weight function for the continuous

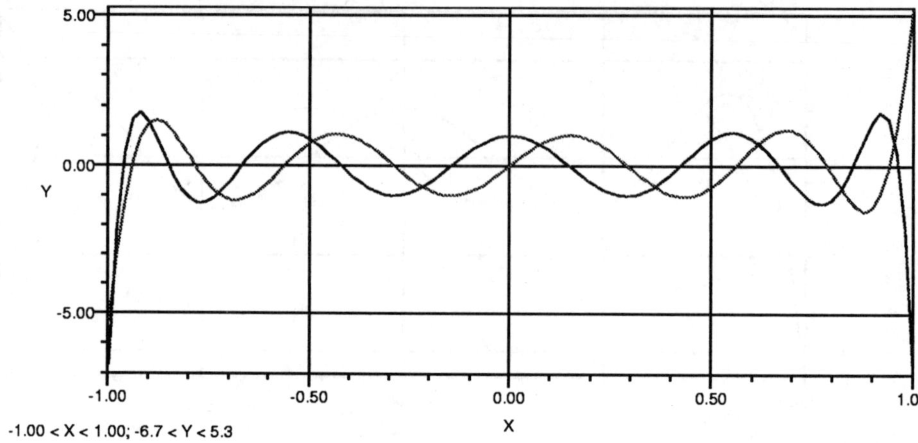

FIGURE 11.13. Functions $S_q(x;\omega_5)$ and $C_q(x;\omega_5)$ for $q=0.1$.

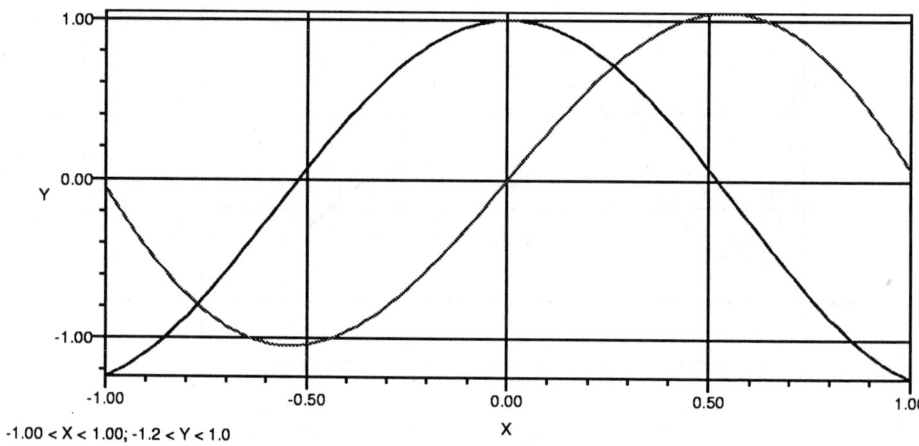

FIGURE 11.14. Functions $S_q(x;\omega_1)$ and $C_q(x;\omega_1)$ for $q=0.5$.

q-ultraspherical polynomials $C_n(x;\beta|q)$ with $\beta = q^{1/2}$, namely,

$$\rho(x) = 4\sqrt{1-x^2}\prod_{k=0}^{\infty}\frac{\left(1+2\left(1-2x^2\right)q^{k+1}+q^{2k+2}\right)}{\left(1+2\left(1-2x^2\right)q^{k+1/2}+q^{2k+1}\right)}. \tag{11.4.1}$$

Substituting $x=0$,

$$\rho(0) = 4\prod_{k=0}^{\infty}\frac{\left(1+q^{k+1}\right)^2}{\left(1+q^{k+1/2}\right)^2}. \tag{11.4.2}$$

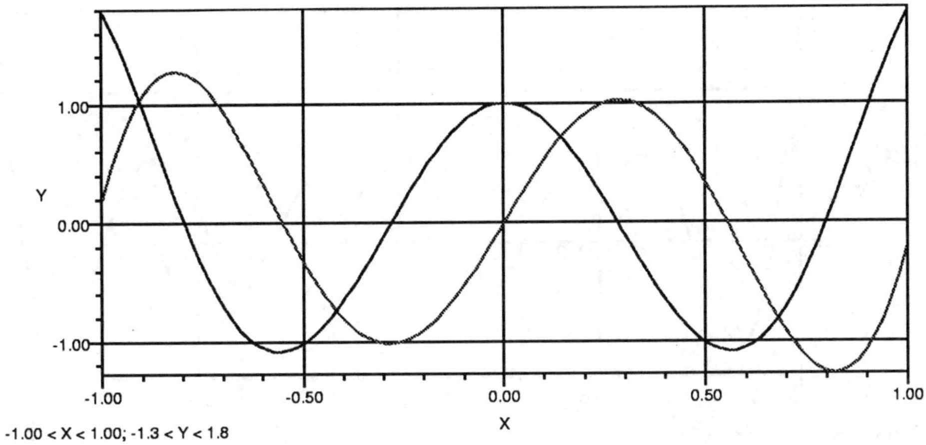

FIGURE 11.15. Functions $S_q(x;\omega_2)$ and $C_q(x;\omega_2)$ for $q=0.5$.

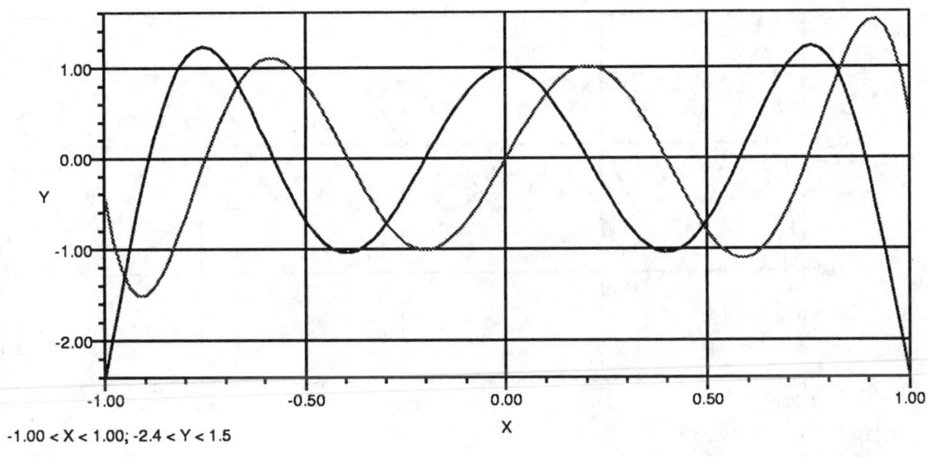

FIGURE 11.16. Functions $S_q(x;\omega_3)$ and $C_q(x;\omega_3)$ for $q=0.5$.

Therefore

$$\frac{\rho(x)}{\rho(0)} \tag{11.4.3}$$

$$= \sqrt{1-x^2}\prod_{k=0}^{\infty}\frac{\left(1+q^{k+1/2}\right)^2\left(1+2\left(1-2x^2\right)q^{k+1}+q^{2k+2}\right)}{\left(1+q^{k+1}\right)^2\left(1+2\left(1-2x^2\right)q^{k+1/2}+q^{2k+1}\right)}.$$

We sketch the graphs of this function for the following values of the parameter $q=0.1,\ 0.2,\ 0.3,\ 0.4,\ 0.5,\ 0.6,\ 0.7,\ 0.8,\ 0.9$ on $[-1,1]$ in Figure 11.19. These numerical results definitely confirm the well-known fact that function (11.4.3), the weight function for the continuous q-ultraspherical

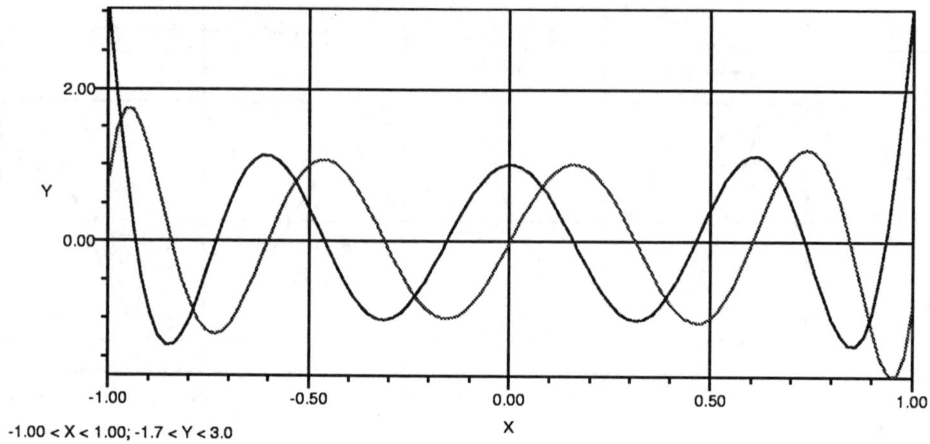

FIGURE 11.17. Functions $S_q(x;\omega_4)$ and $C_q(x;\omega_4)$ for $q = 0.5$.

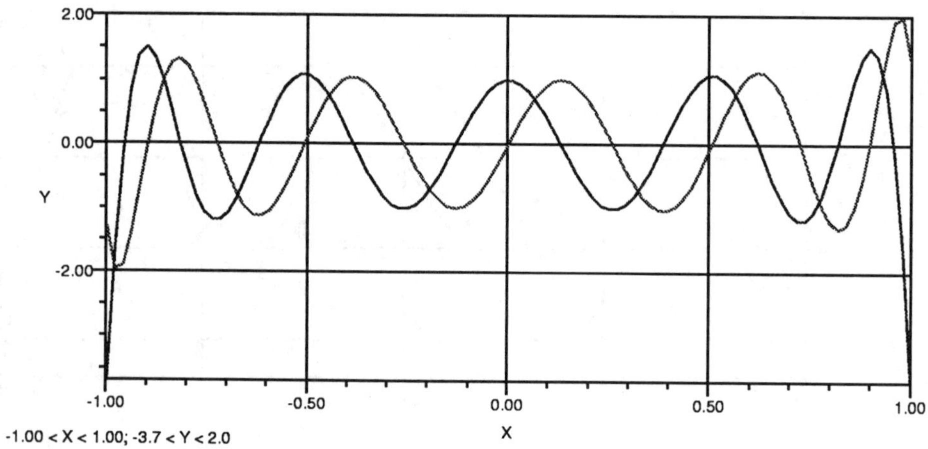

FIGURE 11.18. Functions $S_q(x;\omega_5)$ and $C_q(x;\omega_5)$ for $q = 0.5$.

polynomials $C_n\left(x;q^{1/2}|q\right)$, converges to the uniform distribution on $[-1,1]$ as $q \to 1^-$.

We have also checked numerically the orthogonality relations (5.2.26)–(5.2.28) evaluating the integrals by Simpson's rule.

11.5. Some Examples of Basic Fourier Series and Related Sums

Explicit expansions of elementary and q-functions in basic Fourier series lead naturally to a new interesting class of formulas; see **[26]**, **[146]**, and Chapters 9 to 11. Here we demonstrate convergence of some of these series numerically.

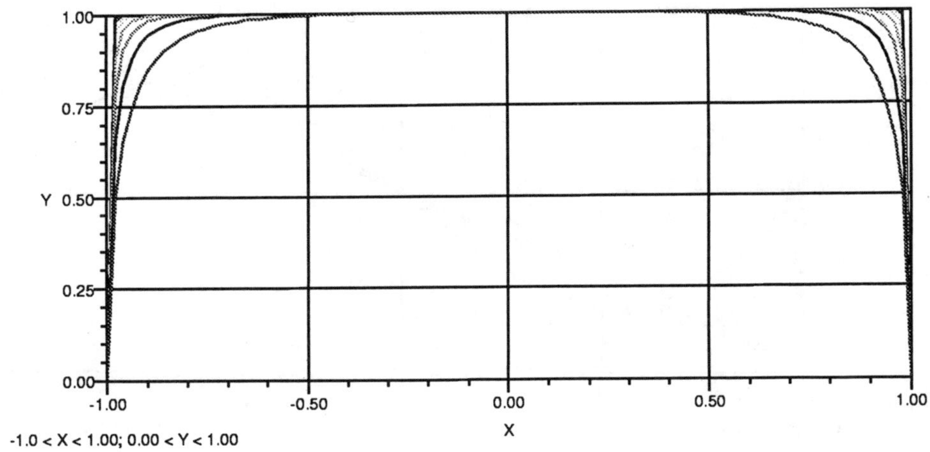

FIGURE 11.19. Function $\rho(x)/\rho(0)$ for $q = 0.1, 0.2, 0.3, 0.4, 0.5, 0.6, 0.7, 0.8, 0.9$ on $[-1, 1]$.

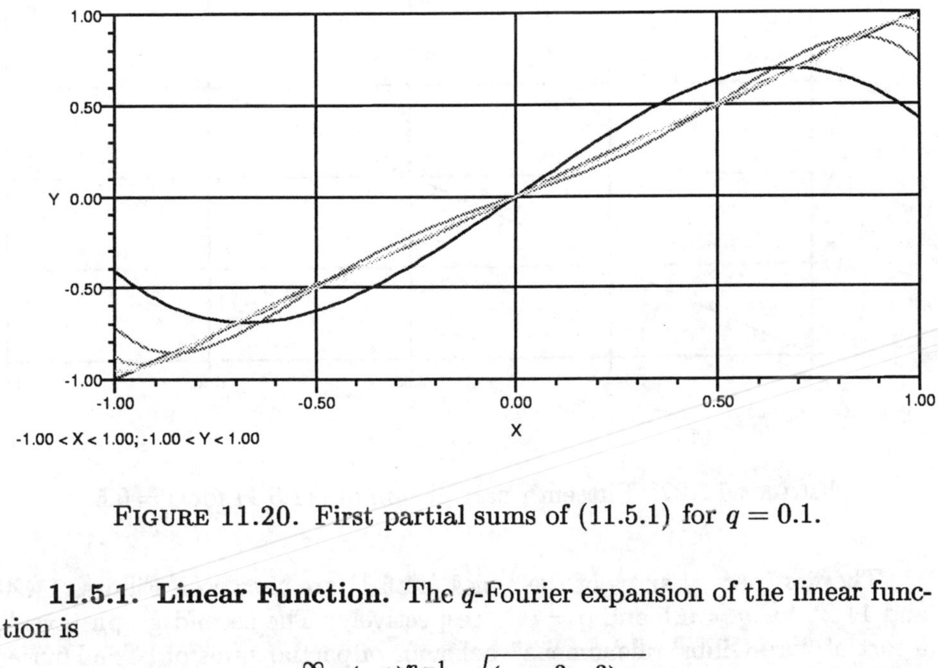

FIGURE 11.20. First partial sums of (11.5.1) for $q = 0.1$.

11.5.1. Linear Function. The q-Fourier expansion of the linear function is

$$x = \left(q^{1/4} + q^{-1/4}\right) \sum_{n=1}^{\infty} \frac{(-1)^{n-1}}{\kappa(\omega_n)\,\omega_n} \sqrt{\frac{(-q\omega_n^2; q^2)_\infty}{(-\omega_n^2; q^2)_\infty}}\; S_q(x; \omega_n), \qquad (11.5.1)$$

where ω_n are the roots of the transcendental equation (11.1.1) and function $\kappa(\omega)$ is given by (5.5.9); see (6.8.6) for more details. This series converge absolutely and uniformly in every closed subinterval of $|x| < \eta = \left(q^{1/4} + q^{-1/4}\right)/2$ and thus on the orthogonality interval $[-1, 1]$.

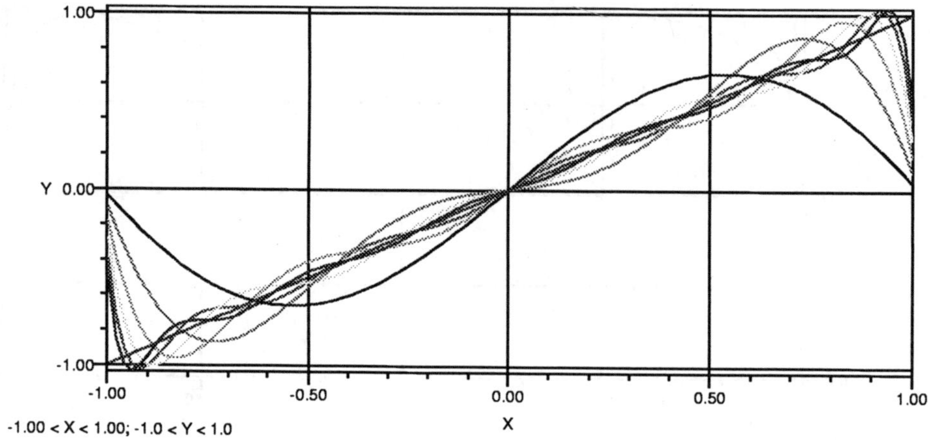

FIGURE 11.21. First partial sums of (11.5.1) for $q = 0.5$.

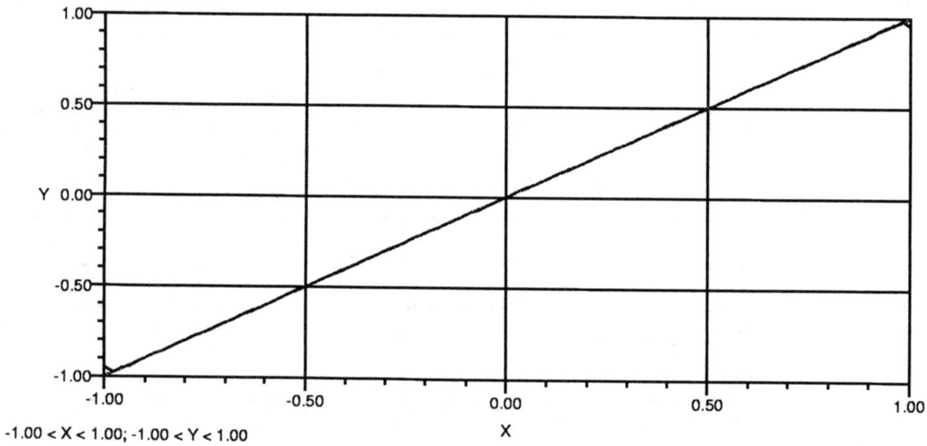

FIGURE 11.22. Fifteenth partial sum of (11.5.1) for $q = 0.5$.

The first partial sums of the series (11.5.1) are shown on Figures 11.20 and 11.21 for $q = 0.1$ and $q = 0.5$, respectively. The second graph reveals a sort of "pre-Gibbs' phenomena" behavior of partial sums of basic Fourier series. It is vanishing when we increase the number of terms in partial sums because of the uniform convergence of the series (11.5.1) on $[-1, 1]$. For instance, Figure 11.22 shows fifteenth partial sum for $q = 0.5$ and one can see that "Gibbs' phenomenon" has disappeared there. We have to go outside the orthogonality interval $(-1, 1)$ in order to investigate this phenomenon at the end points of the uniform convergence interval $(-\eta, \eta)$ with $\eta = \left(q^{1/4} + q^{-1/4}\right)/2$. Numerical investigation of the convergence of this series

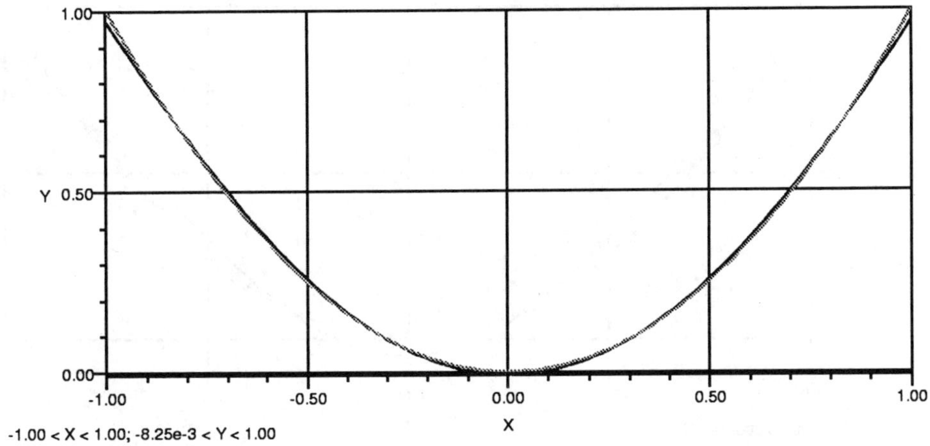

FIGURE 11.23. First partial sums of (11.5.2) for $q = 0.1$.

inside the ellipse $E_{1/4}$ in the complex x-plane is another interesting question; see Section 6.9.

11.5.2. Quadratic Function. Expansion of a quadratic function in the basic Fourier series is

$$x^2 = \frac{\left(1+q^{1/2}\right)^2}{4\left(1+q^{1/2}+q\right)} - \frac{\left(1+q^{1/2}\right)\left(1-q^2\right)}{2q} \times \sum_{n=1}^{\infty} \frac{(-1)^{n-1}}{\kappa\left(\omega_n\right)\ \omega_n^2} \sqrt{\frac{\left(-q\omega_n^2; q^2\right)_\infty}{\left(-\omega_n^2; q^2\right)_\infty}}\ C_q\left(x; \omega_n\right), \tag{11.5.2}$$

where the series converge absolutely and uniformly in every closed subinterval of $|x| < \eta_1 = \left(q^{1/2} + q^{-1/2}\right)/2$; see (6.8.7).

The first partial sums of the series (11.5.2) are shown on Figures 11.23 and 11.24 for $q = 0.1$ and $q = 0.5$, respectively. One can see that in the case $q = 0.1$ just the first two terms give a good approximation. In the second case convergence is slower, but see Figure 11.25 for the graph of the fifth partial sum for $q = 0.5$. It is of interest to investigate numerically convergence of this series in the complex x-plane inside the ellipse $E_{3/4}$; see Section 6.9.

11.5.3. Some Summation Formulas. In Section 10.6 we derived

$$\sum_{n=1}^{\infty} \frac{1}{\kappa\left(\omega_n\right)\ \omega_n^2} = \frac{q^{1/2}}{2\left(1-q^{3/2}\right)}, \tag{11.5.3}$$

$$\sum_{n=1}^{\infty} \frac{(-1)^{n-1}}{\kappa\left(\omega_n\right)\omega_n{}^2} \sqrt{\frac{\left(-q\omega_n^2; q^2\right)_\infty}{\left(-\omega_n^2; q^2\right)_\infty}} = \frac{q}{2\left(1+q\right)\left(1-q^{3/2}\right)} \tag{11.5.4}$$

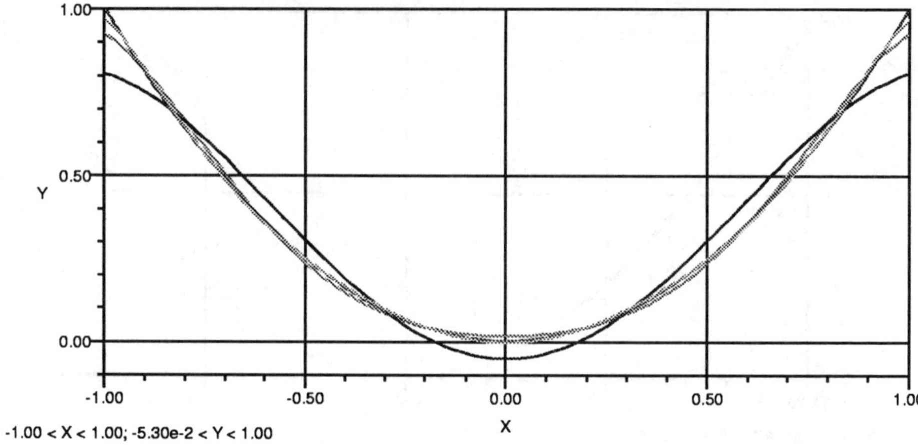

FIGURE 11.24. First partial sums of (11.5.2) for $q = 0.5$.

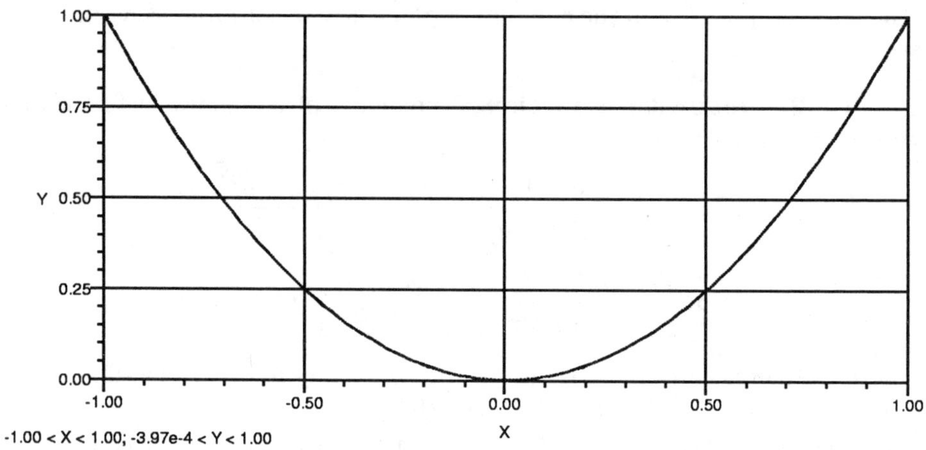

FIGURE 11.25. Fifth partial sum of (11.5.2) for $q = 0.5$.

as analogs of Euler's sums

$$\sum_{n=1}^{\infty} \frac{1}{n^2} = \frac{\pi^2}{6}, \qquad \sum_{n=1}^{\infty} \frac{(-1)^{n-1}}{n^2} = \frac{\pi^2}{12}, \tag{11.5.5}$$

respectively.

The next table shows the convergence of the first partial sums in the left sides of (11.5.3) and (11.5.4) for $q = 0.1$. The values of the right sides are presented here as the case $m = \infty$. One can see that the alternating series (11.5.4) converges faster than (11.5.3).

m	Partial sums of (11.5.3)	Partial sums of (11.5.4)
1	0.14862325246384	0.04822253395912
2	0.16182628926034	0.0468988402049
3	0.16313221634812	0.04694014783294
4	0.16326266721734	0.04693884329082
5	0.16327571088776	0.04693888453862
6	0.16327701524064	0.04693888323428
7	0.1632771456758	0.04693888327552
8	0.1632771587193	0.04693888327422
9	0.16327716002366	0.04693888327426
10	0.16327716015408	0.04693888327426
∞	0.16327716016858	0.04693888327426

In a similar fashion, formulas

$$\sum_{n=1}^{\infty} \frac{1}{\kappa\left(\omega_n\right)\ \omega_n^4} = \frac{q^2}{2\left(1-q^{3/2}\right)^2\left(1-q^{5/2}\right)}, \tag{11.5.6}$$

$$\sum_{n=1}^{\infty} \frac{(-1)^{n-1}}{\kappa\left(\omega_n\right)\omega_n{}^4} \sqrt{\frac{\left(-q\omega_n^2;q^2\right)_\infty}{\left(-\omega_n^2;q^2\right)_\infty}} = \frac{q^{5/2}\left(1+2q+q^{3/2}+2q^2+q^3\right)}{2\left(1+q\right)^2\left(1+q^2\right)\left(1-q^{3/2}\right)^2\left(1-q^{5/2}\right)} \tag{11.5.7}$$

are analogs of the elementary sums

$$\sum_{n=1}^{\infty} \frac{1}{n^4} = \frac{\pi^4}{90}, \qquad \sum_{n=1}^{\infty} \frac{(-1)^{n-1}}{n^4} = \frac{7\pi^4}{720}, \tag{11.5.8}$$

respectively.

The next table demonstrates the convergence of the first partial sums of (11.5.6) and (11.5.7) for $q = 0.1$. The values of the sums are presented as the case $m = \infty$. One can see that the positive series (11.5.6) converges now as fast as the alternating series (11.5.7). Both series converge much faster then those in (11.5.3) and (11.5.4).

m	Partial sums of (11.5.6)	Partial sums of (11.5.7)
1	0.00534457030674	0.00173410767724
2	0.00534879646036	0.00173368397674
3	0.00534880059506	0.00173368410752
4	0.00534880059918	0.00173368410748
∞	0.00534880059918	0.00173368410748

Convergence of other summation formulas derived in Section 10.6 can be analyzed in a similar manner.

11.6. Exercises for Chapter 11

(1) With the notations of Section 11.2.1, suppose that

$$x_1 < 0 < x_2 < x_3 < \dots$$

and that $S_1 < 0$. Show that

$$-\left|S_{2m-1}\right|^{-1/(2m-1)} < x_1 < -\left|S_{2m-1}\right|^{-1/2m}$$

and

$$S_{2m}/S_{2m+1} < x_1 < S_{2m-1}/S_{2m}$$

with $m = 1, 2, 3, \dots$. **[65]**

(2) Numerically interpolate the zeros ω_n and ϖ_n of the basic sine $S_q(\eta;\omega)$ and basic cosine $C_q(\eta;\omega)$ functions, respectively, as functions of q in the interval $0 < q < 1$.

(3) Check numerically Conjectures 8.1 and 8.2.

(4) Investigate numerically asymptotic formulas for Jackson's exponential functions from Ex. 8.4.

(5) *Gibbs' phenomenon.* Investigate numerically convergence of the basic Fourier series (11.5.1) when $x \to \pm\eta = \pm\left(q^{1/4} + q^{-1/4}\right)/2$ and compare the results with Gibbs' phenomenon for the corresponding classical Fourier series.

(6) The *Dirichlet kernel* of order m is defined by $D_0(x) = 1/2$ and

$$D_m(x) = \frac{1}{2} + \sum_{n=1}^{m} \cos nx = \frac{\sin((m+1/2)x)}{2\sin(x/2)},$$

when $m = 1, 2, 3, \dots$. The *Fejér kernel* of order m is defined by $K_0(x) = 1/2$ and

$$\begin{aligned} K_m(x) &= \frac{1}{2} + \sum_{n=1}^{m}\left(1 - \frac{n}{m+1}\right)\cos nx \\ &= \frac{2}{m+1}\left(\frac{\sin((m+1)x/2)}{2\sin(x/2)}\right)^2, \end{aligned}$$

when $m = 1, 2, 3, \dots$. Plot these functions for $m = 1$, 2, 3, 4, and 5. Use the partial sums and Cesàro means for the basic Fourier series in order to introduce q-analogs of these kernels as follows

$$D_0(x, -y; q) = \frac{1}{2k(0)},$$

$$\begin{aligned} D_m(x, -y; q) &= \sum_{n=-m}^{m} \frac{\mathcal{E}_q(x, -y; i\omega_n)}{2k(\omega_n)} \\ &= \frac{1}{2k(0)} + \sum_{n=1}^{m} \frac{C_q(x, -y; i\omega_n)}{k(\omega_n)}, \end{aligned}$$

when $m = 1, 2, 3, \ldots$, and

$$K_0 (x, -y; q) = \frac{1}{2k(0)},$$

$$K_m (x, -y; q) = \frac{D_0 (x, -y; q) + \ldots + D_m (x, -y; q)}{m+1}$$

$$= \frac{1}{2k(0)} + \sum_{n=1}^{m} \left(1 - \frac{n}{m+1}\right) \frac{C_q (x, -y; i\omega_n)}{k(\omega_n)},$$

when $m = 1, 2, 3, \ldots$; see Ex. 5.3. Plot these kernels as functions of two variables for m = 1, 2, 3, 4, 5 and different values of the parameter $0 < q < 1$. Investigate numerically positivity of the q-Fejér kernel.

(7) *Numerical investigation of convergence of the basic Fourier series and quadrature formulas of Gaussian type.* Quadrature formulas of Gaussian type for definite integrals are formulas of the form

$$\int_a^b f(x)\ \rho(x)\ dx \approx \sum_{k=1}^{n} \lambda_k f(x_k),$$

where $\rho(x) > 0$, abscissas x_k with k = 1, 2, ..., n are the zeros of the polynomial $p_n(x)$ that is orthogonal on (a, b) with the weight function $\rho(x)$, and the Christoffel numbers λ_k are given by

$$\lambda_k = \frac{a_n d_{n-1}^2}{a_{n-1} p_n' (x_k)\, p_{n-1} (x_k)},$$

where a_n is the leading coefficient of $p_n(x)$ and d_n^2 is the $\mathcal{L}^2$-norm; see, for example [**113**].

Find numerically the zeros x_k and the Christoffel numbers λ_k with k = 1, 2, ..., n for n = 1, 2, ..., 10 and different values of the parameter $0 < q < 1$ in the case of the q-Legendre polynomials $p_n(x) = C_n\left(x; q^{1/2} \middle| q\right)$.

Study accuracy of the numerical evaluation of the q-Fourier coefficients with the help of quadrature formulas of Gaussian type for the q-Legendre polynomials $C_n\left(x; q^{1/2} \middle| q\right)$ for different functions $f(x)$ in comparison with the evaluation of the corresponding integrals by the Simpson rule. Investigate numerically convergence of the corresponding basic Fourier series.

(8) Verify numerically the uniform bounds (6.1.11)–(6.1.12), (6.1.16) and (6.9.7)–(6.9.8), (6.9.13).

(9) Verify numerically the estimates (8.3.18) and (8.3.19).

(10) Numerically investigate Conjecture 11.1.

CHAPTER 12

Suggestions for Further Work

The study of Fourier series has a long and distinguished history in mathematics. Historically, Fourier series were introduced in order to solve the heat equation, and since then these series were frequently used in various applied problems. Much of modern real analysis including Lebesgue's fundamental theory of integration had its origin in some deep convergence questions in Fourier series. In this book a basic analog of only relatively small part of an extensive theory of the Fourier series has been covered. The author truly believes that the most of the emerging theory of the q-Fourier series and practically all of their applications still remain "Terra Incognita" at the present stage of the investigation. The subject is rapidly changing and it is reasonable to expect that many more interesting results will be discovered in the future. In conclusion, I would like to outline several directions for further work.

Positivity. Positivity of the exponential function has many interesting consequences. Situation with basic exponential function $\mathcal{E}_q(x;\alpha)$ is more complicated. Ismail and Stanton [**73**] observed that the representation (2.3.26) implies that $\mathcal{E}_q(x;\alpha)>0$ when either $0\le\alpha<1$, $x\ge-\eta=-\left(q^{1/4}+q^{-1/4}\right)/2$ or $-1<\alpha\le0$, $x\le\eta$. Almost nothing is known about the positivity of the q-exponential function and/or about its zeros in larger domains. One can use representation (2.3.32) for further investigation of the positivity. There is also an overlap with the study of the completeness of the basic trigonometric systems by using methods of the theory of entire functions.

More explicit q-Fourier expansions – joy of discovering of new formulas! This, obviously, requires more explicit integral evaluations involving the q-exponential functions. The fundamental Askey–Wilson integral [**15**] and its extension, the Nassrallah–Rahman integral [**110**], [**46**], are expected to play a central role in this study.

Completeness of q-trigonometric systems. Many interesting results on the completeness of the trigonometric systems $\left\{e^{i\lambda_n x}\right\}_{n=-\infty}^{\infty}$, where λ_n is a suitable sequence of complex numbers, were established in [**99**]–[**101**] using methods of the theory of entire functions. Some extensions of these results to the case of the basic trigonometric systems have been established; see Chapter 7 and [**145**] for the "first step" in this direction. Extensions of other classical results should be found.

Asymptotics of zeros. It is of interest to establish next terms in asymptotic expansions found in Chapter 8. There are interesting relations with the transformation theory of basic hypergeometric functions and complex analysis. This will be useful in analytic continuation of the q-zeta function. There are also unexpected relations of the coefficients of these expansions with some Lambert series occurring in the theory of elliptic functions and number theory; see, for example, [**102**], [**103**], [**104**] and [**109**]. These results are likely to be extended to the zeros of the q-Bessel functions and to the zeros of basic trigonometric functions with the q-shift which are important for extension of the q-Riemann zeta function to q-Hurwitz zeta function.

The q-Fourier transform. In this book the basic analog of Fourier series on a q-quadratic grid has been studied in detail. Investigation of q-analogs of the classical Fourier integral is another interesting research topic in q-series. See [**13**], [**14**] and references therein for analogs of the Fourier integral related to the Poisson kernels of the Askey–Wilson polynomials and their special cases. This was developed as an extension of Wiener's approach [**165**] to the classical Fourier transform. Another version, related to the q-exponential functions, is to be investigated. One can apply an analog of the Sturm–Liouville type approach as in [**26**] and [**143**] in order to establish another orthogonality relation for the basic exponential functions that seems to be missing at the moment. These results will be important for study q-analogs of the equations of mathematical physics.

Group theoretical interpretation of the q-addition theorems. A "group–theoretic" interpretation of the addition formulas discussed in Chapter 3 should be established. The q-shift operator, see Ex. 3.14, will, possibly, play a significant role in this study. There is a very good chance to find new connections with the modern theory of quantum groups.

Basic analogs of Bernoulli and Euler polynomials. Investigation of q-analogs of the Bernoulli and Euler polynomials and numbers should be continued. The q-Bernoulli and Euler polynomials and numbers were introduced in Chapter 10 on the basis of the q-Fourier series; see also [**146**]. Their properties has to be investigated in detail. A systematic study of another q-version of the Bernoulli polynomials introduced recently by Ismail and Rahman [**67**] by formal q-Fourier expansions needs to be done.

Summability. Methods of summation of basic Fourier series somewhat similar to the classical ones in [**3**], [**20**], and [**166**] should be found. An analog of Abel's summation method has already been analyzed in [**26**] and in Section 6.6 here; see also Ex. 9.12 for yet another q-version of Abel's summation method – a systematic study has not been undertaken.

A detailed theory of summability of basic Fourier series should be developed. This requires, among other things, finding of analogs of several classical kernels. For example, q-analogs of the Dirichlet kernel and the Fejér kernel were discussed in Ex. 11.6. One has to investigate properties of these kernels as functions of two variables; in particular, investigate analytically and numerically positivity of the q-Fejér kernel and corresponding summability results.

Numerical methods. Numerical aspects of the q-Fourier series should be analyzed in detail. It may lead to a new effective method of approximation for certain "arbitrary" functions on the basis of the Gaussian quadrature formulas for the definite integrals; see Ex. 11.7. One has to find numerically the zeros x_k and the Christoffel numbers λ_k with $k = 1, 2, \dots,$ n for $n = 1, 2, \dots, 10$ and different values of the parameter $0 < q < 1$ in the case of the q-Legendre polynomials $p_n(x) = C_n\left(x; q^{1/2} \middle| q\right)$, which have the same weight function as the basic trigonometric system, and to study accuracy of the numerical evaluation of the q-Fourier coefficients with the help of quadrature formulas of Gaussian type. The next step will be to investigate numerically convergence of the corresponding basic Fourier series and to compare the results with other methods of numerical approximation of functions. The obvious advantage of the method of q-Fourier series is a possibility to "optimize" convergence with respect to the basic parameter q and one needs to investigate this question. The end result should be be a computer software for numerical approximation of certain class of functions. It is also of interest to study a possible Gibbs' phenomenon at the ellipse of convergence for different functions. This may lead to interesting results in approximation theory and numerical methods.

The q-zeta function. One goal is a thorough investigation of the q-zeta function introduced in Section 10.6. Analytic continuation and possible "functional equation" are important steps in this direction. Zeros and numerical investigation of the behaviour of this function in the complex plane are other interesting questions. Also, an analog of the Hurwitz zeta function should be introduced and studied in detail.

Theory of q-Fourier–Bessel series and the corresponding integral transforms. The Fourier–Bessel series are natural extensions of the classical Fourier series, their theory is investigated in detail in [**163**]. See [**63**], Exs. 5.6–5.7 and references therein for the new orthogonality relation of the q-Bessel functions on a q-quadratic grid. The detailed theory of these functions somewhat similar to [**163**] has not been developed yet. This study is expected to give a natural extension of the theory of basic Fourier series to the next level of the ${}_2\varphi_1$ q-Bessel functions. It is also interesting to extend these results to the case of the corresponding integral transforms.

Theory of orthogonal basic hypergeometric functions. It is well-known that the q-exponential and q-trigonometric functions discussed in this book are solutions of a very special case of the general difference equation of hypergeometric type on nonuniform lattices [**17**], [**111**] and [**139**]. The Askey–Wilson polynomials [**15**] – and more generally the classical orthogonal polynomials [**9**], [**86**], [**111**] – are well-known as the most important orthogonal solutions of this equation. In the last few years Ismail, Masson and Suslov [**63**], and Suslov [**141**], [**143**] have found another type of orthogonal solutions of this difference equation with respect to continuous weight functions at the higher levels of the nonterminating ${}_2\varphi_1$ and ${}_8\varphi_7$ functions, correspondingly. Recently Koelink and Stokman [**90**] have established an orthogonality relation of the ${}_8\varphi_7$ function under consideration when the measure includes an absolutely continuous part and an infinite series of mass points; see also [**91**]. All these results have to be generalized to the highest level of the nonterminating ${}_{10}\varphi_9$ biorthogonal rational functions and, possibly, to the elliptic hypergeometric functions. See Spiridonov and Zhedanov's paper [**137**] for recent progress in the theory of these functions and bibliography.

Applications to analytic theory of partial difference equations. Many applications of the theory of basic Fourier series are not investigated yet. As an example, we can mention a q-model of the heat equation that uses Askey–Wilson divided difference operators of the first order instead of derivatives. Separation of variables and representation of solution in terms of basic Fourier series give an explicit solution of this model; see [**39**] and Ex. 5.9 for more details. One can apply the method of separation of variables to q-versions of other equations of mathematical physics, like the wave equation and Laplace equation [**39**]. Numerical investigation of these solutions is an interesting research topic.

Conclusion. This is an incomplete list of open problems in the theory of basic Fourier series and related topics from author's perspective. At the moment the theory of basic Fourier series is a new open area of research and this gives an opportunity for any person with good background in classical analysis and q-series, who is interested in developing of the theory, to formulate a more complete list of challenging problems for further study. Research in basic Fourier series deals with fundamental ideas of calculus and classical analysis; it includes also numerical calculations, use of computer algebra, and programming. All this provides an excellent opportunity for integration of these research activities with teaching, training and learning and will be extremely beneficial for graduate students and young researchers who would like to be involved in this project.

APPENDIX A

Selected Summation and Transformation Formulas and Integrals

In this Appendix we collect some summation and transformation formulas from the theory of basic hypergeometric functions and related integrals which are frequently used throughout the book. The reader can find an excellent account on the theory of basic hypergeometric series in Gasper and Rahman's book **[46]** where a more complete list of useful formulas is given in Appendices I–III.

A.1. Basic Hypergeometric Series

We use the standard definition of the *basic hypergeometric series* by

$$ {}_r\varphi_s\left(\begin{matrix} a_1, a_2, ..., a_r \\ b_1, b_2, ..., b_s \end{matrix} ; q, z\right) \tag{A.1.1}$$

$$= \sum_{n=0}^{\infty} \frac{(a_1, a_2, \ldots, a_r; q)_n}{(q, b_1, b_2, \ldots, b_s; q)_n} \left((-1)^n q^{n(n-1)/2}\right)^{1+s-r} z^n,$$

where the *q-shifted factorials* are

$$(a;q)_n := \prod_{k=0}^{n-1}(1-aq^k), \qquad (a_1, a_2, \ldots, a_r; q)_n := \prod_{k=1}^{r}(a_k;q)_n, \tag{A.1.2}$$

$$(a;q)_\infty := \lim_{n\to\infty}(a;q)_n, \qquad (a_1, a_2, \ldots, a_r; q)_\infty := \prod_{k=1}^{r}(a_k;q)_\infty, \tag{A.1.3}$$

$$(a;q)_\alpha := \frac{(a;q)_\infty}{(aq^\alpha;q)_\infty}, \qquad (a_1, a_2, \ldots, a_r; q)_\alpha := \prod_{k=1}^{r}(a_k;q)_\alpha, \tag{A.1.4}$$

provided $|q| < 1$. If $0 < |q| < 1$, the ${}_r\varphi_s$ series converges absolutely for all t if $r \le s$ and for $|z| < 1$ if $r = s+1$.

An ${}_{r+1}\varphi_r$ series called *balanced* if $b_1 b_2 \cdots b_k = q a_1 a_2 \cdots a_{r+1}$. The basic hypergeometric series

$$ {}_{r+1}\varphi_r\left(\begin{matrix} a_1, a_2, ..., a_{r+1} \\ b_1, b_2, ..., b_r \end{matrix} ; q, z\right) \tag{A.1.5}$$

is called *very-well-poised* if the parameters satisfy the relations

$$qa_1 = a_2 b_1 = a_3 b_2 = \ldots = a_{r+1} b_r; \quad a_2 = qa_1^{1/2}, \ a_3 = -qa_1^{1/2}. \tag{A.1.6}$$

A frequently used the more compact notation is

$$ {}_{r+1}W_r\left(a_1;a_4,a_5,...,a_{r+1};q,z\right) \tag{A.1.7}$$
$$= {}_{r+1}\varphi_r\left(\begin{matrix} a_1, qa_1^{1/2}, -qa_1^{1/2}, a_4, ..., a_{r+1} \\ a_1^{1/2}, -a_1^{1/2}, qa_1/a_4, ..., qa_1/a_{r+1} \end{matrix} ; q, z\right).$$

For negative subscripts, the q-shifted factorials are defined by

$$(a;q)_{-n} = \frac{1}{(aq^{-n};q)_n} = \frac{(-q/a)^n q^{n(n-1)/2}}{(q/a;q)_n} \tag{A.1.8}$$

and the *bilateral basic hypergeometric series* is defined by

$$ {}_r\psi_r\left(\begin{matrix} a_1, a_2, ..., a_r \\ b_1, b_2, ..., b_r \end{matrix} ; q, z\right) \tag{A.1.9}$$
$$= \sum_{n=-\infty}^{\infty} \frac{(a_1, a_2, \ldots, a_r; q)_n}{(b_1, b_2, \ldots, b_r; q)_n} z^n$$
$$= \sum_{n=0}^{\infty} \frac{(a_1, a_2, \ldots, a_r; q)_n}{(b_1, b_2, \ldots, b_r; q)_n} z^n$$
$$+ \sum_{n=1}^{\infty} \frac{(q/b_1, q/b_2, \ldots, q/b_r; q)_n}{(q/a_1, q/a_2, \ldots, q/a_r; q)_n} \left(\frac{b_1 \cdots b_r}{a_1 \cdots a_r z}\right).$$

We assume that $|q| < 1$, so that the region of convergence is the annulus

$$\left|\frac{b_1 \cdots b_r}{a_1 \cdots a_r}\right| < |z| < 1. \tag{A.1.10}$$

When $b_k = q$ for some k the second series vanishes and the first series becomes basic hypergeometric series.

A.2. Selected Summation Formulas

The two q-exponential functions,

$$e_q(x) = \sum_{n=0}^{\infty} \frac{x^n}{(q;q)_n} = \frac{1}{(x;q)_\infty}, \quad |x| < 1, \tag{A.2.1}$$

$$E_q(x) = \sum_{n=0}^{\infty} \frac{q^{n(n-1)/2}\, x^n}{(q;q)_n} = (-x;q)_\infty. \tag{A.2.2}$$

The q-binomial theorem,

$$ {}_1\varphi_0(a;-;q,z) = \sum_{n=0}^{\infty} \frac{(a;q)_n}{(q;q)_n} z^n = \frac{(az;q)_\infty}{(z;q)_\infty}. \tag{A.2.3}$$

Jacobi's triple product,

$$\sum_{n=-\infty}^{\infty} q^{n^2} z^n = \left(q^2, -qz, -q/z; q^2\right)_\infty. \tag{A.2.4}$$

Ramanujan's sum,

$$ {}_1\psi_1\left(\begin{matrix} a \\ b \end{matrix};q,z\right) = \frac{(q,\ b/a,\ az,\ q/az;q)_\infty}{(b,\ q/a,\ z,\ b/az;q)_\infty}. \tag{A.2.5} $$

The q-Gauss sum,

$$ {}_2\varphi_1\left(\begin{matrix} a,\ b \\ c \end{matrix};q,\ \frac{c}{ab}\right) = \frac{(c/a,\ c/b;q)_\infty}{(c,\ c/ab;q)_\infty}. \tag{A.2.6} $$

The Bailey–Daum sum,

$$ {}_2\varphi_1\left(\begin{matrix} a,\ b \\ qa/b \end{matrix};q,\ -\frac{q}{b}\right) = \frac{(-q;q)_\infty\,(qa,\ aq^2/b^2;q^2)_\infty}{(-q/b,\ qa/b;q)_\infty}. \tag{A.2.7} $$

The q-Saalschütz sum,

$$ {}_3\varphi_2\left(\begin{matrix} a,\ b,\ q^{-n} \\ c,\ q^{1-n}ab/c \end{matrix};q,\ q\right) = \frac{(c/a,\ c/b;q)_n}{(c,\ c/ab;q)_n}. \tag{A.2.8} $$

Andrews' terminating q-analog of the Watson's ${}_3F_2$ sum,

$$ {}_4\varphi_3\left(\begin{matrix} q^{-n},\ aq^n,\ c,\ -c \\ (aq)^{1/2},\ -(aq)^{1/2},\ c^2 \end{matrix};q,\ q\right) \tag{A.2.9} $$

$$ = \begin{cases} 0 & \text{if } n \text{ is odd,} \\ c^n \dfrac{(q,\ qa/c^2;\ q^2)_{n/2}}{(qa,\ qc^2;\ q^2)_m} & \text{if } n \text{ is even.} \end{cases} $$

Andrews' terminating balanced ${}_4\varphi_3$ sum [**8**],

$$ {}_4\varphi_3\left(\begin{matrix} q^{-2n},\ a^2q^{2n},\ b,\ bq \\ aq,\ aq^2,\ b^2 \end{matrix};q^2,\ q^2\right) = b^n \frac{(1-a)\,(qa/b,\ -q;q)_n}{(1-aq^{2n})\,(a,\ -b;q)_n}. \tag{A.2.10} $$

Bailey's nonterminating extension of Jackson's ${}_8\varphi_7$ sum,

$$ {}_8\varphi_7\left(\begin{matrix} a,\ qa^{1/2},\ -qa^{1/2},\ b,\ c,\ d,\ e,\ f \\ a^{1/2},\ -a^{1/2},\ qa/b,\ qa/c,\ qa/d,\ qa/e,\ qa/f \end{matrix};q,q\right) \tag{A.2.11} $$

$$ -\frac{b\,(qa,\ c,\ d,\ e,\ f,\ qb/a,\ qb/c,\ qb/d,\ qb/e,\ qb/f;q)_\infty}{a\,(qa/b,\ qa/c,\ qa/d,\ qa/e,\ qa/f,\ bc/a,\ bd/a,\ bf/a,\ qb^2/a;q)_\infty} $$

$$ \times {}_8\varphi_7\left(\begin{matrix} b^2/a,\ qba^{-1/2},\ -qba^{-1/2},\ b,\ bc/a,\ bd/a,\ be/a,\ bf/a \\ ba^{-1/2},\ -ba^{-1/2},\ qb/a,\ qb/c,\ qb/d,\ qb/e,\ qb/f \end{matrix};q,q\right) $$

$$ = \frac{(qa,\ b/a,\ qa/cd,\ qa/ce,\ qa/cf,\ qa/de,\ qa/df,\ qa/ef;q)_\infty}{(qa/c,\ qa/d,\ qa/e,\ qa/f,\ bc/a,\ bd/a,\ be/a,\ bf/a;q)_\infty}, $$

where $qa^2 = bcdef$.

A.3. Selected Transformation Formulas

Heine's transformations of ${}_2\varphi_1$ series:

$$ {}_2\varphi_1\left(\begin{matrix} a, \ b \\ c \end{matrix}; q, \ z\right) = \frac{(b, \ az; q)_\infty}{(c, \ z; q)_\infty} \, {}_2\varphi_1\left(\begin{matrix} c/b, \ z \\ az \end{matrix}; q, \ b\right) \tag{A.3.1} $$

$$ = \frac{(c/b, \ bz; q)_\infty}{(c, \ z; q)_\infty} \, {}_2\varphi_1\left(\begin{matrix} abz/c, bz \\ bz \end{matrix}; q, \ c/b\right) \tag{A.3.2} $$

$$ = \frac{(abz/c, \ c; q)_\infty}{(z; q)_\infty} \, {}_2\varphi_1\left(\begin{matrix} c/a, c/b \\ c \end{matrix}; q, \ abz/c\right). \tag{A.3.3} $$

Jackson's transformations of ${}_2\varphi_1$, ${}_2\varphi_2$ and ${}_3\varphi_2$ series:

$$ {}_2\varphi_1\left(\begin{matrix} a, \ b \\ c \end{matrix}; q, \ z\right) = \frac{(az; q)_\infty}{(z; q)_\infty} \, {}_2\varphi_2\left(\begin{matrix} a, \ c/b \\ c, \ az \end{matrix}; q, \ bz\right) \tag{A.3.4} $$

$$ = \frac{(az, \ bz; q)_\infty}{(c, \ z; q)_\infty} \, {}_2\varphi_2\left(\begin{matrix} z, \ abz/c \\ az, \ bz \end{matrix}; q, \ c\right) \tag{A.3.5} $$

$$ = \frac{(b, \ c/a, \ az; q)_\infty}{(b/a, \ c, \ z; q)_\infty} \, {}_3\varphi_2\left(\begin{matrix} a, \ c/b, \ 0 \\ qa/b, \ az \end{matrix}; q, \ q\right) \tag{A.3.6} $$

$$ + \frac{(a, \ c/b, \ az; q)_\infty}{(a/b, \ c, \ z; q)_\infty} \, {}_3\varphi_2\left(\begin{matrix} b, \ c/a, \ 0 \\ qb/a, \ az \end{matrix}; q, \ q\right). $$

Three-term transformation formulas for ${}_2\varphi_1$:

$$ {}_2\varphi_1\left(\begin{matrix} a, \ b \\ c \end{matrix}; q, \ z\right) = \frac{(abz/c, \ q/c; q)_\infty}{(az/c, \ q/a; q)_\infty} \, {}_2\varphi_1\left(\begin{matrix} c/a, \ qc/abz \\ qc/az \end{matrix}; q, \ \frac{qb}{c}\right) $$

$$ - \frac{(b, q/c, c/a, az/q, q^2/az; q)_\infty}{(c/q, qb/c, q/a, az/c, qc/az; q)_\infty} \, {}_2\varphi_1\left(\begin{matrix} qa/c, \ qb/c \\ q^2/c \end{matrix}; q, \ z\right). \tag{A.3.7} $$

$$ {}_2\varphi_1\left(\begin{matrix} a, \ b \\ c \end{matrix}; q, \ z\right) = \frac{(b, \ c/a, \ az, \ q/az; q)_\infty}{(c, \ b/a, \ z, \ q/z; q)_\infty} \, {}_2\varphi_1\left(\begin{matrix} a, \ qa/c \\ qa/b \end{matrix}; q, \ \frac{qc}{abz}\right) $$

$$ + \frac{(a, \ c/b, \ bz, \ q/bz; q)_\infty}{(c, \ a/b, \ z, \ q/z; q)_\infty} \, {}_2\varphi_1\left(\begin{matrix} b, \ qb/c \\ qb/a \end{matrix}; q, \ \frac{qc}{abz}\right). \tag{A.3.8} $$

Bailey's transformations for ${}_2\psi_2$:

$$ {}_2\psi_2\left(\begin{matrix} a, \ b \\ c, \ d \end{matrix}; q, \ z\right) \tag{A.3.9} $$

$$ = \frac{(az, \ d/a, \ c/b, \ qd/abz; q)_\infty}{(\ z, \ d, \ q/b, \ cd/abz; q)_\infty} \, {}_2\psi_2\left(\begin{matrix} a, \ abz/d \\ az, \ c \end{matrix}; q, \ \frac{d}{a}\right). $$

$$ {}_2\psi_2\left(\begin{matrix} a, \ b \\ c, \ d \end{matrix}; q, \ z\right) \tag{A.3.10} $$

$$= \frac{(az,\ bz,\ qc/abz,\ qd/abz;q)_\infty}{(\ q/a,\ q/b,\ c,\ d;q)_\infty}\ {}_2\psi_2\left(\begin{matrix} abz/c,\ abz/d \\ az,\ bz \end{matrix};q,\ \frac{cd}{abz}\right).$$

Transformation (3.4.7) of [**46**]:

$${}_2\varphi_1\left(\begin{matrix} a,\ b \\ qa/b \end{matrix};q,\ \frac{qx}{b^2}\right) = \frac{(qx/b,\ qax^2/b^2;q)_\infty}{(\ qax/b,\ qx^2/b^2;q)_\infty} \tag{A.3.11}$$

$$\times {}_8W_7\left(\frac{ax}{b};x,\ a^{1/2},\ -a^{1/2},\ (qa)^{1/2},\ -(qa)^{1/2};q,\frac{qx}{b^2}\right)$$

provided $|qx/b^2| < 1$ when the two series do not terminate.
Sears' transformations of terminating balanced ${}_4\varphi_3$ series:

$${}_4\varphi_3\left(\begin{matrix} q^{-n},\ a,\ b,\ c \\ d,\ e,\ f \end{matrix};q,\ q\right)$$

$$= \frac{(e/a, f/a;q)_n}{(e,\ f;q)_n}a^n\ {}_4\varphi_3\left(\begin{matrix} q^{-n},\ a,\ d/b,\ d/c \\ d,\ q^{1-n}a/e,\ q^{1-n}a/f \end{matrix};q,q\right) \tag{A.3.12}$$

$$= \frac{(a,\ ef/ab,\ ef/ac;q)_n}{(e,\ f,\ ef/abc;q)_n}$$

$$\times\ {}_4\varphi_3\left(\begin{matrix} q^{-n},\ e/a,\ f/a,\ ef/abc \\ ef/ab,\ ef/ac,\ q^{1-n}/a \end{matrix};q,\ q\right), \tag{A.3.13}$$

where $abcq^{1-n} = def$.
Watson's transformation formula,

$${}_8\varphi_7\left(\begin{matrix} a,\ qa^{1/2},\ -qa^{1/2},\ b,\ c,\ d,\ e,\ f \\ a^{1/2},\ -a^{1/2},\ qa/b,\ qa/c,\ qa/d,\ qa/e,\ qa/f \end{matrix};q,\frac{a^2q^2}{bcdef}\right) \tag{A.3.14}$$

$$= \frac{(qa,\ qa/de,\ qa/df,\ qa/ef;q)_\infty}{(qa/d,\ qa/e,\ qa/f,\ qa/def;q)_\infty}\ {}_4\varphi_3\left(\begin{matrix} qa/bc,\ d,\ e,\ f \\ qa/b,\ qa/c,\ def/a \end{matrix};q,q\right),$$

when the ${}_8\varphi_7$ series converges and the ${}_4\varphi_3$ series terminates.
Transformations of very-well-poised ${}_8\varphi_7$ series:

$${}_8\varphi_7\left(\begin{matrix} a,\ qa^{1/2},\ -qa^{1/2},\ b,\ c,\ d,\ e,\ f \\ a^{1/2},\ -a^{1/2},\ qa/b,\ qa/c,\ qa/d,\ qa/e,\ qa/f \end{matrix};q,\frac{a^2q^2}{bcdef}\right)$$

$$= \frac{(qa,\ qa/ef,\ q\lambda/e,\ q\lambda/f;q)_\infty}{(qa/e,\ qa/f,\ q\lambda,\ q\lambda/ef;q)_\infty} \tag{A.3.15}$$

$$\times {}_8\varphi_7\left(\begin{matrix} \lambda,\ q\lambda^{1/2},\ -q\lambda^{1/2},\ \lambda b/a,\ \lambda c/a,\ \lambda d/a,\ e,\ f \\ \lambda^{1/2},\ -\lambda^{1/2},\ qa/b,\ qa/c,\ qa/d,\ q\lambda/e,\ q\lambda/f \end{matrix};q,\frac{aq}{ef}\right)$$

$$= \frac{(qa,\ \mu bc/a,\ \mu bd/a,\ \mu be/a,\ \mu bf/a;q)_\infty}{(qa/c,\ qa/d,\ qa/e,\ qa/f,\ q\mu,\ \mu b/a;q)_\infty} \tag{A.3.16}$$

$$\times {}_8\varphi_7\left(\begin{matrix} \mu,\ q\mu^{1/2},\ -q\mu^{1/2},\ qa/bc,\ qa/bd,\ qa/be,\ qa/bf,\ \mu b/a \\ \mu^{1/2},\ -\mu^{1/2},\ \mu bc/a,\ \mu bd/a,\ \mu be/a,\ \mu bf/a,\ qa/b \end{matrix};q,b\right)$$

where $\lambda = qa^2/bcd$ and $\mu = q^2a^3/b^2cdef$.

Transformation of an ${}_8\psi_8$:

$$\frac{(qa/b,\ qa/c,\ qa/d,\ qa/e,\ q/ab,\ q/ac,\ q/ad,\ q/ae;q)_\infty}{(af,\ ag,\ f/a,\ g/a,\ qa^2,\ q/a^2;q)_\infty} \tag{A.3.17}$$
$$\times {}_8\psi_8\left(\begin{matrix} qa,\ -qa,\ ab,\ ac,\ ad,\ ae,\ af,\ ag \\ a,\ -a,\ qa/b,\ qa/c,\ qa/d,\ qa/e,\ qa/f,\ qa/g \end{matrix};q,\frac{q^2}{bcdefg}\right)$$
$$= \frac{(q,\ q/bf,\ q/cf,\ q/df,\ q/ef,\ qf/b,\ qf/c,\ qf/d,\ qf/e;q)_\infty}{(af,\ q/af,\ qa/f,\ f/a,\ fg,\ qf^2;q)_\infty}$$
$$\times {}_8\varphi_7\left(\begin{matrix} f^2,\ qf,\ -qf,\ bf,\ cf,\ df,\ ef,\ fg \\ f,\ -f,\ qf/b,\ qf/c,\ qf/d,\ qf/e,\ qf/g \end{matrix};q,\frac{q^2}{bcdefg}\right)$$
$$+ \frac{(q,\ q/bg,\ q/cg,\ q/dg,\ q/eg,\ qg/b,\ qg/c,\ qg/d,\ qg/e;q)_\infty}{(ag,\ q/ag,\ qa/g,\ g/a,\ fg,\ qg^2;q)_\infty}$$
$$\times {}_8\varphi_7\left(\begin{matrix} g^2,\ qg,\ -qg,\ bg,\ cg,\ dg,\ eg,\ fg \\ g,\ -g,\ qg/b,\ qg/c,\ qg/d,\ qg/e,\ qg/f \end{matrix};q,\frac{q^2}{bcdefg}\right)$$

provided $\left|q^2/bcdefg\right| < 1$.

Five-term transformation formula for ${}_4\varphi_3$:

$${}_4\varphi_3\left(\begin{matrix} a_1,\ a_2,\ a_3,\ a_4 \\ b_1,\ b_2,\ b_3 \end{matrix};q,\ z\right) \tag{A.3.18}$$
$$= \frac{(a_2/a_3,\ a_4,\ b_1/a_1,\ b_2/a_1,\ b_3/a_1,\ a_1z, q/a_1;q)_\infty}{(b_1,\ b_2,\ b_3,\ a_2/a_1,\ a_3/a_1,\ a_4/a_1,\ z,\ q/z;q)_\infty}$$
$$\times {}_4\varphi_3\left(\begin{matrix} a_1,\ qa_1/b_1,\ qa_1/b_2,\ qa_1/b_3 \\ qa_1/a_2,\ qa_1/a_3,\ qa_1/a_4 \end{matrix};q,\ \frac{qb_1b_2b_3b_4}{a_1a_2a_3a_4z}\right)$$
$$+ \operatorname{idem}(a_1;a_2,a_3,a_4),$$

where the symbol "idem$(a_1;a_2,a_3,a_4)$" after an expression stands for the sum of the three expressions obtained from the preceding expression by interchanging a_1 with each a_2, a_3, a_4; see the special case $r = 3$ of (4.5.2) in [**46**].

A.4. Some Basic Integrals

The Askey–Wilson integral [**15**],

$$\int_0^\pi \frac{\left(e^{2i\theta},\ e^{-2i\theta};q\right)_\infty\ d\theta}{(ae^{i\theta}, ae^{-i\theta}, be^{i\theta}, be^{-i\theta}, ce^{i\theta}, ce^{-i\theta}, de^{i\theta}, de^{-i\theta};q)_\infty} \tag{A.4.1}$$
$$= \frac{2\pi\,(abcd;q)_\infty}{(q, ab, ac, ad, bc, bd, cd;q)_\infty},$$

where $\max\left(|a|, |b|, |c|, |d|\right) < 1$.

The Nassrallah–Rahman integral [**110**], [**46**],

$$\int_0^\pi \frac{\left(e^{2i\theta}, e^{-2i\theta}, ge^{i\theta}, ge^{i\theta};q\right)_\infty\ d\theta}{(ae^{i\theta}, ae^{-i\theta}, be^{i\theta}, be^{-i\theta}, ce^{i\theta}, ce^{-i\theta}, de^{i\theta}, de^{-i\theta}, fe^{i\theta}, fe^{i\theta};q)_\infty}$$

$$= \frac{2\pi\left(abcf, abdf, acdf, bcdf, fg, g/f; q\right)_\infty}{\left(q, ab, ac, ad, af, bc, bd, bf, cd, cf, df, abcdf^2; q\right)_\infty} \times {}_8W_7\left(abcdf^2q^{-1}; af, bf, cf, df, abcdfg^{-1}; q, g/f\right) \tag{A.4.2}$$

$$= \frac{2\pi\left(ag, bg, cg, abcd, abcf; q\right)_\infty}{\left(q, ab, ac, ad, af, bc, bd, bf, cd, cf, abcg; q\right)_\infty} \times {}_8W_7\left(abcgq^{-1}; ab, ac, bc, g/d, g/f; q, df\right) \tag{A.4.3}$$

$$= \frac{2\pi\left(ag, bg, cg, dg, fg, abcdf/g; q\right)_\infty}{\left(q, ab, ac, ad, af, bc, bd, bf, cd, cf, df, g^2; q\right)_\infty} \times {}_8W_7\left(g^2q^{-1}; g/a, g/b, g/c, g/d, g/f; q, abcdg^{-1}\right), \tag{A.4.4}$$

where $\max\left(|a|, |b|, |c|, |d|, |f|, |q|\right) < 1$ and g is arbitrary. If these series do not terminate, then one must impose the conditions $|g/f| < 1$, $|df| < 1$, $\left|abcdg^{-1}\right| < 1$, respectively, so that they converge. See **[123]** and **[124]** for a complete list of the q-beta integrals and the bibliography.

APPENDIX B

Some Theorems of Complex Analysis

For the benefit of the reader we provide here some definitions and theorems of complex analysis; proofs of these theorems and more details can be found in **[2]**, **[11]**, **[21]**, **[22]**, **[23]**, **[98]**, **[99]**, **[100]**, **[106]**, and **[136]**.

B.1. Entire Functions

A function $f(z)$ analytic in the whole complex plane and, therefore, represented by a power series of the form

$$f(z) = \sum_{n=0}^{\infty} c_n z^n, \qquad \frac{1}{R} = \limsup_{n\to\infty} \sqrt[n]{|c_n|} = 0, \tag{B.1.1}$$

is called an *entire function.* This is the simplest class of all analytic functions containing all polynomials. Polynomials are classified according to their degree, or according to their growth as $|z| \to \infty$. An entire function can grow in various ways along different directions. For a general characterization of the growth of an entire function $f(z)$ as $|z| \to \infty$ the following function

$$M_f(r) = \max_{|z|=r} |f(z)| \tag{B.1.2}$$

is introduced. By the Maximum Principle this function increases monotonically as $|z| \to \infty$.

An entire function $f(z)$ is called a *function of finite order* if

$$M_f(r) < e^{r^k} \tag{B.1.3}$$

for all sufficiently large values of r for some constant $k > 0$. The *order* $\rho = \rho_f$ of an entire function $f(z)$ is the greatest lower bound of all those constants k when this inequality is valid. It can be shown **[99]**, **[100]** that[1]

$$\rho = \limsup_{r\to\infty} \frac{\log\log M_f(r)}{\log r}. \tag{B.1.4}$$

Let ρ be the order of an entire function $f(z)$. The function is said to have a *finite type* if for some constant $A > 0$ the following inequality

$$M_f(r) < e^{Ar^\rho} \tag{B.1.5}$$

[1]Here, as usual, by the definition

$$\limsup_{r\to\infty} \varphi(r) = \lim_{r\to\infty}\left(\sup_{t\geq r}\varphi(t)\right), \qquad \liminf_{r\to\infty} \varphi(r) = \lim_{r\to\infty}\left(\inf_{t\geq r}\varphi(t)\right).$$

holds for sufficiently large values of r. The greatest lower bound for the values of A in the last inequality is called the *type* $\sigma = \sigma_f$ of the entire function $f(z)$. It follows from this definition [**99**], [**100**] that

$$\sigma = \limsup_{r \to \infty} \frac{\log M_f(r)}{r^\rho}. \tag{B.1.6}$$

The following inequalities hold

$$\rho_{fg} \le \max(\rho_f, \rho_g), \qquad \rho_{f+g} \le \max(\rho_f, \rho_g), \tag{B.1.7}$$

$$\sigma_{fg} \le \sigma_f + \sigma_g, \qquad \sigma_{f+g} \le \max(\sigma_f, \sigma_g). \tag{B.1.8}$$

If, for a given $\rho > 0$, the type of an entire function is infinite $\sigma = \infty$, then the function is of a *maximal type*; for $0 < \sigma < \infty$ the type is *normal*; for $\sigma = 0$ the type is *minimal*. In the last case, for any $\varepsilon > 0$ the inequality

$$M_f(r) < e^{\varepsilon r^\rho} \tag{B.1.9}$$

holds for all sufficiently large values of r.

Entire functions of order $\rho = 1$ and normal type σ are called *entire functions of exponential type.*

The growth of an entire function is determined by the decrease of its coefficients in the power series expansion.

THEOREM B.1. *The order and the type of the entire function $f(z)$ given by* (B.1.1) *are determined by the formulas*

$$\rho = \limsup_{n \to \infty} \frac{n \log n}{\log(1/|c_n|)}, \tag{B.1.10}$$

$$\sigma = \frac{1}{\rho e} \limsup_{n \to \infty} \left(n \sqrt[n]{|c_n|^\rho} \right), \tag{B.1.11}$$

respectively.

The following examples show that entire functions of any nonnegative order and type exist. Let $0 < \rho < \infty$ and $0 < \sigma < \infty$. The entire function

$$f(z) = \sum_{n=1}^{\infty} \left(\frac{e\sigma\rho}{n} \right)^{n/\rho} z^n \tag{B.1.12}$$

is of order ρ and type σ. The function

$$f(z) = \sum_{n=2}^{\infty} \left(\frac{e\sigma\rho}{n \log n} \right)^{n/\rho} z^n \tag{B.1.13}$$

is of order ρ and of minimal type, whereas the function

$$f(z) = \sum_{n=2}^{\infty} \left(\frac{e\rho \log n}{n} \right)^{n/\rho} z^n \tag{B.1.14}$$

is of order ρ and of maximal type. The entire function

$$f(z) = \sum_{n=1}^{\infty} \left(\frac{1}{\log n} \right)^{n} z^n \tag{B.1.15}$$

is of infinite order, and the function

$$f(z) = \sum_{n=1}^{\infty} e^{-n^2} z^n \tag{B.1.16}$$

is of zero order.

Let $a_1, a_2, ..., a_n$ be the zeros of an entire function $f(z)$ in the disk $|z| < R$ arranged according to increasing modulus. Denoting $n_f(r)$ the number of points a_n such that $|a_n| < r$, we obtain a continuous from the left, integer-valued and piecewise constant function, which is called a *counting function* of the zeros. The following famous Jensen formula holds.

THEOREM B.2. *If $f(z)$ is holomorphic in a circle of radius R with the center at the origin, and $f(0) \neq 0$, then*

$$\int_0^R \frac{n_f(r)}{r}\, dr = \frac{1}{2\pi} \int_0^{2\pi} \log\left|f\left(Re^{i\theta}\right)\right| \, d\theta - \log|f(0)|, \tag{B.1.17}$$

where $n_f(r)$ is the number of zeros of $f(z)$ in the circle $|z| < r$.

Let $\{a_n\}$ be a sequence of complex numbers not equal to zero and such that for some nonnegative integer p the series

$$\sum_n \frac{1}{|a_n|^{p+1}} < \infty \tag{B.1.18}$$

converges. Introduce the infinite product

$$\Pi(z) = \prod_n G\left(\frac{z}{a_n}, p\right), \tag{B.1.19}$$

where

$$G(u,p) = \begin{cases} 1-u, & p=0 \\ (1-u)\exp\left(u + \dfrac{u^2}{2} + ... + \dfrac{u^p}{p}\right), & p > 0. \end{cases} \tag{B.1.20}$$

The functions $G(u,p)$ are called the *Weierstrass primary factors.*

The inequality

$$|\log G(u,p)| \leq \sum_{k=p+1}^{\infty} \frac{|u|^k}{k} \leq 2|u|^{p+1}, \tag{B.1.21}$$

valid for $|u| < 1/2$, implies that the infinite product $\Pi(z)$ converges absolutely and uniformly in every disk $|z| \leq R < \infty$. This product is called the *Weierstrass canonical product of genus p.*

One of the fundamental theorems of the theory of entire functions is the Hadamard theorem.

THEOREM B.3. *An entire function f of finite order ρ can be represented in the form*

$$f(z) = z^m e^{P_k(z)} \prod_{n=1}^{\infty} G\left(\frac{z}{a_n}, p\right), \tag{B.1.22}$$

where $a_1, a_2, \ldots$ are all zeros of the function $f(z)$ not equal to zero, $p \leq \rho \leq p+1$, $P_k(z)$ is a polynomial in z of degree $k \leq \rho$, and m is the multiplicity of the zero at the origin.

In what follows, p will denote the smallest integer for which the series $\sum_{n=1}^{\infty} |a_n|^{-p-1}$ converges. With this condition the number k is determined uniquely. The integer $g = \max(p, k)$ is called the *genus* of an entire function f. It follows from the Hadamard theorem that the genus of an entire function does not exceed its order.

An entire function of order zero has the form

$$f(z) = Cz^m \prod_{n=1}^{\infty} \left(1 - \frac{z}{a_n}\right) \tag{B.1.23}$$

with

$$\sum_{n=1}^{\infty} \frac{1}{|a_n|} < \infty. \tag{B.1.24}$$

For $\rho < 1$, by the Hadamard theorem a function $f(z)$ of order ρ has exactly the same form. An entire function of genus one has the form

$$f(z) = Cz^m e^{az+b} \prod_{n=1}^{\infty} \left(1 - \frac{z}{a_n}\right) e^{z/a_n}. \tag{B.1.25}$$

Given a sequence $a_1, a_2, \ldots, a_n, \ldots$, $a_n \neq 0$, $\lim_{n\to\infty} a_n = \infty$, the greatest lower bound of λ's such that the series

$$\sum_{n=1}^{\infty} \frac{1}{|a_n|^{\lambda}} < \infty \tag{B.1.26}$$

converges is called the *convergence exponent* of this sequence. The following important theorem due to Borel establishes the relation between the order of an entire function and the convergence exponent of the sequence of its zeros.

THEOREM B.4. *The order ρ of the canonical product is equal to the convergence exponent of the sequence of its zeros.*

The following result **[99]**, **[100]** is very important for investigation of the completeness of basic trigonometric systems in Chapters 6 and 7.

THEOREM B.5. *If the growth of an entire function $f(z)$ is not higher that of the first order and the minimal type, and if this function is bounded $|f(z)| \leq M$ on a line, then $f(z)$ =constant.*

B.2. Lagrange Inversion Formula

In this section we derive a useful inversion formula for analytic functions that is due to Lagrange. Let $f(z)$ be regular at $z = z_0$,

$$w = f(z) = \sum_{n=0}^{\infty} a_n (z - z_0)^n, \qquad f'(z_0) \neq 0 \tag{B.2.1}$$

with $w_0 = f(z_0) = a_0$ and let us find the inversion of this power series as

$$z = h(w) = \sum_{n=0}^{\infty} b_n (w - w_0)^n . \tag{B.2.2}$$

Indeed, in view of $f'(z_0) \neq 0$, by the inverse function theorem there are a disk $D : |z - z_0| < r$ and a disk $D_1 : |w - w_0| < r_1$ such that for all $w \in D_1$ equation $f(z) = w$ has a unique solution $z \in D$. This defines a single-valued function $z = h(w)$ regular in D_1 that can be represented as a power series.

In order to determine the coefficients b_n in the series (B.2.2), consider the integral

$$I(w) = \frac{1}{2\pi i} \int_{\Gamma} \frac{\zeta f'(\zeta)}{f(\zeta) - w} d\zeta, \tag{B.2.3}$$

where Γ is the boundary of the disk D and $w \in D_1$. The integrand

$$F(\zeta) = \frac{\zeta f'(\zeta)}{f(\zeta) - w}$$

is a regular function inside Γ with the exception of the point $z = h(w)$, which is a simple pole of $F(\zeta)$, and by the Cauchy residue theorem

$$\begin{aligned} I(w) &= \operatorname*{Res}_{\zeta = h(w)} F(\zeta) = \lim_{\zeta \to h(w)} (\zeta - h(w)) \frac{\zeta f'(\zeta)}{f(\zeta) - w} \\ &= \lim_{\zeta \to h(w)} \frac{\zeta f'(\zeta)}{(f(\zeta) - w)'} = h(w) = z. \end{aligned}$$

Thus,

$$z = h(w) = \frac{1}{2\pi i} \int_{\Gamma} \frac{\zeta f'(\zeta)}{f(\zeta) - w} d\zeta, \tag{B.2.4}$$

where

$$\begin{aligned} \frac{1}{f(\zeta) - w} &= \frac{1}{f(\zeta) - w_0} \frac{1}{1 - \dfrac{w - w_0}{f(\zeta) - w_0}} \\ &= \sum_{n=0}^{\infty} \frac{(w - w_0)^n}{(f(\zeta) - w_0)^{n+1}}. \end{aligned} \tag{B.2.5}$$

The series (B.2.5) converges uniformly with respect to $\zeta \in \Gamma$ because $|w - w_0| < r_1$ as $w \in D_1$ and $|f(\zeta) - w_0| \geq r_1$ as $\zeta \in \Gamma$.

Hence, multiplying Eq. (B.2.5) by

$$\frac{1}{2\pi i} \zeta f'(\zeta)$$

and integrating over Γ, we arrive at the power series (B.2.2)

$$z = h(w) = \sum_{n=0}^{\infty} b_n (w - w_0)^n , \tag{B.2.6}$$

with the coefficients of the form

$$b_n = \frac{1}{2\pi i}\int_\Gamma \frac{\zeta\, f'(\zeta)}{\left(f(\zeta) - w_0\right)^{n+1}}\, d\zeta, \qquad n = 0, 1, 2, \dots . \tag{B.2.7}$$

By (B.2.4) and (B.2.7) one gets $b_0 = z_0$. Integration by parts for $n \geq 1$,

$$\int_\Gamma \frac{\zeta\, f'(\zeta)}{\left(f(\zeta) - w_0\right)^{n+1}}\, d\zeta = -\frac{1}{n}\frac{\zeta}{\left(f(\zeta) - w_0\right)^n}\Bigg|_{\zeta=\zeta_0}^{\zeta_0} + \frac{1}{n}\int_\Gamma \frac{d\zeta}{\left(f(\zeta) - w_0\right)^n},$$

gives

$$b_n = \frac{1}{2\pi n i}\int_\Gamma \frac{d\zeta}{\left(f(\zeta) - w_0\right)^n}. \tag{B.2.8}$$

The function

$$\frac{1}{\left(f(\zeta) - w_0\right)^n}$$

has obviously a pole of order n at the point $\zeta = z_0$ but no other singularities inside Γ. The residue is

$$\operatorname*{Res}_{\zeta=z_0} \frac{1}{\left(f(\zeta) - w_0\right)^n} = \frac{1}{(n-1)!}\lim_{z\to z_0}\left(\frac{d^{n-1}}{dz^{n-1}}\left(\frac{z - z_0}{f(z) - w_0}\right)^n\right)$$

and we arrive at

$$b_n = \frac{1}{n!}\lim_{z\to z_0}\left(\left(\frac{z - z_0}{f(z) - w_0}\right)^n\right)^{(n-1)}. \tag{B.2.9}$$

This completes the proof of the *Lagrange inversion formula*:

$$z = h(w) = z_0 + \sum_{n=1}^{\infty} b_n (w - w_0)^n, \tag{B.2.10}$$

where

$$\begin{aligned} b_n &= \frac{1}{2\pi i}\int_{|z-z_0|=r} \frac{z f'(z)}{\left(f(z) - w_0\right)^{n+1}}\, dz \\ &= \frac{1}{n!}\lim_{z\to z_0}\frac{d^{n-1}}{dz^{n-1}}\left(\frac{z - z_0}{f(z) - w_0}\right)^n. \end{aligned} \tag{B.2.11}$$

More details and applications can be found in [**10**], [**21**], [**106**], and [**136**].

B.3. Dirichlet Series

The Dirichlet series of the general form [**11**], [**21**], [**106**] are

$$\sum_{n=1}^{\infty} a_n e^{-\lambda_n z}, \tag{B.3.1}$$

where the a_n are complex numbers and the exponents λ_n are nonnegative real numbers satisfying the conditions

$$\lambda_n < \lambda_{n+1}, \quad n = 1, 2, 3, \dots, \qquad \lim_{n\to\infty} \lambda_n = \infty. \tag{B.3.2}$$

If we make the change of variable $w = e^{-z}$, the Dirichlet series (B.3.1) becomes

$$\sum_{n=1}^{\infty} a_n w^{\lambda_n}, \tag{B.3.3}$$

which is a series expansion in arbitrary nonnegative powers of w. In general, the powers may be noninteger, but they must increase without limit as $n \to \infty$. From this point of view, a Dirichlet series can be regarded as a generalization of a power series to the case of arbitrary powers w^{λ_n} satisfying the conditions (B.3.2).

Dirichlet series have many properties similar to those of power series but the region of convergence is now half-plane instead of a disk.

THEOREM B.6. *If the Dirichlet series* (B.3.1) *converges at the point* $z_0 = x_0 + iy_0$, *then it converges at every point of the half-plane* Re $z = x > x_0$. *Moreover, the convergence is uniform on every closed domain D of the form*

$$|\arg(z - z_0)| \leq \theta < \frac{\pi}{2}. \tag{B.3.4}$$

See [21] and [106] for the proof of this and the following theorems.

A Dirichlet series can behave in any of the following ways:

(1) The series may be convergent nowhere;
(2) The series may be convergent everywhere;
(3) There may exist a line Re $z = C$ such that the series is convergent at every point of the half-plane of Re $z > C$ but not at any point of the half-plane of convergence Re $z < C$.

In the third case, by the definition, the Dirichlet series (B.3.1) is said to have *abscissa of convergence* C, or *half-plane of convergence* Re $z > C$, if the series converges at every point of the half-plane Re $z > C$, but diverges at any point of the half-plane Re $z < C$. The first two cases can be regarded as limiting cases of the third, corresponding to $C = +\infty$ and $C = -\infty$, respectively.

THEOREM B.7. *The sum of the Dirichlet series* (B.3.1) *is an analytic function in the half-plane of convergence* Re $z > C$ *when* $C < \infty$.

One might expect that any function analytic in a half-plane of Re $z > C$ can be represented by a Dirichlet series (B.3.1). However, this conjecture, which is the converse of Theorem B.7 is not true; see [106] for the details.

Now we turn our attention to the absolute convergence of Dirichlet series.

THEOREM B.8. *If the Dirichlet series* (B.3.1) *converges absolutely at the point* $z_0 = x_0 + iy_0$, *then it converges absolutely and uniformly in the closed half-plane* Re $z = x \geq x_0$.

A Dirichlet series (B.3.1) can behave in one of the following three ways:

(1) The series may be absolutely convergent nowhere;
(2) The series may be absolutely convergent everywhere;

(3) There may exist a line $\operatorname{Re} z = A$ such that the series is absolutely convergent at every point of the half-plane of $\operatorname{Re} z > A$ but not at any point of the half-plane of convergence $\operatorname{Re} z < A$.

In the third case, the Dirichlet series is said to have *abscissa of absolute convergence* A, or *half-plane of absolute convergence* $\operatorname{Re} z > A$. The first two cases can be regarded as limiting cases of the third, corresponding to $A = +\infty$ and $A = -\infty$, respectively. Note that the series is either absolutely convergent at every point of the line $\operatorname{Re} z = A$ or at no point of at no point of $\operatorname{Re} z = A$.

THEOREM B.9. *If the exponents* λ_n *satisfy the condition*

$$L = \limsup_{n\to\infty} \frac{\log n}{\lambda_n} < \infty, \tag{B.3.5}$$

as well as the conditions (B.3.2), *then*

$$0 \le A - C \le L, \tag{B.3.6}$$

where C *is the abscissa of convergence and* A *the abscissa of absolute convergence of the Dirichlet series* (B.3.1).

According to (B.3.6), $A = C$ when $L = 0$, e. g., when the Dirichlet series corresponds to a power series. More generally, the abscissa of convergence and the abscissa of absolute convergence coincide for the Dirichlet series whose exponents satisfy the condition

$$\limsup_{n\to\infty} \frac{\log n}{\lambda_n} = 0, \tag{B.3.7}$$

as well as the conditions (B.3.2). The following generalization of the Cauchy–Hadamerd theorem holds for a Dirichlet series of this kind.

THEOREM B.10. *For the Dirichlet series* (B.3.1) *satisfying the conditions* (B.3.2) *and* (B.3.7), *the numbers* A *and* C *are given by the formula*

$$A = C = \limsup_{n\to\infty} \frac{\log |a_n|}{\lambda_n}. \tag{B.3.8}$$

One can look at **[21]**, **[106]** for the proof of this theorem.

B.4. Asymptotics

Let complex-valued functions $f(z)$ and $g(z)$ be defined on a set E in the complex z-plane and let a be a cluster point of E. Throughout the book we use the symbols $\sim$, o, and O in a usual sense:

Asymptotic Formula	Definition
$f(z) \sim g(z), \quad z \to a, z \in E.$	$\lim\limits_{z \to a,\ z \in E} \dfrac{f(z)}{g(z)} = 1.$
$f(z) = o(g(z)), \quad z \to a, z \in E.$	$\lim\limits_{z \to a,\ z \in E} \dfrac{f(z)}{g(z)} = 0.$
$f(z) = O(g(z)), \quad z \in E.$	The quotient $f(z)/g(z)$ is bounded on E : $\lvert f(z)/g(z) \rvert \leq M, \ z \in E.$
$f(z) = O(g(z)), \quad z \to a, z \in E.$	The quotient $f(z)/g(z)$ is bounded on an intersection of a neighborhood of a and the set E.

In particular, if $\lim\limits_{z \to a,\ z \in E} f(z)/g(z) < \infty$, then $f(z) = O(g(z))$ as $z \to a$, $z \in E$.

APPENDIX C

Tables of Zeros of Basic Sine and Cosine Functions

We used Newton's method to find first 10 to 15 zeros of the basic sine $S_q(\eta;\omega)$ and basic cosine $C_q(\eta;\omega)$ functions for different values of parameter q with the double precision. Numerical results are represented below for the following values of the parameter $q = 0.1,\ 0.2,\ 0.25,\ 0.5,\ 0.75,\ 0.9$.

These are zeros of the basic sine $S_q(\eta;\omega)$:

n	ω_n for $q = 0.1$	ω_n for $q = 0.2$
1	5.27335469027306	2.90042982204458
2	55.893882379295	16.2987904328376
3	562.002001518	83.177038985378
4	5623.07402037004	417.548017711194
5	56233.7932967102	2089.3989447214
6	562340.985968945	10448.6527898134
7	5623412.91268218	52244.921857416
8	$5.62341321798136 \times 10^7$	261226.267163863
9	$5.62341324851126 \times 10^8$	1306132.99368979
10	$5.62341325156426 \times 10^9$	6530666.62631813
11		$3.26533347894596 \times 10^7$
12		$1.63266675605166 \times 10^8$

n	ω_n for $q = 0.25$	ω_n for $q = 0.5$
1	2.34744308318202	1.03665575431726
2	10.8635366494448	2.78262866580542
3	44.8113923829444	6.17199290693112
4	180.577524671584	12.9107915913224
5	723.635937047936	26.3706668081984
6	2895.86806976414	53.2820520205418
7	11584.7962222	107.100756434088
8	46340.5087373844	214.736160530784
9	185363.358774485	430.005973321244
10	741454.758916978	860.54510292812
11	2965820.35948547	1721.62311458506
12	$1.1863282761759 \times 10^7$	3443.77901422794
13	$4.74531323708534 \times 10^7$	6888.09075170494
14	$1.8981253080723 \times 10^8$	13776.7141957612
15	$7.5925012455274 \times 10^8$	27553.9610684268

n	ω_n for $q = 0.75$	ω_n for $q = 0.9$
1	0.43035133535186	0.16170612298962
2	0.91671492533346	0.32634395186574
3	1.51365313041482	0.49680932157838
4	2.27856366955664	0.6759429891091
5	3.27895991411962	0.86653632394786
6	4.6000413029874	1.071360004545
7	6.35280631145732	1.29320814382726
8	8.6837804284664	1.53494991950668
9	11.7874516856602	1.7995832583507
10	15.9225866579814	2.09028792587322
11	21.4338483890054	2.41047738220166
12	28.780545919667	2.7638498802646
13	38.5749244608914	3.15443977153218
14	51.6331934919836	3.5866701300521
15	69.0435483468518	4.06540781404834

These are zeros of the basic cosine $C_q(\eta; \omega)$:

n	ϖ_n for $q = 0.1$	ϖ_n for $q = 0.2$
1	1.39953705981838	1.00483939787922
2	17.440279553054	7.05020895442214
3	177.488395589304	36.9669325529092
4	1777.94015637718	186.503670036848
5	17782.4548758522	934.178429102712
6	177827.601782267	4672.55045571158
7	1778279.07081759	23364.4102356826
8	$1.77827937611678 \times 10^7$	116823.70906495
9	$1.7782794066467 \times 10^8$	584120.203197169
10	$1.7782794096997 \times 10^9$	2920602.67385544
11		$1.46030150271462 \times 10^7$
12		$7.30150767936 \times 10^7$
13		$3.65075385625868 \times 10^8$
14		$1.82537692978722 \times 10^9$
15		$9.12688465059396 \times 10^9$

n	ϖ_n for $q = 0.25$	ϖ_n for $q = 0.5$
1	0.87903921114982	0.4747323848864
2	5.1971495113873	1.77314109908414
3	22.1817697308248	4.19140497789744
4	90.0673154950768	8.96538332724022
5	361.59713041096	18.4870254827466
6	1447.7133482657	37.5181960345504
7	5792.1774623152	75.5747194388514
8	23170.0337293626	151.68491466547
9	92681.4587502764	303.903893340048
10	370727.158822114	608.341148265578
11	1482909.95910651	1217.21530776662
12	5931641.16024335	2434.96345180858
13	$2.37265659647904 \times 10^7$	4870.45965246608
14	$9.49062651829792 \times 10^7$	9741.45201008128
15	$3.79625062055734 \times 10^8$	19483.436703465

n	ϖ_n for $q = 0.75$	ϖ_n for $q = 0.9$
1	0.21163295654602	0.08066936320478
2	0.6631110346124	0.24347640531324
3	1.1979229666883	0.41066929631968
4	1.87121785185694	0.58511723965354
5	2.74450551750258	0.76963400875336
6	3.8931856660269	0.96699587317768
7	5.4142679993379	1.17997908079938
8	7.43515498817398	1.41140860247754
9	10.1245860329766	1.66421128336434
10	13.706854888959	1.9414694124004
11	18.4805771697354	2.24647318704016
12	24.8436134908758	2.58277208896086
13	33.3262444242814	2.9542259509929
14	44.635371037976	3.36505678143784
15	59.7134289358994	3.81990247086372

More extensive tables of zeros can be found on author's webpage at `http://hahn.la.asu.edu/~suslov/index.html`.

APPENDIX D

Numerical Examples of Improved Asymptotics

Here we present numerical examples illustrating the accuracy of asymptotic formulas (8.4.24) and (8.4.25) for the zeros of the basic sine and cosine functions, respectively.

These are the sums,

$$\omega_n^{(3)} = \omega_n^{(0)} - c_1(q) - \frac{c_1^2(q)}{2\omega_n^{(0)}} - \left(c_1^2(q)\left(2c_2(q)+3\right) - \frac{2q}{1+q^{1/2}+q}\right)\frac{c_1(q)}{6\left(\omega_n^{(0)}\right)^2},$$

of the first four terms in the right side of the asymptotic formula (8.4.24) for the zeros ω_n of the basic sine function $S_q(\eta;\omega)$ and the relative errors $\left(\omega_n^{(3)} - \omega_n\right)/\omega_n$ when $q = 0.5$:

n	$\omega_n^{(3)}$	ω_n	$\left(\omega_n^{(3)} - \omega_n\right)/\omega_n$
1	1.04337757573196	1.03665575431726	0.00648414035876
2	2.78339752684992	2.7826286658054	$2.76307454881782 \times 10^{-4}$
3	6.17208408141604	6.17199290693112	$1.47722925632098 \times 10^{-5}$
4	12.9108026835328	12.9107915913224	$8.59142554979188 \times 10^{-7}$
5	26.3706681759636	26.3706668081984	$5.18669229227938 \times 10^{-8}$
6	53.2820521903502	53.2820520205418	$3.18697375650674 \times 10^{-9}$
7	107.100756455242	107.100756434088	$1.97518339725236 \times 10^{-10}$
8	214.736160533424	214.736160530784	$1.22994850318212 \times 10^{-11}$
9	430.005973321574	430.005973321244	$7.68565224115714 \times 10^{-11}$
10	860.545102928162	860.54510292812	$4.84844656042604 \times 10^{-14}$
11	1721.62311458508	1721.62311458506	$1.12258962190646 \times 10^{-14}$
12	3443.77901422794	3443.77901422794	0

These are the sums,

$$\varpi_n^{(3)} = \varpi_n^{(0)} - c_1(q) - \frac{c_1^2(q)}{2\varpi_n^{(0)}} - \left(c_1^2(q)\left(2c_2(q)+3\right) - \frac{2q}{1+q^{1/2}+q}\right)\frac{c_1(q)}{6\left(\varpi_n^{(0)}\right)^2},$$

of the first four terms in the right side of the asymptotic formula (8.4.25) for the zeros ϖ_n of the basic cosine function $C_q(\eta;\omega)$ and the relative errors $\left(\varpi_n^{(3)} - \varpi_n\right)/\varpi_n$ when $q = 0.5$:

n	$\varpi_n^{(3)}$	ϖ_n	$\left(\varpi_n^{(3)} - \varpi_n\right)/\varpi_n$
1	0.49445085268924	0.4747323848864	0.04153596516816
2	1.7754058144102	1.77314109908414	0.00127723356434
3	4.19166863426672	4.19140497789744	$6.29040550055418 \times 10^{-5}$
4	8.96541505582614	8.96538332724022	$3.5390105204586 \times 10^{-6}$
5	18.4870293732778	18.4870254827466	$2.10446581640882 \times 10^{-7}$
6	37.5181965162004	37.5181960345504	$1.2837770709017 \times 10^{-8}$
7	75.5747194987678	75.5747194388514	$7.92812001944104 \times 10^{-10}$
8	151.684914672942	151.68491466547	$4.92638243987756 \times 10^{-11}$
9	303.90389334098	303.903893340048	$3.0699542000224 \times 10^{-12}$
10	608.341148265694	608.341148265578	$1.92486474385494 \times 10^{-13}$
11	1217.21530776664	1217.21530776662	$1.62514467550172 \times 10^{-14}$
12	2434.96345180858	2434.96345180858	0

One can see that our asymptotic formulas give a very good approximation even for the small zeros.

APPENDIX E

Numerical Examples of Euler–Rayleigh Method

Here we present numerical examples demonstrating the accuracy of Euler–Rayleigh's method.

Consider, for instance, the case $q = 0.1$ when the approximate values of the first zeros of $S_q(\eta; \omega)$ and $C_q(\eta; \omega)$ found in Appendix C with the double precision are $\omega_1 = 5.27335469027306$ and $\varpi_1 = 1.39953705981838$, respectively. The following two tables give an idea how fast the lower and upper bounds converge to these approximate values. Here we have used the general representation (11.2.8) in the expressions (11.2.16) for the upper and lower bounds of these zeros. We checked also the explicit formulas for these bounds found in Section 11.2.2 when $m = 1, 2, 3$.

These are the bounds for the first zero of $S_q(\eta; \omega)$ when $q = 0.1$:

m	Lower Bounds for ω_1	Upper Bounds for ω_1	Average
1	5.24980941954232	5.29679571148216	5.27330256551224
2	5.27325023297116	5.2735617515226	5.27340599224688
3	5.2733540704436	5.27335653321032	5.27335530182696
4	5.27335468613518	5.27335470667728	5.27335469640622
5	5.2733546902436	5.27335469041908	5.27335469033134
6	5.27335469027284	5.27335469027436	5.2733546902736
7	5.27335469027306	5.27335469027308	5.27335469027306
8	5.27335469027306	5.27335469027306	5.27335469027306

These are the bounds for the first zero of $C_q(\eta; \omega)$ when $q = 0.1$:

m	Lower Bounds for ϖ_1	Upper Bounds for ϖ_1	Average
1	1.39500892260642	1.4040507804411	1.39952985152376
2	1.39952254948174	1.39956589406876	1.39954422177524
3	1.39953699752826	1.39953724548542	1.39953712150684
4	1.39953705951754	1.39953706101402	1.39953706026578
5	1.39953705981684	1.39953705982608	1.39953705982146
6	1.39953705981838	1.39953705981844	1.3995370598184
7	1.39953705981838	1.39953705981838	1.39953705981838

As another example, let us consider the case $q = 0.5$ when approximate values of the first zeros of $S_q(\eta; \omega)$ and $C_q(\eta; \omega)$ are $\omega_1 = 1.03665575431726$ and $\varpi_1 = 0.4747323848864$, respectively; see Appendix C. One can observe that in this case the Euler–Rayley method gives a "more accurate" value for the first zero of q-cosine, namely, $\varpi_1 = 0.47473238488638$.

These are the bounds for the first zero of $S_q(\eta;\omega)$ when $q = 0.5$:

m	Lower Bounds for ω_1	Upper Bounds for ω_1	Average
1	0.95614515758492	1.11281598065148	1.0344805691182
2	1.0315103543751	1.0456150381798	1.03856269627744
3	1.03619064753468	1.0378594662099	1.03702505687228
4	1.03660760046932	1.03682168586534	1.03671464316732
5	1.03665041401184	1.0366787511926	1.03666458260222
6	1.03665513682152	1.03665894511092	1.03665704096622
7	1.036655680862	1.03665619714084	1.03665593900142
8	1.03665574539684	1.03665581577602	1.03665578058642
9	1.03665575321676	1.0366557628471	1.03665575803192
10	1.03665575417978	1.03665575550112	1.03665575484044
11	1.0366557542999	1.03665575448156	1.03665575439072
12	1.03665575431504	1.03665575434006	1.03665575432754
13	1.03665575431696	1.03665575432042	1.03665575431868
14	1.03665575431722	1.0366557543177	1.03665575431746
15	1.03665575431724	1.03665575431732	1.03665575431728
16	1.03665575431726	1.03665575431726	1.03665575431726

These are the bounds for the first zero of $C_q(\eta;\omega)$ when $q = 0.5$:

m	Lower Bounds for ϖ_1	Upper Bounds for ϖ_1	Average
1	0.45508986056222	0.49391281196484	0.47450133626352
2	0.4741041159144	0.47590327812534	0.47500369701986
3	0.47470307962652	0.47481403698708	0.4747585583068
4	0.47473081653556	0.4747382090177	0.47473451277662
5	0.47473229502278	0.47473280200132	0.47473254851204
6	0.47473237951916	0.4747324147813	0.47473239715022
7	0.47473238455662	0.47473238702924	0.47473238579292
8	0.4747323848657	0.47473238503998	0.47473238495284
9	0.47473238488506	0.47473238489738	0.47473238489122
10	0.4747323848863	0.47473238488716	0.47473238488672
11	0.47473238488636	0.47473238488644	0.4747323848864
12	0.47473238488638	0.47473238488638	0.47473238488638

Our numerical analysis shows that the convergence become very slow when parameter q approaches to 1.

APPENDIX F

Numerical Examples of Lower and Upper Bounds

Here we present numerical examples of the monotone sequences of the lower and upper bounds for the zeros of basic sine and cosine functions constructed in Theorems 11.2 and 11.3. We consider only the cases of the first zeros in order to compare our results with the corresponding sequences of the lower and upper bounds for the first zeros available from the Euler–Rayleigh method in Appendix E; see also [**48**].

These are the lower $L_n^{(k)}$ and upper $U_n^{(k)}$ bounds given by (11.3.1) and (11.3.2), respectively, for the first zero $\omega_1 = 1.03665575431726$ of $S_q(\eta;\omega)$ when $q = 0.5$:

k	$L_1^{(k)}$	$U_1^{(k)}$	$\frac{1}{2}\left(L_1^{(k)} + U_1^{(k)}\right)$
0		1.5	
1	0.66154494749078	1.20277795015158	0.93216133117676
2	1.0209636974334	1.04768180922342	1.03432273864746
3	1.03662674267916	1.03668345189868	1.03665509728892
4	1.03665575415158	1.0366557544829	1.03665575431724
5	1.03665575431726	1.03665575431726	1.03665575431726

These are the lower $\overline{L}_n^{(k)}$ and upper $\overline{U}_n^{(k)}$ bounds given by (11.3.3) and (11.3.4), respectively, for the first zero $\varpi_1 = 0.47473238488638$ of $C_q(\eta;\omega)$ when $q = 0.5$:

k	$\overline{L}_1^{(k)}$	$\overline{U}_1^{(k)}$	$\frac{1}{2}\left(\overline{L}_1^{(k)} + \overline{U}_1^{(k)}\right)$
0		0.75	
1	0.3710966356424	0.52409262492642	0.4475946302844
2	0.47390457472684	0.4753606746592	0.47463260333478
3	0.47473231616194	0.47473245319626	0.4747323846791
4	0.47473238488638	0.47473238488638	0.47473238488638

These are the lower $L_n^{(k)}$ and upper $U_n^{(k)}$ bounds given by (11.3.17) and (11.3.18), respectively, for the first zero $\omega_1 = 1.03665575431726$ of $S_q(\eta;\omega)$ when $q = 0.5$:

k	$L_1^{(k)}$	$U_1^{(k)}$	$\frac{1}{2}\left(L_1^{(k)}+U_1^{(k)}\right)$
0		1.5	
1	0.66154494749078	1.07645596218222	0.8690004548365
2	1.0361841364819	1.0366605636724	1.03642235007714
3	1.03665575431226	1.03665575431726	1.03665575431476
4	1.03665575431726	1.03665575431726	1.03665575431726

These are the lower $\overline{L}_n^{(k)}$ and upper $\overline{U}_n^{(k)}$ bounds given by (11.3.19) and (11.3.20), respectively, for the first zero $\varpi_1 = 0.47473238488638$ of $C_q(\eta;\omega)$ when $q = 0.5$:

k	$\overline{L}_1^{(k)}$	$\overline{U}_1^{(k)}$	$\frac{1}{2}\left(\overline{L}_1^{(k)}+\overline{U}_1^{(k)}\right)$
0		0.75	
1	0.3710966356424	0.48748870155732	0.42929266859986
2	0.47469735288332	0.4747324713825	0.4747149121329
3	0.47473238488638	0.47473238488638	0.47473238488638

These results indicate that the convergence is up to three or four times faster then convergence in the Euler–Rayleigh method discussed in Appendix E.

Bibliography

[1] M. Abramowitz and I. A. Stegun, *Handbook of Mathematical Functions*, Dover, New York, 1972.

[2] L. V. Ahlfors, *Complex Analysis: An Introduction to the Theory of Analytic Functions of One Complex Variable*, Third Edition, McGraw–Hill, Inc., New York, 1979.

[3] N. I. Akhiezer, *Theory of Approximation*, Frederick Ungar Publishing Co., New York, 1956.

[4] N. I. Akhiezer, *The Classical Moment Problem and Some Related Questions in Analysis*, Oliver and Boyed, Edinburgh, 1965.

[5] N. I. Akhiezer and I. M. Glazman, *Theory of Linear Operators in Hilbert Space*, Dover, New York, 1993.

[6] W. A. Al-Salam, *q-Bernoulli numbers and polynomials*, Math. Nachr. **17** (1959), 239–260.

[7] W. A. Al-Salam and T. D. Chihara, *Convolutions of orthogonal polynomials*, SIAM J. Math. Anal. **7** (1976), 16–28.

[8] G. E. Andrews, *Connection coefficient problems and partitions*, Proc. Symp. Pure Math **34** (1979), 1–24.

[9] G. E. Andrews and R. A. Askey, *Classical orthogonal polynomials*, in: *"Polynômes orthogonaux et applications"*, Lecture Notes in Math. **1171**, Springer-Verlag, 1985, pp. 36–62.

[10] G. E. Andrews, R. A. Askey, and R. Roy, *Special Functions*, Cambridge University Press, Cambridge, 1999.

[11] T. M. Apostol, *Modular Functions and Dirichlet Series in Number Theory*, Springer–Verlag, New York, 1976, pp. 161–189.

[12] R. A. Askey, *Orthogonal Polynomials and Special Functions*, CBMS–NSF Regional Conferences Series in Applied Mathematics, SIAM, Philadelphia, Pennsylvania, 1975.

[13] R. A. Askey, N. M. Atakishiyev, and S. K. Suslov, *An analog of the Fourier transformation for the q-harmonic oscillator*, Preprint IAE-5611/1, Kurchatov Institute, Moscow, 1993; in: *"Symmetries in Science VI"*, B. Gruber, ed., Plenum Press, New York, 1993, pp. 57–63.

[14] R. A. Askey, M. Rahman, and S. K. Suslov, *On a general q-Fourier transformation with nonsymmetric kernels*, J. Comp. Appl. Math. **68** (1996), 25–55.

[15] R. A. Askey and J. A. Wilson, *Some basic hypergeometric orthogonal polynomials that generalize Jacobi polynomials*, Memoirs Amer. Math. Soc., Number **319** (1985).

[16] N. M. Atakishiyev, M. Rahman, and S. K. Suslov, *On the classical orthogonal polynomials*, Constr. Approx. **11** (1995), 181–226.

[17] N. M. Atakishiyev and S. K. Suslov, *Difference hypergeometric functions*, in: *"Progress in Approximation Theory: An International Perspective"*, A. A. Gonchar and E. B. Saff, eds., Springer Series in Computational Mathematics, Vol. 19, Springer–Verlag, 1992, pp. 1–35.

[18] N. M. Atakishiyev and S. K. Suslov, *On the Askey–Wilson polynomials*, Constr. Approx. **8** (1992), 363–369.

[19] R. Ayoub, *Euler and the zeta function*, Amer. Math. Month. **81** (1974), 1067–1086.

[20] N. K. Bary, *A Treatise on Trigonometric Series*, two volumes, Macmillan, New York, 1964.

[21] C. A. Berenstein, R. Gay, *Complex Variables: An Introduction*, Springer–Verlag, New York, 1991, pp. 168–170, 482–495.

[22] C. A. Berenstein, R. Gay, *Complex Analysis and Special Topics in Harmonic Analysis*, Springer–Verlag, New York, 1995.

[23] R. P. Boas, *Entire Functions*, Academic Press, New York, 1954.

[24] D. Bressoud, *Linearization and related formulas for q-ultraspherical polynomials*, SIAM J. Math. Anal. **12** (1981), 161–168.

[25] J. Bustoz and J. L. Cordoso, *Basic analog of Fourier series on a q-linear grid*, J. Approx. Theory **112** (2001), 134–157.

[26] J. Bustoz and S. K. Suslov, *Basic analog of Fourier series on a q-quadratic grid*, Methods Appl. Anal. **5** (1998), 1–38.

[27] L. Carlitz, *q-Bernoulli numbers and polynomials*, Duke Math. J. **15** (1948), 987–1000.

[28] L. Carlitz, *q-Bernoulli and Eulerian numbers*, Trans. Amer. Math. Soc. **76** (1954), 332–350.

[29] L. Carlitz, *Expansions of q-Bernoulli numbers* , Duke Math. J. **25** (1958), 355–364.

[30] I. Cherednik, *From double Hecke algebra to analysis*, Documenta Mathematica, Extra Volume ICM 1998; Preprint math.QA/9806097, v2, 13 Jul 1998.

[31] I. Cherednik, *On q-analogues of Riemann's zeta*, Preprint math.QA/9804099, v3, 8 Oct 1998.

[32] I. Cherednik, *Calculations of zeros of a q-zeta function numerically*, Preprint math.QA/9804100, v2, 12 May 1998.

[33] S. Cooper, private communication.

[34] S. Cooper, *On sums of an even number of squares, and an even number of triangle numbers: an elementary approach based on Ramanujan's $_1\psi_1$ summation formula*, in: *"q-Series with Applications to Combinatorics, Number Theory, and Physics"*, B. C. Berndt and K. Ono, eds., Contemporary Mathematics, Vol. 291, Amer. Math. Soc., 2001, pp. 115–137.

[35] C. Dunkl and Y. Xu, *Orthogonal Polynomials of Several Variables*, Cambridge University Press, Cambridge, 2001.

[36] H. M. Edwards, *Rieman's Zeta Function*, Dover, New York, 1974.

[37] A. Erdélyi, *Higher Transcendental Functions*, Vols. I–III, A. Erdélyi, ed., McGraw–Hill, 1953.

[38] H. Exton, *q-Hypergeometric Functions*, Ellis Horwood, 1983.

[39] K. Ey, A. L. Ruffing, and S. K. Suslov, *Method of separation of the variables for basic analogs of equations of mathematical physics*, preprint, under preparation.

[40] R. Floreanini and L. Vinet, *On the quantum group and quantum algebra approach to q-special functions*, Lett. Math. Phys. **27** (1993), 179–190.

[41] R. Floreanini and L. Vinet, *A model for the continuous q-ultraspherical polynomials*, J. Math. Phys. **36** (1995), 3800–3813.

[42] R. Floreanini and L. Vinet, *More on the q-oscillator algebra and q-orthogonal polynomials*, J. Phys. A: Math. Gen. **A28** (1995), L287–L293.

[43] R. Floreanini, J. LeTourneux, and L. Vinet, *Symmetry techniques for the Al-Salam–Chihara polynomials*, J. Phys. A: Math. Gen. **A30** (1997), 3107–3114.

[44] R. Floreanini, J. LeTourneux, and L. Vinet, *Symmetries and continuous q-orthogonal polynomials*, in: *"Algebraic Methods and q-Special Functions"*, J. F. van Diejen and L. Vinet, eds., CRM Proceedings & Lecture Notes, Vol. 22, Amer. Math. Soc., 1999, pp. 135–144.

[45] G. Gasper, *Elementary derivations of summation and transformation formulas for q-series*, in: *"Special Functions, q-Series and Related Topics"*, M. E. H. Ismail,

D. R. Masson, and M. Rahman, eds., Fields Institute Communications, Vol. 14, Amer. Math. Soc., 1997, pp. 55–70.

[46] G. Gasper and M. Rahman, *Basic Hypergeometric Series*, Cambridge University Press, Cambridge, 1990.

[47] R. W. Gosper, *Experiments and discoveries in q-trigonometry*, in: "Symbolic computation, Number Theory, Special Functions, Physics and Combinatorics", F. G. Garvan and M. E. H. Ismail, eds., Developments in Mathematics, Vol. 4, Kluwer Academic Publishers, Dordrecht – Boston – London, 2001, pp. 79–105.

[48] R. W. Gosper, Jr. and S. K. Suslov, *Numerical investigation of basic Fourier series*, in "q-Series from a Contemporary Perspective", M. E. H. Ismail, D. R. Stanton, eds., Contemporary Mathematics, Vol. 254, Amer. Math. Soc., 2000, pp. 199–227.

[49] F. A. Grünbaum and L. Haine, *Some functions that generalize the Askey–Wilson polynomials*, Comm. Math. Phys. **106** (1997), 173–202.

[50] L. Haine and P. Iliev, *Askey–Wilson type functions, with bound states*, Duke Math. J., to appear.

[51] J. R. Higgins, *Completeness and Basic Properties of Sets of Special Functions*, Cambridge University Press, Cambridge, 1977.

[52] E. Hille, *Lectures on Ordinary Differential Equations*, Addison–Wesley, Reading, 1969, pp. 479–491.

[53] E. Hille, *Analytic Function Theory*, Vol. 1, Chelsea Publishing Company, New York, 1973.

[54] Y. Chen, M. E. H. Ismail, and K. A. Muttalib, *Asymptotics of basic Bessel functions and q-Laguerre polynomials*, J. Comp. Appl. Math. **54** (1994), 263–272.

[55] M. E. H. Ismail, *The basic Bessel functions and polynomials*, SIAM J. Math. Anal. **12** (1981), 454–468.

[56] M. E. H. Ismail, *The zeros of basic Bessel functions, the functions $J_{\nu+ax}(x)$, and associated orthogonal polynomials*, J. Math. Anal. Appl. **86** (1982), 1–19.

[57] M. E. H. Ismail, *Some properties of Jackson's third q-Bessel functions*, preprint.

[58] M. E. H. Ismail, *The Askey–Wilson operator and summation theorems*, in "Mathematical Analysis, Wavelets, and Signal Processing", M. E. H. Ismail, M. Z. Nashed, A. Zayed, and A. Ghaleb, eds., Contemporary Mathematics, Vol. 190, Amer. Math. Soc., 1995, pp. 171–178.

[59] M. E. H. Ismail, *Lectures on q-orthogonal polynomials*, in: "Special Functions 2000: Current Perspective and Future Directions", J. Bustoz, M. E. H. Ismail and S. K. Suslov, eds., NATO Science Series II: Mathematics, Physics and Chemistry, Vol. 30, Kluwer Academic Publishers, Dordrecht – Boston – London, 2001, pp. 179–219.

[60] M. E. H. Ismail, *Orthogonality and completeness of q-Fourier type systems*, preprint, March 1, 2001; revised version in Z. Anal. Anwendungen **20** (2001), 761–775.

[61] M. E. H. Ismail, *An operator calculus for the Askey–Wilson operator*, Annals of Combinatorics, **5** (2001), 347–362.

[62] M. E. H. Ismail, D. R. Masson, and S. K. Suslov, *Some generating functions for q-polynomials*, in: "Special Functions, q-Series and Related Topics", M. E. H. Ismail, D. R. Masson, and M. Rahman, eds., Fields Institute Communications, Vol. 14, Amer. Math. Soc., 1997, pp. 91–108.

[63] M. E. H. Ismail, D. R. Masson, and S. K. Suslov, *The q-Bessel functions on a q-quadratic grid*, in: "Algebraic Methods and q-Special Functions", J. F. van Diejen and L. Vinet, eds., CRM Proceedings & Lecture Notes, Vol. 22, Amer. Math. Soc., 1999, pp. 183–200.

[64] M. E. H. Ismail and M. E. Muldoon, *On the variation with respect to a parameter of zeros of Bessel and q-Bessel functions*, J. Math. Anal. Appl. **135** (1988), 187–207.

[65] M. E. H. Ismail and M. E. Muldoon, *Bounds for the small real and purely imaginary zeros of Bessel and related functions*, Methods Appl. Anal. **2** (1995), 1–21.

[66] M. E. H. Ismail and M. Rahman, *The associated Askey–Wilson polynomials*, Trans. Amer. Math. Soc. **328** (1991), 201–237.

[67] M. E. H. Ismail and M. Rahman, *Inverse operators, q-fractional integrals, and q-Bernoulli polynomials*, J. Approx. Theory, **114** (2002), 269–307.

[68] M. E. H. Ismail, M. Rahman, and D. Stanton, *Quadratic q-exponentials and connection coefficient problems*, Proc. Amer. Math. Soc. **127** (1999) #10, 2931–2941.

[69] M. E. H. Ismail, M. Rahman, and R. Zhang, *Diagonalization of certain integral operators II*, J. Comp. Appl. Math. **68** (1996), 163–196.

[70] M. E. H. Ismail and D. Stanton, *On the Askey–Wilson and Rogers polynomials*, Canad. J. Math. **40** (1988), 1025–1045.

[71] M. E. H. Ismail and D. Stanton, *Addition theorems for the q-exponential functions*, in *"q-Series from a Contemporary Perspective"*, M. E. H. Ismail, D. R. Stanton, eds., Contemporary Mathematics, Vol. 254, Amer. Math. Soc., 2000, pp. 235–245.

[72] M. E. H. Ismail and D. Stanton, *q-Integral and moment representations for q-orthogonal polynomials*, Canad. J. Math. **54** # 4 (2002), 709–735.

[73] M. E. H. Ismail and D. Stanton, *Applications of q-Taylor theorems*, preprint, September 2001.

[74] M. E. H. Ismail and D. Stanton, private communication.

[75] M. E. H. Ismail, D. Stanton, and G. Viennot, *The combinatorics of q-Hermite polynomials and the Askey–Wilson integral*, European J. Combinatorics, **8** (1987), 379–392.

[76] M. E. H. Ismail and R. Zhang, *Diagonalization of certain integral operators*, Advances in Math. **109** (1994), 1–33.

[77] F. H. Jackson, *The application of basic numbers to Bessel's and Legendre's functions*, Proc. London Math. Soc. (2) **2** (1905), 192–220.

[78] F. H. Jackson, *The application of basic numbers to Bessel's and Legendre's functions (Second paper)*, Proc. London Math. Soc. (2) **3** (1905), 1–23.

[79] F. H. Jackson, *Theorems relating to a generalization of Bessel's function*. II, Trans. Roy. Soc. Edin. **41** (1905), 399–408.

[80] T. Kakehi, *Eigenfunction expansion associated with the Casimir operator on the quantum $SU(1,1)$ group*, preprint.

[81] T. Kakehi, T. Masuda, and K. Ueno, *Spectral analysis of q-difference operator which arises from the quantum $SU(1,1)$ group*, J. Operator Theory **33** (1995), 159–196.

[82] E. G. Kalnins and W. Miller, *Models of q-algebra representations: q-integral transforms and "addition theorems"*, J. Math. Phys. **35** (1994), 1951–1975.

[83] E. G. Kalnins and W. Miller, *q-Algebra representations of the Euclidean, Pseudo-Euclidean and oscillator algebras, and their tensor products*, in: "Symmetries and Integrability of Difference Equations", D. Levi, L. Vinet, and P. Winternitz, eds., CRM Proceedings & Lecture Notes, Vol. 9, Amer. Math. Soc., 1996, pp. 173–183.

[84] K. Knopp, *Theory of Functions*, Parts I and II, Dover, New York, 1996.

[85] N. Koblitz, *On Carlitz's q-Bernoulli numbers*, J. Number Theory **14** (1982), 332–339.

[86] R. Koekoek and R. F. Swarttouw, *The Askey scheme of hypergeometric orthogonal polynomials and its q-analogues*, Report 94–05, Delft University of Technology, 1994.

[87] H. T. Koelink, *On Quantum Groups and q-Special Functions*, Ph. D. thesis, University of Leiden, 1991, §3.2 and §3.6.

[88] H. T. Koelink, *On q-Bessel Functions Related to the Quantum Group of Plane Motions*, Report W–91–26, University of Leiden, 1991, §2 and §6.

[89] H. T. Koelink, *The quantum group of plane motions and the Hahn–Exton q-Bessel function*, Duke Math. J. **76** (1994), 483–508.

[90] E. Koelink and J. V. Stokman, *The Askey–Wilson function transform*, International Mathematics Research Notices (2001), No. 22, 1203–1227.

[91] E. Koelink and J. V. Stokman, *The Askey–Wilson function transform scheme*, in: "*Special Functions 2000: Current Perspective and Future Directions*", J. Bustoz, M. E. H. Ismail and S. K. Suslov, eds., NATO Science Series II: Mathematics, Physics and Chemistry, Vol. 30, Kluwer Academic Publishers, Dordrecht – Boston – London, 2001, pp. 221–241.

[92] H. T. Koelink and R. F. Swarttouw, *On the zeros of the Hahn–Exton q-Bessel Functions and associated q-Lommel polynomials*, J. Math. Anal. Appl. **186** (1994), 690–710.

[93] A. N. Kolmogorov and S. V. Fomin, *Introductory Real Analysis*, Dover, New York, 1970.

[94] T. Koornwinder, *Special functions and q-commuting variables*, in: "*Special Functions, q-Series and Related Topics*", M. E. H. Ismail, D. R. Masson, and M. Rahman, eds., Fields Institute Communications, Vol. 14, Amer. Math. Soc., 1997, pp. 131–166.

[95] T. Koornwinder and R. F. Swarttouw, *On q-analogues of the Fourier and Hankel transforms*, Trans. Amer. Math. Soc. **333** (1992), 445–461.

[96] C. Krattenthaler, *HYP and HYPQ: Mathematica packages for the manipulation of binomial sums and hypergeometric series, respectively q-binomial sums and basic hypergeometric series*, Symbolic Computation in Combinatorics Δ_1 (Ithaca, 1993), J. Symbolic Comput. **20** (1995) #5–6, 737–744.

[97] L. D. Landau and E. M. Lifschitz, *Quantum Mechanics: Non-relativistic Theory*, Third Edition, Butterworth-Heinemann, 1998.

[98] S. Lang, *Complex Analysis*, Fourth Edition, Springer–Verlag, New York, 1999.

[99] B. Ya. Levin, *Distribution of Zeros of Entire Functions*, Translations of Mathematical Monographs, Vol. 5, Amer. Math. Soc., Providence, Rhode Island, 1980.

[100] B. Ya. Levin, *Lectures on Entire Functions*, Translations of Mathematical Monographs, Vol. 150, Amer. Math. Soc., Providence, Rhode Island, 1996.

[101] N. Levinson, *Gap and Density Theorems*, Amer. Math. Soc. Colloq. Publ., Vol. 36, New York, 1940.

[102] Z.-G. Liu, *Residue theorem and theta function identities*, Ramanujan J. **5** (2001), 129–151.

[103] Z.-G. Liu, *Some Eisenstein series identities associated with the Borwein functions*, in: "*Symbolic computation, Number Theory, Special Functions, Physics and Combinatorics*", F. G. Garvan and M. E. H. Ismail, eds., Developments in Mathematics, Vol. 4, Kluwer Academic Publishers, Dordrecht – Boston – London, 2001, pp. 147–169.

[104] Z.-G. Liu, *On the representation of integers as sums of squares*, in: "*q-Series with Applications to Combinatorics, Number Theory, and Physics*", B. C. Berndt and K. Ono, eds., Contemporary Mathematics, Vol. 291, Amer. Math. Soc., 2001, pp. 163–176.

[105] A. Magnus, *Associated Askey–Wilson polynomials as Laguerre–Hahn orthogonal polynomials*, in: "*Orthogonal Polynomials and Their Applications*", M. Alfaro et al., eds., Lecture Notes in Math. **1329**, Springer-Verlag, New York, 1988, pp. 261–378.

[106] A. I. Markushevich, *Theory of Functions of a Complex Variable*, Second Edition, Chelsea Publishing Company, New York, 1985, Vol. II, pp. 86–89, 26–34.

[107] T. Masuda, K. Mimachi, Y. Nakagami, M. Noumi, Y. Saburi, and K. Ueno, *Unitary representations of the quantum group $SU_q(1,1)$: II—Matrix elements of unitary representations and basic hypergeometric functions*, Lett. Math. Phys. **19** (1990), 195–204.

[108] J. N. McDonald and N. A. Weiss, *A Course in Real Analysis*, Academic Press, New York, 1999.

[109] S. C. Milne, *Infinite families of exact sums of squares formulas, Jacobi elliptic functions, continued fractions, and Schur functions*, Ramanujan J. **6** (2002), 7–149.

[110] B. Nassrallah and M. Rahman, *Projection formulas, a reproducing kernel and a generating function for q-Wilson polynomials*, SIAM J. Math. Anal. **16** (1985), 211–229.

[111] A. F. Nikiforov, S. K. Suslov, and V. B. Uvarov, *Classical Orthogonal Polynomials of a Discrete Variable*, Nauka, Moscow, 1985 [in Russian]; English translation, Springer–Verlag, Berlin, 1991.

[112] A. F. Nikiforov, S. K. Suslov, and V. B. Uvarov, *Classical orthogonal polynomials in a discrete variable on nonuniform lattices*, Soviet Math. Dokl. **34** (1987), 576–579.

[113] A. F. Nikiforov and V. B. Uvarov, *Special Functions of Mathematical Physics*, Nauka, Moscow, 1984 [in Russian]; English translation, Birkhäuser, Boston, 1988.

[114] M. Rahman, *The linearization of the product of continuous q-Jacobi polynomials*, Canad. J. Math. **33** (1981), 255–284.

[115] M. Rahman, *A q-extension of Feldheim's bilinear sum for Jacobi polynomials and some applications*, Canad. J. Math. **37** (1985), 551–576.

[116] M. Rahman, *An integral representation and some transformation properties of q-Bessel functions*, J. Math. Anal. Appl. **125** (1987), 58–71.

[117] M. Rahman, *An addition theorem and some product formulas for q-Bessel functions*, Canad. J. Math. **40** (1988), 1203–1221.

[118] M. Rahman, *The q-exponential functions, old and new*, preprint.

[119] M. Rahman, *q-Wilson functions of the second kind*, SIAM J. Math. Anal. **17** (1986), 1280–1286.

[120] M. Rahman, *Askey–Wilson functions of the first and second kind: series and integral representations of* $C_n^2(x;\beta|q) + D_n^2(x;\beta|q)$, J. Math. Anal. Appl. **164** (1992), 263–284.

[121] M. Rahman, *A q-extension of a product formula of Watson*, Quaestiones Mathematicae **22** (1999), 27–42.

[122] M. Rahman, private communication.

[123] M. Rahman and S. K. Suslov, *The Pearson equation and beta integrals*, SIAM J. Math. Anal. **25** (1994), 646–693.

[124] M. Rahman and S. K. Suslov, *Barnes and Ramanujan-type integrals on the q-linear lattice*, SIAM J. Math. Anal. **25** (1994), 1002–1022.

[125] M. Rahman and S. K. Suslov, *Singular analogue of the Fourier transformation for the Askey–Wilson polynomials*, in: *Symmetries and Integrability of Difference Equations*, D. Levi, L. Vinet, and P. Winternitz, eds., CRM Proceedings & Lecture Notes, Vol. 9, Amer. Math. Soc., 1996, pp. 289–302.

[126] E. D. Rainville, *Special Functions*, The Macmillan Company, New York, 1960.

[127] F. Riesz and B. Sz.-Nagy, *Functional Analysis*, Dover, New York, 1990.

[128] W. Rudin, *Principles of Mathematical Analysis*, Third Edition, McGraw–Hill, Inc., New York, 1976.

[129] S. N. M. Ruijsenaars, *First order analytic difference equations and integrable quantum systems*, J. Math. Phys. **38** (1997), 1069–1146.

[130] S. N. M. Ruijsenaars, *Special functions associated with Calogero–Moser type quantum systems*, in: "*Proceedings of the 1999 Montréal Séminaire de Mathématiques Supérieures*", J. Harnad, G. Sabidussi, and P. Winternitz, eds., CRM Proceedings & Lecture Notes, Amer. Math. Soc., 2000, pp. 1–38.

[131] S. N. M. Ruijsenaars, *On Barnes' multiple zeta and gamma functions*, Advances in Math, **156** (2000), 107–132.

[132] S. N. M. Ruijsenaars, *Special functions defined by analytic difference equations*, in: "*Special Functions 2000: Current Perspective and Future Directions*", J. Bustoz, M. E. H. Ismail and S. K. Suslov, eds., NATO Science Series II: Mathematics,

Physics and Chemistry, Vol. 30, Kluwer Academic Publishers, Dordrecht – Boston – London, 2001, pp. 281–333.

[133] J. Satoh, *q-Analogue of Riemann's ζ-function and q-Euler numbers*, J. Number Theory **31** (1989), 346–362.

[134] A. Sharma, *q-Bernoulli and Euler numbers of higher order*, Duke Math. J. **25** (1958), 343–353.

[135] J. Shohat and J. D. Tamarkin, *The Problem of Moments*, Revised Edition, Amer. Math. Soc., Providence, 1950.

[136] Yu. V. Sidorov, M. V. Fedoryuk, and M. I. Shabunin, *Lectures on the Theory of Functions of a Complex Variable*, [in Russian], Nauka, Moscow, 1976.

[137] V. P. Spiridonov and A. S. Zhedanov, *Generalized eigenvalue problem and a new family of rational functions biorthogonal on elliptic grids*, in: "*Special Functions 2000: Current Perspective and Future Directions*", J. Bustoz, M. E. H. Ismail and S. K. Suslov, eds., NATO Science Series II: Mathematics, Physics and Chemistry, Vol. 30, Kluwer Academic Publishers, Dordrecht – Boston – London, 2001, pp. 365–388.

[138] J. V. Stokman, *An expansion formula for the Askey–Wilson function*, J. Approx. Theory **114** (2002), 308–342.

[139] S. K. Suslov, *The theory of difference analogues of special functions of hypergeometric type*, Russian Math. Surveys **44** (1989), 227–278.

[140] S. K. Suslov, *"Addition" theorems for some q-exponential and q-trigonometric functions*, Methods Appl. Anal. **4** (1997), 11–32.

[141] S. K. Suslov, *Some orthogonal very-well-poised ${}_8\varphi_7$-functions*, J. Phys. A: Math. Gen. **30** (1997), 5877–5885.

[142] S. K. Suslov, *Another addition theorem for the q-exponential function*, J. Phys. A: Math. Gen. **33** (2000), L375–380.

[143] S. K. Suslov, *Some orthogonal very-well-poised ${}_8\varphi_7$-functions that generalize Askey–Wilson polynomials*, Ramanujan J. **5** (2001), 183–218.

[144] S. K. Suslov, *Basic exponential functions on a q-quadratic grid*, in: "*Special Functions 2000: Current Perspective and Future Directions*", J. Bustoz, M. E. H. Ismail and S. K. Suslov, eds., NATO Science Series II: Mathematics, Physics and Chemistry, Vol. 30, Kluwer Academic Publishers, Dordrecht – Boston – London, 2001, pp. 411–456.

[145] S. K. Suslov, *Completeness of basic trigonometric system in $\mathcal{L}^p$*, in: "*q-Series with Applications to Combinatorics, Number Theory, and Physics*", B. C. Berndt and K. Ono, eds., Contemporary Mathematics, Vol. 291, Amer. Math. Soc., 2001, pp. 229–241.

[146] S. K. Suslov, *Some expansions in basic Fourier series and related topics*, J. Approx. Theory, **115** # 2 (2002), 289–353.

[147] S. K. Suslov, *Asymptotics of zeros of basic sine and cosine functions*, J. Approx. Theory, to appear.

[148] S. K. Suslov, *A note on completeness of of basic trigonometric system in $\mathcal{L}^2$*, Rocky Mountain J. Math. **33** #1 (Spring 2003), to appear.

[149] S. K. Suslov, *An Introduction to Special Functions*, Lecture Notes, under preparation.

[150] R. F. Swarttouw, *The Hahn–Exton q-Bessel Functions*, Ph. D. thesis, Delft Technische Universiteit, 1992.

[151] G. Szegö, *Orthogonal Polynomials*, Amer. Math. Soc. Colloq. Publ., Vol. 23, Rhode Island, 1939.

[152] N. M. Temme, *Special Functions: An Introduction to the Classical Functions of Mathematical Physics*, John Wiley & Sons, Inc., New York, 1996.

[153] J. Thomae, *Beiträge zur Theorie der durch die Heinesche Reihe ...*, J. reine angew. Math., **70**, 258–281.

[154] E. C. Titchmarsh, *The Theory of Function*, Second Edition, Press, Oxford University Press, 1939.

[155] E. C. Titchmarsh, *The Theory of the Riemann Zeta-function*, Second Edition, Claredon Press, Oxford, 1986.

[156] G. P. Tolstov, *Fourier Series*, Dover, New York, 1962.

[157] H. Tsumura, *On the values of a q-analogue of the q-adic L-functions*, Mem. Fac. Sci. Kyushu Univ. **44** (1990), 49–60.

[158] H. Tsumura, *A note on q-analogues of the Dirichlet series and q-Bernoulli numbers*, J. Number Theory **39** (1991), 251–256.

[159] H. Tsumura, *On evaluation of the Dirichlet series at positive integers by q-calculation*, J. Number Theory **48** (1994), 383–391.

[160] H. Tsumura, *A note on q-analogues of the Dirichlet series*, Proc. Japan Acad. **75** (1999), 23–25.

[161] W. R. Wade, *An Introduction to Analysis*, Second Edition, Prentice–Hall, Upper Saddle River, New Jersey, 1999.

[162] M. Ward, *A calculus of sequences*, Amer. J. Math. **58** (1936), 255–266.

[163] G. N. Watson, *A Treatise on the Theory of Bessel Functions*, Second Edition, Cambridge University Press, Cambridge, 1944.

[164] E. T. Whittaker and G. N. Watson, *A Course of Modern Analysis*, Fourth Edition, Cambridge University Press, Cambridge, 1952.

[165] N. Wiener, *The Fourier Integral and Certain of Its Applications*, Cambridge University Press, Cambridge, 1933; Dover edition published in 1948.

[166] A. Zygmund, *Trigonometric Series*, Second Edition, Cambridge University Press, Cambridge, 1968.

Index

Author and Subject Index

Symbol Index

Developments in Mathematics

1. Alladi et al. (eds.): *Analytic and Elementary Number Theory.* 1998
ISBN 0-7923-8273-0
2. S. Kanemitsu and K. Györy (eds.): *Number Theory and Its Applications.* 1999
ISBN 0-7923-5952-6
3. A. Blokhuis, J.W.P. Hirschfeld, D. Jungnickel and J.A. Thas (eds.): *Finite Geometries.* Proceedings of the Fourth Isle of Thorns Conference. 2001 ISBN 0-7923-6994-7
4. F.G. Garvan and M.E.H. Ismail (eds.): *Symbolic Computation, Number Theory, Special Functions, Physics and Combinatorics.* 2001 ISBN 1-4020-0101-0
5. S.C. Milne: *Infinite Families of Exact Sums of Squares Formulas, Jacobi Elliptic F. Continued Fractions, and Schur Functions.* 2002 ISBN 1-4020-0491-5
6. C. Jia and K. Matsumoto (eds.): *Analytic Number Theory.* 2002
ISBN 1-4020-0545-8
7. J. Martínez (ed.): *Ordered Algebraic Structures.* 2002 ISBN 1-4020-0752-3
8. S. Kanemitsu and C. Jia (eds.): *Number Theoretic Methods..* Future Trends. 2002
ISBN 1-4020-1080-X
9. S.K. Suslov: *An Introduction to Basic Fourier Series.* 2003 ISBN 1-4020-1221-7

KLUWER ACADEMIC PUBLISHERS – DORDRECHT / BOSTON / LONDON